Proterozoic Lithospheric Evolution

Geodynamics Series

Geodynamics Series

Proterozoic Lithospheric Evolution

Edited by A. Kröner

Geodynamics Series
Volume 17

American Geophysical Union
Washington, D.C.

Geological Society of America
Boulder, Colorado
1987

Publication No. 0130 of the International Lithosphere Program

Published under the aegis of AGU Geophysical Monograph Board.

Library of Congress Cataloging-in-Publication Data

Proterozoic lithospheric evolution.

(Geodynamics series, ISSN 0277-6669 ; v. 17)
(Publication no. 0130 of the International Lithosphere
Program)
"Based on a symposium entitled 'Proterozoic crustal
dynamics and lithospheric evolution' that was held
during the special session on the International
Lithosphere Programme at the 27th International
Geological Congress in Moscow, USSR, in August 1984"—
Pref.
   1. Geology, Stratigraphic—Pre-Cambrian—Congresses.
2. Geodynamics-Congresses.   3. Sedimentation and
deposition—Congresses.   1. Kröner, A. (Alfred)
II. International Geological Congress (27th : 1984 :
Moscow, R.S.F.S.R.)   III. Series.   IV. Series:
Publication . . . of the International Lithosphere
Program ; no. 0130.
QE653.P798   1987   551.7'15   86-28803
ISBN 0-87590-517-X
ISSN 0277-6669

CONTENTS

FOREWORD

Raymond A. Price

Past-President, International Lithosphere Program
and
Director General, Geological Survey of Canada,
601 Booth Street, Ottawa, Ontario, K1A OE8

The International Lithosphere Program was launched in 1981 as a ten-year project of inter-disciplinary research in the solid earth sciences. It is a natural outgrowth of the Geodynamics Program of the 1970's, and of its predecessor, the Upper Mantle Project. The Program — "Dynamics and Evolution of the Lithosphere: The Framework of Earth Resources and the Reduction of Hazards" — is concerned primarily with the current state, origin and development of the lithosphere, with special attention to the continents and their margins. One special goal of the program is the strengthening of interactions between basic research and the applications of geology, geophysics, geochemistry and geodesy to mineral and energy resource exploration and development, to the mitigation of geological hazards, and to protection of the environment; another special goal is the strengthening of the earth sciences and their effective application in developing countries.

An Inter-Union Commission on the Lithosphere (ICL) established in September 1980, by the International Council of Scientific Unions (ICSU), at the request of the International Union of Geodesy and Geophysics (IUGG) and the International Union of Geological Sciences (IUGS), is responsible for the overall planning, organization and management of the program. The ICL consists of a seven-member Bureau (appointed by the two unions), the leaders of the scientific Working Groups and Coordinating Committees, which implement the international program, the Secretaries-General of ICSU, IUGG and IUGS, and liaison representatives of other interested unions or ICSU scientific committees. National and regional programs are a fundamental part of the International Lithosphere Program and the Chairman of the Coordinating Committee of National Representatives is a member of the ICL.

The Secretariat of the Commission was established in Washington with support from the U.S., the National Academy of Sciences, NASA, and the U.S. Geodynamics Committee.

The International Scientific Program initially was based on nine International Working Groups.

WG-1 Recent Plate Movements and Deformation
WG-2 Phanerozoic Plate Motions and Orogenesis
WG-3 Proterozoic Lithospheric Evolution
WG-4 The Archean Lithosphere

WG-5 Intraplate Phenomena
WG-6 Evolution and Nature of the Oceanic Lithosphere
WG-7 Paleoenvironmental Evolution of the Oceans and Atmosphere
WG-8 Subduction, Collision, and Accretion
WG-9 Process and Properties in the Earth that Govern Lithospheric Evolution

Eight Committees shared responsibility for coordination among the Working Groups and between them and the special goals and regional groups that are of fundamental concern to the project.

CC-1 Environmental Geology and Geophysics
CC-2 Mineral and Energy Resources
CC-3 Geosciences Within Developing Countries
CC-4 Evolution of Magmatic and Metamorphic Processes
CC-5 Structure and Composition of the Lithosphere and Asthenosphere
CC-6 Continental Drilling
CC-7 Data Centers and Data Exchange
CC-8 National Representatives

Both the Bureau and the Commission meet annually, generally in association with one of the sponsoring unions or one of their constituent associations. Financial support for scientific symposia and Commission meetings has been provided by ICSU, IUGG, IUGS, and UNESCO. The constitution of the ICL requires that membership of the Bureau, Commission, Working Groups, and Coordinating Committees change progressively during the life of the project, and that the International Lithosphere Program undergo a mid-term review in 1985. As a result of this review there has been some consolidation and reorganization of the program. The reorganized program is based on six International Working Groups:

WG-1 Recent Plate Movements and Deformation
WG-2 The Nature and Evolution of the Continental Lithosphere
WG-3 Intraplate Phenomena
WG-4 Nature and Evolution of the Oceanic Lithosphere
WG-5 Paleoenvironmental Evolution of the Oceans and the Atmosphere
WG-6 Structure, Physical Properties, Composition and Dynamics of the Lithosphere-Asthenosphere System

and six Coordinating Committees:

CC-1 Environmental Geology and Geophysics
CC-2 Mineral and Energy Resources
CC-3 Geosciences Within Developing Countries
CC-4 Continental Drilling
CC-5 Data Centers and Data Exchanges
CC-6 National Representatives
      Sub-Committee 1 - Himalayan Region
      Sub-Committee 2 - Arctic Region

This volume is one of a series of progress reports published to mark the completion of the first five years of the International Geodynamics Project. It is based on a symposium held in Moscow on the occasion of the 26th International Geological Congress.

Further information on the International Lithosphere Program and activities of the Commission, Working Groups and Coordinating Committees is available in a series of reports through the Secretariat and available from the President — Prof. K. Fuchs, Geophysical Institute, University of Karlsruhe, Hertzstrasse 16, D–7500 Karlsruhe, Federal Republic of Germany; or the Secretary-General — Prof. Dr. H.J. Zwart, State University Utrecht, Institute of Earth Sciences, P.O. Box 80.021, 3508 TA Utrecht, The Netherlands.

R.A. Price, President
Inter-Union Commission on the Lithosphere, 1981-85

PREFACE AND DEDICATION

There have been significant new results in many fields of the earth sciences as concerns the Proterozoic, and many of these originated from new techniques, new concepts and from increasing multidisciplinary research. The most significant result, perhaps, is the recognition of a major crust-forming event of global proportions some 1.7 to 2 Ga ago when juvenile crust was generated in a variety of tectonic settings that are still vigorously debated. Arguments for subduction-related horizontal accretion during the mid-Proterozoic are presented for North America and the Baltic Shield while the Australian crust does not seem to fit such a pattern, and vertical accretion is favored there. In northeast Africa arc and microplate accretion with extensive ophiolite obduction in the late Proterozoic is remarkably similar to modern tectonic processes such as in the SW Pacific, while crust-generation rates seem to have been abnormally high at that time and in this region. In contrast, models for intracrustal orogeny are preferred to explain the evolution of some late Proterozoic African terrains. Proterozoic foredeeps have been recognized adjacent to several major thrust belts in North America and contain major iron formations that may be genetically related to foredeep magmatism.

Geophysical data have shown major differences between Archean and Proterozoic crust that may be related to different crust-forming processes, and seismic profiles together with petrological and structural features in high-grade terrains seem to argue for extensive horizontal tectonism and extension of major shearzones deep into the lower crust. The role of fluids during granulite formation remains a topic of wide interest, and the origin of CO_2 in lower crustal granulites is still a matter of controversy. While chemical and physical heterogeneities are now well established for the modern oceanic upper mantle we still know very little about the ancient sub-crustal mantle lithosphere, and new results from long-range seismic profiles in Siberia seem to indicate that large-scale alternating low- and high-velocity layers characterize this region to depths of more than 200 km.

This volume summarizes recent advances in research to elucidate the evolution of the Proterozoic continental crust and is based on a symposium entitled "Proterozoic Crustal Dynamics and Lithospheric Evolution" that was held during the Special Session on the International Lithosphere Programme at the 27th International Geological Congress in Moscow, USSR, in August 1984.

This volume does not claim to provide a comprehensive coverage of Proterozoic terrains, and there are gaps from Asia, South America, Antarctica and parts of Africa. However, the present contributions present an insight into new approaches that try to understand the evolution of the Proterozoic crust on the basis of better-studied Phanerozoic regions or in applying new field and laboratory data. One further aspect not covered in this volume is the often cited apparent global change in crust-production and tectonics from the Archean to the Proterozoic, and further insights into this and other major problems will hopefully emerge during the second half of the International Lithosphere Program.

I gratefully acknowledge the assistance of numerous reviewers who are listed overleaf and and whose constructive comments and assessments were appreciated by all authors.

Finally, it is my sad duty to report that Mike R. Wilson, major author of a paper in this volume and one of the most active researchers on the Proterozoic evolution of the Baltic Shield, tragically died at the age of 42 years after completion of his manuscript in 1985. His work on granitoid genesis in Sweden has contributed significantly to the new concept of Proterozoic accretion tectonics in Scandinavia, and I should like to honor his research in dedicating this volume to his memory.

Alfred Kröner
Editor and Chairman of Working Group 3,
1982 - 1985
Mainz, Federal Republic of Germany

PHANEROZOIC AND PRECAMBRIAN CRUSTAL GROWTH

Arthur P.S. Reymer[1] and Gerald Schubert

Department of Earth and Space Sciences, University of California, Los Angeles, CA 90024

Abstract. The continental crust grows at present at a small rate of about 1 km^3a^{-1}, mainly by the formation and accretion of island arcs, but also by basaltic magmatism above hotspots. Newly formed and accreted crust forms marginal slivers along some (circumPacific) continents. Nd isotopic data identify certain time intervals in which large Precambrian terrains formed at rates far in excess of modern arc production/accretion rates. The geometry and thermal history of these Precambrian terrains are also different from modern sites of crust formation/accretion. Continental freeboard allows 10 to 40% growth of the continental crust since the Archean, but it seems probable that the actual growth will be close to the value required to maintain freeboard at a constant level, i.e. 20 to 30%. A more precise determination of crustal addition and subtraction rates over geological time is required.

Introduction

Detailed studies of the oceans have led to major improvements in our understanding of physical and chemical geodynamics. However, the young age of the oceanic lithosphere (less than 0.2 Ga) puts a limitation on the reliability of modelling geodynamic processes farther back in time. The former existence of ocean basins, ridges and subduction zones must be deduced from geologic data (e.g. ophiolites, blueschists, eclogites). While evidence for former ocean basins is still reasonably well preserved in Paleozoic orogenic belts (Urals, Caledonides), it is rare and enigmatic in Precambrian rocks older than ca. 0.8 Ga (only tectonic patterns in some Precambrian orogenic sections could be interpreted as consistent with a plate tectonic regime similar to the present one). Determination of continental crustal growth rates through geologic time is not only an important task in itself but its comparison with modern crustal growth rates can elucidate whether plate tectonic crustal growth

____

[1] New address: Dept. of Marine, Earth & Atm. Sciences, North Carolina State Univ., Raleigh, NC 27905-8208, USA.

mechanisms comparable to the Phanerozoic have operated in the Precambrian.

This report reviews our earlier work along these lines [Reymer and Schubert, 1984; 1985; Schubert and Reymer, 1985]; it is organized in three parts. The first deals with modern crustal growth processes. We define the various processes which lead to crustal addition and subtraction and define crustal growth as the difference between addition and subtraction. We calculate modern rates of addition, subtraction, and growth and assess the geometry of Phanerozoic accreted margins. The second part deals with Precambrian crustal growth. We apply our modern growth rates to certain Precambrian orogenic terrains. We use the constancy of freeboard and its uncertainty to assess constraints on crustal growth since 2.5 Ga, and investigate the possibilities of crustal thickening and thinning and changes in ocean volume. In the last part, we discuss the constraints that can presently be applied to model crustal growth curves and the implications of the shapes of these curves for geotectonic and geodynamic models.

Mass Transfer Processes Between Crust and Mantle

Crustal addition is defined as the flux of crust-like material from the mantle to the outer shell of the Earth (crust) [Dewey and Windley, 1981]. This crust-like material may vary from felsic granites to mafic flood basalts and ophiolites. The body of material, which segregates from the mantle source rock, may be emplaced directly into existing continents, or after a relatively short lifespan as intraoceanic masses, may be tectonically accreted onto existing continents. The sum of the compositions of these added materials should equal the average composition of the continental crust which, however, is imperfectly known [Kay & Kay, 1981]. Sites and rates of crustal addition are directly connected to the geometry and nature of modern (Mesozoic-Cenozoic) plate tectonics. Magmatic arcs form at subduction zones and basalts intrude along rifts and above hotspots.

Although virtually all the oceanic crust is returned to the mantle, its formation, alteration

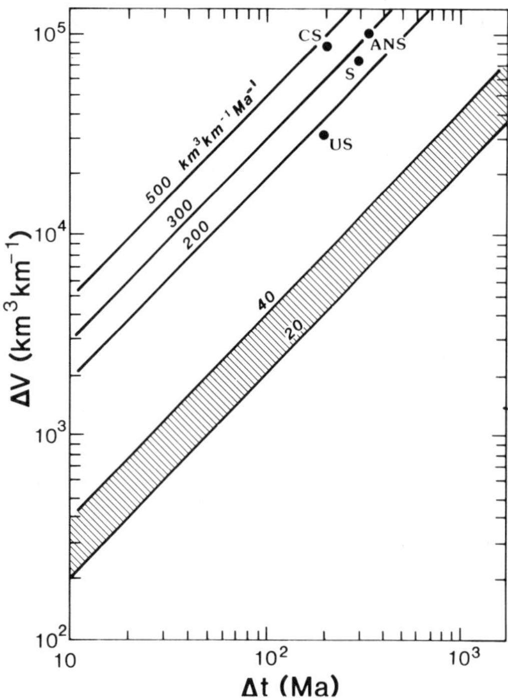

Fig. 1. Addition volumes V of the regions listed in Table 1 versus the growth times of the regions. V is expressed in km^3km^1 by assuming that each region formed by accretion of one arc along the long dimension of the region. The stippled band between addition rates of 20 and 40 $km^3km^1Ma^1$ represents the range of Mesozoic Cenozoic magmatic arc addition rates determined by Reymer and Schubert [1984]. CS--Canadian Shield; S Svecokarelian; US West Central USA; ANS-- Arabian Nubian Shield.

by interaction with seawater, and subsequent heating during subduction is an important contribution to the generation of magmas along arcs [Ringwood, 1982; DeVore, 1983]. Thus, the presence of oceans on the Earth and their capacity to hydrate basalts has been held responsible for the fact that the Earth has sialic continents, whereas the other terrestrial planets may lack both sialic continents and oceans, and do not show signs of plate tectonics [Campbell and Taylor, 1983; Walker and Dennis, 1983].

Isotopic geochemical models of crust and mantle require that fractionated crustal material is returned to the mantle in order to explain observed isotopic patterns in oceanic basalts [e.g. Allegre, 1982; DePaolo, 1983]. Physical support for subtraction of crustal material comes from sediment-starved trenches (e.g. Marianas), where most or all of the ocean floor sediments carried into the trench are apparently subducted together with the oceanic lithosphere. At other sites (e.g. Antilles), ocean floor sediments are scraped off and accreted to form a sediment wedge or fore-arc, but part of these sediments are also subducted [White and Dupré, 1985]. Tectonic erosion [Karig, 1974] may occur at a few places, notably along the Peruvian trench, where the subducting plate removes material from the overlying continental plate. Part of the subducted sediments or eroded material is returned directly to the crust when the material becomes involved in arc magmas [White and Dupré, 1985].

Net crustal growth is the difference between crustal addition and crustal subtraction. Addition and subtraction are both quasi-continuous processes in modern plate tectonics, but the difference may vary depending on changes in the rates of each of these two processes. For example, the addition rate may be a function of convergence rate [Reymer and Schubert, 1984], or of the angle of subduction and age of the descending lithosphere [Abbott and Hoffman, 1984]. The subtraction rate will be influenced by the supply of material to the ocean basins by major rivers [e.g. McLennan and Taylor 1983] (erosion rate in turn depending on latitude of the continents and climate) and the plate velocities, as well as the mode of subduction, e.g. strong coupling versus weak coupling [Uyeda, 1982]. Variations in crustal growth can therefore be expected, but the variations required during some Precambrian time intervals are clearly far in excess of what can be reasonably expected on the basis of modern crustal growth processes [Reymer & Schubert, 1985], a point which will be addressed below. Nevertheless, the process of crustal growth is not unidirectional [Moorbath, 1977]. It allows, at least theoretically, crustal growth to be positive, zero, or negative.

Mesozoic-Cenozoic Crustal Addition

Mesozoic-Cenozoic crustal addition along magmatic arcs was determined by us to proceed with a rate of 20–40 $km^3km^{-1}Ma^{-1}$ [Reymer & Schubert, 1984] as shown in Fig. 1 by the shaded band. We also showed that volcanic eruption rates are a factor of 5-10 less than total addition rates, and that individual hotspots produce basalts at a much higher rate, although the average addition rate over a whole hotspot chain is comparable to that of magmatic arcs. By including contributions from both ophiolites and the oceanic basement of accreting arcs, we arrived at a total worldwide addition rate of 1.7 km^3a^{-1} (Table 1). Addition along magmatic arcs alone is estimated at 1.1 km^3a^{-1}.

We considered in detail crustal addition in the SW Pacific [Reymer & Schubert, 1985], because this region is sometimes used as an analog for the evolution of certain Paleozoic and Precambrian terrains, e.g. the Arabian-Nubian shield [Kroner, 1985]. Fig. 2 shows the study region with a land and shelf area totaling about 6×10^6 km^2. This region is about the size of the Arabian-Nubian shield and is comparable in area to other Precambrian age provinces. We calculated

TABLE 1. Mesozoic-Cenozoic Crustal Growth Rate[a]

	Value
ADDITION:	
Average arc accretion rate (30 $km^3km^{-1}Ma^{-1}$) x length of arcs (37,000 km)	1.1 km^3a^{-1}
Accreted arc basement	0.05
Ophiolites	0.07
Intraplate volcanism, oceanic	0.2
Continental	0.1
Under- and overplating	0.18
Total addition	1.7
SUBTRACTION[b]:	
Subducted continental material	0.6
GROWTH = ADDITION - SUBTRACTION	1.1 km^3a^{-1}

[a] After Reymer and Schubert [1985].
[b] Average of estimates quoted in the text.

the length and lifespan of the various subduction systems that are, or have been, operative in this region since the Upper Paleozoic, a roughly 300 Ma time interval (also comparable to the growth interval of many of the Precambrian and Paleozoic provinces). We applied our average arc-addition rate of 30 $km^3km^{-1}Ma^{-1}$ and found that a maximum of 30% of this SW Pacific region can be composed of newly added crust. This figure compares well with the amounts of newly added and pre-existing crust shown on the geological map (Fig. 2) and supports the application of our average modern arc-addition rate to this type of analysis. Our conclusions are further supported by Sm-Nd data on Malaysian granites [Liew and McCulloch, 1985] which indicate the presence of mid-Proterozoic crust in this region. Average crustal thickness in this region is about 25 km. The consequences of this calculation for Precambrian tectonic/age provinces will be discussed later, but it is already clear, that large Precambrian provinces such as the Arabian-Nubian shield should consist largely of preexisting crust, if the analogy is valid, and this is at variance with the interpretation of isotopic geochemical data.

Mesozoic-Cenozoic Net Crustal Growth

In order to obtain a measure of net crustal growth over the last 0.2 Ga, one must subtract from the added volume whatever continental mate rial has been either returned to the mantle or incorporated in arc magmas. As far as we know, mantle recycling occurs mainly by sediment subduction along oceanic trenches. Our estimate of about 0.6 km^3a^{-1} of subtracted material [Reymer & Schubert, 1984], is close to estimates of less than 0.8 km^3a^{-1} by McLennan & Taylor [1983], 0.5+0.2 km^3a^{-1} by Veizer and Jansen [1984, pers. comm.], and about 0.7 km^3a^{-1} as recalculated from

data by White and Dupré [1985] from the Lesser Antilles. More work should be done to constrain this figure, because isotope geochemists claim subtraction rates of 2.5+0.5 km^3a^{-1} [DePaolo, 1983] or 1-3 km^3a^{-1} [Armstrong, 1981]. Either one or both of these estimates need revision, or other mechanisms exist for returning continental crust back to the mantle reservoir, e.g. delamination of the continental lithosphere including the lower crust [Bird, 1978; Dewey & Windley, 1981].

Upon combining addition and subtraction rates we calculate a net Mesozoic-Cenozoic crustal growth rate of about 1 km^3a^{-1}. Since the average growth rate of the continental crust over geologic time is about 1.7 km^3a^{-1} (total volume of continental crust divided by 4.56 Ga [Reymer & Schubert, 1984]), crustal growth in the Precambrian must have been faster than at present.

Redistribution of Continental Crust

Continental crust, once in existence, may collide with other segments of continental crust or break up due to intracontinental rifting. Thus, modern continental tectonics, while responsible for the main Phanerozoic rift zones and orogenic belts and perhaps also some Precambrian orogenic belts, does not produce or destroy continental crustal material. Addition and subtraction occurs along subduction zones. However, some overlap in time and space exists: fairly large amounts of basalts may intrude the continental crust during initial stages of rifting; Andean-type orogenic margins involve the addition of new material along the magmatic arc axis. Nevertheless, it is useful to discriminate between processes which involve mantle-crust interaction and mass transfer from processes which merely redistribute existing continental crustal material.

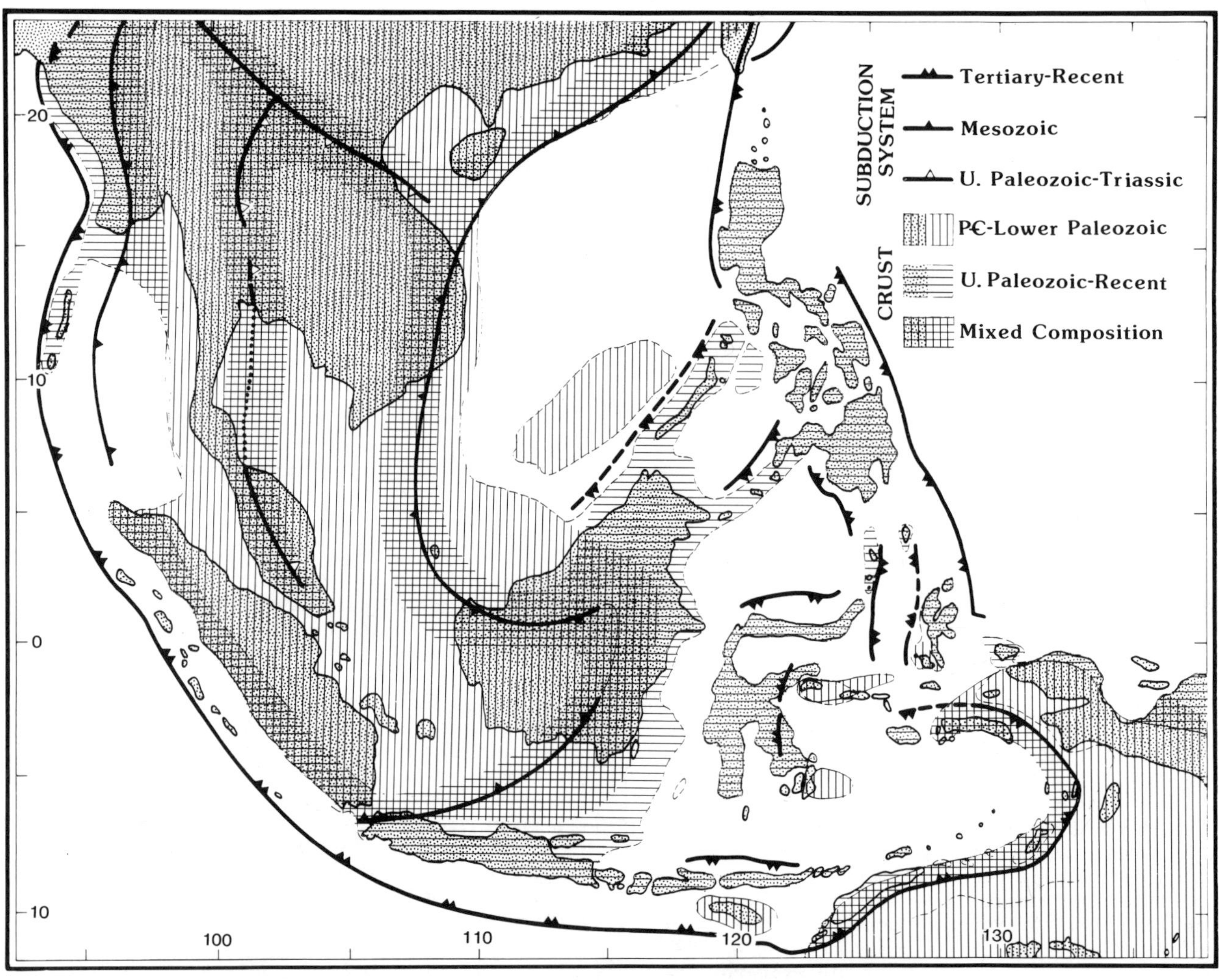

Fig. 2. Map of the area of the SW Pacific. Subduction systems active in different time intervals are indicated. Dark or hachured continental crust newly added during the last 300 Ma either as accreted arcs or onto preexisting crust. Cross hatched Precambrian and Lower Paleozoic crust.

The original configuration of Precambrian age provinces can be restored once worldwide mapping of the age of the continents has sufficiently progressed and can be combined with paleomagnetic data. The main problem is more general in nature, namely extracting information from the deeper continental crust. Seismic studies, deep continental drilling, xenolith studies and new geochemical techniques for the analysis of granitoid rocks will all help to constrain the age, structure and composition of the deeper crust.

Precambrian Crustal Growth

Although it is not yet possible to map out all additions to the continental crust over Earth's history, we may obtain a constraint on the growth of the continental crust during the last 2.5 Ga by looking at the freeboard of the continents through time.

Continental Freeboard

Constancy of freeboard since the Archean has been used to argue for no-growth of the continental crust during the Proterozoic and Phanerozoic [Armstrong, 1968; 1981].

Temporal variations in freeboard (transgressions and regressions of sealevel) can arise from numerous causes including melting of polar ice, post-glacial rebound, changes in the number and areal distribution of continental segments [Harrison et al., 1981], changes in the volume of midocean ridges and continents, etc. Some of these effects produce only short term changes in freeboard. Here, we consider only the secular

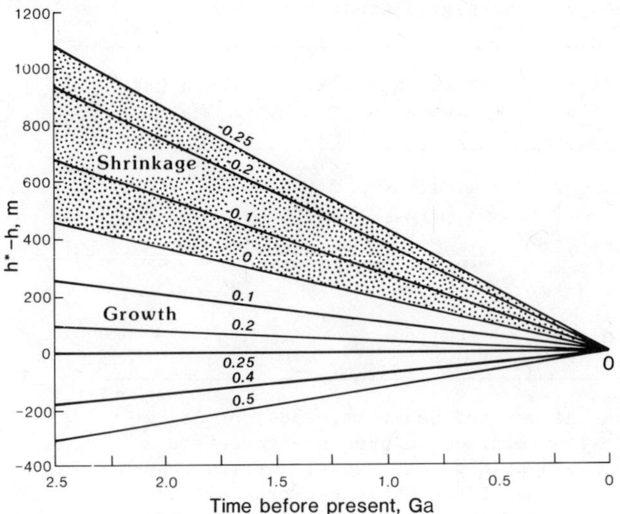

Fig. 3. Freeboard variation with time during the Proterozoic Phanerozoic for prescribed percentage changes in continental volume between 2.5 Ga ago and the present. Both crustal growth and shrinkage are considered. Crustal thickness is assumed constant.

changes in freeboard due to changes in the area and thickness of the continents, and changes in the volume of the oceans particularly as a consequence of the deepening of the ocean basins with time on a cooling Earth.

Equations relating the volume, area, thickness and density of the continental crust to the volume, area, mean depth and density of the oceans are given in Schubert and Reymer [1985] (see also Wise, 1974). If no continental growth has occurred since the Archean, and if crustal thickness has remained the same (around 38 km), freeboard must have increased by 400 m. Conversely, constant freeboard requires about 25% crustal growth since 2.5 Ga ago, due to the deepening of the ocean basins on a cooling Earth. However, measures of freeboard now and 2.5 Ga ago are imperfectly known and an uncertainty of ± 200 m translates into a possible variation in crustal growth of 10 to 40% (Fig. 3). Crustal shrinkage can virtually be ruled out because this would require an increase of more than 500 m in freeboard, an unlikely large amount.

Changes in crustal thickness of more than 10% must be considered unlikely. For example, 10% crustal thickening since the Archean would mean that the continents did not rise above the oceans until at least the mid-Proterozoic. Geological data show evidence of erosion in the Late Archean and Early Proterozoic, contradicting even a small change in crustal thickness. Thus, crustal thickness is considered to have remained essentially constant, in agreement with conclusions by Condie [1973] and Wise [1974].

There are a large number of other effects which may have influenced freeboard. These include variations in ocean volume (assumed to be constant in our previous calculations) and sediment thickness in the ocean basins. Lack of data makes a useful discussion of these points difficult. However, we can invert the problem and consider data on the actual growth of the continental crust, as deduced from arc production rates and Sm-Nd age determinations, and compare these with the curves of Fig. 3.

Crustal Growth Since 2.5 Ga

Based on our Mesozoic-Cenozoic crustal growth rate of about 1 km^3a^{-1} and assuming that Phanerozoic crustal growth mechanisms (arcs, hotspots) did not differ from those observed today, we calculate about 6% growth of the continental crust during the last 0.5 Ga. Presently available Proterozoic Sm-Nd crustal formation ages [e.g. Duyverman et al., 1982; Patchett and Bridgwater, 1984; Patchett et al., 1984; Nelson and DePaolo, 1985; Liew and McCulloch, 1985; Wilson et al., 1985] show major episodes of crust formation 1.9 to 1.7 Ga ago (West-Central USA and Svecokarelian in northern Europe, Ketidilian in Greenland) and 0.9 to 0.6 Ga ago (the formation of the Arabian-Nubian shield). At least 10% growth of the continental crust occurred in these two periods, and it seems highly probable that this growth will turn out to be even larger once a more complete dataset is obtained. Thus, minimum crustal growth since the end of the Archean can be estimated at 16%, and a number on the order of 20 to 30% may be more realistic.

Such a growth rate would suggest that freeboard remained within 200 m, and possibly within 100 m, of its present value (Fig. 3). The relative importance of a number of unknown effects on freeboard can now be assessed within this perspective. A secular variation of less than ± 5% in ocean volume (corresponding to a ± 200 m change in sealevel) cannot be resolved. If a high imbalance were to be found between degassing rates and water subduction rates along trenches, then other processes must be identified to offset this imbalance. An increase in sediment thickness on the ocean floor would be countered by a slight increase in mean depth of the oceans due to the increased load. Erosion of the continents would tend to reduce freeboard, but this would be compensated by the isostatic adjustment of the continents. The combined effect of all this requires detailed analysis, but the above growth rate estimate suggests that it is small.

Other processes which would further tend to complicate the effect of deepening of the ocean basins are possible, and these add uncertainty to the case for positive crustal growth since 2.5 Ga ago. Examples include secular variations in the depth to ridge crests, in oceanic crustal thickness, and in the development of deep mantle roots under the continents. However, without the crustal growth estimate above it would be diffi-

TABLE 2. Growth Rates of Segments of Continental Crust

	Area (km^2)	Volume[a] (km^3)	Age[b] (Ga)	Growth Rate (km^{3}a^{-1})	Arc Addition Rate[c] (km^{3}km^{-1}Ma^{-1})
Canadian Shield	8x10^6	4x10^8	3.0-2.7	1.33	295
Svecokarelian	5x10^6	2.5x10^8	2.2-1.9	0.83	160
West-central USA	4x10^6	2x10^8	1.9-1.7	1.00	185
Arabian-Nubian Shield	6x10^6	3x10^8	0.9-0.6	1.00	310

[a] Calculated using an average crustal thickness [Reymer and Schubert, 1984] of 38 km with the addition of 12 km to account for erosion (minimum volumes are therefore also minimum). Large accretionary wedges composed of old continental detritus, such as Barbados (Lesser Antilles) could account for some of the estimated crustal volume. However, such large wedges are not abundant and their presence either at the surface or in isotopic signature has not been demonstrated for any of the listed terrains.
[b] Based on Sm-Nd crustal formation ages.
[c] Taking longest dimension of each area as the length of the arc.

cult to explain the known substantial addition of new crust since 2.5 Ga, because an equivalent amount of crust would have to be removed. We have argued earlier that net reduction in areal size of existing continental crust is difficult and probably minor.

We conclude that the approximate constancy of freeboard model can be used with simple crustal growth models to constrain crustal evolution since the Archean. Crustal thickness has probably remained virtually constant, and we would argue for positive growth of the crust on the order of 20 to 30% since 2.5 Ga ago. Small variations in ocean volume and other effects are not resolvable. This suggests that 70 to 80% of the crust was in existence by the end of the Archean, an often quoted number. In the next section we show how values of modern arc addition rates, derived above, can be used to elucidate mechanisms of crust formation in the Precambrian.

Rapid Growth of Some Major Segments of Continental Crust

The combination of our average modern arc addition rate of 30 km^{3}km^{-1}Ma^{-1} with the Sm-Nd age dating technique gives us a new method to investigate the growth rate of Precambrian and Paleozoic tectonic provinces and to test the applicability of modern plate tectonic processes to the ancient record. This is based on the calculation of crustal formation ages [McCulloch and Wasserburg, 1978] which date the time of separation of continental crustal material from a mantle reservoir. Since this process corresponds to arc magmatism and hotspot activity, Nd ages reflect the time of arc magmatism if a strict uniformitarian approach applies.

Care must be taken with the interpretation of Sm-Nd model ages. Such ages should be calculated with respect to the same depleted mantle curve if comparisons among various laboratories are made. More important, model ages only define true "crustal formation ages" if the measured Nd values plot on or close to the depleted mantle curve, or if they define a linear array intersecting the depleted mantle curve at T_{DM}. In other cases, Nd model ages may represent "mixed ages", reflecting older basement components and new addition at the time of magma formation. In general, detailed studies are required to define the Nd systematics of a certain crustal segment [e.g. DePaolo, 1981; Farmer & DePaolo, 1983, 1984; Nelson & DePaolo, 1985].

We analyzed a number of orogenic provinces: the Canadian Shield, 3.0-2.7 Ga in age [McCulloch and Wasserburg, 1978], the West-Central USA (1.9-1.7 Ga), the Svecokarelian in Northern Europe (2.2-1.9 Ga), and the Arabian-Nubian Shield (0.9-0.6 Ga) (Proterozoic ages referenced above).

Table 2 expresses the growth rates of the various regions in two formats. One is simply the result of dividing total produced rock volume by the time interval, the other expresses the growth rate in units of km^{3}km^{-1}Ma^{-1}. The latter allows rapid comparison with modern arc addition rates and allows an estimate of how many arcs would have been involved in the build-up of a particular segment. The total addition rates are close to 1 km^{3}a^{-1}, or equal to the present day worldwide addition rate concentrated in one region. Expressed as a number of arcs, Table 2 and Fig. 4 show that 6-12 "superarcs" each operating over the entire time span (about 300 Ma) and over the entire length of each crustal segment are necessary to build these crustal seg-

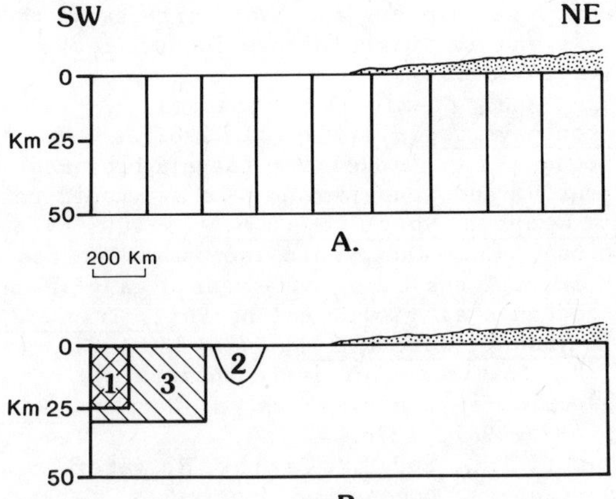

Fig. 4. A Cross section through the Arabian-Nubian shield. The thickness of 50 km represents present crust (approximately 38 km thick) plus a minimum estimate for erosion [Reymer and Schubert, 1984] of about 12 km. The volumes of the 9 "superarcs" which would be necessary to build the crust on the basis of the arc accretion model are indicated (see text). The section of the shield now covered with Phanerozoic deposits is also indicated. B--Same cross-section. 1, Volume of a 100 Ma old arc; 2, volume of a present-day arc (Marianas). Twenty-seven type 1 arcs or 55 type 2 arcs would be necessary to build the crust if the process consisted only of arc accretion. 3, Volume of accreted margin of the North American Cordillera, which took about 250 Ma to accumulate (volume of 1 included). All arcs are assumed to extend 2500 km along the longest dimension of the shield. A volume corresponding to 6 km of oceanic basement is included in the arcs in A and B. From Reymer and Schubert [1985].

ments. Fig. 4a shows this arrangement for the Arabian-Nubian shield. More realistic arc volumes, shown in Fig. 4b, imply that tens of arcs are necessary to build these segments.

Because arc formation/accretion is not rapid enough to build large crustal segments that formed within a few hundred million years, these segments must either contain large amounts of undetected older crustal material, or other mechanisms of crustal addition operated, possibly together with arc accretion. The undetected presence of large amounts (at least 80%) of older preexisting crust seems improbable, despite the geographically sparse data base. Therefore, we conclude that arc accretion has been of subordinate importance for the formation of these crustal segments.

The same conclusion can be reached in a qualitative way by comparing Mesozoic-Cenozoic continental growth patterns with the Precambrian terrains listed here. Instead of long narrow slivers of new crust accreted to preexisting cratons, we observe wide, more or less square Precambrian terrains with equal growth intervals.

The Shape of the Crustal Growth Curve

Compilation of a crustal growth curve from geochronologic data was pioneered by Hurley and Rand [1969], but new data have substantially changed their original curve [Reymer and Schubert, 1984; Nelson and DePaolo, 1985]. Fig. 5 shows two hypothetical examples of crustal addition and subtraction rates over geologic time. The subtraction rate curve S consists of two parts: one covers the period 4.5 to about 4 Ga, when wholesale subduction of any existing crust occurred (to explain the absence of rocks older than ca. 4 Ga [e.g. Miller and O'Nions, 1985]), and a second covers the rest of the time until the present, during which sediment subduction is the most important subtraction mechanism. The increase in crust survivability is correlated

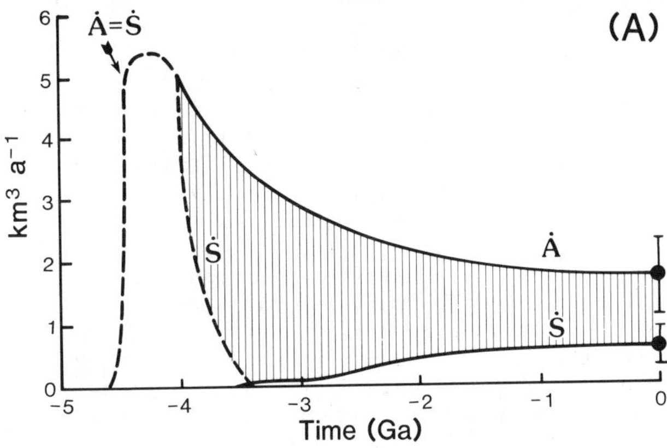

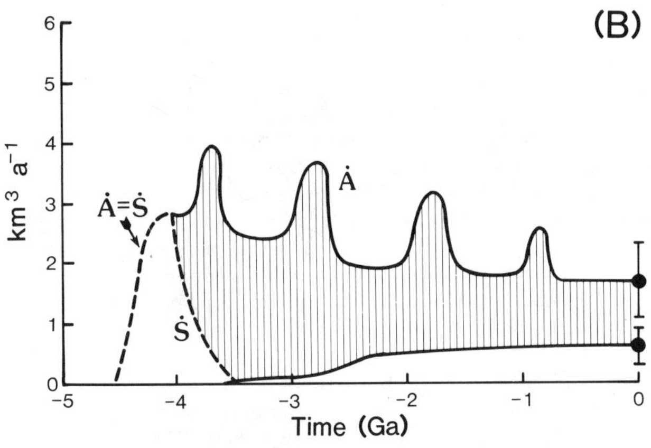

Fig. 5. (a) Hypothetical addition rate curve (A) proportional to the decline in heat flow from the mantle. S – subtraction rate curves. (b) Hypothetical addition rate curve showing several short periods of high crustal addition rates. Other curves identical to (a). See text for discussion.

with an increase in sediment subduction. The addition rate curve A in Fig. 5a represents a model in which addition (e.g by arc formation/accretion) is proportional to the decline in heat flow from the mantle. Fig. 5b shows an addition rate curve that contains a number of spikes, arbitrarily placed at 3.83.5 Ga, 3.02.6 Ga, 1.9-1.6 Ga and 0.9-0.6 Ga. The curve in Fig. 5b suggests major addition events in the mantle during certain periods in Earth's history. It requires an episodic modulation of crustal addition by either arc accretion or hot spot mechanisms. Integration of A-S over geologic time yields the present day volume of continental crustal material, about 7.76×10^9 km^3 [Reymer and Schubert, 1984]. Both curves are hypothetical, because we have as yet only two data points: the addition and subtraction rates for the last 0.2 Ga, discussed earlier in this report. However, the curves illustrate that a quantitative assessment of crustal growth with time would be able to put constraints on geodynamic models of the Earth's evolution.

Conclusions

Mesozoic-Cenozoic crustal addition and subtraction rates indicate that the amount of continental crust has grown relatively slowly during the past 200 Ma at a rate of about 1 km^{3}a^{-1}. The main mechanisms of crust formation during this period are arc magmatism and arc accretion, and the intrusion of hotspot derived basalts.

Continental freeboard calculations suggest that the continental crust has grown 10 to 40% since the Archean. Determination of the actual growth over this period by geochemical means will allow inversion of the freeboard model to constrain other variables, such as the water volume of the oceans, about which little is known. Preliminary estimates suggest that growth may have actually been close to that required to maintain constant freeboard, i.e. around 25%. Addition of material to the continental crust has occurred episodically [Moorbath, 1977], based on new Nd isotopic data. Major periods occurred at 3.0-2.6 Ga, 1.9-1.6 Ga and 0.9-0.6 Ga.

Future work should be directed towards constraining the crustal addition curve, and towards developing more efficient models of crust formation. Crust-mantle exchange processes at the crust-mantle boundary (delamination, underplating) could have been important.

Acknowledgments. We thank N. Arndt, M.J. Bickle and one anonymous reviewer for helpful comments. This study was done under NASA grant NAG-9-76.

References

Abbott, D.H. and S.E. Hoffman, Archean plate tectonics revisited, part 1: Heat flow, spreading rate, and the age of subducting oceanic lithosphere and their effects on the origin and evolution of continents, Tectonics, 3, 429-448, 1984.

Allegre, C.J., Chemical geodynamics, Tectonophysics, 81, 109-132, 1982.

Armstrong, R.L., A model for the evolution of strontium and lead isotopes in a dynamic earth, Rev. Geophys. Space Phys., 6, 175-200, 1968.

Armstrong, R.L., Radiogenic isotopes: The case for crustal recycling on a near-steady-state no-continental growth earth, Phil. Trans. R. Soc. London, Ser. A, 301, 443-472, 1981.

Bird, P., Initiation of intracontinental subduction in the Himalaya, J. Geophys. Res., 83, 4975-4987, 1978.

Campbell, I.H., and S.R. Taylor, No water, no granites - no oceans, no continents, Geophys. Res. Lett., 10, 1061-1064, 1983.

Condie, K.C., Archean magmatism and crustal thickening, Geol. Soc. Am. Bull., 84, 2981-2992, 1973.

DePaolo, D.J., Nd in the Colorado Front Range and Implications for crust formation and mantle evolution in the Proterozoic, Nature, 291, 193-196, 1981.

DePaolo, D., The mean life of continents: Estimates of continental recycling rates from Nd and Hf isotopic data and implications for mantle structure, Geophys. Res. Lett., 10, 705-708, 1983.

DeVore, G.W., Relations between subduction, slab heating, slab dehydration and continental growth, Lithos, 16, 255-263, 1983.

Dewey, J.F., and B.F. Windley, Growth and differentiation of the continental crust, Phil. Trans. R. Soc. London, Ser. A, 301, 189-206, 1981.

Duyverman, H.J., N.B.W. Harris, and C.J. Hawkesworth, Crustal accretion in the Pan-African: Nd and Sr isotope evidence from the Arabian Shield, Earth Planet. Sci. Lett., 59, 315-326, 1982.

Farmer G.L., and D.J. DePaolo, Origin of Mesozoic and Tertiary granite in the western United States and implications for pre-Mesozoic crustal structure - 1. Nd and Sr isotopic studies in the geocline of the northern Great Basin, J. Geophys. Res., 88, 3379-3401, 1983.

Farmer, G.L. and D.J. DePaolo, Origin of Mesozoic and Tertiary granite in the western United States and implications for pre-Mesozoic crustal structure - 2. Nd and Sr isotopic studies of unmineralized and Cu- and Mo-mineralized granite in the Precambrian craton, J. Geophys. Res., 89, 10141-10160, 1984.

Harrison, C.G.A., G.W. Brass, E. Saltzman, J. Sloan II, J. Southam, and J.M. Whitman, Sea level variations, global sedimentation and the hypsographic curve, Earth Planet. Sci. Lett., 54, 1-16, 1981.

Hurley, P.M., and J.R. Rand, Pre-drift continental nuclei, Science, 164, 1229-1242, 1969.

Karig, D.E., Tectonic erosion at trenches, Earth Planet. Sci. Lett., 21, 209-212, 1974.

Kay, R.M., and S.M. Kay, The nature of the lower crust: inferences from geophysics, surface geology, and crustal xenoliths, Rev. Geophys. Space Phys., 19, 271-297, 1981.

Kroner, A., Ophiolites and the evolution of tectonic boundaries in the Late Proterozoic Arabian-Nubian shield of Northeast Africa and Arabia, Precambrian Res., 27, 277-300, 1985.

Liew, T.C., and M.T. McCulloch, Genesis of granitoid batholiths of Peninsular Malaysia and implications for models of crustal evolution: Evidence from a Nd-Sr isotopic and U-Pb zircon study, Geochem. Cosmochim. Acta, 49, 587-600, 1985.

McCulloch, M.T., and G.J. Wasserburg, Sm-Nd and Rb-Sr chronology of continental crust formation, Science, 200, 1003-1011, 1978.

McLennan S.M., and S.R. Taylor, Continental freeboard, sedimentation rates and growth of the continental crust, Nature, 306, 169-172, 1983.

Miller, R.G., and R.K. O'Nions, Source of Precambrian chemical and clastic sediments, Nature, 314, 325-330, 1985.

Moorbath, S., Ages, isotopes and evolution of Precambrian continental crust, Chemical Geology, 20, 155-187, 1977.

Nelson, B.K., and D.J. DePaolo, Rapid production of continental crust 1.7-1.9 Ga ago: Nd and Sr isotopic evidence from the basement of the North American midcontinent, Geol. Soc. Am. Bull., 96, 746-754, 1985.

Patchett, J., and D. Bridgwater, Origin of continental crust of 1.9-1.7 Ga age defined by Nd isotopes in the Ketidilian terrain of S. Greenland, Contrib. Mineral. Petrol., 87, 311-318, 1984.

Patchett, J., R. Gorbatschev, O. Kouvo, and W. Todt, Origin of continental crust of 1.9-1.7 Ga age: Nd isotopes in the Svecokarelian terrain of Sweden and Finland, Geol. Soc. Am. Abstracts with Programs, 16, 619, 1984.

Reymer, A., and G. Schubert, Phanerozoic addition rates to the continental crust and crustal growth, Tectonics, 3, 63-77, 1984.

Reymer, A., and G. Schubert, Rapid growth of major segments of continental crust, Geology, 14, 299-302, 1986.

Ringwood, A.E., Phase transformations and differentiation in subducted lithosphere: implications for mantle dynamics, basalt petrogenesis and crustal evolution. J. Geol., 90, 611-643, 1982.

Schubert, G. and A.P.S. Reymer, Continental volume and freeboard through geologic time, Nature, 316, 336-339, 1985.

Uyeda, S., Subduction zones: An introduction to comparative subductology, in Geodynamics Final Symposium, edited by A.L. Hales, Tectonophysics, 81, 133-159, 1982.

Veizer, J., and S.L. Jansen, Basement and sedimentary recycling and continental evolution, J. Geol., 87, 341-370, 1979.

Walker, C.T., and J.G. Dennis, Exogenic processes and the origin of the sialic crust, Geol. Rundschau, 72, 743-755, 1983.

White, W.M., and B. Dupre, Sediment subduction and magma genesis in the Lesser Antilles: isotopic and trace element constraints, J. Geophys. Res., 91, 5927-5941, 1986.

Wilson, M.R., P.J. Hamilton, A.E. Fallick, M. Aftalion, and A. Michard, Granites and early Proterozoic crustal evolution in Sweden: evidence from Sm-Nd, U-Pb, and O isotope systematics, Earth Planet. Sci. Lett., 72, 376-388, 1985.

Wise, D.U., Continental margins, freeboard, and the volumes of continents and oceans through time, in Geology of Continental Margins, edited by C.A. Burk and C.L. Drake, pp. 45-58, Springer, New York, 1974.

PETROLOGIC ASPECTS OF PRECAMBRIAN GRANULITE FACIES TERRAINS BEARING ON THEIR ORIGINS

Robert C. Newton

Department of the Geophysical Sciences, University of Chicago, Chicago, Illinois 60637

Abstract. Several petrologic features of
Precambrian and younger granulite facies terrains may be of fundamental importance in deduc-
ing the mechanisms of high-grade metamorphism.
These include:
Early-metamorphic horizontal tectonism. Great
crustal thickening accompanied high-grade metamorphism in virtually all well-described terrains. Pressures of 8 ± 2 kbar, corresponding
to burial under one continental thickness, were
generated in the majority of metamorphic events.
*High-temperature metamorphism outlasting horiz-
ontal tectonics*. Temperatures of 650°–950° were
generated over large terrains. Recrystallization
partially obliterated early-formed flat foliation. These temperatures exceed the normal geotherms of late-Archaean continents. Thermal perturbation accompanying crustal thickening is
clearly implied.
Low H_2O activity. Desiccation of whole terrains
prior to or during metamorphism was necessary for
initially H_2O-rich crustal materials, to sustain
the high temperatures without wholesale melting.
Fluid inclusions in minerals in granulite facies
terrains are commonly carbonic.
*Granulite facies transition zones with mappable
isograds*. The zones of progressive metamorphism
are analogous in many ways to those of younger
orogenic-metamorphic belts.
*Massif anorthosites with high-pressure metamor-
phic overprint*. Pre-orogenic igneous rocks were
emplaced in a rifting continental setting at
shallow depths, creating dry, high-temperature
aureoles. The shallow thermal effects were sometimes preserved through subsequent high-pressure
recrystallization.

A consistent, oft-repeated sequence of events
is implied by these features. First, there was
stretching and thinning of a continental interior
by subcrustal activity. Shallow marine sediments, many highly carbonated and some evaporitic, were deposited in rift basins and intruded
by anorogenic magmas, including anorthosites.
Closure of the abortive rifts took place with one
crustal segment overriding the shallow marine
basin and trapping the supracrustals at the flat
continent-continent interface. Temperatures were
increased by augmented radioactivity of the
thickened crust, with additional thermal action
from upward transport by magmas and volatiles
generated in the deep lower continental plate.
The high CO_2 content of buried sediments or
mantle CO_2 tapped by crustal shearing lowered
H_2O activities and prevented extensive melting.
Subsequent uplift and erosion, either rapid or
slow, restored the continent to normal thickness,
and often returned the supracrustal horizon to
the surface.

The above sequence of events could be explained by opening and closing of a small ocean basin
during operation of the Wilson Cycle. However,
evidence of entirely ensialic orogenesis in some
terrains indicates that rifting commonly stopped
short of opening of an ocean. Kröner's A-subduction hypothesis, wherein old dense subcrust
decoupled and foundered beneath an abortive rift,
triggering closure, could be a viable alternative. Static models of crustal thickening and
heating by magmatic underplating or overplating
are less capable of explaining the petrologic
features.

Introduction

Granulite facies terrains have come under intensive study in recent years because of their
bearing on crustal evolution and the nature of
the deep crust. The most generally recognized
fact of the large Precambrian high-grade terrains
is that they were recrystallized in high-temperature, low-H_2O metamorphic environments. Another
salient feature is that some of the highest-grade
terrains show marked depletion of the large-ion
lithophile (LIL) elements, such as Rb, U and Th,
relative to average upper-crustal rocks. This
fact, more than any other, suggests granulite
terrains as possible models of the deep continental crust (Fountain and Salisbury, 1981), because
of the low radioactive heat generation required
by crustal heat flow studies (Heier, 1973).

Other aspects of granulite terrains are less
generally agreed upon or are subjects of active
debate. Much of the debate centers around the
possible role of Precambrian plate tectonics in
growth of the continents. Several authors have
found evidence of meta-ophiolites and calc-alka-

line plutonic suites which could represent
ancient collapsed ocean basins and volcanic arc
magmatic rocks. Kröner (1982), however, pointed
out that several examples of Precambrian high-
grade terrains display continental interior environments, with deformed shallow platform sediments unconformably overlying older sialic crust.
The major evidence for the existence of old
oceanic crust and arc plutonics in the older Precambrian terrains is almost entirely geochemical
and may be suspect because of subsequent metamorphic alteration and limited chemical criteria
divorced from field observations. Another source
of controversy is the Precambrian geothermal regimes. A common assumption has been that the
high grade of metamorphism displayed by some
Precambrian terrains as compared to more modern
metamorphic belts results from secular decrease
of the average geothermal gradient (Ernst, 1972).
Recent heat-flow modelling, on the other hand,
has led some authors to the conclusion that normal continental geothermal regimes were not markedly higher in the late Archaean than today, despite the substantially higher radioactive heat
production (Bickle, 1978; Davies, 1979). Differences in metamorphic grade and tectonic style
of Precambrian metamorphic belts from those of
the Phanerozoic have been explained by deeper
levels of erosional exposure of the oldest areas.
Such an inference does not explain what replaced
the continental crust at its roots as material
was eroded off the top over thousands of millions
of years, nor why some of the highest grade terrains, such as the Ptarmigan Complex of southern
Labrador (Emslie et al., 1979), are a billion
years younger than other extensive terrains of
lower grade.

Additional facts about Precambrian high-grade
terrains have become apparent from recent geochronology, structural and petrographic analysis,
experimental petrology, and geothermometry-geo-
barometry. Some of these facts may be general or
even fundamental, but here-to-fore have not been
used extensively in discussions of crustal evolution. Table 1 presents significant facts
about a number of granulite terrains. A digest
of important generalizations follows:
1) Virtually all granulite-facies terrains show
evidence of large-scale horizontal tectonics, such
as recumbent folding, usually early in the metamorphic cycle. High-grade metamorphism commonly
set in near the end of the period of horizontal
tectonics and outlasted it, with obliteration of
early-formed foliation.
2) The temperature range of granulite metamorphism was 650° to over 900°C in some areas. The
implied metamorphic geotherms are too high in
some cases to be normal for Precambrian continents, and require some sort of thermal anomaly.
The discrete radiometric ages found for many
granulite terrains supports the concept of
granulite formation during specific thermal events, rather than under ambient conditions of
the Precambrian lower crust.

3) The typical range of pressures of granulite
metamorphism was 8 ± 3 kbar, suggesting that some
control operated to give the onset of granulite
metamorphism at a crustal depth of 17-25 km. The
high pressures were generated in rocks which in-
variably included layers of surficial origin.
The extreme burial implies great crustal thickening attendant on granulite facies metamorphism.
4) Water activities during the initial production of orthopyroxene in several lithologies were
of the order of 0.3 to 0.1, based on thermodynamic analysis of mineral assemblages. For metamorphism involving a vapor phase, this activity
corresponds approximately to the molar fraction
of H_2O in the vapor. The precursors of many of
the orthopyroxene-bearing rocks were supracrustals, with abundant initial H_2O. Therefore, a
specific drying mechanism was needed. The characteristic fluid inclusions of granulite terrains
are CO_2-dominated, in contrast to the H_2O-domi-
nated fluid inclusions of lower-grade rocks
(Touret, 1981). Some agency operated either to
extract H_2O partially from the pore fluids, or to
contribute overwhelming amounts of CO_2. Desiccation and carbonic fluids are two outstanding
differences between ancient metamorphism and
younger metamorphism.
5) Many granulite facies terrains show transitional regions of progressive metamorphism, with
more or less mappable isograds, principally the
incoming of orthopyroxene. Some of these transitional terrains show variations of paleotemperature and/or paleopressure across them, increasing in the direction of increasing amount of
granulite-grade rocks. Apparent $P(H_2O)$ steadily
decreases across the transition zones, and CO_2
fluid inclusions have been shown to become more
abundant in some cases.
6) Massif-type anorthosites are conspicuous in
several Proterozoic high-grade terrains, leading
various authors to postulate a fundamental link
between anorthosite petrogenesis and granulite
metamorphism. The most recent work on the Adirondack and Grenville anorthosites strongly indicates that some of the largest bodies were initially emplaced at shallow depths, probably in a
rifted continental setting. Involvement of older continental crust in the genesis of the anorthosite suite is implied by geochemical criteria. Subsequent high-pressure metamorphism
occurred 100-300 million years after igneous
emplacement. The genetic link between anorthosites and granulites is less direct than formerly thought, but may be fundamental nevertheless.

The present paper reviews, in turn, the bearing of these aspects of granulite terrains on
models of their origin.

Regional and Structural Aspects

A tendency for increasing area of granulite
terrains with increasing age may be deduced from
Table 1. The largest terrains are Archaean. The

Table 1. Ages, Outcrop Areas, Physical and Petrologic Characters of Precambrian Granulite Facies Terrains

Complex	Age m y	Area $10^3 km^2$	Temp °C	Press. kbar	Deformation related to Metamorphism**	Timing of Metamorphism	Lithologies, LIL Depletion	Refs.
NAPIER Antarctica	3100-3000	100	900-950	7-10	D_1: Flat isoclines D_2: Tight assym. folds	Late D_1 through D_2	Mostly(?) undepleted metaigneous, metasedimentary	1, 2, 3
BUKSEFJORDEN Greenland	2800-2700	1.2+	650-800	7-8	D_3: Recumbent folds D_4: Vertical shearing	D_3, outlasting D_4	Undepleted orthogneiss, minor metasediments	4, 5, 6
SCOURIE Scotland	2800-2600	0.5+	820±20	7-9*	F_1: Flat-lying foliation F_2: Intrafolial folds	Late F_1 through F_2	Dominantly depleted orthogneiss	7, 8
NEW QUEBEC N. Canada	> 2500	120	700-800*	6-8*	Unknown	Unknown	Mostly(?) undepleted ortho-, paragneiss	9, 10
WHEAT BELT W. Australia	> 2600	60	700-800*	6-7*	Unknown	Unknown	Mostly(?) undepleted orthogneiss, supracrustals	11
S. INDIA	2650-2500	70	650-850	6-9	Early nappes, later shearing, tight folds	Late-nappe thru shearing	Depleted orthogneiss, undepleted supracrustals	12, 13
HIGHLANDS Sri Lanka	> 2100	28	650-800*	6-7.5*	Early great recumbent folds, later tight folds	Outlasted overfolding	Dominantly undepleted supracrustals	14
INARI Finland	2150-1900	12	700-760	5-7	Early overthrusting, flat foliation	Outlasted overthrusting	Dominantly undepleted supracrustals	15
WILLYAMA N.S. Wales	1700	5	650-800	4.5-6.5	F_1: Isoclinal recumbent folds F_2: Tight upright folds	Late F_1, through F_2	Dominantly undepleted supracrustals	16, 17
PTARMIGAN Labrador	1660	2.6	900-1000	9-10	Unknown	Unknown	Undepleted plutonic, sedimentary	18, 19
NAMAQUALAND S. Africa	1200	4.3	750-800	5-7	F_2: Isoclinal folding	Accompanied, outlasted F_2	Dominantly undepleted supracrustals	20,21
SW GRENVILLE Canada	1200-1000	30	650-800	6-8	Early overturned isoclinal folds	Synchronous w/ folding	Undepleted gneisses, plutonics, supracrustals	22,23
ADIRONDACKS New York	1100-1000	27	650-800	6-8.5	D_2: Isoclinal folds, nappes D_3: Upright folds	Outlasted D_3	Undepleted plutonic, sedimentary	24, 25

References: 1) Sheraton et al. (1980); 2) Grew (1980); 3) Harley (1983); 4) Wells (1979); 5) Chadwick and Nutman (1979); 6) Newton and Perkins (1982); 7) Bowes (1975); 8) Rollinson (1981); 9) Eade and Fahrig (1971); 10) Herd (1978); 11) Wilson (1978); 12) Drury et al. (1984); 13) Harris et al. (1982); 14) Cooray (1962); 15) Hörmann et al. (1980); 16) Laing et al. (1978); 17) Phillips and Wall (1981); 18) Emslie et al. (1979); 19) Emslie (1981); 20) Zelt (1980); 21) Waters (1984); 22) Wynne-Edwards (1972); 23) Bourne (1978); 24) Wiener et al. (1983); 25) Bohlen et al. (1984)
*   Estimate of the present author from published or personal mineral data.
** Original D and F terminology of cited authors.

New Quebec terrain of northern Canada is the
largest continuously exposed high-grade province
(Herd, 1978). Some decrease in the scope of
granulite facies metamorphism seems implied by
the secular decrease in the outcrop areas.

A striking aspect of Table 1 is the nearly
ubituitous occurrence of early-metamorphic large-
scale horizontal deformation often in the form of
recumbent overfolds. This appears to be a definitive feature of most well-described granulite
terrains, including the few post-Precambrian examples, such as the Cretaceous Big Sur granulites
(Compton, 1966, p. 281) and the Eocene Prince
Rupert British Columbia, granulites (Hutchinson,
1970, p. 382; Hollister, 1975).

Early flat deformation is commonly overprinted
by other deformations. A second deformation may
also be recumbent, or may be upright open to
tight folding with later structures visible on
smaller scales or by interference fold patterns
with the earlier flat deformations. Large-scale
shear strain along steep planes followed early
recumbent deformation in SW Greenland (Chadwick
and Nutman, 1979). This has been suggested for
southern India also (Drury et al., 1984), but
field data are not yet adequate to demonstrate
this relationship there. High-grade metamorphism started relatively late in the horizontal
deformation cycle and continued through subsequent deformation. Chadwick and Nutman (1979)
suggested that the change of tectonic style in
SW Greenland was attendant on drying out, and
hence, embrittlement, of the infrastructure in
granulite metamorphism, causing a different
response to the same deforming forces.

Extensive crustal shortening must have accompanied the subhorizontal deformation. The suggestion of Gastil (1979) that flat foliation is
formed at deep levels in response to escape of
voluminous acid magmas from the lower crust may
have some validity, but cannot account for the
consistent directionality of overfolding or for
the crustal shortening evident in many high-grade
terrains. Similarly, the suggestion that horizontal deformations may be caused commonly by deep
extensional movements, as over a crustal magma
underplate (Gastil, 1979), could not account for
the overfolding, which must be a compressional
feature.

An analogy of structural style of the ancient
granulite terrains with younger orogenic areas
seems evident. The phenomena of early recumbent
folds followed by more brittle deformation with
increasing metamorphic intensity is recorded in
the Southern Highlands of Scotland (Elles and
Tilley, 1930). Metamorphism outlasted major
deformation and left simple isograds superposed
on complex nappe structures.

Temperatures and Pressures of Metamorphism

The temperature ranges of Table 1 were established from continuous mineralogic geothermometry
or by use of experimental univariant equilibria
(the "petrogenetic grid"). The most generally
useful continuous temperature scale for granulites based on the chemistry of major minerals is
the Fe-Mg exchange between garnet and clinopyroxene. The calibration most widely used is the
experimental scale of Ellis and Green (1979).
Garnet-orthopyroxene exchange is also useful
(Harley, 1984). Other temperature scales in
current use are Na distribution between coexist-
ing alkali feldspar and plagioclase (Stormer,
1975) and the exchange of Mg, Fe and Ca between
orthopyroxene and clinopyroxene, the most recent
calibration of which is that of Lindsley (1983).
The generally most useful continuous geobarometer
for granulites is based on the assemblage garnet-
orthopyroxene-plagioclase-quartz. This assemblage was first used to calculate equilibrium
pressures of granulites in South Harris, Outer
Hebrides, by Wood (1975). Three nearly convergent calibrations of this temperature-insensitive
geobarometer scale are those of Wells (1979),
based entirely on experimental work, of Newton
and Perkins (1982), based entirely on thermodynamic measurements, and of Bohlen et al. (1983),
based entirely on experimental work. Application of these "charnockite barometers" has established the typical pressure range of granulite
metamorphism.

Figure 1 shows representative temperatures and
pressures of granulite metamorphism in a number
of terrains based on a number of geothermometers
and on the temperature-insensitive charnockite
geobarometers. The large temperature range extends from the hydrous melting curve of alkali
feldspars and quartz ("granite minimum") nearly
up to the dry granite solidus, which effectively
limits the possible P-T range of crustal materials. There is no consistent age control apparent in the distribution. The pressure spread is
generally from 6 to 10 kbar. The results for
individual terrains contrast with earlier extreme estimates of 15 $\pm$ 3 kbar for Scourie
(O'Hara and Yarwood, 1978) and 2-3 kbar for the
Adirondacks (Saxena, 1977). Fig. 1 also shows
that the deduced temperatures and pressures are
consistent with the experimental Al_2SiO_5 diagram.

The major information about metamorphic
temperatures from Fig. 1 is that they represent
perturbed thermal conditions, rather than ambient
geothermal regimes, despite the fact that radioactive heat production in a late Archaean crust
was nearly three times that in the present crust.
Metamorphic P,T conditions in Precambrian granulite terrains all lie to the high-temperature side
of geothermal curves in 30-km-thick late Archaean
continents (Davies, 1979; Baer, 1981). Therefore, additional sources of heat are needed. The
simplest means of augmenting temperatures of the
deep crust is mechanical thickening. Fig. 1
shows Baer's (1981) calculated geothermal curve
for a 46-km late Archaean continent. Within the
uncertainties inherent in geothermal calculations, it appears possible that substantial
thickening of the relatively radioactive crust

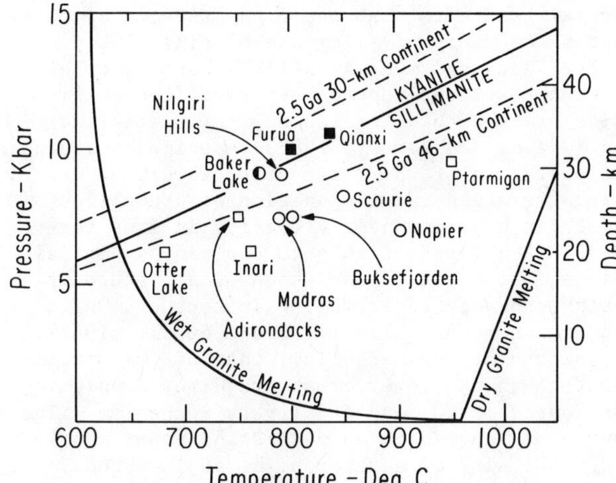

Fig. 1. Representative temperatures and pressures of a number of Precambrian granulite terrains at peak metamorphic conditions. <u>Symbols</u>: <u>square</u> = Proterozoic terrain; <u>circle</u> = late Archaean terrain; <u>open symbol</u> = sillimanite-bearing terrain; <u>filled symbol</u> = kyanite-bearing terrain; <u>half-filled symbol</u> = terrain with both kyanite and sillimanite. P-T data of Newton and Perkins (1982), up-dated, with addition of data for late-Archaean Baker Lake (NW Canada) terrain (Schau, 1982). Shown also are two late-Archaean steady-state geothermal curves calculated by Baer (1981). Kyanite-sillimanite relations from Holdaway (1971). Granite melting relations as in Fig. 2.

may have been sufficient to elevate temperatures at 30-km depth to the granulite facies range, without additional heat sources, provided that subsequent uplift and erosion took place on time scales longer than 100 million years (England and Bickle, 1984). Additional heat sources may be necessary to explain the highest-temperature terrains, such as the Napier Complex of Enderby Land and the Ptarmigan Complex of Labrador. Other possible heat sources which have been suggested are a basaltic underplate (Kröner, 1982), tonalitic mid-crustal intrusions (Wells, 1979), and upward transport of heat over large terrains by CO_2 from a degassing mantle diapir (Harris et al., 1982).

The range of pressures of granulite metamorphism evident in Fig. 1 corresponds nearly to the pressure at the base of a continent of 20-30 km thickness. This coincidence, and the near uniformity of paleopressures over large areas containing abundant shallow-water sediments, as in southern India and Sri Lanka, suggest that wholesale crustal underthrusting of one segment under another has been postulated by many authors for situations ranging from modern times to the Archaean. Some authors proposing this mechanism are Powell and Conaghan (1975) for the Himalayas, Allis (1981) for the New Zealand Southern Alps,

Radain et al. (1981) for the western Arabian shield, Cuthbert et al. (1983) for the southern Norwegian Basal Gneisses, Hodges et al. (1982) for the northern Norwegian Caledonides, and Rubie (1984) for the western Alps. Thickening by thrust-stacking of thin nappes (Bridgewater et al., 1974), magmatic underplating (O'Hara, 1977) or overplating (Wells, 1979) all seem unlikely to have produced 8 kbar repeatedly over broad terrains.

H_2O Activities of Metamorphism

The melting curves of feldspar-quartz under reduced H_2O activities shown in Fig. 2 are indicative of the relative dryness necessary for the formation of charnockites. In some terrains, such as South India, there is evidence that a vapor phase was present during deep-seated metamorphism (Hansen et al., 1984). In order for a vapor to have coexisted with the charnockitic assemblage of feldspars plus quartz, the mol fraction of H_2O must have been considerably less than unity, and less than 0.5 in the majority of cases.

Thermodynamic calculations on the stabilities of biotite and amphibole relative to orthopyroxene have provided more exact knowledge of H_2O pressures in granulite metamorphism. Based on these methods, Phillips (1980) found a steady decrease of H_2O pressure across the amphibolite facies to granulite facies transition in the

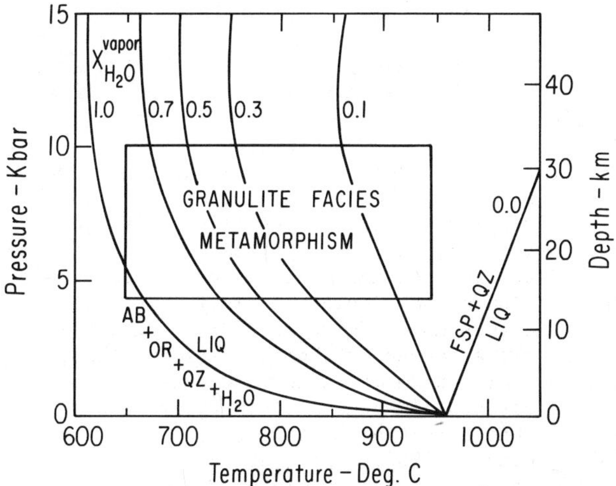

Fig. 2. Fractional isopleths of H_2O ($X[H_2O]$) in a H_2O-CO_2 vapor phase which could coexist with feldspar and quartz, projected from the data for albite-H_2O-CO_2 of Bohlen et al. (1982) to the system albite-sanidine-quartz (Huang and Wyllie, 1975). Symbols: AB = $NaAlSi_3O_8$; OR = $KAlSi_3O_8$; QZ = SiO_2; LIQ = liquid; FSP = hypersolvus alkali feldspar. The figure shows that a granitic charnockite can coexist with a vapor phase at granulite facies P-T conditions only at reduced H_2O concentrations.

Broken Hill area, New South Wales. According to
the biotite stability calculations, the critical
H_2O activity for the appearance of orthopyroxene
was 0.5, and according to the amphibole stability
calculations it was considerably lower. Hansen
et al. (1984) calculated a critical H_2O activity
for orthopyroxene stability of 0.3 or less at
6-8 kbar and 750°C in southern India. Valley et
al. (1983) calculated that metasedimentary rocks
in the Adirondacks containing orthopyroxene,
clinopyroxene and quartz instead of tremolite
were formed at water activities less than 0.2.
If a vapor phase was present during metamorphism,
these activity values correspond approximately to
the molar fractions of H_2O in the vapors.

Granulite facies metamorphism operated in many
cases on rocks of surficial origin which had abundant initial H_2O. The temperatures and pressures of metamorphism were well within the hydrous melting regions of common lithologies. Although migmatization may be present near the orthopyroxene isograd, as in southern India (Condie
et al., 1982), melting does not become extensive
at higher grades, as many workers have noted (for
instance, Sheraton and Collerson, 1984 in the
Vestvold Block of Antarctica). Water cannot escape as a pure vapor phase during dehydration
reactions at granulite facies conditions without
causing extensive melting. Therefore, some
specific mechanism of desiccation is required to
explain the low H_2O activities. There are three
outstanding possibilities:
1) Absorption of H_2O into anatectic melts on a
regional scale with upward removal of the H_2O-
bearing liquid. This hypothesis has been cited
for a number of specific terrains, as, for example, the Namaqualand terrain of South Africa
(McCarthy, 1976; Waters, 1984) and the Willyama
Complex, New South Wales (Phillips, 1980).
2) Purging of the deep crust by voluminous CO_2
or other low-H_2O volatiles. This is suggested by
the nearly ubiquitous occurrence of dense CO_2-
rich fluid inclusions in granulite-facies rocks
(Touret, 1981), and by the field evidence in some
granulite transitional zones of the subsolidus
dehydration of amphibole and biotite to orthopyroxene (for example, Schreurs, 1984). There is
now nearly a consensus that Rb depletion in granulite metamorphism is better accounted for by
partition into and removal by a vapor phase,
rather than a partial melt (Tarney and Windley,
1977; Weaver, 1980; Okeke et al., 1983;
Smalley et al., 1983; Sheraton and Collerson,
1984).
3) Baking out of H_2O at shallow depths by contact with plutons prior to high-pressure metamorphism. If the pressure is low enough, H_2O
can escape to the surface in a dehydration event
without forming siliceous melts. This is an important possibility suggested by McLelland and
Husain (1986) for the Adirondack granulites in
the vicinity of the Marcy anorthosite, which show
strong gradients of oxygen fugacity and oxygen
isotope ratios suggestive of the strong local

thermal gradients that would be produced by contact metamorphism (Valley and O'Neill, 1984).

The fact that many granulites have major element compositions appropriate for silicate liquids, rather than for residues of partial melting (Weaver and Tarney, 1982) is evidence against
the operation of the first of these drying mechanisms as a general explanation of granulites,
though it may have been effective in some terrains. Equally difficult to explain by partial
melting processes is the isochemical nature of
patchy charnockite formation in southern India
(Janardhan et al., 1982; Condie et al., 1982).
The second drying mechanism, that of CO_2 streaming through the lower crust, requires a voluminous source of CO_2 and a delivery mechanism. The
source may have been an outgassing upper mantle
under the site of metamorphism, as inferred for
southern India by Harris et al. (1982), exsolu-
tion from a crystallizing basaltic underplate or
deep-crustal intrusions (Touret, 1971), exsolu-
tion from mid-crustal tonalitic intrusions of
mantle origin (Wells, 1979), or deeply buried
carbonaceous sediments (Glassley, 1983). It
would have been necessary to bury sediments rapidly, in order to release the CO_2 by decarbona-
tion reactions at depth. This could conceivably
have been done by entrainment of carbonate sediments including, perhaps, evaporites, by overthrusting during closure of a continental margin
basin or an infracontinental basin. The third
possible mechanism, that of shallow out-baking
prior to high-pressure metamorphism, also has
important implications for some terrains like the
Adirondacks. Modern thought on the shallow, anorogenic origin of massif anorthosites (Emslie,
1978; Morse, 1982) is closely related to the
concept of shallow desiccation of sediments by
contact metamorphism in a rift environment as a
common forerunner to high-pressure metamorphism.

Amphibolite Facies to Granulite Facies
Transition Zones

Many classic granulite facies terrains have
well-defined transitional borders with lower-
grade terrains, including the Adirondack Highlands (Wiener et al., 1983), the Bamble area of
southern Norway (Touret, 1971), the Broken Hill
area (Willyama Complex) of New South Wales
(Phillips and Wall, 1981) and southern India
(Condie et al., 1982; Hansen et al., 1984).
These transition zones usually display mappable
isograds, such as the first appearance of orthopyroxene in quartzofeldspathic gneisses. The
pyroxene gneisses at the orthopyroxene isograd
are not depleted in LIL elements (Condie, 1982;
Field et al., 1980). Somewhere in the grade
progression rocks of pronounced LIL depletion may
appear. Not many detailed geochemical traverses
across granulite facies transition zones have
been made. In the Bamble terrain, pronounced
depletion begins abruptly at the Island of
Tromoy, the southern extremity of the granulite

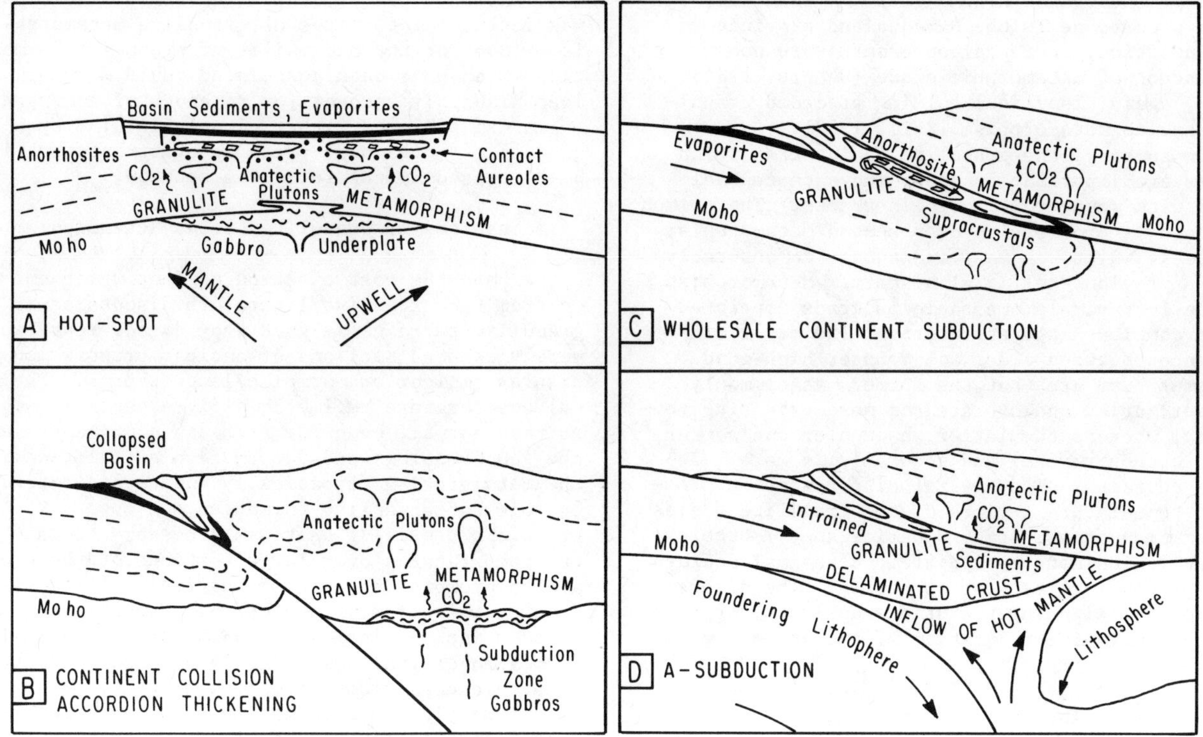

Fig. 3. Various hypotheses of granulite metamorphism. a) Hot-spot hypothesis with crustal thinning and underplating over a mantle upwell. b) Continental collision with accordion-style thickening (Dewey and Burke, 1973). c) Continental subduction (Hodges et al., 1982). A-subduction, with decoupling of the crust from the mantle (Kroner, 1982).

terrain (Smalley et al., 1983). A majority of the gneisses there are depleted. A Rb traverse across the transition zone of southern India showed that depletion sets in at about the point where charnockite (quartzofeldspathic granulite) becomes dominant (Hansen et al., 1984). Only a few rocks there are depleted, however; many still show Rb levels characteristic of upper-crustal rocks. There is no apparent correlation of Rb depletion with Mg/Fe ratio, which might indicate that removal of an anatectic melt was responsible for the depletion, nor with the disappearance of biotite, which is a major-mineral host for Rb. The causes of pronounced Rb depletion remain unknown.

It is likewise not certain yet whether temperature or pressure increases are generally characteristic across the transition zones. Phillips and Wall (1981) inferred a temperature increase of 650°C to 800°C across the Broken Hill zone. Paleopressure there increases concomitantly with paleotemperature, from about 4.5 kbar in the amphibolite facies to 6.5 kbar at the highest granulite grade reached. Hansen et al. (1984) were unable to discern a definite temperature increase across the South Indian transition zone. The mean temperature is close to 750°C. Paleopressures based on nearly temperature-independent

barometers show a definite increase from about 5 kbar at the orthopyroxene isograd to about 7.5 kbar in the depleted massif area. The relations there are suggestive that the amphibolite facies to granulite facies transition is a crustal depth-zone profile. Particularly significant are the paleopressures calculated from the densities of CO_2 fluid inclusions in charnockites, which agree quantitatively with the mineralogical geo-barometry, and give direct evidence that CO_2 was the major fluid species during peak metamorphic conditions.

By contrast, Schreurs (1984) deduced a rather sudden increase of temperature, from about 700° to 800°C over a 5-km lateral interval, in the W. Uusimaa area of southern Finland. The paleo-pressure does not rise with temperature, but is constant at about 4 kbar across the granulite facies transition interval. CO_2 fluid inclusion barometry is again in quantitative agreement with mineralogical barometry. Structural and litholo-gic features are amazingly continuous across the S. Finland transition zone, which would appear to be simply a thermal overprint, were it not for the presence of early-metamorphic isoclinal folds, which groups the W. Uusimaa area with most other granulite facies areas. The orthopyroxene isograd is very well-defined and cuts across complex structures at high angles. A seemingly similar occurrence is the Namaqualand province of South Africa, where paleopressures are nearly independent of metamorphic grade (Waters, 1984). Here, again, isoclinal folding preceded granulite-facies metamorphism (Zelt, 1980).

In summary, the granulite facies transition zones are important in providing a conceptual link with younger metamorphic tracts. The common feature is progressive deep-seated metamorphism of rocks which, in many instances, demonstrably had a previous surficial history. Metamorphism often left simple, mappable isograds imprinted over complex nappe structures. The major differences between older and younger high-grade metamorphism are that the former was commonly hotter, drier and operated on more extensive terrains, whereas the latter was cooler and wetter (sub-orthopyroxene) and generally in more belt-like terrains. The several differences in pressure-temperature regimes of the granulite facies transition zones, some isobaric, some isothermal, probably indicate that different thermal sources and/or time-scales were operative. The diversity of P-T signatures cautions against overgeneralization in interpretation of ancient metamorphism.

The Anorthosite Problem

The presence of massif-type anorthosite bodies in several Proterozoic granulite-facies terrains has long been a mystery. The anorthosites are sometimes enormous accumulations of andesine or labradorite not obviously related to any large gabbro bodies, whereas Archaean anorthosites are more calcic and many are manifestly segregations in layered gabbro complexes, as in the Stillwater intrusion of Montana. The presence of massif anorthosites in the Adirondacks, western Grenville and SW Norway high-grade terrains has led several authors to presume that anorthositic magmatism and emplacement has something fundamental to do with the granulite metamorphism there. Recent work on massif anorthosites has resulted, however, in the picture of shallow emplacement in a continental rift environment (Berg, 1977; Emslie, 1978; Morse, 1982). This revised viewpoint is strengthened by recent age determinations, particularly those based on Nd-Sm, which have shown that igneous ages predate metamorphic ages by 50-300 million years for the South Harris anorthosite (Cliff et al., 1983), the Adirondack anorthosite (Ashwal and Wooden, 1983) and the anorthosite of the Wilmington Complex (Foland and Muessig, 1978). The earlier suggestion by Emslie (1978) that the Grenville anorthosites may have igneous ages as much as 400 million years older than the metamorphic ages is supported.

Thus, there may be an evolutionary relationship between massif anorthosites and Proterozoic granulite metamorphism, but it is less direct than supposed by earlier workers, who thought that anorthosites crystallized from orogenic magmas intruded into sites of granulite metamorphism. Some of the mechanisms of granulite formation which have been postulated could more or less plausibly account for an epoch of anorogenic magmatism followed rather closely in time by a granulite facies high-pressure overprint, as will be discussed in the following section.

Mechanisms of Granulite Facies Metamorphism

Perhaps the most outstanding fact which emerges from the extensive literature discussion of granulite terrains is that they do not represent merely exhumed sections of ancient crust which display ambient or normal paleogeotherms. The paleotemperature regime at a given depth is higher than normal, even for Archaean continents, and the geochronology and deformation associated with the metamorphism emphatically indicate discrete episodes of granulite formation. A powerful transient source of heat was necessary to raise the temperatures over large portions of the crust and to promote dehydration, partial melting, and incompatible element depletion. Possible thermal sources include influx of magmas or hot volatiles from a subcrustal upwell, enhanced radioactivity of a thickened crust, or, less plausibly, mechanical working in deformation. The various possibilities figure prominently in several hypotheses of high-grade infracrustal metamorphism, summarized below.

"Hot-spot" Hypothesis

A magmatic crustal underplate which emanates from a zone of increased thermal activity in the mantle has been suggested many times as a thermal source for high-grade metamorphism. The magma must have been mafic, in order to have been ponded in large quantities at the base of the crust (Herzberg et al., 1983). This hypothesis provides crustal thickening, which was necessary for the subsequent elevation of a deep-seated terrain. The basic magmas could provide a source of low-P(H_2O) volatiles, which could have purged the lower crust of water and created a carbonic pore-fluid milieu (Touret, 1971). The model is diagrammed in Fig. 3A.

A few Precambrian high-grade terrains show rapid lateral increase of paleotemperature in the granulite transition zones without significant change of paleopressure. Examples are the West Uusimaa area of southern Finland (Schreurs, 1984) and the Namaqualand terrain (Waters, 1984). This relationship could very plausibly be the result of local deep-seated emplacement of a gabbro body. The extremely high-temperature Napier Complex of Enderby Land has a pronounced positive free-air anomaly (Wellman and Tingey, 1982), suggesting dense mafic material underneath.

In general, however, an unmodified "hot-spot" theory falls short of providing a comprehensive explanation of most granulite terrains. The ma-

jority do not show positive gravity anomalies.
The South India regional gravity map shows a
featureless free-air pattern (Subrahmanyam,
1983), and that of Sri Lanka is nearly so
(Hatherton et al., 1975). The common involvement
of nappe structures, continental shelf or basin
sediments and evaporites, and post-igneous high-
pressure metamorphism of anorthosites during the
Proterozoic are unaccounted for.

"Nappe-stacking"

The evidence for pronounced crustal shortening
in nearly all well-documented granulite-facies
events requires a compressional regime, and the
time relations of metamorphism and deformation
often indicate that crustal thickening took place
before the full effects of the thermal event were
felt. There may have been an earlier widespread
high-level contact metamorphism, as deduced by
Valley (1984) for the Adirondacks, but this was
distinct from the main period of regional metamorphism, which overprinted it. Residual heat
from the earlier contact metamorphism is not
likely to have contributed significantly to the
regional metamorphism in the Adirondacks, which
came 300 million years after the anorthosite emplacement (Ashwal and Wooden, 1983). Mechanical
work during nappe-stacking is also inadequate to
explain the elevation to temperatures near 800°C
over hundreds of kilometers in the Adirondacks.
The increased radioactivity and thermal blanket
of a thickened continent could have raised temperatures at the 30-km depth level substantially
if uplift and erosion took place on a time scale
longer than 100 million years (England and
Bickle, 1984), and additional heat could have
been supplied by synorogenic magmas.

Piling up of nappes is conceptually the simplest means of crustal thickening. This mechanism has difficulty in explaining the early uniform paleopressures of 6-9 kbar found across
large terrains like the Adirondacks or southern
India. It seems unlikely that successive superposition of thin slices could have produced an
increased crustal thickness of 20-30 km under
large areas. A more coherent form of crustal
thickening may be called for.

Accordion-style Thickening

Dewey and Burke (1973) applied plate-tectonic
concepts to orogenesis of continental interiors.
They envisioned continental collision, with ac-
cordion-like thickening of a continental foreland, as shown in Fig. 3B. Magmas rising from a
subduction zone could have supplied heat to the
base of a thickened continent, promoting granulite metamorphism. Escape upward of deep-crustal
anatectic melts could have left dried granulite
residues, and it is conceivable that CO_2 released from subducted limestone could have aided
the desiccation process.

The Dewey and Burke hypothesis is appealing in

providing means of crustal thickening and heating. The common involvement of continental shelf
or basin sediments, including evaporites, in the
deep-crustal milieu of granulite metamorphism is
not explained by this model, nor the prevalence
of early-metamorphic large-scale horizontal
structures. The deformation style of the crustal
roots inferred from Fig. 3B would be dominantly
vertical.

Continental-scale Underthrusting

A number of reconstructions of Phanerozoic
continental-margin orogenesis have suggested subduction of nearly continental thicknesses of
crustal material underneath another crustal segment, with effective doubling of the continental
thickness. The continental segments may have
been formerly separate continents, island arcs
subducted or obducted at a continental foreland,
or marginal splinters of continents separated
previously by the opening of a marginal basin.
An example is the collision of the Greenland and
Baltic plates in the Caledonide orogeny of northern Norway, in which Baltica drove under Greenland at a very small angle for a distance of at
least 250 km (Hodges et al., 1982). Metamorphism
in the lower plate achieved high amphibolite
grade, and postorogenic uplift took place in
about 100 million years with assemblages remaining in the kyanite field of stability in the upward transport (Royden and Hodges, 1984). A considerable amount of continental shelf sediment
was entrained at the flat continent-continent
interface in the collision. A very similar reconstruction was put forth by Wagner (1982) to
explain the relations of the Baltimore Gneiss
(western plate), Wissahickon Schist (cover sequence) and Wilmington Complex (eastern plate)
of SE Pennsylvania. Whole-continent underthrusting may have been localized by a lubricated horizon of shallow-water sediments and/or evaporites, as shown in Fig. 3C. Behr and Horn (1981)
suggested localization of large-scale thrusting
in the Pan-African Damara Orogen by an evaporite-
rich sedimentary sequence, which provided lubri-
cating horizons and high fluid pressures at the
soles of thrusts. Ortega-Gutiérrez (1983) described the abundant metaevaporites associated
with granulites and anorthosites in the Grenville-age Oaxaca Complex of southern Mexico, and
listed many other literature references to metaevaporites in Precambrian high-grade terrains.
The metaevaporites need not have resulted from a
world-wide arid climate during the Proterozoic,
but may have been supratidal and intertidal deposits of shallow seas. Lougheed (1983) suggested ankeritic muds, gypsum and salt as the
original deposits of the Lake Superior iron
formations.

Large-scale continental underthrusting has
many appealing aspects in explaining granulite
facies metamorphism. In addition to explaining
the horizontally-deformed cover sequence and

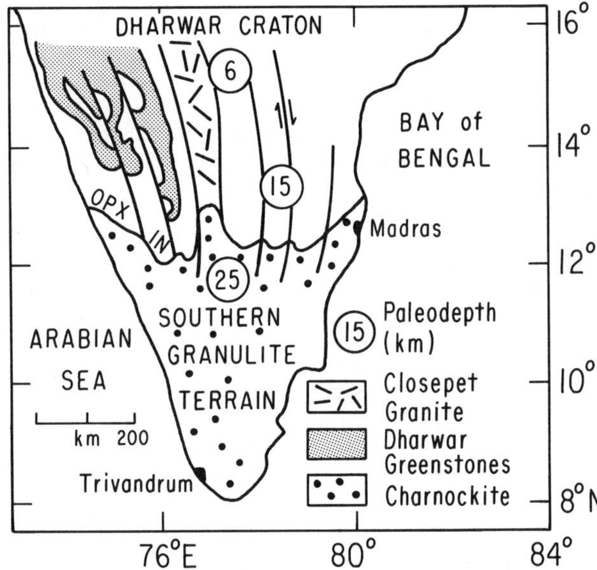

Fig. 4. South India Shield, with gradient of
progressive metamorphism from the Dharwar Peninsular Gneiss granite-greenstone terrain southward
through an orthopyroxene isograd to the charnockitic terrain of southernmost India. Unpatterned
area is tonalitic-trondhjemitic Peninsular
Gneiss. Shown also are deep vertical shears
postulated by Drury and Holt (1980) and paleo-
depths exposed, as indicated by geobarometer
assemblages.

basement associations common in Precambrian terrains, this hypothesis also explains the approximately 8 kbar paleopressures and provides an uplift mechanism in the isostatic rebound of a
doubly-thick continent. The increased radioactivity of a thickened crust may be a sufficient
potential source of heat to elevate temperatures
from normal Proterozoic or late Archaean geotherms into the granulite facies range, commonly
700-800°C, if erosion and uplift took place on
time-scales of 100 million years or more. Carbonic fluids may have protected deeply-buried
crustal rocks against extensive melting, and
these fluids may have been supplied by highly
carbonated and evaporitic basin sediments. This
volatile source would also explain the frequent
reports of F in biotites and amphiboles
(Leelanandam, 1970), sulfate and carbonate in
scapolite (Von Knorring and Kennedy, 1958) and of
B in sillimanite (Grew and Hinthorne, 1983).
Continental collisions are discrete events, occurring over limited time intervals, which would
explain the often well-defined radiometric age
dates associated with granulite-facies rocks.
Self-heating of a thickened crust would necessarily involve a time lag after crustal duplex
formation, which could explain why granulite
metamorphism tended to occur late in the cycle of
early horizontal-style deformation, with common

obliteration of flat foliation by mineral growth.
A major problem of an unmodified continental
collision model is that in many large granulite
terrains there is little evidence of the remains
of a collapsed ocean basin. Anorogenic magmatism, sedimentation, deformation and metamorphism
appear to have taken place in an entirely infracontinental milieu in a number of instances
(Kroner, 1982). Another problem is how a considerable portion of subcontinental upper mantle
was disposed of as the crust was shortened above
it. According to conventional plate tectonic
models, some subduction zone remote from the
scene of crustal shortening must accommodate the
outflow of lithosphere. Such major subduction
zones coupled with Precambrian high-grade terrains have not been often identified. However,
it should be emphasized that detachment of the
crust from the underlying mantle is not a necessary feature of crustal doubling. Very flat subduction, as is taking place under northern Chile
(Jordan et al., 1983), would be sufficient. It
is possible that such flat subduction results in
doubling of the lithosphere itself, as pictured
by Vlaar (1982).
The model of infracontinental subduction or
"A-subduction" of Kröner (1981, 1982) can account
for most of the petrologic constraints considered
above, and has the further advantage that it
could have operated entirely within a continental
setting. According to this hypothesis, a tecton-
ic-metamorphic cycle began with a sub-continental
hot-spot, perhaps an incipient rift zone. The
crust was thinned and distended, with the emplacement of high-level and deep-level anorogenic
magmas. If this process slackened before complete continental disruption, cooling and gravitational instability of a subcontinental mantle
peridotite layer would occur. Foundering of this
slab would trigger crustal shortening above it,
the simplest means of which might be simple
crustal duplex formation, as shown in Fig. 3D. A
subduction zone remote from the metamorphic-tec-
tonic site is not necessary.
It is possible that anorogenic magmas introduced into the crust were over-emphasized as a
source of heat for metamorphism by Kröner (1982).
The early magmatic episode may have commonly
produced high-level contact metamorphism and
drying out of the upper layers of the crust, and
melting in the deep parts of the crust, with
hybridization of basalt, as seems to be impli-
cated in the chemistry of the anorthosite suite
(Duchesne and Demaiffe, 1978). However, the time
separation between anorogenic magmatism and regional metamorphism was as long as 300 million
years in the Adirondacks (Ashwal and Wooden,
1983), which would be ample time for cooling, and
the petrographic evidence indicates a discrete
second episode of heating in subsequent regional
metamorphism, which could have been provided by
mere thickening of the crust. If crustal thickening followed closely upon rifting, the early
and late thermal effects might be indecipherably

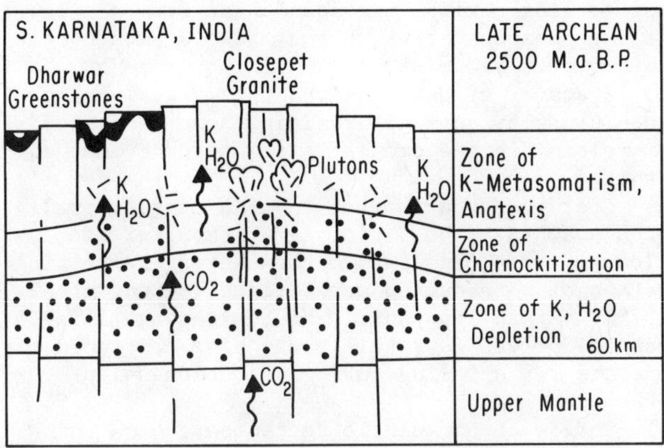

Fig. 5. Late–Archaean CO_2 streaming through thickened S. India crust, as postulated as an object of high-grade metamorphism, metasomatism and anatexis by Harris et al. (1982) and Friend (1983).

associated, as seems to have been the case with anorthositic magmatism and very high-temperature metamorphism in the Rogaland terrain of SW Norway (Kars et al., 1980).

Infracontinental Transform Zones and Shear Belts

The term "mobile belt" has come into wide use in connection with the Precambrian high-grade terrains. The areas occupied by granulite facies rocks have, in many instances, been zones of crustal mobility which have been active long after the epoch of granulite metamorphism. Examples are the Limpopo Belt of southern Africa, which underwent periods of lateral and vertical faulting subsequent to the late Archaean-early Proterozoic granulite-facies events (DuToit et al., 1983), and the granulite terrain of southern India (Drury and Holt, 1980), where numerous "megashears", some of probable large transcurrent motion, post-date the late Archaean granulite metamorphism. The continued mobility of the South Indian terrain suggested to Katz (1976) that the granulite metamorphism itself was a consequence of infracontinental transform motion, whereby crustal spreading, driven by the accretion of the cratonal Dharwar volcanics in southern India and by their pre-drift correlatives, the Kalgoorlie greenstones of SW Australia, was accommodated by intense shearing in the toe of India and in Sri Lanka.

This hypothesis does not provide ready explanations for the involvement of shelf or basin-type supracrustals nor their burial to depths of 20+ kilometers, sometimes uniformly over large areas. The origins of early-metamorphis nappes and flat foliation are not apparent in this hypothesis.

Large-scale shearing of the continent could have had an important role in the South Indian granulite facies metamorphism, according to Drury and Holt (1980) and Friend (1983), though not in the manner envisioned by Katz (1976). Fig. 4 shows the essential relations of the charnockitic terrain, the late Archaean supracrustals (Dharwar greenstones) and the 2500 Ma Closepet Granite. According to Drury and Holt (1980), N–S shearing of the lower-grade craton and the northern part of the high-grade belt took place after thickening of the latter by overthrusting, perhaps by a continuation of the same tectonic forces. The deep shear zones gave access of mantle CO_2 to the crust and thus could have been an important agency of lower-crustal dehydration and depletion. In the picture of Friend (1983), summarized in Fig. 5, the CO_2 dehydrated biotite and amphibole in the deep crust and carried H_2O and K upwards. At the midcrustal level of the thickened crust, K-metasomatism occurred and the tonalitic-trondhjemitic Peninsular Gneiss was brought closer to the hydrous "granite minimum" composition, providing for the anatexis which formed the Closepet Granite. This hypothesis provides for the N–S linear aspect of the Closepet, the closely associated timing of K-metasomatism, granite and charnockite emplacement observed by Friend (1983) in the southern Karnataka transition zone, the abundant CO_2 inclusions of granites and charnockites in this area (Hansen et al., 1984), and the necessity of metasomatic transfer of LIL elements in the transition zone deduced by the geochemical studies of Weaver (1980) and Condie et al. (1982). Upward streaming of mantle CO_2 could have been an important factor in the thermal budget, as suggested by Harris et al. (1982).

The mantle source of CO_2 has not yet been elucidated in this hypothesis. Mantle carbonate of long-term stable residency might have been disturbed by the deep shearing, or CO_2 could have been enriched in the mantle under the craton by limestone subduction prior to continental collision.

Summary

The petrologic constraints described above suggest a possible sequence of events in Precambrian granulite facies metamorphism. A tectono-metamorphic cycle started with diapiric upwelling of the mantle under a continental interior. Stretching of the crust took place, with subsidence of rift-basins. Shallow platform sediments were deposited, which could have been generally more carbonated than shelf sediments of the present day, and contained abundant evaporites. Early rifting of continents in Phanerozoic times has been characterized by thick evaporite deposits as slump-basins were periodically flooded and desiccated (Rona, 1982). The near-surface expression of deep-crustal basaltic magmatism included the anorthosite suite. The elevated initial Sr ratios and alkaline chemistry of associated acid plutonics imply considerable crustal involvement in the magmatic products (Duchesne and Demaiffe, 1978; Taylor et al., 1984). The

andesine anorthosites could be products of remelting of more mafic rock at depth (Gromet and
Dymek, 1981). Longer deep-crustal residence time
in a slowly-rifting environment, and hence, alkali contamination, of the lower-crustal mafic accumulation could account for the more sodic anorthosites of the later Proterozoic.

Incipient rifting could have terminated within
a relatively short time after its inception. A
period of static cooling might have followed,
lasting perhaps 50-400 million years. This could
have been followed in turn by a period of regional compression, caused by collapse of a small
marginal basin in the normal operation of the
Wilson Cycle (Dewey and Burke, 1973), foundering
of a cooled, dense segment of uppermost mantle
under an abortively-rifted continent, as proposed by Kröner (1982), or simply from a change
of the stress regimes operating in different
parts of the earth's crust, transmuting disten-
tion into thickening. The supracrustal sequence
may itself have localized tectonic transport,
facilitating continental doubling.

Closure of the rift resulted in the generation of 6-10 kbar pressure on the supracrustal-
rich section and anorogenic igneous rocks. Temperatures ultimately generated at this horizon
would be near 800°C simply from self-heating of
a doubly-thick late Archaean crust, if erosion
and rebound took longer than about 100 million
years (England and Bickle, 1984). Faster uplift
would result in a steep dP/dT path of ascent,
similar to that inferred from coronal mineral
reactions for the Central Limpopo Belt (Windley
et al., 1984; Harris and Holland, 1984). Possibly the duration of high-grade metamorphism, the
rate of uplift, the temperatures ultimately realized, and the presence or absence of mineral zon-
ing and coronas were strongly determined by the
width of penetration of one continent under another. Lateral penetration of several hundred
kilometers may have resulted in a very stable
duplex crust which remained thickened for several hundred million years.

Some terrains show evidence of paleotemperature of 900°C or more, such as the Napier Complex
of Enderby Land and the Ptarmigan Complex of
southern Labrador, where the sapphirine-quartz
assemblage, a high-temperature, high-pressure
equivalent of cordierite, is regionally distributed. The late Proterozoic Rogaland terrain of
SW Norway shows a metamorphic zonation reaching
a pigeonite isograd in orthogneiss. The isograds are well-defined and spatially-related to
the anorthosite-mangerite intrusive bodies.
These relations strongly suggest either preservation of the effects of earlier high-temperature
shallow contact metamorphism, as has been demonstrated for the Adirondacks by Valley and O'Neill
(1984), or synmetamorphic heat from igneous
bodies. Igneous heat sources could have originated in several possible ways:
1) Closure and crustal thickening immediately
after anorogenic magmatism, without an interven-

ing cooling period. Residual heat from shallow
or deep-seated intrusions may have augmented the
metamorphic temperatures.
2) Transfer of heat upwards in a greatly thickened crust by copious granitic magma generated
by melting in the deepest part of the thickened
crust.
3) Continued weak operation of a mantle upwell
with associated basaltic magmatism after the
closure phase, similar to the thermal perturbing
action of a subducted ocean ridge (DeLong et al.,
1979). If an important heat source was rising
mantle-derived CO_2, this could have continued to
operate for a period during the compression
phase.
4) Inflow of hot mantle to the base of a thickened crust replacing foundered older mantle
(Kröner, 1982).
5) Subduction of a still-active island arc.

Some terrains show very little evidence of
melting despite metamorphic temperatures in the
range 800-900°C. This can only mean a very dry
metamorphic environment. Desiccation could have
been accomplished by baking out of the terrain
about shallow intrusions during the anorogenic
phase, absorption of H_2O into anatectic melts,
or purging of H_2O from the crust by rising CO_2-
rich vapors. High-density CO_2 inclusions in
minerals in many granulite terrains strongly
suggest the latter as an important drying mechanism. The CO_2 could have come from an upper
mantle previously carbonated by subducted limestone, with access facilitated by deep shearing
motions, or from an entrained sedimentary source.
Deeply buried sediment-rich horizons could
account for the zone of high electrical conductivity found under the Adirondack Highlands
(Klemperer et al., 1984).

If continent-scale thrusting was commonly
lubricated and localized by platform sediments
and evaporites, the interesting possibility suggests itself that the underthrusting may have
been ultimately self-limiting. As temperature
increased in the underthrust sediments, devola-
tilization reactions would occur, and some of
the lower part of the thickened crust might
undergo a change of deformation style from overthrusting to dominantly shear movements, as
postulated by Chadwick and Nutman (1979) for the
late Archaean deformation of SW Greenland. Increasing friction might have slowed down underthrusting, allowing lateral plastic outflow to
compensate for convergence, as pictured by Molnar
and Tapponier (1975) for the Himalayan-Tibet
collision.

Acknowledgements. The author's research is
supported by National Science Foundation grants
EAR 82-19248 and 84-11192. I benefitted from
conversations with many people, including Mike
Bickle, Brian Chadwick, Jean-Clair Duchesne,
Bill Glassley, Ed Hansen, Simon Harley, Nigel
Harris, Alfred Kröner, Jim McLelland, Michael

Raith, Frank Richter, Jacques Touret, John Valley
and Dave Waters.

References

Allis, R.G., Continental underthrusting beneath
the Southern Alps of New Zealand. Geology, 9,
303-307, 1981.

Ashwal, L.D. and J.L. Wooden, Sr and Nd isotope
geochronology, geologic history, and origin of
the Adirondack Anorthosite. Geochim. et
Cosmochim. Acta, 47, 1875-1886, 1983.

Baer, A.J., Geotherms, evolution of the lithosphere, and plate tectonics. Tectonophysics,
72, 203-227, 1981.

Behr, H.J. and E.E. Horn, Fluid inclusion systems in metaplaya deposits and their relationships to mineralization and tectonics. Chem.
Geol., 37, 173-190, 1982.

Berg, J., Regional geobarometry in the contact
aureoles of the anorthositic Nain Complex,
Labrador. J. Petrol., 18, No. 3, 399-430,
1977.

Bickle, M.J., Heat loss from the earth: a constraint on Archaean tectonics from the relation
between geothermal gradients and the rate of
plate production. Earth, Plan. Sci. Lett., 40,
301-315, 1978.

Bohlen, S.R., A.L. Boettcher and V.J. Wall, The
system albite-H_2O-CO_2: a model for melting and
activities of water at high pressures. Amer.
Mineral, 67, 451-462, 1982.

Bohlen, S.R., V.J. Wall and A.L. Boettcher, Experimental investigation and application of
garnet granulite equilibria. Contr. Min. Pet.,
83, 52-61, 1983.

Bourne, J.H., Metamorphism in the eastern and
southern portions of the Grenville Province.
Geol. Surv. Canada Pap. 78-10, 315-328, 1978.

Bowes, D.R., Archaean crustal history in northwestern Britian. In B.F. Windley, Ed., The
Early History of the Earth. Wiley, 469-479,
1975.

Bridgwater, D., V.R. McGregor and J.S. Myers,
A horizontal tectonic regime in the Archaean of
Greenland and its implications for early crust-
al thickening. Grønlands Geologiske
Undersøgelse Miscellaneous Papers no. 142, 1974.

Chadwick, B. and A.P. Nutman, Archaean structural
evolution in the northwest of the Buksefjorden
Region, southern West Greenland. Precamb.
Res., 9, 199-226, 1979.

Cliff, R.A., C.M. Gray and H. Huhma, A Sm-Nd
isotopic study of the South Harris igneous
complex, the Outer Hebrides. Contr. Min. Pet.,
82, 91-116, 1983.

Compton, R.R., Granitic and metamorphic rocks of
the Salinian Block, California Coast Ranges.
Calif. Div. Mines and Geol. Bull., 190, 277-
287, 1966.

Condie, K.C., P. Allen and B.L. Narayana, Geochemistry of the Archean low- to high-grade
transition zone, southern India. Contr. Min.
Pet., 81, 157-167, 1982.

Cooray, P.G., Charnockites and their associated
gneisses in the Pre-cambrian of Ceylon. Quart.
J. Geol. Soc. Lond., 118, 239-266, 1962.

Cuthbert, S.J., M.A. Harvey and D.A. Carswell,
A plate tectonic model for the metamorphic ev-
olution of the Basal Gneiss Complex in western
Norway. J. Metamorphic Geol., 1, 63-90, 1983.

Davies, G.F., Thickness and thermal history of
continental crust and root zones. Earth, Plan.
Sci. Lett., 44, 231-238, 1979.

DeLong, S.E., W.M. Schwarz and R.N. Anderson,
Thermal effects of ridge subduction. Earth,
Plan. Sci. Lett., 44, 239-246, 1979.

Dewey, J.F. and K.C.A. Burke, Tibetan, Variscan,
and Precambrian basement reactivation: products of continental collision. J. Geol., 81,
683-692, 1973.

Drury, S.A., N.B.W. Harris, R.W. Holt, G.J.
Reeves-Smith and R.T. Wightman, Precambrian
tectonics and crustal evolution in South India.
J. Geol., 92, 1-20, 1984.

Drury, S.A. and R.W. Holt, The tectonic framework of the South India Craton: a reconnaissance involving LANDSAT imagery. Tectonophysics, 65, T1-T15, 1980.

DuToit, M.C., D.D. Van Reenen and C. Roering,
Some aspects of the geology, structure and me-
tamorphism of the southern marginal zone of the
Limpopo metamorphic complex. Spec. Publ. Geol.
Soc. S. Africa, 8, 121-142, 1983.

Eade, K.E. and W.F. Fahrig, Geochemical evolutionary trends of continental plates - preliminary study of the Canadian Shield. Geol.
Surv. Canada Bull., 179, 1-51, 1971.

Elles, G.L. and C.E. Tilley, Metamorphism in relation to structure in the Scottish Highlands.
Trans. Roy. Soc. Edinburgh, 56, 621-646, 1930.

Ellis, D.J. and D.H. Green, An experimental study
of the effect of Ca upon garnet-clinopyroxene
Fe-Mg exchange equilibria. Contr. Min. Pet.,
71, 13-22, 1979.

Emslie, R.F., Anorthosite massifs, rapakivi granites, and late Proterozoic rifting of North
America. Precamb. Res., 7, 61-98, 1978.

Emslie, R.F., L.J. Hulbert, C.P. Brett and D.F.
Garson, Geology of the Red Wine Mountains,
Labrador: the Ptarmigan Complex. Geol. Surv.
Canada Pap. 78-1A, 129-134, 1979.

England, P.C. and M.J. Bickle, Continental thermal and tectonic regimes during the Archaean.
J. Geol., 92, 353-368, 1984.

Ernst, W.G., Occurrence and mineralogic evolution
of blueschist belts with time. Amer. J. Sci.,
272, 657-668, 1972.

Field, D., S.A. Drury and D.C. Cooper, Rare-earth
and LIL element fractionation in high-grade
charnockitic gneisses, south Norway. Lithos,
13, 281-289, 1980.

Foland, K.A. and K.W. Muessig, A Paleozoic age
for some charnockitic-anorthositic rocks.
Geology, 6, 143-146, 1978.

Fountain, D.M. and M.H. Salisbury, Exposed cross-
sections through the continental crust: implications for crustal structure, petrology and

evolution. Earth, Plan. Sci. Lett., 56, 263-277, 1981.

Friend, C.R.L., The link between charnockite formation and granite production: evidence from Kabbaldurga, Karnataka, southern India. In M.P. Atherton and C.D. Gribble, Eds., Migmatites, Melting and Metamorphism. Nantwich (U.K.): Shiva Publishing Co., 246-276, 1983.

Gastil, R.G., A conceptual hypothesis for the relation of differing tectonic terranes to plutonic emplacement. Geology, 7, 542-544, 1979.

Glassley, W.E., Deep crustal carbonates as CO_2 fluid sources: evidence from metasomatic reaction zones. Contr. Min. Pet., 84, 15-24, 1983.

Grew, E.S., Sapphirine + quartz association from Archean rocks in Enderby Land, Antarctica. Amer. Mineral., 65, 821-836, 1980.

Grew, E.S. and J.R. Hinthorne, Boron in sillimanite. Science, 217, 547-549, 1983.

Gromet, L.P. and R.F. Dymek, Petrological and geochemical characterization of the St. Urbain anorthosite massif, Quebec: Summary of initial results. Abstr., Lunar, Plan. Sci. Conf. on Layered Intrusives, 1981.

Hansen, E.C., R.C. Newton and A.S. Janardhan, Pressures, temperatures and metamorphic fluids across an unbroken amphibolite-facies to granulite-facies transition in southern Karnataka, India. In A. Kröner, A.M. Goodwin and G.N. Hanson, Eds., Archaean Geochemistry. Springer, 161-181, 1984.

Harley, S.L., Regional geobarometry-geothermometry and metamorphic evolution of Enderby Land, Antarctica. In R.L. Oliver, P.R. James and J. Jago, Eds., Antarctic Earth Science. Australian Acad. Sci., 25-30, 1983.

Harley, S.L., An experimental study of the partitioning of Fe and Mg between garnet and orthopyroxene. Contr. Min. Pet., 86, 359-373, 1985.

Harris, N.B.W. and T.J.B. Holland, The significance of cordierite-hypersthene assemblages from the Beitbridge region of the Central Limpopo Belt; evidence for rapid decompression in the Archaean? Amer. Mineral., 69, 1036-1049, 1985.

Harris, N.B.W., R.W. Holt and S.A. Drury, Geobarometry, geothermometry, and late Archean geotherms from the granulite facies terrain of South India. J. Geol., 90, 509-528, 1982.

Hatherton, T., D.B. Pattiaratchi and V.V.C. Ranasinghe, Gravity Map of Sri Lanka. Sri Lanka Geol. Surv. Dept. Prof. Pap., 3, 1-39, 1975.

Heier, K.S., Geochemistry of granulite facies rocks and problems of their origin. Phil. Trans. R. Soc. Lond., A-273, 429-442, 1973.

Herd, R.K., Notes on metamorphism in New Quebec. In Metamorphism in the Canadian Shield, Geol. Surv. Canada Pap. 78-10, 79-83, 1978.

Herzberg, C.T., W.S. Fyfe and M.J. Carr, Density constraints on the formation of the continental Moho and crust. Contr. Min. Pet., 84, 1-5, 1983.

Hodges, K.V., J.M. Bartley and B.C. Burchfiel, Structural evolution of an A-type subduction zone, Lofoten-Rombak area, Northern Scandinavian Caledonides. Tectonics, 1, 441-462, 1982.

Holdaway, M.J., Stability of andalusite and the aluminum silicate phase diagram. Amer. J. Sci., 271, 97-131.

Hollister, L.S., Granulite facies metamorphism in the Coast Range Crystalline Belt. Canad. J. Earth Sci., 12, 1953-1955, 1975.

Hörmann, P.K., M. Raith, P. Raase, D. Ackermand and F. Seifert, The granulite complex of Finnish Lappland: petrology and metamorphic conditions in the Ivalojoki-Inarijärvi area. Geol. Surv. Finland Bull., 308, 1-95, 1980.

Huang, W.-L. and P.J. Wyllie, Melting reactions in the system $NaAlSi_3O_8-KAlSi_3O_8-SiO_2$ to 35 kilobars, dry and with excess water. J. Geol., 83, 737-748, 1975.

Hutchison, W.W., Metamorphic framework and plutonic styles in the Prince Rupert region of the Central Coast Mountains. Canad. J. Earth. Sci., 7, 376-405, 1970.

Janardhan, A.S., R.C. Newton and E.C. Hansen, The transformation of amphibolite facies gneiss to charnockite in southern Karnataka and northern Tamil Nadu, India. Contr. Min. Pet., 79, 130-149, 1982.

Jordan, T.E., B.L. Isacks, R.W. Allmendinger, J.A. Brewer, V.A. Ramos and C. Ando, Andean tectonics related to geometry of subducted Nazca plate. Geol. Soc. Amer. Bull., 94, 341-361, 1983.

Kars, H., J.B.H., Jansen, A.C. Tobi and R.P.E. Poorter, The metapelitic rocks of the polymetamorphic Precambrian of Rogaland, SW Norway. Part II: Mineral reactions between cordierite, hercynite and magnetite within the osumilite-in isograd. Contr. Min. Pet., 74, 235-244, 1980.

Katz, M.B., Early Precambrian granulites-greenstones, transform mobile belts and ridge-rifts on early crust? In B.F. Windley, Ed., The Early History of the Earth. Wiley-Interscience, 147-158, 1976.

Klemperer, S.L, L.D. Brown, J.E. Oliver, C.J. Ando, B.L. Czuchra and S. Kaufman, Some results of COCORP seismic reflection profiling in the Grenville-age Adirondack Mountains, New York State. Canad. J. Earth Sci. (in press), 1985.

Kröner, A., Precambrian plate tectonics. In A. Kröner, Ed., Precambrian Plate Tectonics. Elsevier, 58-90, 1981.

Kröner, A., Archaean to early Proterozoic tectonics and crustal evolution: a review. Rev. Brasileira de Geosciencias, 12, 15-31, 1982.

Laing, W.P., R.W. Marjoribanks and R.W.R. Rutland, Structure of the Broken Hill Mine area and its significance for the genesis of the ore bodies. Econ. Geol., 73, 1112-1136, 1978.

Leelanandam, C., Chemical mineralogy of hornblendes and biotites from the charnockitic rocks of Kondapalli, India. J. Petrol., 11, 475-505.

Lindsley, D.H., Pyroxene thermometry. Amer. Mineral., 68, 477-493, 1983.

Lougheed, M.S., Origin of Precambrian iron-formations in the Lake Superior region. Geol. Soc. Amer. Bull., 94, 325-340, 1983.

McCarthy, T.S., Chemical interrelationships in a low-pressure granulite terrain in Namaqualand, South Africa, and their bearing on granite genesis and the composition of the lower crust. Geochim. et Cosmochim. Acta., 40, 1057-1068, 1976.

McLelland, J.M. and J. Husain, Nature and timing of anatexis in the eastern and southern Adirondack Highlands. J. Geol. (in press), 1986.

Molnar, P. and P. Tapponier, Cenozoic tectonics of Asia: effects of a continental collision. Science, 189, 419-426, 1975.

Morse, S.A., A partisan review of Proterozoic anorthosites. Amer. Mineral., 67, 1087-1100, 1982.

Newton, R.C. and D. Perkins, Thermodynamic calibration of geobarometers based on the assemblages garnet-plagioclase-orthopyroxene (clinopyroxene)-quartz. Amer. Mineral., 67, 203-222, 1982.

O'Hara, M.J., Thermal history of excavation of Archaean gneisses from the base of the continental crust. J. Geol. Soc. Lond., 134, 185-200, 1977.

O'Hara, M.J. and Yarwood, G., High pressure-temperature point on an Archaean geotherm, implied magma genesis by crustal anatexis and consequences for garnet pyroxene thermometry and barometry. Phil. Trans. Roy. Soc. Lond., A-228, 441-456, 1978.

Okeke, P.O., G.D. Borley and J. Watson, A geochemical study of Lewisian metasedimentary granulites and gneisses in the Scourie-Laxford area of the north-west Scotland. Mineral. Mag., 47, 1-10, 1983.

Ortega-Gutiérrez, F., Evidence of Precambrian evaporites in the Oaxacan Granulite Complex of southern Mexico. Precamb. Res., 23, 377-393, 1984.

Phillips, G.N., Water activity changes across an amphibolite-granulite facies transition, Broken Hill, Australia. Contr. Min. Pet., 75, 377-386, 1980.

Phillips, G.N. and V.J. Wall, Evaluation of prograde regional metamorphic conditions: their implications for the heat source and water activity during metamorphism in the Willyama Complex, Broken Hill, Australia. Bull. Minéral., 104, 801-810, 1981.

Powell, C.M. and P.J. Conaghan, Tectonic models for the Tibetan Plateau. Geology, 3, 727-731, 1975.

Radain, A.A.M., W.S. Fyfe and R. Kerrich, Origin of peralkaline granites of Saudi Arabia. Contr. Min. Pet., 78, 358-366, 1981.

Rollinson, H.R., Garnet-pyroxene thermometry and barometry in the Scourie granulites, NW Scotland. Lithos, 14, 225-238, 1981.

Rona, P.A., Evaporites at passive margins. In R.A. Scrutton, Ed., Dynamics of Passive Margins. Amer. Geophys. Union, Washington, 116-132, 1982.

Royden, L. and Hodges, K.V., A technique for analyzing the thermal and uplift histories of eroding orogenic belts: a Scandinavian example. J. Geophys. Res., 89, 7091-7106, 1984.

Rubie, D.C., A thermal-tectonic model for high-pressure metamorphism and deformation in the Sesia Zone, western Alps. J. Geol., 92, 21-36, 1984.

Saxena, S.K., The charnockite geotherm. Science, 198, 614-617, 1977.

Schau, M., Two sapphirine localities in the Kramanituar Complex, Baker Lake region, District of Keewatin. Geol. Surv. Canada Pap. 82-1C, 99-102, 1982.

Schreurs, J., The amphibolite-granulite facies transition in West-Uusimaa, S.W. Finland. A fluid inclusion study. J. Meta. Geol., 2, 327-342, 1985.

Sheraton, J.W. and Collerson, K.D., Geochemical evolution of Archaean granulite-facies gneisses in the Vestfold Block and comparisons with other Archaean gneiss complexes in the East Antarctic Shield. Contr. Min. Pet., 87, 51-64, 1984.

Smalley, P.C., D. Field, R.C. Lamb and P.W.L. Clough, Rare earth, Th-Hf-Ta and large-ion lithophile element variation in metabasalts from the Proterozoic amphibolite-granulite transition zone at Arendal, South Norway. Earth, Plan. Sci. Lett., 63, 446-458, 1983.

Stormer, J.C., A practical two-feldspar geothermometer. Amer. Mineral., 60, 667-674, 1975.

Subrahmanyam, C., An overview of gravity anomalies, Precambrian metamorphic terrains and their boundary relationships in the southern Indian shield. In S.M. Naqvi and J.J.W. Rogers, Eds., Precambrian of South India. Geol. Soc. India Memoir, 4, 553-566, 1983.

Tarney, J. and B.F. Windley, Chemistry, thermal gradients and evolution of the lower continental crust. J. Geol. Soc., 134, 153-172, 1977.

Taylor, S.R., I.H. Campbell, M.T. McCullough and S.M. McLennan, A lower crustal origin for massif-type anorthosites. Nature, 311, 372-374, 1984.

Touret, J., Le faciès granulite en Norwège Méridionale. Lithos, 4, 239-249; 423-436, 1971.

Touret, J., Fluid inclusions in high grade metamorphic rocks. In L.S. Hollister and M.L. Crawford, Eds., Short Course in Fluid Inclusions: Application to Petrology. Min. Assoc. of Canada, 182-208.

Valley, J.W., Polymetamorphism in the Adirondacks. Volume of Abstracts, NATO Advanced Study Institute on the Deep Proterozoic Crust in the North Atlantic Provinces, Moi, Norway, July 1984, 29, 1984.

Valley, J.W., J. McLelland, E.J. Essene and W. Lamb, Metamorphic fluids in the deep crust: evidence from the Adirondacks. Nature, 301, No. 5897, 226-228.

Valley, J.W. and J.R. O'Neil, Fluid heterogeneity

during granulite facies metamorphism in the Adirondacks: stable isotope evidence. <u>Contr. Min. Pet.</u>, <u>85</u>, 158-173, 1984.

Vlaar, N.J., Lithospheric doubling as a cause of intracontinental tectonics. <u>Geophys.</u>, <u>B 85</u>, (4), 469-483, 1982.

von Knorring, O. and W.Q. Kennedy, The mineral paragenesis and metamorphic status of garnet-hornblende-pyroxene-scapolite gneiss from Ghana (Gold Coast). <u>Mineral. Mag.</u>, <u>31</u>, 846-859, 1958.

Wagner, M.E., Taconic metamorphism at two crustal levels and a tectonic model for the Pennsylvania-Delaware Piedmont. <u>Geol. Soc. Amer. Prog. w/ Abstr.</u>, <u>14</u>, 640, 1982.

Waters, D.J., Dehydration melting and the granulite transition in metapelites from southern Namaqualand, S. Africa. <u>Proc. of the Conf. on Middle to Late Proterozoic Lithosphere Evolution</u>, Cape Town, July 1984, 1984.

Weaver, B.L. and J. Tarney, Andesitic magmatism and continental growth. In R.S. Thorpe, Ed., <u>Andesites: Orogenic Andesites and Related Rocks</u>. John Wiley, 639-661, 1982.

Wellman, P. and R.J. Tingey, A gravity survey of Enderby and Kemp Lands, Antarctica. In C. Craddock, Ed., <u>Antarctic Geoscience</u>. Univ. of Wisconsin, Madison, 937-940, 1982.

Wells, P.R.A., Chemical and thermal evolution of

Archaean sialic crust, southern West Greenland. <u>J. Petrol.</u>, <u>20</u>, 187-226, 1979.

Wiener, R.W., J.M. McLelland, Y.W. Isachsen and L.M. Hall, Stratigraphy and structural geology of the Adirondack Mountains, New York: review and synthesis. <u>Geol. Soc. Amer. Spec. Pap.</u>, <u>194</u>, 1-55, 1983.

Wilson, A.F., Comparison of some of the geochemical features and tectonic setting of Archaean and Proterozoic granulites, with particular reference to Australia. In B.F. Windley and S.M. Naqvi, Eds., <u>Archaean Geochemistry</u>. Elsevier, 241-268, 1978.

Windley, B.F., D. Ackermand and R.K. Herd, Sapphirine/kornerupine-bearing rocks and crustal uplift history of the Limpopo Belt, southern Africa. <u>Contr. Min. Pet.</u>, <u>86</u>, 342-358, 1984.

Wood, B.J., The influence of pressure, temperature and bulk composition on the appearance of garnet in orthogneisses - an example from South Harris, Scotland. <u>Earth, Plan. Sci. Lett.</u>, <u>26</u>, 299-211, 1975.

Wynne-Edwards, H.R., The Grenville Province. Geol. Assoc. Canada Spec. Pap 11, 263-334, 1972.

Zelt, G.A.D., Granulite-facies metamorphism in Namaqualand, South Africa. <u>Precamb. Res.</u>, <u>13</u>, 253-274, 1980.

FLUID DISTRIBUTION IN THE CONTINENTAL LITHOSPHERE

Jacques Touret

Earth Science Institute, Free University, Amsterdam (The Netherlands)

Abstract. Synmetamorphic fluid inclusions trapped in rock forming minerals indicate a systematic change in the fluid regime at depth from H_2O- to CO_2-dominant. The ubiquity of CO_2 in the lower crust is explained by selective removal of H_2O in anatectic melts and related to the emplacement, at depth, of magmatic, mantle-derived carbonate phases.

Introduction

Most metamorphic and metasomatic processes in the lower crust and in the upper mantle involve a fluid phase, which may participate in mineral reactions (H_2O for hydration, CO_2 for carbonation) or, if not directly participating, exert a profound influence on the density, heat balance, and fugacities of other volatiles in the enclosing rocks. Its composition is generally estimated indirectly, from experimental data or thermo-dynamical calculations, but it has become evident during the last decade that minute quantities of volatiles trapped in minerals as fluid inclusions may contain valuable direct information. The study of fluid inclusions is as old as petrography (Sorby, 1858). However, only in recent years have decisive advances in technology (freezing and heating microscopic stages, Micro-Raman spectral analysis, etc.) and in the understanding of fluid behaviour at high P-T permitted the interpretation of data from lower crustal and upper mantle rocks. Despite considerable µvariation in rock types and in the location of investigated samples, some general rules have emerged, which have led to a general model of fluid distribution in the earth's crust and upper mantle (Touret, 1974).

In presenting this model it will be tacitly assumed that the reader is aware of the specific techniques used in fluid inclusion study, notably microthermometry (melting and homogenization temperatures), which have been reviewed in a number of recent publications (e.g. Roedder, 1984, Weisbrod et al., 1976; Touret, 1977; Hollister and Crawford, 1981; Tomilenko and Chupin, 1983). The following discussion will therefore emphasize the recent possibilities offered by laser microraman analysis, which now permits the routine, non-destructive analysis of inclusions in the micrometer range and which is as important for fluid inclusions as the electron microprobe has been for solid minerals.

Properties of Fluid Inclusions

Fluid inclusions investigations have shown two remarkable and constrasting aspects:
At the scale of the thin section, there are considerable interpretative complications such as number of fluid inclusions, variability in the physicochemical properties and difficulties to correlate the fluid evolution with the rest of the rock history. On a regional scale, however, things are often much more simple. Granulites and mantle-derived ultrabasic xenoliths from all over the world contain always the same type of inclusions (dense CO_2). Fundamental types of fluid inclusions amount to very few species: H_2O, H_2O + dissolved ions (brines), CO_2, CH_4 and N_2.

Despite many causes for fluid variations and errors in interpretation (non-representativity of the trapped fluids, reactivity of the mixture, leakage, changes that may occur after trapping, etc., see the abovementioned specialized literature) there are now some well documented cases (alpine metamorphism, granulites), where peak metamorphic fluids can be characterized with a reasonable degree of confidence. The method, always the same, involves three successive steps, sometimes not easy to carry out:

(1) Determination of the chronology of the trapped fluid inclusions in order to recognize peak metamorphic fluids that are present during the growth of the diagnostic metamorphic paragenesis.

(2) Determination of the P-T conditions for formation of the solid phases from one of the many thermometers or barometers now available for metamorphic rocks (see also Newton, this volume).

(3) Comparison of the P-T data for the solids and for the fluids. With a few exceptions (e.g. boiling, immiscible fluids trapped simultaneously), fluids do not give a unique P and T, but define a relation between P and T (isochore). In most cases, solid phase geothermometry and barometry do not indicate a point, but a

surface (P-T "box"). If isochores do not intersect this box, the corresponding fluid cannot have been trapped at peak metamorphic conditions (e.g. Touret and Dietvorst 1983).

Even if isochores do intersect the P-T box there is still the possibility for the fluid to have been trapped somewhere else along the isochore. This possibility, however, is limited by two sets of arguments:

(1) Fluid inclusions often give minimum trapping temperatures, limiting considerably the possible part of the isochore. This is less true for high-grade metamorphic rocks than for most other varieties: For high density CO_2, the homogenization temperature is so low (well below 0^oC) that no significant information is derived other than that trapping must occur at higher T than homogenization. However, fluid pressures at low T are so high (about 4 kb at 200^oC for CO_2 which homogenizes near the triple point) that only the part of the isochore corresponding to T above several 100^oC can be considered seriously.

(2) Different rock types metamorphosed at the same P-T conditions contain fluids of similar composition and density. This is especially true for granulites for which many arguments indicate that the early, high-density CO_2 inclusions are representative of peak metamorphic conditions (Coolen, 1980, Hollister and Crawford, 1981). Since the first estimates of fluid entrapment conditions were made (Touret, 1974), simultaneous rapid advances in seemingly unrelated fields (such as P-T estimates for solid phases, technological developments of fluid inclusion techniques and knowledge of the isochores of pure and natural systems) have resulted in a drastic decrease of the uncertainty. The correspondence between data for fluid and solid phases should now satisfy even the most critical mind. It is remarkable that these developments have not modified fundamentally the initial conclusions, in spite of very significant changes in the representative data (e.g. in Southern Norway, where changes in estimates of the slopes of the isochores have been cancelled by the discovery of higher density fluids as shown by e.g. the comparison of Fig. 5, Touret 1971 and Fig. 6, Touret 1985.

Model of Fluid Distribution in the Continental
Lithosphere

The model, illustrated in Fig. 1, simply corresponds to the vertical juxtaposition of several domains of variable depth in which the synmetamorphic fluids are now reasonably well characterized. In fact, a few well documented cases are available: the Western Alps for upper crustal diagenesis and low to medium grade metamorphism, Southern Norway and a few other places (Tanzania, India) for the lower crust. There are also a number of fragmental indications suggesting the general validity of the model, including deep crustal and upper mantle xenoliths. The model is based on the succession, from top to bottom, of three successive major fluid regimes: CH_4, H_2O and CO_2. The picture is very complicated near the surface as a result of widespread immiscibility and the dominating influence of the lithology of the host rock. It ends to be simpler or at least more continuous, at depth where the same fluid (CO_2) occurs in very different rocks ranging from acid to basic or ultrabasic in composition. In addition to previous descriptions (Touret, 1974; Dietvorst and Touret, 1983), a significant new feature is now introduced by the recent discovery of granulites containing NaCl brines (about 30% wt NaCl) instead of CO_2. These rocks indicate that CO_2, if dominant, is not ubiquitous and that it can coexist with other fluids in an apparently immiscible way. All granulites observed so far to contain brines are metasediments: metapelites, former evaporites and skarns. The brines are probably inherited from the sedimentary stage and can be preserved throughout the metamorphic history. Very significant seems to be the fact that the density of the brine is of the same order of magnitude as that of the neighbouring CO_2. This is a further argument to support the principle of a density controlled fluid distribution in the deeper part of the continental crust (Touret and Dietvorst, 1983).

Attention must be drawn to the artificial and static character of the model. I do not think that there is any part of the continental crust where the complete section is exposed continuously and has been formed during a single metamorphic episode. Much of the continental crust is continuously reworked and recycled, and any vertical section certainly passes through segments of very different age. However, there is no relation between age and fluid type; Archaean metamorphites contain exactly the same fluid inclusions than younger rocks of comparable metamorphic grade. The proposed model therefore expresses the fact that there seems to be a constant relation between P-T and fluid composition in the continental crust since the Archaean. Critical parts of the model can also be verified from exposed sections at the surface, such as the transition from a water-dominant to a CO_2-dominant regime in Southern Norway (N, Fig. 1).

Fluid Buffering Mechanisms: Internal Versus
External Buffering

From the preceeding discussion it appears that the fluid composition in the earth's continental crust and upper mantle may be represented in the C-O-H-N-S system, with addition of NaCl (and other dissolved ions) for some selected compositions. N_2 has recently been discovered in a variety of environments (Swanenberg, 1980, Kreulen and Schuiling, 1982), but it does not participate in the chemical equilibrium, at least at a first approximation. Sulfur species might be more fre-

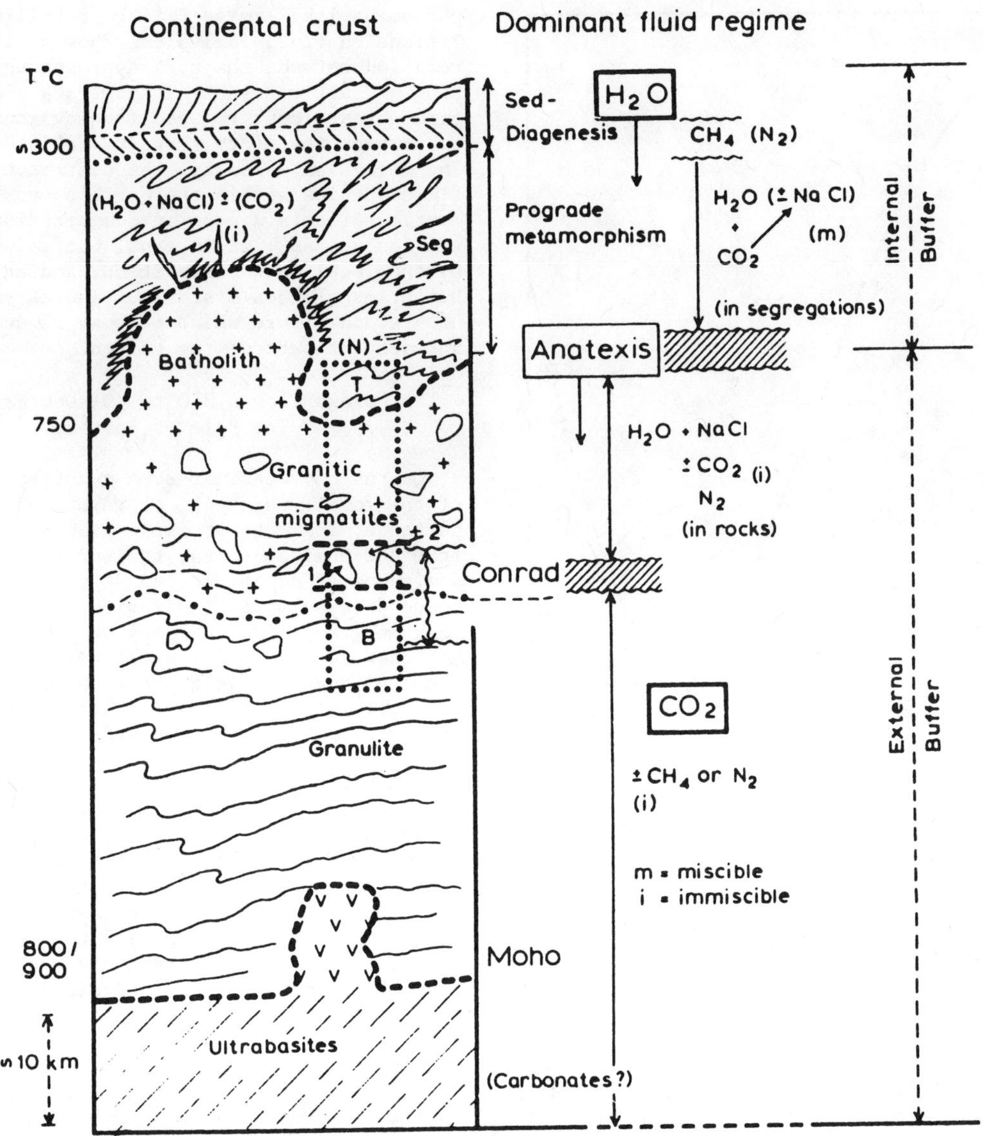

Fig. 1. Model of fluid distribution in the continental lithosphere (Touret and Dietvorst 1983). Left: Petrological section; the lower crust is shown to consist entirely of granulite. N (dotted rectangle): Section now exposed at the surface in Southern Norway (T = Telemark, B = Bamble). Right: Dominant fluid regime as inferred from fluid inclusions studies.

quent than commonly assumed (Touray et Guilhaumou, 1984), but they remain restricted to specific environments, notably metacarbonates. The most important volatile components are therefore C-H-O, a system first discussed by French (1966) and later by Eugster and Skippen (1967) and Holloway (in Hollister and Crawford, 1981), among others.

At variable P-T, the composition of the system depends on a number of hypothetical conditions which I shall not attempt to review here. Most important is the fact that, if graphite is present, the fugacity of one component fixes the fugacity of all other components. Since Eugster and Skippen (1967) it has become customary to consider that the fugacity of oxygen is fixed by mineral assemblages, notably by oxide or silicate-oxide "buffers". Fig. 2 (Holloway, 1981) gives results for graphite saturation at a total pressure of 2 kb and an oxygen fugacity of the buffer Q-F-M (Quartz-Fayalite-Magnetite) minus two log units (slightly more reducing than the Q-F-M buffer). We observe an increasing temperature sequence of dominant fluids: CH_4-H_2O-CO_2, which is the same as that observed in the metamorphic column at comparable (slightly higher) tempera-

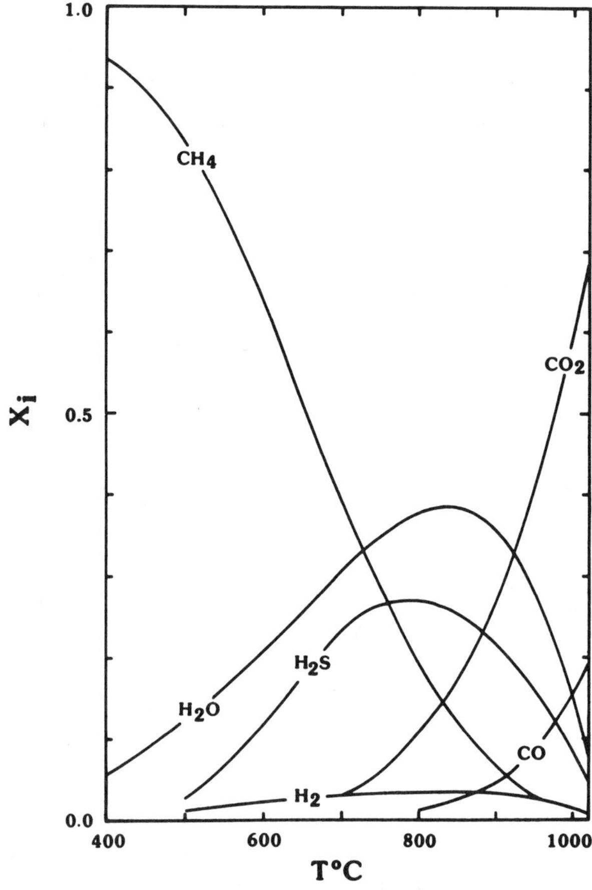

Fig. 2. Composition of fluids in equilibrium with graphite at log f_{O_2} = QFM - 2 and total P = 2kb (Holloway 1981).

tures. More extended calculations would show that the order of succession of the dominant fluid would remain the same at higher pressure (up to about 4 kb) also for a limited variation in f_{O_2}. It is, however, far from evident that the preceeding results may be immediately applicable to natural rocks, for at least two reasons. One is that graphite may not always be present. It might be more abundant than commonly assumed, but it is obvious that there are many rock types, notably at upper crustal levels, where graphite is not stable. The other is that the inferred oxygen fugacity is compatible with most natural parageneses only in the lower crust and in the upper mantle. Closer to the surface, many indications suggest that the oxygen fugacity is commonly much higher than the Q-F-M buffer. The overall results are, however, suggestive that some kind of large-scale buffering might exist, controlling successively CH_4, H_2O and CO_2 at increasing depth. This obviously calls for an external buffering mechanism, independent of the local rock composition.

A different picture is given for fluids in the ionic systems (H_2O + NaCl). Besides specific phenomena like immiscibility or boiling, observations on fluid inclusions show a distinct relation between the rock type and the concentration of salt : brines either trace old evaporites or are restricted to specific environments; their occurrence therefore suggests an internal buffering mechanism. Finally, the distribution of the fluid species within the earth's crust and upper mantle reflects a constant competition between two tendencies : an internal buffering, imposed by the local rock composition, and an external buffering of more mysterious and as yet little understood nature which becomes predominant in the deepest part of the column.

The Change From a H_2O to CO_2 Dominant Regime:
The Role of Anatexis

Let us now consider some quantitative aspects of the model. CH_4 being of relatively minor importance, the most striking aspect is the change from H_2O-dominance near the surface to CO_2-dominance at depth. The quantity of CO_2 is not known precisely, but inclusions are so abundant in some granulites that may be quite important, at least locally. If the change corresponds to a chemical reaction (e.g. between water and graphite), it would involve large quantities of graphite and produce much hydrogen. Diffusion could be a reason for the absence of hydrogen, but some graphite should remain in the system. Graphite is present in small amount (typically less than 1%) in some granulites, but is normally absent in most upper crustal rocks, except in very specific environments (metapelites). It seems therefore necessary to search for other processes in order to explain the transition from H_2O to CO_2 at depth. Direct observations and theoretical consideration (Kadik et al., 1973; Kadik, 1975) suggest that anatexis plays a major role.

One of the most striking results of fluid inclusions studies has been to recognize the rarity of pure CO_2 inclusions where one would have normally expected them (carbonate-bearing sediments) and their abundance in rocks where their presence might be considered unlikely: granitic and anatectic rocks. The common occurrence of CO_2 in granites has been known since the early days of petrography (e.g. Fig. 52 in Sorby, 1858). But its significance remained obscure, as most inclusions in granitic rocks do not relate to early, magmatic processes but to later hydrothermal overprinting (Weisbrod, in Hollister and Crawford, 1981).

More interesting is the case where in situ "granite" formation can be seen at the hand specimen scale, namely at the onset of anatexis in metamorphic rocks. When studying a prograde metamorphic sequence, a sudden change is observed in fluid inclusion distribution as soon as granitic mobilisates appear, most commonly near the second sillimanite isograd (Evans and Guidotti, 1966): the coarser grained "mobilisate" contains

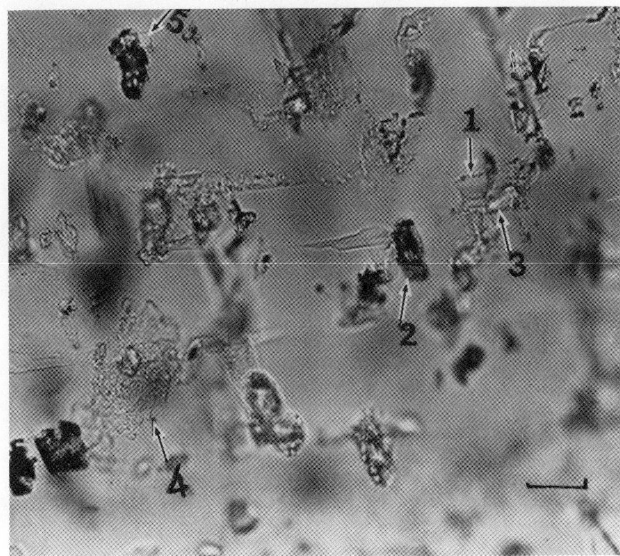

Fig. 3. Carbonate inclusions in a plagioclase
of charnockitic gneiss, Tromoy, Norway (Touret
1985). 1 and 5: unknown isotropic phase, 2: gas,
3: well crystallized carbonate (ferroan dolomite), 4: cryptocrystalline calcite (probably
late alteration). The inclusions are interpreted
as former melt-carbonatite droplets immiscible
in the silicate magma and trapped during plagioclase growth (Touret 1985). Length of bar
25 μm.

many more inclusions than the enclosing rock
which, on the other hand, is richer in hydrous
minerals, notably micas. This observation
supports the idea that most volatiles trapped in
the mobilisate are locally derived and provide a
clue as to the influence of melting on fluid
distribution.

A detailed study has been undertaken in one
case (Songe amphibolite, Norway, Touret and
Dietvorst, 1983) indicating that the earliest
fluids, probably in equilibrium with the melt
during the first stage of mineral crystallization, are CO_2 rich. Subsequent successive pulses
of different fluids are recorded, namely N_2, CH_4
and H_2O. The interpretation of pressure and temperature for the trapping is much more complicated than initially assumed (Touret and Olsen
1985), but this is mainly due to the difficulty
of establishing a precise chronology of fluid
entrapment. The case of H_2O in particular is
hopelessly complicated: some aqueous inclusions, notably those which homogenize at low
temperature (dense inclusions, T_h close to
$100^{\circ}C$), are very late, but others (high T_h) might
be earlier than the CH_4-bearing ones, and this
induces very significant variations in possible
post-metamorphic P-T paths. (Touret and Olsen,
1985). In spite of such complications, the
relation between pure CO_2 and local melting
remains valid and is one of the most obvious

inferences from the study of high-grade metamorphic rocks.

Origin of Deep Crustal CO_2:
The Carbonatite Connection

Although anatexis provides a mechanism to
separate CO_2 and H_2O, it cannot by itself produce
CO_2. We may envisage that the deep CO_2 is residual, but this does not agree very well with the
large number of carbonic inclusions found in most
granulites. No direct relation can be inferred
between the number of inclusions and the quantity
of fluid actually present when the rocks were
buried, but the general increase in the quantity
of inclusions when reaching the granulite facies
is so spectacular that a reintroduction of CO_2 is
most probable. The first idea that comes to the
mind is to invoke progressive decarbonization of
CO_2-bearing sediments. This is possible, even
probable in some specific cases (Glassley 1983),
but not generally applicable for two major
reasons:

(1) In many cases the protolith for rocks now
in granulite facies terrains can be determined
with reasonable certainty. Metasediments are
common, but not always predominant, and they are
far more often clastic (metapelite) than CO_2-
bearing.

If carbonates were originally present, they
are still there in most cases: marbles cover wide
areas in some classical granulite provinces
(Adirondacks, Canada, Madagascar) and, except
very locally, they do not show any sign of
destabilisation. Moreover, CO_2-bearing minerals
(e.g. scapolite) are more common and typical for
the granulite domains than for the neighbouring
areas.

Thus direct evidence is not in favour of CO_2
mobilisation by decarbonization. In the only case
which, to my knowledge, has been investigated in
detail (South Norway, Touret 1985), inclusions
related to former carbonate-bearing sediments
("skarns") are not CO_2-rich, but contain NaCl
brines. Carbonic inclusions are far more abundant
near basic, synmetamorphic intrusives.

(2) The carbon isotopic data (δC^{13}) are
complicated but also tend to eliminate a large
scale participation of metasedimentary carbonate:
carbonic inclusions have very low δC^{13} (-20 $^o/oo$
or lower relative to PDB), whose precise significance remains a matter of debate (Hoefs and
Touret 1975, Pineau et al. 1981), but which can
hardly derive from former limestones.

For all these reasons I have always favoured a
juvenile, mantle-derived origin for the CO_2 of
the lower crust. An important confirmation has
been obtained recently in southern Norway by the
discovery of minute carbonate inclusions in
plagioclase of intermediate granitoid intrusives
(Touret, 1985) (Figs. 3 and 4). The core of the
granulite area of southeastern Norway (region of
Arendal) is characterized by the occurrence of
highly LIL-depleted rocks of enderbitic composi-

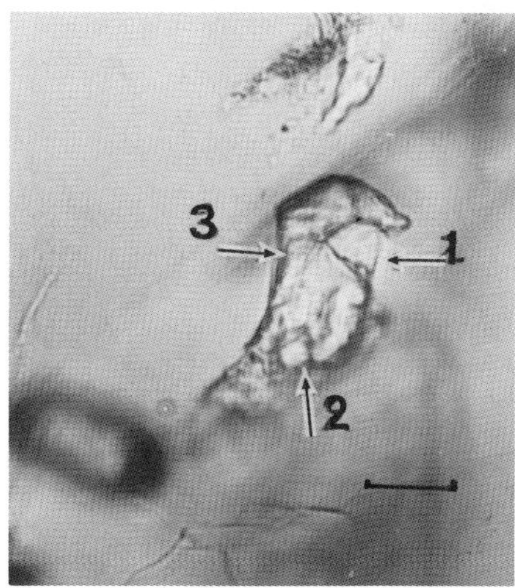

Fig. 4. Detail of a carbonate inclusion: 1 and 2: euhedral ferroan dolomite, 3: calcite. Length of bar 10 µm.

tion ("arendalite" in the sense of Bugge, 1943). From major and trace element geochemistry it was found (Field et al., 1980; Smalley et al., 1983) that the LIL-depletion was not acquired during metamorphic evolution, but by magmatic differentiation and direct crystallization under granulite facies conditions. Many typical magmatic features are indeed still present in the rocks, notably in idiomorphic inclusions preserved in some minerals (especially feldspars, see Fig. 7 in Touret, 1985). But most important are carbonate-bearing inclusions, probably crystallized from melts which predate the CO_2 fluids and which bear obvious affinities with carbonatites. In southern Norway it has consequently been argued that most of the CO_2 was initially introduced as immiscible carbonate droplets in granitoid intrusions of intermediate, trondjhemitic composition. These magmas were emplaced during granulite facies conditions and crystallized directly at a depth of about 25 km (pressure $\sim$ 8 kb). There are reasons to believe that the case of Norway is not unique since similar carbonate-melt inclusions have been found in other granulite areas, notably in high pressure granulites (Tanzania, China) (Touret, unpubl. data). More detailed studies are required, however, in order to determine whether observations made in southern Norway are applicable to all granulite terrains.

Acknowledgements. Constructive remarks and suggestions by A. Kroner and an anonymous reviewer are gratefully acknowledged. Thanks are due to Mr. Sion and Mr. Van der Bliek for the photos and the drawings, respectively. This work forms part of the research program IvA GEFY 83/2 of the Free University, Amsterdam.

References

Bugge, J. A. W., Geological and petrological investigations in the Kongsberg-Bamble formation, Norges Geol. Unders., 150, 150 pp., 1943.

Coolen, J. J. M. M. M., Chemical geology of the Furua Granulite Complex, southern Tanzania. GUA Ser. (Univ. van Amsterdam, Ser. 1, 13, 258 pp, 1980.

Eugster, H. P. and G. B. Skippen, Igneous and metamorphic reactions involving gas equilibria, in : Researches in Geochemistry, 2, edited (P. H. Abelson), pp. John Wiley Sons, NY., 1967.

Evans, B. W. and C. V. Guidotti, The sillimanite-potash feldspar isograd in Western Main, USA, Contrib. Mineral. Petrol., 12, 25 -62, 1966.

Field, D., A. Drury, and D. C. Cooper, Rare earth and LIL Fractionation in high grade charnockitic gneisses, South Norway, Lithos, 13, 281-289, 1980.

French, B. M., Some geological implications of equilibrium between graphite and a C-H-O gas phase at high temperature and pressure, Rev. Geophys., 4, 223-253, 1966.

Glassley, W. E., Deep crustal carbonates as CO_2 fluid sources. Evidence from metasomatic reaction zones, Contrib. Mineral. Petrol., 84, 15-24, 1983.

Hoefs, J. and J. Touret, Fluid inclusion and carbon isotope studies from Bamble granulites, Southern Norway. A preliminary investigation, Contrib. Mineral. Petrol., 52, 165-174, 1975.

Hollister, L. S. and M. L. Crawford (Eds), Short Course in Fluid Inclusions: Applications to Petrology. Min. As. Canada, vol. 6, Calgary, 304 p.p., 1981.

Holloway, J. R., Compositions and volumes of supercritical fluids in the Earth's Crust, pp. 13-36 in Hollister, L.S. and M. L. Crawford, op. cit., 1981.

Kadik, A. A., Influence of basic magmas degassing on H_2O and CO_2 regimes in the Crust and Upper Mantle (in russian), Acad. Sci. SSSR, Int. Geophys. Project, 3, 67-86, 1975.

Kadik, A.A. and O.A. Lukanin, The solubility dependant behaviour of water and carbon dioxyde in magmatic processes. Geochim. Inter., 10, 115-129, 1973.

Kreulen, R. and R. D. Schuiling, N_2-CH_4-CO_2 fluids during formation of the Dome de l'Agout, France. Geochim. Cosmo. Acta, 46, 193-203, 1982.

Pichavant, M., C. Ramboz, and A. Weisbrod, Fluid immiscibility in natural systems: Use and misuse in fluid inclusion data, Chem. Geol., 37, 1-27, 1982.

Pineau, F., M. Javoy, F. Behar and J. Touret, La géochimie isotopique du faciès granulite du Bamble (Norvège) et l'origine des fluides carbonés de la croûte profonde, Bull. Mineral., 104, 630-641, 1981.

Roedder, E., Fluid Inclusions, Rev. in Mineral., <u>Am. Min. Soc.</u>, <u>12</u>, 644 pp, 1984.

Smalley, P. C., D. Field, R. C. Lamb and P. W. L. Clough, Rare earth, Th-Hf-Ta and LIL element variations in metabasites from the Proterozoic amphibolite - granulite transition zone at Arendal, south Norway, <u>Earth Plan. Sci. Let.</u>, <u>63</u>, 446-458, 1983.

Sorby, H. C., On the microscopic structure of crystals, indicating the origin of minerals and rocks. <u>Quat. Jl. geol. Soc. London</u>, <u>14</u>, 453-500, 1858.

Swanenberg, H., Fluid inclusions in high grade metamorphic rocks from southern Norway. Ph D Thes., <u>Geologica Ultraiectina</u>, Utrecht, <u>25</u>, 147 pp, 1980.

Tomilenko, A. A. and V.P. Chupin, Thermobarochemistry of metamorphic complexes (in Russian). <u>Acad. Sci. SSSR Siberian Branch</u>, <u>524</u>, 200 pp., 1983.

Touray, J. C. and N. Guilhaumou, Characterization of H_2S bearing inclusions. <u>Bul. Mineral.</u>, 107, 181-188, 1984.

Touret, J., Le facies granulite et Norvege meridionale. Associations minerales et inclusions fluides. <u>Lithos</u>, <u>4</u>, 239-249 et 423-435, 1971.

Touret, J., Facies granulite et fluides carboniques. Geol. Domaines cristallins (vol. P. Michot), <u>Soc. Geol. Belgique</u>, h.s., 267-287, 1974.

Touret, J. The significance of fluid inclusions in metamorphic rocks in D. Fraser, ed., <u>Thermodynamics in Geology</u>, pp. 203-227, Reidell, Dordrecht, 1977.

Touret, J., Fluid regime in southern Norway : the record of fluid inclusions, in Tobi, A. C. and Touret, J. (eds)., <u>The deep proterozoic crust in the North Atlantic provinces</u>, NATO Ad. Stud. Inst., 603 pp., 1985.

Touret, J., and P. Dietvorst, Fluid inclusions in high grade anatectic metamorphites. <u>Jl. Geol. Soc. London</u>, <u>140</u>, 635-649, 1983.

Touret, J., and S. Olsen, Fluid inclusions in migmatites, 265-286 in Ashworh, J. R., ed. <u>Migmatites</u>, Blakie pub., 1985.

Weisbrod, A., B. Poty et J. Touret, Les inclusions fluides: Tendances actuelles. <u>Bul. Mineral.</u>, <u>99</u>, 140-152, 1976.

____

J. Touret, Instituut voor Aardwetenschappen, Vrije Universiteit, Postbus 7161, 1007 MC Amsterdam

PRECAMBRIAN CARBONACEOUS FORMATIONS: THEIR EVOLUTION AND METAL CONTENT

N.A. Sozinov and O.V. Gorbachev

Institute of the Lithosphere, USSR Academy of Sciences, 22 Staromonetny per., 109180 Moscow, USSR

Abstract. The guiding factor in carbon accumulation throughout earth history is the evolution of sedimentary basins which, in turn, reflects the evolution in the style of tectonics of the lithosphere. The distribution of ore deposits in various carbonaceous formations in the age interval from the Archean to the Lower Paleozoic is closely related to the evolution of the upper continental crust during that period. The carbonaceous deposits can be specified as a group of "through-going" formations.

Formational Types of Carbonaceous Deposits

Of principal importance for the analysis of sediments is the comparison of formations of the same type but of differing ages which emerged in the course of geological history. Such comparison helps to reveal the characteristic features related to evolution of these formations.

One of the best objects for research under discussion may be provided by deposits enriched in carbonaceous matter and a complex of ores and trace elements which were formed in a specific geological setting throughout the history of the Earth. Four formational types of carbonaceous deposits can be singled out on the basis of their paragenesis: terrigenous-carbonaceous, siliceous-carbonaceous, carbonate-carbonaceous, and volcanic-siliceous-(carbonate)-carbonaceous.

Since the role of volcanism for carbonaceous deposits is not always known, and the fourth type is often polyformational, we have arbitrarily divided carbonaceous deposits into three main formational types according to the predominant component: terrigenous-carbonaceous, carbonate-carbonaceous and siliceous-carbonaceous.

The Evolution of Carbonaceous Formations

To trace the evolution of carbon-bearing sequences in the Precambrian, we have chosen the eastern part of the Baltic Shield where these accumulations are persistent from the Lopian (early Archean) to the Riphean (late Proterozoic).

Data on the distribution of major rock types in formations and age divisions were provided by Negrutzy et al. (1981).

Already during the earliest Archean, two types of basins can be delineated (Fig. 1). One is a series of narrow greenstone basins bordered by granite-gneiss domes (Fig. 1-1,2), the other is a trough-like basin of the Keivian sequences formed in the axial, most stable part of the Kola megablock (Fig. 1-3). The Archean carbonaceous deposits are characterized by their association with ferruginous-siliceous sediments in the greenstone basins and with highly aluminiferous sediments in the Keivian association.

The early Proterozoic (Sariolian, 2800-2300 Ma) (Fig. 2-4,5) was marked by a rifting stage in the reactivation of the previously stabilized craton and thus did not favor mass accumulation of organic matter because of coarse terrigenous clastics and volcanics in trough-type basins.

The peak in carbonaceous deposition is associated with the Jatulian cycle (2300-2150 Ma) and the Karelian tectonogenesis which was accompanied by subsidence in the Kola-Karelian region.

The differential nature of this subsidence is reflected within Karelian mobile belts (Fig. 2) in a series of geosynclinal structures which accommodated highly carbonaceous accumulations. Abundant carbonate and siliceous, organic-rich rocks reveal the high bioproductivity of this basin type. It is this level that shows the largest stratiform deposits of sulphide ores on the Baltic Shield (Outukumpu, Vikhanti, Vuonos, Boliden) which were formed with a significant involvement of volcanic processes. Within the areas of cratonized basement (Figs. 2-10), probably at the close of this cycle, the specific trough-like basins (Onega and Suoyarvi troughs) were formed and characterized by slow subsidence and thick accumulation of highly carbonaceous rocks (schungites) (Fig. 2-10).

In Kalevian times (2100-1950 Ma) (Fig. 2-12-14) the Pechenga basin accumulated thick sequences of effusive-sedimentary rocks with up to 12% carbon in the upper horizons; at that time the stabilized zones near Lake Ladoga and Lake Onega already showed epi-platform terrigeneous carbon-poor sediments which, in the Vepsian (1960-1600 Ma), underwent redeposition under typically continental conditions (Fig. 2-15).

The Riphean is the final stage in Precambrian carbonaceous sediment accumulation in the eastern

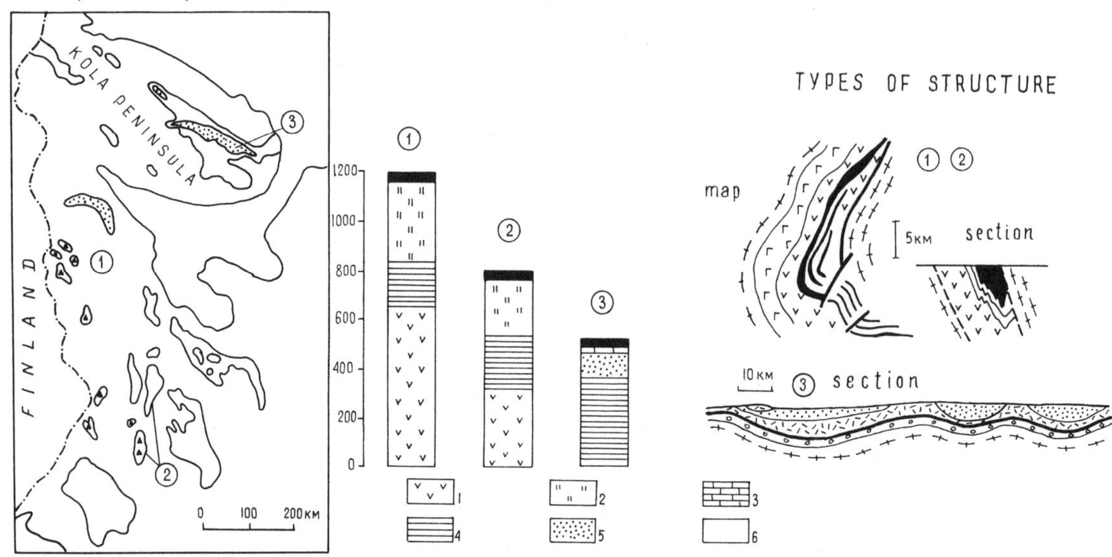

Fig. 1. Structural position and compostion of carbonaceous formations of the Eastern Baltic Shield (Archean). 1-3 (numbers in circles), names of sequences: 1-Gimolian, 2-Parandovian, 3-Keivian. 1-6, rocks: 1-volcanics, 2-quartzite, 3-carbonate, 4-metapelite, 5-metapsammite, 6-carbonaceous.

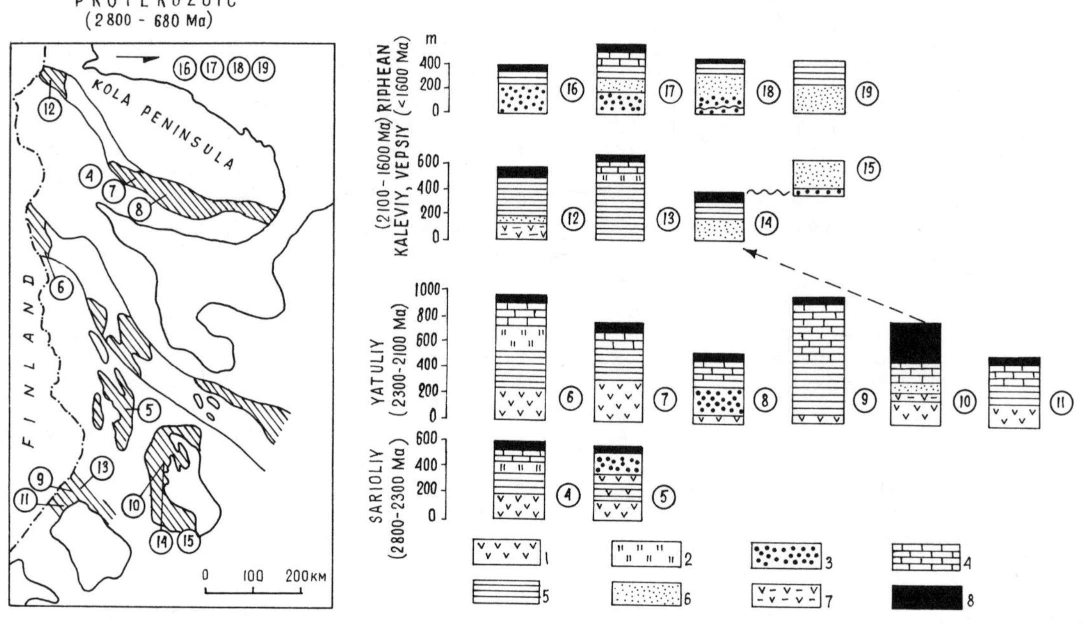

Fig. 2. Structural position and composition of carbonaceous formations of the Eastern Baltic Shield (Proterozoic). 4-19 (numbers in circles), age levels, types of sequences: 4-5 (Sarioly): 4-Sydoreshenskiy, 5-Seletzky; 6-11 (Yatuly): 6-Sovayarvinsky, 7-Prichibinsky, 8-Varzugsky, 9-Soanlachtisky, 10-Zaonezhsky, 11-Sortovalsky, 12-14 (Kalevy): 12-Pilgujarvinsky, 13-Ladozhsky, 14-Nigozersky, 15 (Vepsy)-Kamennoborsky, 16-Kildinsky, 17-Skarbeevsky, 18-Volkvoy, 19-Borgoutsky. 1-8, rocks: 1, 2, 4-6, 8 (see Fig. 1), 3-conglomerate, 7-tuffite.

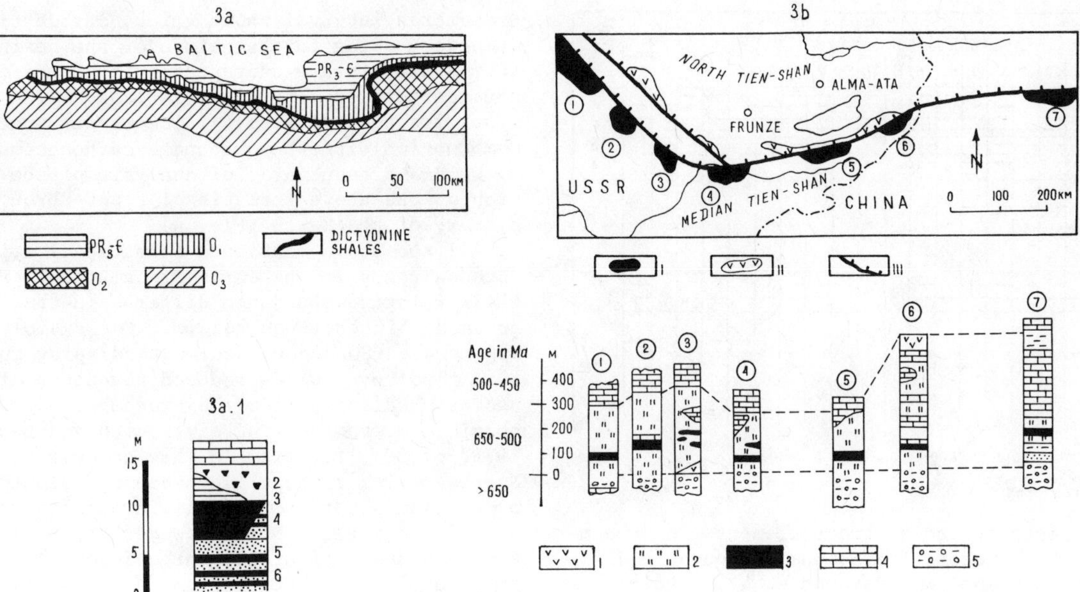

Fig. 3. Structural position and composition of carbonaceous formations (late Proterozoic-early Paleozoic). (a) - Platform type of Baucov (1973), dictyonine shales of the Baltic basin. 3a.1, typical column Lower Ordovician. 1-limestone, dolostone, 2-glauconite, sandstone, 3-clay, 4-dictyonine shales, 5-sandstone with obolus, 6-conglomerate. (3b)-Geosynclinal type of Orlov (1981) and Kholodov (1970). I - Region of carbonaceous formations, II - Zone of volcanicity, III - Continental margin; 1-7 (numbers in circles): 1-Ulu-Tau, 2-Karatau, 3-Sandalash-Pskem, 4-Kok-Jyrym-Too, 5-Jetym-Too, 6-Sary-Jas, 7-Kuruk-Tag. 1-5, rocks: 1-volcanics, 2-siliceous, 3-carbonaceous, 4-carbonate, 5-pebble clay ("tilloid").

Baltic Shield. These deposits are limited to narrow depressions and are represented mostly by terrigenous sediments poor in dispersed organic matter.

Thus, the guiding factor in carbon accumulation within the studied period of time is the evolution of sedimentary basins which, in turn, reflects the evolution in the style of tectonics of the region,

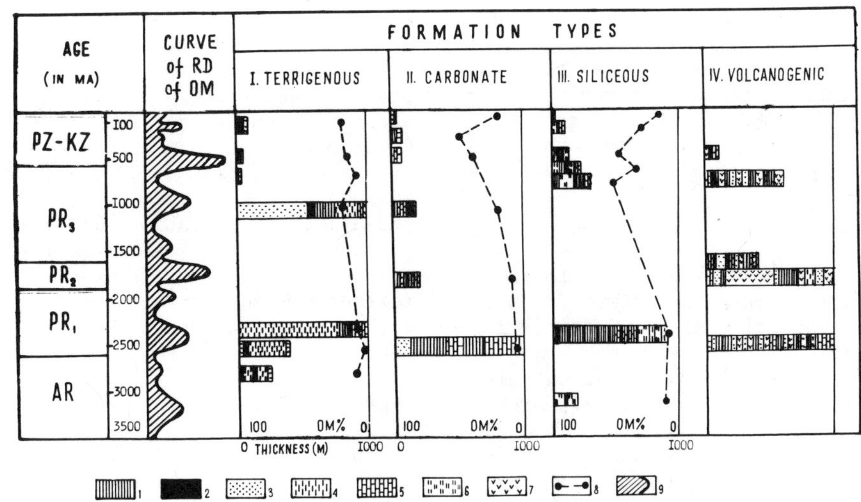

Fig 4. Relative distribution (RD) of different carbonaceous formation types in Precambrian and Phanerozoic times. 1-7, rocks: 1-clayey, 2-carbonaceous, 3-sandy, 4-aleurolitic, 5-carbonate, 6-siliceous, 7-volcanogenic, 8-organic atter (OM), 9-curve of relative distribution of OM.

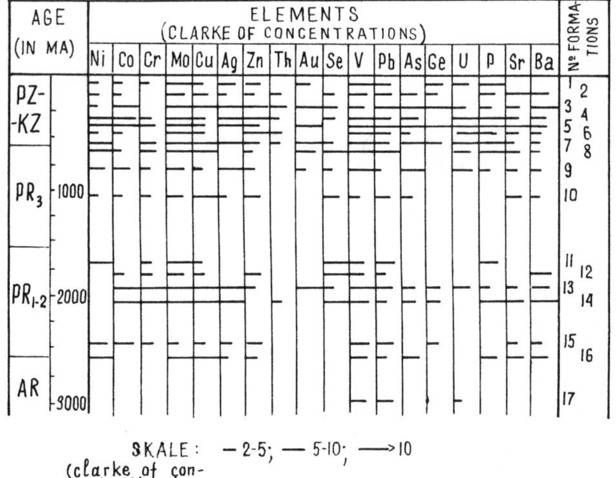

Fig. 5. Distribution of trace elements in Precambrian and Phanerozoic formations: 1-Recent sediments enriched in C_{org} (Baturin et al., 1967); 2-Green River Formation, U.S.A. (Vine and Tourletot, 1970); 3-Kupferschiefer carbonaceous deposits (Germany, Wedepohl, 1964); 4-black shales of the Phosphoria Formation, U.S.A. (Vine and Tourletot, 1970); 5-black shales of the Chattanooga Formation, U.S.A. (Vine and Tourletot, 1970); 6-Kolm of Sweden (Bain, 1960); 7-black shales of the Sino-Korean shield; 8-black shales of the Shinsay sutie, Kazakhstan; 9-black shales of the Altai-Sayany area (Zhdanova, 1971); 10-black shales of the Nunsach suite, U.S.A. (Vine and Tourletot, 1970); 11-the Vinkhya black shales, India (Murty et al., 1962); 12-black shales of the Sangilen Upland, Tuva (Borovskaya et al., 1966); 13-black shales of the Outokumpu area, Finland (Peltola, 1960); 14-anthracite coal of the Iron River District, U.S.A. (Tyler et al., 1975); 15-black shales of the Timskaya suite of the Kursk magnetic anomaly area; 16-graphitic shales of the Olkhon series, Baikal area (Ostapenko et al., 1970); 17-black shales of the Soudan series, Canada (Could et al., 1965). Clarke of concentrations: the ratio of the local or regional average content of a given chemical element to its global average content.

that is, successive cratonization of mobile areas and formation of platform-type structures.

We may argue, irrespective of the Baltic region evidence, that differentiation between basin types on blocks with different protocrust, evident already in the Archean, led to formation of large basins of the platform-type proper and of the geosyncline type by the start of the Paleozoic. Such early Paleozoic basins with thick accumulation of biogenic carbon (Fig. 3) are represented by dictyonine shales of the Baltic basin on the one hand and by the geosynclinal belts of the Ural-Tien-Shan fold area on the other, where the Vendian-Ordovician interval shows thick, carbon-rich sequences. Owing to the nature of the basins, platform sediments are characteristic of the terrigenous-carbonaceous type, while geosynclinal sediments are predominantly siliceous-carbonaceous in combination with the carbonate-carbonaceous type.

We now present a brief analysis of the evolution of the above formational types through the history of the Earth (Fig. 4).

All the above mentioned formational types existed already in the early Precambrian. However, their relative abundance differed in the Precambrian and in the Phanerozoic. For example, terrigenous-carbonaceous sediments display the general tendency towards reduced abundance of carbonaceous sediments from the Precambrian to the Cenozoic. W. Dawson (in Voiytkevich and Lebedko, 1975, p. 39), for example, has calculated that North America's Grenville sequences alone contain more organic carbon than all Carboniferous deposits of the world. Note, however, that the relative abundance of highly carbonaceous deposits in Phanerozoic sediments is generally higher than in the Precambrian.

Biogenic carbon associations with highly aluminiferous and predominantly kaolinitic sediments in the Precambrian give way to clay sediments of mixed composition (hydromica, kaolinite, montmorillonite) in the Phanerozoic (Gorbachev and Sozinov, 1985). There are also drastic changes in the morphology of the basins where carbonaceous sediments were accumulating. Local shallow Precambrian basins on a rigid basement become large shallow basins in the platforms of the Phanerozoic.

A similar tendency to changing thickness and organic matter content in rocks is traced in carbonate-carbonaceous sediments. Clay components of this formational type change in composition from predominantly montmorillonitic in the Precambrian to largely illitic in the Phanerozoic. Instead of basins of the miogeosynclinal type typical of the late Precambrian, the Paleozoic shows inland sea and lacustrine sedimentary basins, such as bituminous shales of Eastern Siberia or the Green River Formation in the USA.

A specific feature of siliceous-carbonaceous sediments in the early Precambrian is their close paragenetic association with BIF. Siliceous-carbonaceous sediments attained their maximum development in the late Precambrian and Paleozoic (extensive sequences of metal-bearing siliceous-carbonaceous shales in Eurasia). Their Meso- and Cenozoic extent is considerably less.

The volcano-carbonaceous type, equally well represented in the Precambrian (Yatulian schungites, USSR, Keewatin, USA, Transvaal, South Africa, Birrimian, West Africa, carbonaceous shales), reaches its widest distribution in lower Paleozoic eugeosynclinal sediments as in the Altai-Sayan fold belt of the USSR. Its role in Mesozoic geosynclinal sediments is considerably lower (Sozinov and Gorbachev, 1984).

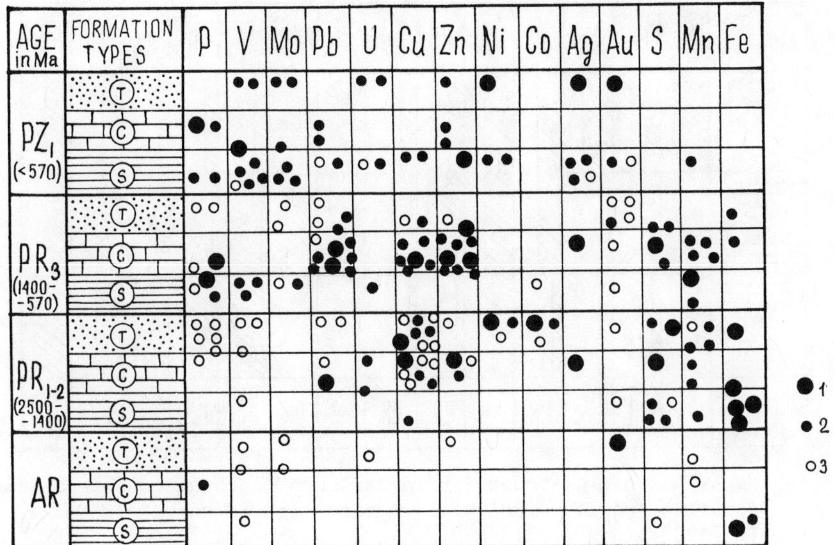

Fig. 6. Ore deposits and occurrences of various carbonaceous formation types in age intervals from the Archean to the Lower Paleozoic. 1-large ore deposits and ore belts, 2-median and small ore deposits, 3-ore occurrence and potential ore-bearing horizons. Carbonaceous formation types: T-terrigenous, C-carbonate, S-siliceous.

Metal-bearing Features of Black Shale Formations

The biogenic origin of metamorphosed carbonaceous matter, involving the vital activity and subsequent burying of ancient marine organism, should predetermine the metal-bearing character of the corresponding rocks. Data on the ability of marine organisms to concentrate some elements are well known and do not require special consideration (Vernadsky, 1940; Manskaya and Drosdova, 1964). Nevertheless, a comparative study of the composition and content of the Precambrian and Phanerozoic helps to reveal a directed evolutionary change in the composition of fossil organic matter in the course of the geological history of the Earth.

This question has been studied using both bibliographic and the authors' data on the distribution of trace elements in carbon-rich (>10%) black shale formations ranging in age from Recent sediments to the Archean, with an absolute age interval of more than 3000 Ma (Fig. 5).

The diagram pattern shows a constant spectrum of elements of higher concentrations, regardless of the age of formations. These are, as a rule, the same elements that are noted in Precambrian and Phanerozoic formations. All the elements characteristic of Phanerozoic formations are also typical of the entire Precambrian. The absolute contents of elements vary considerably, depending upon depositional conditions and subsequent metamorphism. Nevertheless, the characteristic composition of elements remains constant.

These data also indicate that whatever the age, black shale formations contain such elements as Co, Ni, Cr, Mo, Co, Ag, Zn, V, Pb, U, P, Sr and Ba in higher concentrations. Their concentration level varies within a wide range and depends upon conditions of formation, sediment composition and intensity of epigenetic changes. A dependence of changes in element content upon geological age and geotectonic setting has not been established. Moreover, in unmetamorphosed sediments the content of some elements (V, U, Ag, Mo, and some others) is quantitatively related to the organic matter. These data support the opinion that there were no significant changes in the chemical composition of living organisms during the geological history of the Earth (Vernadsky, 1940).

Secondary changes and the metamorphism of organic matter greatly influence the content of trace elements. The ratio of Pb, As, Ge, U, P, Sr, Ba, Zn and Mo contents to their global averages (clarke of concentrations), in metamorphosed Precambrian formations are considerably lower than those in unmetamorphosed Phanerozoic rocks. Moreover, formations subjected to amphibolite and granulite facies metamorphism have minimum concentrations of these elements. A noticeable increase in the content of these elements is observed in black shales which have undergone greenschist facies metamorphism, and a maximum increase is noted in unmetamorphosed formations of the Phanerozoic. These data apparently indicate that in the course of epigenetic and metamorphic transformation of sediments, the above elements become separated from the organic component. Metamorphism leads to destruction of organometallic compounds and breaks up sorption links in the course of organic matter transformation.

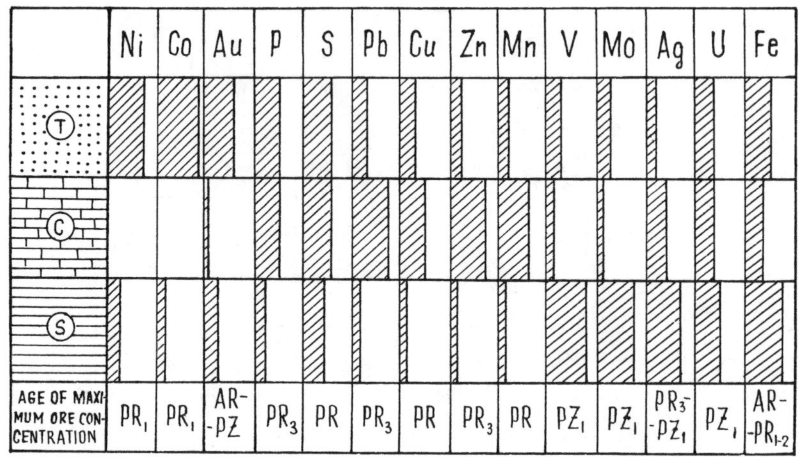

Fig. 7. Metallogeny and ore-distribution in different carbonaceous formation types (Fig. 6) in Precambrian and lower Paleozoic rocks. Age in Ma: AR >2500; PR 2500-570; PR_1 2500-1650, PR_3 1400-570; PZ_1 570-450.

Other elements (Ni, Co, Cr, Cu, Ag, Th, V) are less affected by metamorphic transformation. These elements first appear in higher concentrations in carbonaceous deposits of the Archean some 3000-3200 Ma ago and can be traced through the entire stratigraphic column up to Recent sediments, usually without a noticeable decrease or increase in their content. However, in specific formations of some regions there may be local departures.

Economically important mineral deposits are associated with Precambrian black shale formations. Many deposits of Cu, Au, Mn and P are confined to terrigenous black shale formations. Associations with deposits of V, W, U and Au are characteristic of siliceous black shale formations. Pb and Zn, phosphorite, shungite and graphite are confined to carbonate black shale formations, while Au, U, W, Mn and S (pyrite) are associated with volcanogenic black shale formations.

As a result of our analysis of the distribution of ore concentrations associated with carbonaceous sediments in more than 70 regions of the world from the Archean to the Lower Paleozoic (Fig. 6), a number of regularities have been established at a quantitative level showing the distribution of major ore components in different formational types. Data on the ore-producing potential of the Precambrian and Lower Paleozoic carbonaceous formations of the Soviet Union and some other regions of the world, used to construct Fig. 6, were taken from the compilations devoted to the problems of Precambrian geology (Sidorenko, 1978, 1982; Borodaevskaya et al., 1979; Gezeva et al., 1981; Formozova, 1973; Krauskopf, 1956). It was found (Fig. 7) that terrigenous-carbonaceous sediments are marked by ore occurrences of Ni, Co, Au and P; for carbonate-carbonaceous sediments it is Pb, Cu and Zn. Manganese and sulphide ores are typical of terrigenous-carbonate formations. Siliceous-carbonaceous deposits are marked by concentrations of V, Mo, Ag and Fe. The revealed epochs demonstrating the maximum concentration of ore occurrences (Fig. 8) associated with definite types of formations are: Lower-Middle Proterozoic (terrigenous-carbonaceous: Ni, Cu, Co), Upper Proterozoic (carbonate-carbonaceous: Pb, Zn, Cu, sulphide ores), and Lower Paleozoic (siliceous-carbonaceous: V, Mo, Ag); they reflect the most typical stages in the

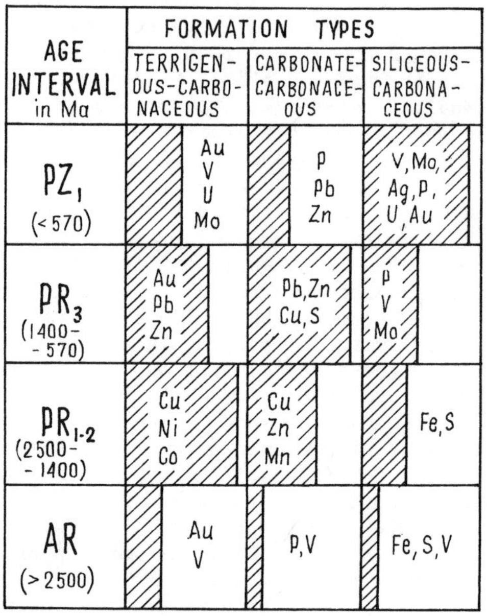

AGE INTERVAL in Ma	FORMATION TYPES		
	TERRIGENOUS-CARBONACEOUS	CARBONATE-CARBONACEOUS	SILICEOUS-CARBONACEOUS
PZ_1 (< 570)	Au V U Mo	P Pb Zn	V, Mo, Ag, P, U, Au
PR_3 (1400--570)	Au Pb Zn	Pb, Zn Cu, S	P V Mo
PR_{1-2} (2500--1400)	Cu Ni Co	Cu Zn Mn	Fe, S
AR (> 2500)	Au V	P, V	Fe, S, V

Fig. 8. Relative distribution of ore deposits and sulfide mineral occurrences according to different carbonaceous formation types (Fig. 6) and ages.

tectonic development of Precambrian to Lower Paleozoic sedimentary basins.

The Archean-Proterozoic stage is characterized
by the formation of carbonaceous sequences in
local but sufficiently deep sea basins related to
block tectonics within the stabilized cratons.
The Upper Proterozoic stage relates to the epoch
of widespread creation of epi-continental basins
on the border of pericratonic depressions and
platforms with accumulation of carbonate-carbonaceous sequences where polymetallic deposits of
the stratiform type are located (eastern Siberia,
margins of the Baltic Shield). And, finally, the
third stage is related to the formation of extended strips of metal-bearing chert-(carbonate)-
carbonaceous shales within the continental margin
of the paleo-Tethys (Ural, Kazakhstan, Middle
Asia, China).

Conclusions

There are two groups of factors which affect
the evolution of carbonaceous formations. The
first group comprises factors related to the general development of geological processes within
the earth crust: the style of tectonic development of geostructures, sedimentation and volcanism. The second group includes factors which
determine the specific uniformity of carbonaceous
sediments throughout geological history, so that
we can refer to them as a group of "through-going"
formations. These are: (1) accumulation of
carbonaceous sediments in basins with a stable
regime in epochs which are synchronous with the
final stages of peneplanation or that develop
immediately after them (platformal inland or epi-
continental basins, passive margins of geosynclines); (2) persistency of the main geochemical
system which led to accumulation of carbonaceous
matter in sediments and resulted in associations
of carbonate-carbon, amorphous silica - carbon,
highly colloidal systems (clay minerals) - carbon;
(3) the universal character of biochemical processes which result in the concentration of chemical elements in biogenic sediments is probably
related to the domination in these processes of
primitive marine organisms the evolution of which
ended with the Archean history of the Earth.

Acknowledgments. The authors are grateful to
A. Kröner for assistance in improving the English
and for useful comments.

References

Bain, G.W., Patterns to ores in layered rocks.
Econ. Geology 55: 695-731, 1960.
Baturin, G.N., Kochenov, A.V. and Shimkus, K.M.,
Uran i redkie metally v kolonkah donnykh osad-
kov Chernogo i Sredizemnogo morey (Uranium and
rare metals in cores of bottom sediments of
Black and Mediterranean seas). Geokhimiya, No.
1, 41-49, 1967.

Baukov, S.S., Ordovikskie slantzenosnye formazii
(Pribaltiyskiy basseyn) (Ordovician oil shale
formations) - In: Formazii goryuchikh slantzev,
Tallin, Valus, 17-39, 1973.
Borodaevskaya, M.B., Gorzhevskiy, D.Y. and
Kryvzov, A.Y., Kolchedannye mestorozhdeniya
mira (Sulphide deposits of the world) Moscow,
"Nedra", 284 pp., 1979.
Borovskaya, I.S., Mycyakina, V.S. and Volkova,
L.V., Rifeiskie fosfority Sangilena (Yugovostok
Tuvy) (Riphean phosphorites of Sanfilen, South-
East Tuva). Metallogeniya osadochnykh i osa-
dochno-metamorficheskikh porod. Moscow, "Nauka"
91-100, 1966.
Cloud, P.E., Gruner, J.W. and Hagen, H., Carbona-
ceous rocks of the Soudan iron formation.
Science 148, 278-279, 1965.
Formozova, L.N., Formatsionnye tipy zheleznykh
rud i ikh evolutsiya (Formational types of iron
ores and their evolution). Moscow, "Nauka",
172 pp., 1973.
Geologiya atomnych syrievych materialov (Geology
of the Atomic-source materials), Moscow, Gos-
geoltehizdat, 1956.
Gezeva, R.V., Derjagin, A.A., Sozinov, N.A. and
Sidorenko, Sv. A.,Geologicheskie osobennosty i
uranosnost formazii chernych slanzev (Geological features of the black shales formations and
their uraniferous). Moscow, "Nauka", 120 pp.,
1981.
Gorbachev, O.V. and Sozinov, N.A., Neckotorye
petrochimicheskie i geochimicheskie aspecty
typyzazyi uglerodistych otlozheniy docembriya
(Some petrochemical and geochemical aspects of
the typification of carbonaceous formation).
Problems of Sediment. Geol. of the Precambr.,
10, Moscow, "Nauka", 55-62, 1985.
Kholodov, V.N., O metallogenii venda i cembriya
Evrazil (On metallogeny of Vendian and Cambrian
of Eurasia) - Lithologiya i poleznye isokopaemye,
No. 4, 24-44, 1970.
Krauskopf, K.B., Factors controlling the concen-
trations of thirteen rare metals in seawater;
Geochim. et Cosmochim. Acta, 9, 1-32b, 1956.
Lithologiya i osadochnaya geologiya docembriya
(Lithology and Sedimentary Geology of the Pre-
cambrian). "Nauka" Cazach. SSR, 200 pp., 1981.
Manskaya, S.D. and Drozdova, T.V., Geokhimiya
organicheskogo veshestva (Geochemistry of or-
ganic matter). Moscow, "Nauka", 315 pp., 1964.
Murty, P.S.M., Aswathanarayana, U. and Nahadevan,
C., Geochemistry of the siliceous black shales
at Nagaryuna Sagar damsite, India. Econ. Geology, 57, 614-619, 1962.
Negrutza, V.Z., Shurgin, S.S. and Zhuravlyov,
V.A., Dokembriyskie uglerodsoderzhaschie porody
vestochnoy chasti Baltiyskogo schita. (Precambrian carbon-bearing rocks in the Eastern Baltic Shield). In: Problemy osadochnoy geologii
dokembriya, 7, 66-79, Moscow, 1981.
Orlov, L.N., Nisznepaleosoyskaya uglerodisto-
kremnistro-slantzevaya formatziya Kokirimtau i
nekotorye voprosy eyo genezisa. (The lower Pa-

leozoic carbonaceous-cherty-schistose formation of Kokirimtau and problems bearing on its genesis) Problemy osadochnoy geologii dokembriya, 7, 135-138, Moscow, 1981.

Ostapenko, Y.P. and Kardash, V.T., Vanadienosnye porody v arkheyskikh otloyheniyakh Tsentralnogo Pribaikalya (Vanadium-bearing rocks in Archean deposits of central part of west-of-Baikal region) Trudy Irkutskogo politekhnicheskogo instituta, Irkutsk, 51, 80-85, 1970.

Peltola, E., On the black schists in the Outokumpu region in Eastern Finland. Bull. Comm. Geol. Finlande, 192, 192 pp., 1960.

Problemy osadochnoy geologii docembriya (Problems of sedimentary Geology of the Precambrian), 7, Moscow, "Nauka", 260 pp., 1981.

Sozinov, N.A. and Gorbachev, O.V., Uglerodistye formazyi i ich evoluziya v istoryi zemly (Carbonaceous formations and their evolution through the history of the Earth). In: Evoluziya osadochnogo rudoobrasovaniya v istoryi Zemli. Moscow, "Nauka", 214-224, 1984.

Sidorenko, A.V. (ed.) Dokembry i problemy fomirovaniya zemnow cory (Precambrian and problems of the earth crust formations). Moscow, "Nauka", 312 pp., 1978.

Sidorenko, A.V. (ed.) Osadochnaja geologiya docembriya (Sedimentary geology of the Precambrian) Moscow, "Nauka", 248 pp., 1982.

Tyler, S.A., Barghorn, E.S. and Barrett, L.P., Anathracitic coal from Precambrian upper Huronian black shale of the Iron River Districts, Northern Michigan. Bull. Geol. Soc. America 68: 1293-1304, 1957.

Wedepol, K.G., Geokhimicheskoe i petrograficheskoe issledovaniye "medistogo slantsa" v severo-zapadnoy Germanii (Geochemical and petrographic investigations of "cupriferous schists" in north-west Germany). In: Khimiya zemnoy kory. Moscow, "Nauka", 2: 11-17, 1964.

Vernadsky, V.I., Biokhimicheskie ocherki 1922-1932 (Essays on biochemistry, 1922-1932). Moscow, Publishing House of the USSR Academy of Sciences, 160 pp., 1940.

Vine, J.D. and Tourtelot, E.B., Geochemistry of black shale deposits - a summary report. Econ. Geol. 65, 253-272, 1970.

Voytkevich, G.V. and Lebedko, G.I., Poleznye isopaemye i metallogeniya docembriya (Mineral Resources and Metallogeny of the Precambrian, Moscow, "Nedra", 232 p., 1975.

Zhdanova, L.V., Raspredelenie radioaktivnykh elementov i C_{org} v nizhneproterozoyskikh otlozheniyakh Zhayminskoy svity (Distribution of radioactive elements and C_{org} in Lower Proterozoic deposits of the Zhayminskaya suite). In: Voprosy geologii i geokhimii Sibiri. Novosibirsk, 128-131, 1971.

EARLY AND MIDDLE PROTEROZOIC PROVINCES IN THE CENTRAL UNITED STATES

W. R. Van Schmus[1], M. E. Bickford[1], and I. Zietz[2]

[1]Department of Geology, University of Kansas, Lawrence, Kansas

[2]The Phoenix Corporation, McLean, Virginia

Abstract. Petrographic, geochemical, and U–Pb
geochronologic studies of sub-surface samples
obtained from drill holes to basement have been
combined with interpretation of recent
geophysical maps to yield better understanding
of Lower and Middle Proterozoic continental
basement in the central U. S. The Archean
Superior Craton is truncated south of Lake
Superior by 1.83 to 1.89 Ga old orogenic suites
of the Penokean Province. To the west, the
Archean Wyoming Craton is truncated on the south
by the Cheyenne Belt and bordered by 1.70 to 1.78
Ga old units of the Colorado Province.
Subsurface samples between the Archean cratons
and between the Penokean and Colorado provinces
indicate general continuity of units about 1.8 Ga
old. Data from the southwest U. S. and from
midcontinent basement indicate that the 1.7 to
1.9 Ga old orogenic provinces are flanked to the
south by igneous and metamorphic rocks 1.6 to 1.7
Ga old. These Lower Proterozoic terranes
represent lateral accretion to the south of the
Archean cratons. The Lower Proterozoic terranes
were intruded by Middle Proterozoic plutons and,
to the south, the older terranes are covered by
large areas of rhyolite and epizonal granite. The
granite and rhyolite represent two distinct
suites, 1.45 to 1.50 Ga and 1.34 to 1.40 Ga old.
The older suite is more extensive, extending from
California to Labrador; the younger suite is only
known from the south-central U. S. Both suites
appear to have been derived from partial melting
of Lower Proterozoic continental crust. Thus,
they represent vertical redistribution of older
Proterozoic continental crustal material, rather
than lateral growth. Subsequently, the
Midcontinent Rift System developed in the
interior of the continent, and the Grenville
Province and similar Llano Province of southern
Texas formed along the eastern and southern
flanks of the continent. These events are
probably related to tectonic regimes associated
with major continental collision about 1.1 Ga
ago.

Introduction

It has long been known [e.g., Muehlberger et
al., 1967] that the Precambrian basement of the
United States, and hence the southern half of
North America, consists primarily of Proterozoic
rocks, with Archean units present only in the
Lake Superior region and in the Wyoming region
[Condie, 1976; Sims and Peterman, 1981].
However, much of it is buried beneath Phanerozoic
sedimentary rocks, especially in the central
United States where Phanerozoic cover is nearly
continuous. Consequently, details of ages,
lithologies, and subdivisions into genetic
terranes for Proterozoic basement in the United
States are still sparse. Nonetheless, if we are
to understand the evolution of North American
continental lithosphere during the Proterozoic,
it is essential that data be obtained wherever
and whenever possible. This report summarizes
the results of our efforts to obtain and study
subsurface samples, our study of recent gravity
and aeromagnetic maps, and our interpretation of
the data in terms of regional Proterozoic
geologic provinces and their genesis.

New Sources of Information

Since the last general report on the
Proterozoic crust of the continental interior of
the United States [Van Schmus and Bickford,
1981], several new and varied sources of
information have become available which
substantially aid in interpretation of the
Precambrian basement geology of the midcontinent
region. In addition, new studies in other
Proterozoic orogenic terranes in North America
[e.g. Hoffman and Bowring, 1984; Premo, 1984;
Bickford et al., 1984; Bowring et al., 1984] have
helped considerably in understanding tectonic and
petrologic evolution of continental crust during
the Proterozoic. Figure 1 shows our current
interpretation of Precambrian basement in the
continental interior of the United States.

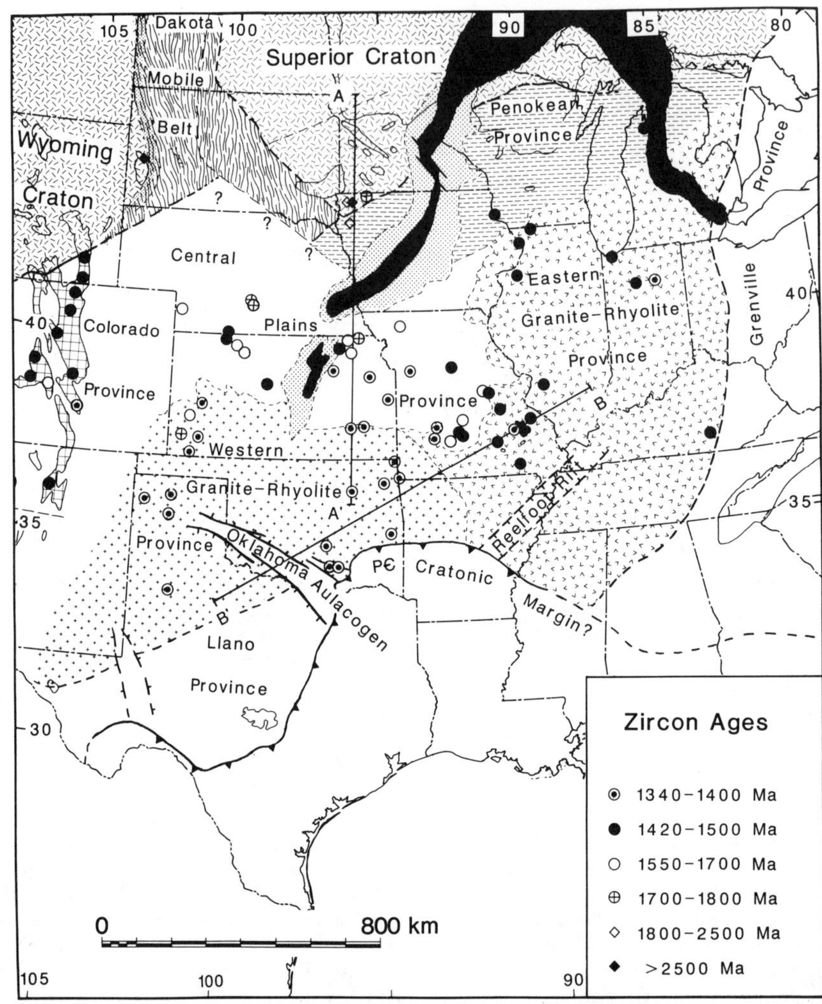

Fig. 1. Generalized geologic map for Precambrian basement of the central United States, showing major tectonic and/or petrologic provinces. Based on petrologic, geophysical, and geochronologic data summarized in the text.

Geophysical Data

Some major contributions to data available for interpretation of buried continental basement are (a) the aeromagnetic map of the United States [Zietz, 1982], (b) the gravity map of the United States [Soc. Explor. Geoph., 1982], and (c) processed versions of gravity data [e.g., Hildenbrand et al., 1982; Arvidson et al, 1984] and aeromagnetic data [Yarger, 1985]. These maps and graphic presentations (Figures 2 and 3) reveal many features that are manifestations of Precambrian continental crust underlying Phanerozoic cover. Furthermore, these data are useful in interpreting both shallow features (represented primarily by magnetic data) and deeper features (represented in much gravity data, particularly on a regional scale). Detailed discussion of the geophysical data is beyond the scope of this paper, but they are essential to refinement of models of the structure and composition of North American continental lithosphere. Because magnetic data tend to reflect the nature of the crust near the Precambrian surface, we regard them as the most useful and have relied upon them more than other types of geophysical data.

U-Pb Ages on Zircons

One of the most important aspects of deciphering the geology of Precambrian basement is knowing primary formation ages and distribution of rock units. Because many samples retrieved from drill holes are very small, limited to single sites, and often weathered or altered, they are not suitable for accurate dating by Rb-Sr or K-Ar techniques. However, we have been able to extract zircons from many samples, including drill cuttings weighing less

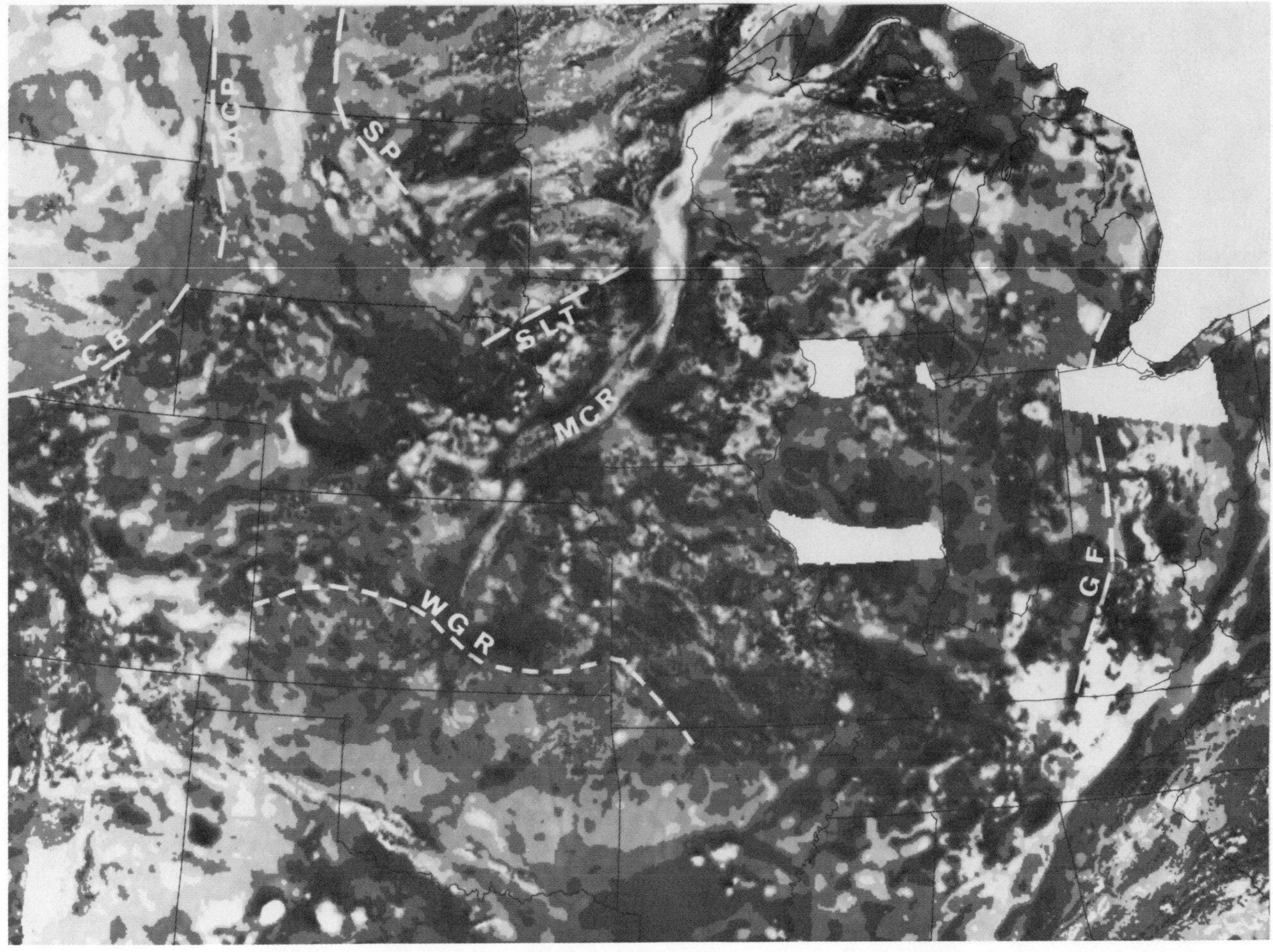

Fig. 2. Aeromagnetic anomaly map for the central part of the United States. Positive anomalies are represented by light areas and negative anomalies are represented by darker areas. White areas represent no data. Based on digital version of U.S. magnetic anomaly map (Zietz, 1982) as produced by the Phoenix Corporation, McLean, Virginia. See Fig. 1 for corresponding geologic interpretation. NACP = North American Central Plains conductive anomaly; SP = western margin of undisturbed Superior Province; SLT = Storm Lake Trend in northwestern Iowa; CB = Cheyenne Belt in southeastern Wyoming; MCR = Mid-Continent Rift system; GF = approximate of Greville Front; WGR = northern limit of contiguous Western Granite-Rhyolite Province.

than 0.5 kg. The resulting zircon concentrates, some less than 5 mg total, can be split into two or more fractions and analysed with modern techniques of U-Pb geochronology to give accurate and precise ages. Although small sample sizes frequently prevent processing samples in order to obtain optimum degrees of concordance, the results are still better than can be obtained by other methods. Since the report of Van Schmus and Bickford [1981], more data have been reported for subsurface samples in the midcontinent region [Thomas et al., 1984; Hoppe et al., 1983; Bickford et al., 1981a], for exposed Precambrian terranes surrounding the midcontinent region [Bickford et al., 1984; Premo, 1984; Condie and Bowring, 1984], and new data, reported here, have

been obtained. Tables 1 and 2, and Figures 4 to 6, summarize the new information; analytical techniques and results are given in the next section.

Sm-Nd Isotopic Data and Chemical Petrology

Nelson and DePaolo [1985] have obtained Sm-Nd isotopic data on several samples from the midcontinent region that were dated by zircons, and DePaolo [1981] reported data from Proterozoic units in Colorado. Their results yield two major conclusions. First, mantle separation ages for Lower Proterozoic units are about 1.8 Ga and indicate clearly that Archean continental basement does not extend very far south of the

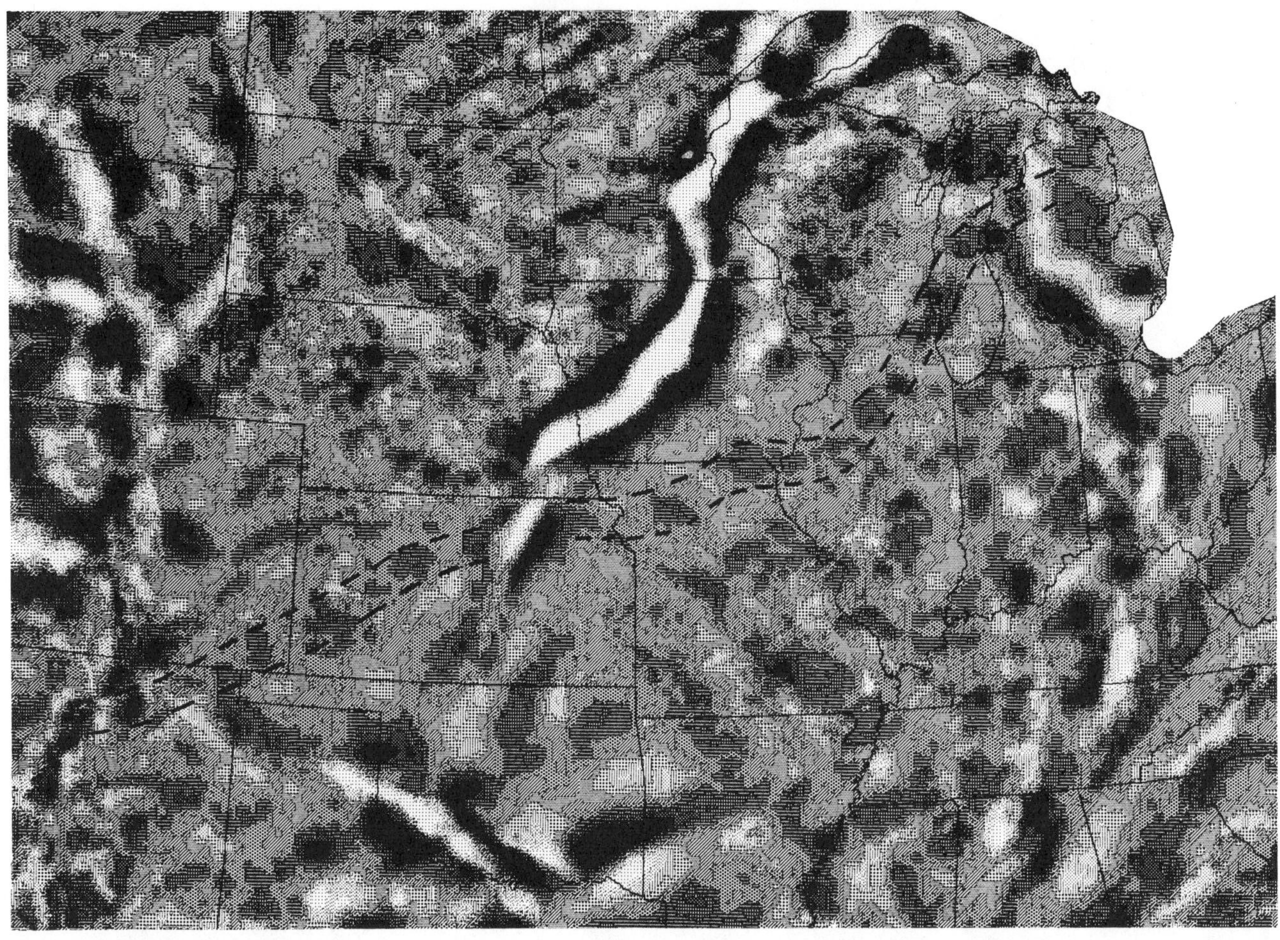

Fig. 3. A portion of black-and-white version of the filtered gravity map of the United
States, for wavelengths less than 250 km. Provided by T. Hildenbrand, U.S. Geological
Survey. Dark colors are gravity lows, light areas are gravity highs. The Midcontinent
Geophysical Anomaly is particularly noticeable, extending south-southwestward from Lake
Superior into central Kansas. Note also the correspondence between gravity features
and magnetic features in the north-central part of the area shown. The dashed line
traces the north edge of a belt of subtle gravity lows that may represent the northern
edge of 1650 Ma accretionary terranes (see text).

Wyoming and Superior cratons. Second, Middle
Proterozoic plutons that intrude the Lower
Proterozoic terranes and rhyolites and epizonal
granites that underlie the southern part of the
midcontinent region also have mantle separation
ages of 1.6 to 1.8 Ga and were derived by melting
of older, early Proterozoic, continental crust.
These conclusions are fully consistent with
petrologic studies on these granites [Anderson,
1983] that indicate they were derived by partial
melting of lower continental crust.
Unfortunately, the Sm-Nd data base is still very
limited for the midcontinent region, and there
are several instances where interpretation is
ambiguous.

New U-Pb Ages

Table 1 lists subsurface samples for which we
have obtained U-Pb ages on zircon over the past
few years and which have not been published
previously. The sample numbers correspond to the
numbered sites shown in Figure 4; the ages are
summarized from Table 2, which lists full
analytical data and results of regression
analysis.

Analytical Techniques

For most samples it was necessary to separate
zircons from small pieces of drill core or

purified drill cuttings of the crystalline
basement; in many instances beginning sample size
was less than 1 kilogram. Many samples were
first ground to -60 mesh and a split was set
aside for future chemical studies; for samples
processed more recently, the chemical splits were
taken before final grinding. The remaining
sample was washed with water to remove fine rock
flour, dried, and processed with heavy liquids
and magnetic separation to concentrate zircons.
Zircon concentrates were purified using heavy
liquids, nitric acid washes, magnetic separation,
and hand-picking as necessary. Purified
concentrates were split into various fractions
using magnetic susceptibility and, in some cases,
size. In some instances only a few very small
fractions could be obtained, with the result that
control on the resulting ages was adequate,
though not ideal.

Analytical data (Table 2) were obtained at the
Isotope Geochemistry Laboratory, Department of
Geology and KU Center for Research, University of
Kansas. Zircons were dissolved and Pb and U
separated using procedures modified after Krogh
[1973]. Aliquots of dissolved samples were
spiked with mixed Pb-208 - U-235 tracer solution.
Isotopic ratios were measured on a 23 cm radius
Nier-type 60 deg. sector, single filament, solid-
source mass spectrometer. Pb and U analyses were
carried out using silica gel and phosphoric acid
on single Re filaments, and mass fractionation of
0.15 percent bias per mass unit was applied to
all Pb data. Radiogenic Pb-207 and Pb-206 were
calculated by correcting measured data for modern
blank lead (207/204 = 15.6, 206/204 = 18.7) and
for non-radiogenic original Pb corresponding to
Stacey and Kramers [1975] model Pb for the age of
the sample. All samples are sufficiently
radiogenic that uncertainties in composition of
non-radiogenic Pb used do not contribute
significantly to uncertainties in the ages
obtained. During the course of these analyses
analytical blanks were 0.25 ng or less for Pb-206
and less than 0.1 ng for U. Decay constants used
were 0.155125 x 10^{-9}/yr for U-238 and 0.98485 x
10^{-9}/yr for U-235 [Steiger and Jäger, 1977].

Uncertainties for measured 207/206 and 208/206
ratios are estimated at 0.1 percent (2-sigma);
relative uncertainties on 204/206 ratios varied,
depending on the magnitude of the ratio, but
absolute uncertainty in the ratio is estimated at
± 0.000010 (1 percent for a sample having a
204/206 ratio of 0.00100). The U/Pb ratios are
considered accurate to ± 1.0 percent at 2-sigma.
After correction for various sources of
uncertainty, radiogenic Pb-207/Pb-206 ratios are
considered to be accurate to ± 0.25 percent at 2-
sigma.

Concordia intercepts (Table 2 and Figures 5
and 6) were calculated using regression and error
analysis methods of Ludwig [1980, 1983], which
are based on regressions of York [1966, 1969].
Data were assigned 2-sigma uncertainties of ± 1.0
percent for Pb-207/U-235 and Pb-206/U-238 ratios

with a correlation coefficient of 0.98
(uncertainty in Pb-207/Pb-206 = 0.20 percent).
Results are reported using Ludwig's Model 1
solution [essentially York, 1969] for
probablilities of fit of 30 percent or better.
If the probability of fit was less than 30
percent, implying sources of scatter outside of
analytical error (e.g., "geological" scatter),
Ludwig's Model 2 or Model 4 solution [Ludwig,
1983] was used. The Model 2 solution is
essentially a York [1966] fit, where errors are
based on the scatter of the data. The Model 4
solution follows the method of Davis [1982], in
which greater weight is given to more concordant
points. Ages are quoted at the 2-sigma level for
the model used.

Results

Archean

Only one subsurface sample (IA-LY-9) yielded
an Archean age. It is from northwestern Iowa and
was encountered in a drill hole (Matlock C-10) as
part of extensive exploration of aeromagnetic
anomalies in Lyon and Sioux counties by New
Jersey Zinc Co. [Tvrdik, 1983]. The rock is
granulated and deformed quartz monzodiorite, and
the age of 2523 Ma (Figure 5A) is interpreted as
the age of crystallization for the associated
pluton. Contact relationships of this unit with
others encountered in nearby drill holes of the
Matlock project are unknown, so that this age can
not be used to constrain the ages of other units.
However, these results clearly indicate that
Archean basement extends this far south.

Early Lower Proterozoic (2000-2500 Ma)

Two samples from northwestern Iowa yield
apparent ages between 2000 Ma and 2500 Ma. The
older of these is a rhyolite porphyry (IA-LY-7)
that apparently overlies Archean basement 5 km.
west of IA-LY-9. Zircons were extracted from
several feet of core, but the yield was small and
mixed with microcrystalline sphene that could not
be removed except by hand-picking. Results on
four fractions of -200 mesh zircons yield an
apparent age of 2280 ± 110 Ma (Figure 5A). The
large uncertainty exists because the four
fractions are moderately discordant and grouped
close together. The zircons are sharp and clear,
so the oldest Pb-Pb age of ca. 2022 Ma indicates
that this unit is significantly older than
Penokean volcanic units found in Wisconsin or
Michigan. This rhyolite may be roughly Huronian
in age (2.1 to 2.4 Ga) or may be correlative with
Lower Proterozoic volcanics about 1.9 to 2.1 Ga
old found in the Homestake Mine in the Black
Hills of South Dakota [Peterman and Zartman,
1985].

The other sample, IA-PL-1, is badly weathered
feldspathic gneiss that appears to be
metamorphosed arkose, although an igneous origin

TABLE 1. Information on Subsurface Samples

Number[1]	Well Name, County, State	Location[2]	Depth[3] (m)	Description
1: IA-LY-7	New Jersey Zinc, Matlock C-5 Lyon County, Iowa	28-98N-44W 43:17N 96:03W	140 215	Quartz-porphyry rhyolite (core)
2: IA-LY-9	New Jersey Zinc, Matlock C-10 Lyon County, Iowa	25-98N-44W 43:16N 95:59W	213 214	Altered monzodiorite (core)
3: IA-OS-1	Harris State Test, D-13 Osceola County, Iowa	17-100N-39W 43:27N 96:37W	278 280	Medium-grained granite
4: IA-PL-1	Iowa Geological Survey, D-21 Plymouth County, Iowa	02-92N-45W 42:49N 96:07W	323 328	Feldspathic gneiss (core)
5: KS-GH-42	Petrol. Mgt., #3 Eva Desbien Graham County, Kansas	15-06S-21W 39:31N 99:40W	484 486	Recrystallized granite (cuttings)
6: KS-GH-43	Petrol. Mgt., #1 Lappin Graham County, Kansas	01-06S-21W 39:32N 99:37W	1100 1101	Recrystallized granite (cuttings)
7: KS-KE-1	AMOCO, #2 Longwood Kearny County, Kansas	35-23S-37W 38:01N 101:20W	1987 1992	Granite (core)
8: KS-MS-44	Houston Int. Min., Vermillion Marshall County, Kansas	01-04S-09E 39:43N 96:27W	369 378,93	Granite (core)
9: KS-MS-46	Texas Gulf, #1 Gerstner Marshall County, Kansas	12-04S-09E 39:48N 96:28W	356 397	Granite (core)
10: KS-NM-18	Transocean, #1 Ukele Nemaha County, Kansas	27-01S-14E 39:54N 95:52W	1204 1212	Granite (cuttings)
11: KS-NM-21	Houston Int. Min., Baileyville Nemaha County, Kansas A: dike, d=262 m; B: dike, d=284 m; C: dike, d=289 m; D: gneiss, d=329 m; E: gneiss, d=340 m	03-02S-11E 39:54N 96:12W	255 d	(Core)
12: KS-NT-334	Hummon Oil, #1 Frederick Norton County, Kansas	35-03S-23W 39:45N 99:57W	1140 1141-51	Recrystallized granite (cuttings)
13: KS-ST-1	AMOCO, #1 McPherson Stanton County, Kansas	28-29S-39W 37:30N 101:33W	2261 2263	Altered granite (core)
14: MO-CM-2	Exxon Minerals, DB-1 Camden County, Missouri	14-37N-17W 37:58N 92:49W	367 368	Granitic gneiss (core)
15: MO-CM-3	Exxon Minerals, TD-1 Camden County, Missouri	28-37N-17W 37:56N 92:46W	483 486	Granitic gneiss (core)
16: MO-LC-3	St. Joe Minerals, 63W29 Laclede County, Missouri	09-35N-14W 37:48N 92:27W	502 518	Felsic paragneiss (core)
17: MO-PO-2	Union Carbide, MHR-1 Polk County, Missouri	18-35N-21W 37:46N 93:12W	496 d	Granite (core) A: d=503 m; B: d=518 m
18: MO-RI-2	Gulf Oil, PBW-2 Ripley County, Missouri	29-25N-04E 36:47N 90:39W	742 743	Granite (core)
19: MO-WE-1	Union Carbide, M1J1 Webster County, Missouri	27-31N-18W 37:23N 92:55W	576 604-49	Granite (core)

Number[1]	Well Name, County, State	Location[2]	Depth[3] (m)	Description
20: NB-BF-1	Ohio Oil, #1 Pettett Buffalo County, Nebraska	20-09N-18W 40:44N 99:23W	<u>1181</u> 1182	Foliated tonalite (core)
21: NB-CS-2	Ohio Oil, #1 Bremer Chase County, Nebraska	05-07N-39W 40:36N 101:46W	<u>1635</u> 1641	Augen gneiss (core)
22: NB-DA-2	Ohio Oil, #1 Dunse Dawson County, Nebraska	01-10N-19W 40:52N 99:26W	<u>1225</u> 1228	Granite (core)

[1]Numbers refer to locations in Fig. 4.
[2]Section, township, and range numbers, respectively, are separated by dashes; degrees and
 minutes of latitude and longitude, respectively, are separated by colons (degrees:minutes).
[3]Depths give top of Precambrian/sample depth in meters.

for its protolith cannot be ruled out at present. Two small zircon fractions yielded very discordant results with an apparent upper intercept age of about 2.07 Ga (Figure 5A). However, this apparent age could represent a hybrid age between Archean detrital zircons and ca. 1.8 Ga metamorphic overgrowths or 1.8 Ga detrital zircons. Alternatively, it could reflect the provenance age and indicate that the source of the detritus was similar in age to the volcanic unit mentioned above. An additional 240 feet of core from this locality, including a substantial amount of unweathered material, was recently recovered, and it will be studied in more detail to determine the nature of the Precambrian bedrock, its origin, and its age at this locality.

Late Lower Proterozoic (1600 - 1800 Ma)

Several subsurface samples yielded ages in the range of 1600 to 1800 Ma. The oldest ages were obtained from granite from northwest Iowa (IA-OS-1) and from two gneissic samples from central Nebraska (NB-BF-1 and NB-DA-2). The ages of these samples, 1804, 1802, and 1787 Ma, respectively (Figure 5B), fall into a gap between ages found in the Penokean Orogen of Wisconsin [1830 to 1880 Ma: Van Schmus, 1980 and unpublished] and in early Proterozoic orogenic suites of southern Wyoming and northern Colorado, south of the Wyoming craton [1750 to 1780 Ma: Premo, 1984]. We believe that the ages represent crystalization ages for the rocks concerned, so that correlation of these units is uncertain, as discussed further in a later section.

Granitic gneiss from northeastern Kansas (KS-NM-21) yielded complex results several rock units occur in a single 100 meter-long core. Three intervals represent cross-cutting dikes about 1.44 Ga old (KS-NM-21A,B,C; Table 2). Two gneissic portions of core yielded older, but different results: zircons from interval 21E yielded an apparent age of 1552 Ma, while zircons from interval 21D yielded ages close to 1800 Ma. In the latter case, two distinct types of zircons could be distinguished in the separates: an euhedral variety and a round, anhedral variety. The euhedral zircons have an approximate age of 1780 Ma, whereas the round fraction appears to be distinctly older (ca. 1820 Ma; Figure 5C). Assuming the gneiss is meta-igneous, we interpret the euhedral zircons as giving the age of crystallization of the protolith of the gneiss, whereas the round variety may reflect a xenocrystic component. In any case, the data indicate that crust about 1800 Ma old extended as far south as northeastern Kansas. We interpret the apparent age of about 1550 Ma from interval 21E as a hybrid age due to metamorphic alteration of 1800 Ma old zircons during intrusion of 1440 Ma old dikes (see below).

The only other age greater than 1700 Ma was found for gneissic granite from southwestern Kansas (KS-ST-1; Figure 5C). Although the zircons are moderately discordant, they still define a reasonable chord with an upper intercept age of 1716 ± 29 Ma, which we interpret as the crystallization age of the granite. This age is similar to many of those found in the Proterozoic terrane of southern and central Colorado [Bickford et al., 1984] and is consistent with extension of that terrane eastward into Kansas.

Granite core recovered from another recent drill hole in southwestern Kansas (KS-KE-1) yielded an age of 1671 ± 11 Ma (Figure 5C), slightly younger than the 1700 to 1780 age range found throughout most of Colorado. However, plutons 1660 to 1690 Ma old intrude early Proterozoic complexes in central andsouthern Colorado [Bickford and Boardman, 1984; Bickford, unpub.; Van Schmus, unpub.], and the results for this sample, in conjunction with KS-ST-1 (above), are fully consistent with projected continuation

TABLE 2. Analytical Data for Zircon Fractions

Fraction[1]	Concentrations[2]		Pb Isotopic Composition[3]			Radiogenic Ratios[4]			Ages[5]		
	U (ppm)	Pb (ppm)	Pb-204/Pb-206	Pb-207/Pb-206	Pb-208/Pb-206	Pb-206/U-238	Pb-207/U-235	Pb-207/Pb-206	Pb-206/U-238	Pb-207/U-235	Pb-207/Pb-206
IA-LY-7 (New Jersey Zinc, Matlock C-5): Rhyolite.											
NM(-1)F	188	65	0.00080	0.13508	0.1681	0.3079	5.288	0.12454	1730	1867	2022
M(-1)F	196	66	0.00089	0.13256	0.1783	0.2984	4.971	0.12080	1684	1814	1968
M(0)F	207	68	0.00136	0.13577	0.2016	0.2823	4.585	0.11780	1603	1747	1923
M(1)F	221	72	0.00187	0.14028	0.2229	0.2755	4.382	0.11535	1569	1709	1885
Regression: U.I. = 2280 + 110 Ma; L.I. = 1104 Ma; P = 0.03. See text.											
IA-LY-9 (New Jersey Zinc, Matlock C-10): Syenite.											
NM(2)	979	489	0.00143	0.18230	0.2149	0.4070	9.226	0.16440	2201	2361	2502
M(2)	1038	464	0.00082	0.17346	0.1942	0.3762	8.464	0.16320	2058	2282	2489
M(3)	1203	481	0.00092	0.17304	0.1961	0.3355	7.471	0.16150	1865	2170	2472
M(4)	1216	441	0.00094	0.17130	0.2017	0.3019	6.636	0.15942	1701	2064	2450
Regression: U.I. = 2523 + 5 Ma; L.I. = 287 Ma; P = 0.74											
IA-OS-1 (Harris State Test; IGS D-13): Granite.											
NM(0)	227	66	0.00111	0.12474	0.2287	0.2488	3.763	0.10972	1432	1585	1795
M(0)	254	68	0.00061	0.11714	0.2123	0.2333	3.502	0.10886	1352	1528	1780
M(1)	323	76	0.00080	0.11980	0.2240	0.2024	3.039	0.10892	1188	1418	1782
M(2)	413	89	0.00134	0.12637	0.2531	0.1790	2.668	0.10809	1062	1320	1767
Regression: U.I. = 1804 + 17 Ma, L.I. = 84 Ma; P = 0.00											
IA-PL-1 (Iowa Geol. Surv., D-21, Camp Quest): Feldspathic gneiss.											
M(2)	470	62	0.00118	0.13875	0.1478	0.1182	2.005	0.12307	720	1117	2001
M(5)	769	65	0.00086	0.13115	0.1336	0.0783	1.292	0.11970	486	842	1952
Regression: U.I. = 2065 + 10 Ma, L.I. = 61 Ma; P = "1". Uncertain age; see text											
KS-GH-42 (Petroleum Management, #3 Eva Desbien): Granite.											
NM(-1)	361	94	0.00058	0.10581	0.1139	0.2473	3.335	0.09780	1425	1489	1583
M(-1)	364	101	0.00136	0.11657	0.1798	0.2450	3.306	0.09786	1413	1482	1584
M(0)	475	123	0.00177	0.12166	0.1646	0.2283	3.060	0.09721	1326	1423	1571
M(1)	544	131	0.00184	0.12249	0.1737	0.2104	2.816	0.09708	1231	1360	1569
Regression: U.I. = 1593 + 10 Ma, L.I. = 139 Ma; P = 0.06. Probably false age; see text.											
NM(-1)AA	329	95	0.00058	0.10690	0.1237	0.2708	3.696	0.09898	1545	1570	1605
M(-1)AA	332	94	0.00053	0.10656	0.1238	0.2659	3.641	0.09933	1520	1558	1611
M(0)AA	426	114	0.00044	0.10494	0.1230	0.2534	3.453	0.09883	1456	1517	1602
M(1)AA	592	143	0.00058	0.10561	0.1241	0.2277	3.066	0.09766	1323	1424	1580
Regression: U.I. = 1618 + 11 Ma, L.I. = 242 Ma; P = 0.00. See text.											
KS-GH-43 (Petroleum Management, #1 Lappin): Granite.											
NM(4)	623	117	0.00079	0.10872	0.1029	0.1801	2.430	0.09785	1067	1251	1584
M(4)	722	106	0.00094	0.11125	0.1157	0.1382	1.872	0.09824	835	1071	1591
M(6)	788	101	0.00117	0.11430	0.1258	0.1181	1.599	0.09819	720	970	1590
Regression: U.I. = 1579 + 9 Ma, L.I. = -18 Ma; P = 0.11. Probably false age; see text.											

[1]M = nonmagnetic, M = magnetic; numbers in parentheses indicate side tilt used on Franz separator at 1.5 amp power. F = fine size (-200 mesh). AA = air abraded.

[2]Total U and Pb, corrected for analytical blank. Concentrations in parentheses uncertain due to weighing errors of very small samples. Age calculations unaffected, since mixed spike used.

[3]Measured ratios, uncorrected for blank or non-radiogenic Pb; see text for corrections.

[4]Pb corrected for blank and non-radiogenic Pb (see text).

[5]Ages given in Ma. See text for decay constants used. Also see text for details on regression methods. P = probability of fit (1.00 = 100 %); U.I. = upper intercept, L.I. = lower intercept. Uncertainties at 2σ level.

TABLE 2. (continued)

	Concentrations[2]		Pb Isotopic Composition[3]			Radiogenic Ratios[4]			Ages[5]		
Fraction[1]	U (ppm)	Pb (ppm)	Pb-204/Pb-206	Pb-207/Pb-206	Pb-208/Pb-206	Pb-206/U-238	Pb-207/U-235	Pb-207/Pb-206	Pb-206/U-238	Pb-207/U-235	Pb-207/Pb-206

KS-KE-1 (AMOCO. #2 Longwood): Granite.

Fraction	U (ppm)	Pb (ppm)	Pb-204/Pb-206	Pb-207/Pb-206	Pb-208/Pb-206	Pb-206/U-238	Pb-207/U-235	Pb-207/Pb-206	Pb-206/U-238	Pb-207/U-235	Pb-207/Pb-206
NM(-1)AA	137	45	0.00041	0.10809	0.2697	0.2803	3.959	0.10245	1593	1626	1669
NM(-1)	166	46	0.00084	0.11285	0.2688	0.2342	3.274	0.10136	1357	1475	1649
M(-1)	190	53	0.00156	0.12234	0.2988	0.2223	3.094	0.10094	1294	1431	1641
M(0)	232	58	0.00114	0.11625	0.2833	0.2044	2.837	0.10064	1199	1365	1636
M(1)	253	56	0.00100	0.11379	0.2743	0.1825	2.519	0.10011	1081	1278	1626

Regression: U.I. = 1671 + 6 Ma, L.I. = 141 Ma; P = 0.04

KS-MS-44 (Houston Int. Mineral Co., "Vermillion"): Granite.

Fraction	U (ppm)	Pb (ppm)	Pb-204/Pb-206	Pb-207/Pb-206	Pb-208/Pb-206	Pb-206/U-238	Pb-207/U-235	Pb-207/Pb-206	Pb-206/U-238	Pb-207/U-235	Pb-207/Pb-206
B: M(1)	248	67	0.00031	0.09439	0.1552	0.2508	3.114	0.09004	1443	1436	1426
B: M(0)	310	82	0.00023	0.09326	0.1454	0.2485	3.088	0.09011	1431	1430	1428
B: NM(-1)	292	77	0.00027	0.09382	0.1371	0.2467	3.065	0.09011	1421	1424	1428
B: M(2)	328	86	0.00021	0.09296	0.1492	0.2450	3.041	0.09002	1413	1418	1426
A: M(0)	149	37	0.00035	0.09471	0.1131	0.2371	2.936	0.08984	1371	1391	1422
A: M(2)	266	64	0.00063	0.09854	0.1262	0.2304	2.853	0.08980	1337	1370	1421
A: M(4)	320	73	0.00057	0.09703	0.1214	0.2177	2.674	0.08910	1270	1321	1406

Regression: (A+B) U.I. = 1428 + 3 Ma, L.I. = 216 Ma; P = 0.19

KS-MS-46 (Texas Gulf, #1 Gerstner): Granite.

Fraction	U (ppm)	Pb (ppm)	Pb-204/Pb-206	Pb-207/Pb-206	Pb-208/Pb-206	Pb-206/U-238	Pb-207/U-235	Pb-207/Pb-206	Pb-206/U-238	Pb-207/U-235	Pb-207/Pb-206
NM (-2)	182	46	0.00055	0.09762	0.1365	0.2386	2.959	0.08995	1379	1397	1425
M (-2)	170	44	0.00087	0.10250	0.1519	0.2377	2.960	0.09034	1375	1398	1433
M (-1)	195	49	0.00057	0.09791	0.1371	0.2355	2.920	0.08993	1363	1387	1424
M (0)	194	49	0.00075	0.10026	0.1444	0.2336	2.893	0.08985	1353	1380	1422
M (+2)	193	48	0.00137	0.10924	0.1672	0.2239	2.782	0.09011	1302	1351	1428
M (+4)	179	44	0.00085	0.10133	0.1446	0.2273	2.804	0.08946	1320	1357	1414

Regression: U.I. = 1431 + 18 Ma, L.I. = 151 Ma; P = 0.00

KS-NM-18 (Transocean, #1 Ukele): Granite.

Fraction	U (ppm)	Pb (ppm)	Pb-204/Pb-206	Pb-207/Pb-206	Pb-208/Pb-206	Pb-206/U-238	Pb-207/U-235	Pb-207/Pb-206	Pb-206/U-238	Pb-207/U-235	Pb-207/Pb-206
NM(1)	552	129	0.00168	0.12148	0.1304	0.2127	2.884	0.09832	1243	1378	1593
M(1)	670	123	0.00142	0.11683	0.1195	0.1697	2.276	0.09724	1011	1205	1572
M(2)	854	127	0.00148	0.11682	0.1223	0.1369	1.821	0.09644	827	1053	1556
M(4)	1033	124	0.00164	0.11757	0.1291	0.1103	1.441	0.09481	674	906	1524

Regression: U.I. = 1609 + 6 Ma, L.I. = 92 Ma; P = 0.08

KS-NM-21ABC (Houston Int. Mineral Co., "Baileyville"): Dikes.

Fraction	U (ppm)	Pb (ppm)	Pb-204/Pb-206	Pb-207/Pb-206	Pb-208/Pb-206	Pb-206/U-238	Pb-207/U-235	Pb-207/Pb-206	Pb-206/U-238	Pb-207/U-235	Pb-207/Pb-206
A: M(-1)	344	113	0.00300	0.13258	0.3268	0.2494	3.120	0.09074	1435	1438	1441
A: NM(-1)	345	99	0.00076	0.10151	0.2421	0.2448	3.067	0.09088	1411	1425	1444
A: M(0)	443	124	0.00072	0.10053	0.2422	0.2408	3.004	0.09050	1391	1409	1436
B: NM(-2)	597	151	0.00141	0.10953	0.2001	0.2188	2.713	0.08994	1275	1332	1424
B: M(0)	759	180	0.00191	0.11601	0.2239	0.2000	2.466	0.08943	1176	1262	1413
C: NM(-1)	1035	208	0.00041	0.09460	0.1791	0.1825	2.237	0.08887	1081	1193	1402
B: M(-2)	671	144	0.00203	0.11745	0.2291	0.1799	2.211	0.08914	1066	1185	1407
C: M(-1)	1191	226	0.00077	0.09883	0.2070	0.1673	2.033	0.08813	997	1127	1385
C: M(0)	1130	182	0.00043	0.09377	0.1716	0.1472	1.783	0.08781	886	1039	1379

Regression: (A+B+C) U.I. = 1441 + 4 Ma, L.I. = 161 Ma; P = 0.00

KS-NM-21D (HIMCO, "Baileyville"): Granitic Gneiss.

Fraction	U (ppm)	Pb (ppm)	Pb-204/Pb-206	Pb-207/Pb-206	Pb-208/Pb-206	Pb-206/U-238	Pb-207/U-235	Pb-207/Pb-206	Pb-206/U-238	Pb-207/U-235	Pb-207/Pb-206
NM(-1)r	(1729)	(589)	0.00077	0.12147	0.1007	0.3251	4.978	0.11105	1815	1816	1817
NM(-1)e	(600)	(156)	0.00034	0.11269	0.0793	0.2582	3.845	0.10802	1480	1602	1766
M(-1)e	(1372)	(338)	0.00075	0.11863	0.0936	0.2399	3.586	0.10841	1386	1546	1773

(e = hand-picked euhedral; r = hand-picked round)
Average Pb-Pb age = 1770 + 10 Ma; Est. age = 1780 + 20 Ma

KS-NM-21E (HIMCO, "Baileyville"): Granitic Gneiss.

Fraction	U (ppm)	Pb (ppm)	Pb-204/Pb-206	Pb-207/Pb-206	Pb-208/Pb-206	Pb-206/U-238	Pb-207/U-235	Pb-207/Pb-206	Pb-206/U-238	Pb-207/U-235	Pb-207/Pb-206
NM(-1)	392	105	0.00115	0.11182	0.1189	0.2518	3.330	0.09592	1448	1488	1546
M(-1)	425	112	0.00080	0.10744	0.1063	0.2498	3.318	0.09633	1438	1485	1554

TABLE 2. (continued)

Fraction[1]	Concentrations[2]		Pb Isotopic Composition[3]			Radiogenic Ratios[4]			Ages[5]		
	U (ppm)	Pb (ppm)	Pb-204/Pb-206	Pb-207/Pb-206	Pb-208/Pb-206	Pb-206/U-238	Pb-207/U-235	Pb-207/Pb-206	Pb-206/U-238	Pb-207/U-235	Pb-207/Pb-206
M(0)	832	218	0.00211	0.12519	0.1545	0.2307	3.054	0.09604	1338	1421	1548

Regression: U.I. = 1552 + 21 Ma; L.I. = 32 Ma; P = 0.00. Probably false age; see text.

KS-NT-334 (Hummon Oil, #1 Frederick): Granite.

Fraction[1]	U (ppm)	Pb (ppm)	Pb-204/Pb-206	Pb-207/Pb-206	Pb-208/Pb-206	Pb-206/U-238	Pb-207/U-235	Pb-207/Pb-206	Pb-206/U-238	Pb-207/U-235	Pb-207/Pb-206
NM (0)	505	125	0.00083	0.10884	0.1221	0.2400	3.223	0.09740	1387	1463	1575
M (0)	664	153	0.00153	0.11856	0.1597	0.2062	2.771	0.09749	1208	1348	1577
M (+3)	779	102	0.00197	0.12496	0.1847	0.1621	2.184	0.09774	968	1176	1582
M (+6)	984	118	0.00201	0.12391	0.1849	0.1050	1.392	0.09610	644	885	1550

Regression: U.I. = 1580 + 12 Ma, L.I. = 23 Ma; P = 0.00. Probably false age; see text.

KS-ST-1 (AMOCO, #1 McPherson): Granitic Gneiss.

Fraction[1]	U (ppm)	Pb (ppm)	Pb-204/Pb-206	Pb-207/Pb-206	Pb-208/Pb-206	Pb-206/U-238	Pb-207/U-235	Pb-207/Pb-206	Pb-206/U-238	Pb-207/U-235	Pb-207/Pb-206
NM (0)	414	66	0.00123	0.11861	0.2136	0.1369	1.921	0.10179	827	1088	1657
M (0)	615	81	0.00163	0.12256	0.2389	0.1100	1.520	0.10022	673	938	1628
M (+1)	685	86	0.00255	0.13524	0.2773	0.1006	1.389	0.10015	618	884	1627
M (+2)	752	86	0.00215	0.12815	0.2691	0.0924	1.256	0.09863	570	826	1598

Regression: U.I. = 1716 + 29 Ma, L.I. = 87 Ma; P = 0.00

MO-CM-2 (Exxon Minerals, DB-1): Gneiss.

Fraction[1]	U (ppm)	Pb (ppm)	Pb-204/Pb-206	Pb-207/Pb-206	Pb-208/Pb-206	Pb-206/U-238	Pb-207/U-235	Pb-207/Pb-206	Pb-206/U-238	Pb-207/U-235	Pb-207/Pb-206
NM(1)	704	82	0.00047	0.10377	0.1254	0.1102	1.479	0.09733	674	922	1574
M(1)	929	80	0.00077	0.10420	0.1320	0.0801	1.033	0.09358	497	720	1500
M(3)	1316	85	0.00076	0.10215	0.1254	0.0608	0.768	0.09160	381	579	1459
M(4)	2244	103	0.00095	0.10210	0.1319	0.0425	0.521	0.08885	269	426	1401

Regression: U.I. = 1633 + 50 Ma, L.I. = 75 Ma; P = 0.00

MO-CM-3 (Exxon Minerals, TD-1): Gneiss.

Fraction[1]	U (ppm)	Pb (ppm)	Pb-204/Pb-206	Pb-207/Pb-206	Pb-208/Pb-206	Pb-206/U-238	Pb-207/U-235	Pb-207/Pb-206	Pb-206/U-238	Pb-207/U-235	Pb-207/Pb-206
NM(1)	662	121	0.00185	0.11936	0.1595	0.1624	2.098	0.09371	970	1148	1502
M(1)	848	125	0.00181	0.11682	0.1523	0.1313	1.659	0.09160	796	993	1459
M(3)	1159	141	0.00221	0.12034	0.1664	0.1062	1.310	0.08948	651	850	1415
M(5)	1706	152	0.00269	0.12388	0.1821	0.0758	0.901	0.08628	471	652	1345

Regression: U.I. = 1552 + 20 Ma, L.I. = 137 Ma; P = 0.00. Probably false age; see text.

MO-LC-3 (St. Joe Minerals, 63W29): Felsic Paragneiss.

Fraction[1]	U (ppm)	Pb (ppm)	Pb-204/Pb-206	Pb-207/Pb-206	Pb-208/Pb-206	Pb-206/U-238	Pb-207/U-235	Pb-207/Pb-206	Pb-206/U-238	Pb-207/U-235	Pb-207/Pb-206
NM (−1)	410	81	0.00032	0.10048	0.0722	0.1963	2.600	0.09608	1155	1301	1549
M (−1)	339	74	0.00031	0.10108	0.0784	0.2158	2.881	0.09679	1260	1377	1563
M (0)	486	89	0.00048	0.10157	0.0770	0.1805	2.361	0.09486	1070	1231	1525
M (1)	667	106	0.00049	0.10067	0.0766	0.1554	2.012	0.09392	931	1120	1506

Regression: U.I. = 1598 + 17 Ma, L.I. = 198 Ma; P = 0.00. Probably false age; see text.

MO-PO-2 (Union Carbide, MHR-1): Granite.

Fraction[1]	U (ppm)	Pb (ppm)	Pb-204/Pb-206	Pb-207/Pb-206	Pb-208/Pb-206	Pb-206/U-238	Pb-207/U-235	Pb-207/Pb-206	Pb-206/U-238	Pb-207/U-235	Pb-207/Pb-206
B: NM(−2)	243	60	0.00290	0.12754	0.2416	0.2004	2.402	0.08695	1177	1243	1359
B: M(−2)	265	58	0.00104	0.10109	0.1863	0.1961	2.340	0.08654	1155	1225	1350
B: M(−1)	281	61	0.00090	0.09903	0.1758	0.1966	2.343	0.08645	1157	1226	1348
B: M(0)	279	66	0.00186	0.11302	0.2282	0.1985	2.379	0.08692	1167	1236	1359
A: M(−1)	182	45	0.00405	0.14270	0.2894	0.1876	2.222	0.08588	1108	1188	1336
B: M(1)	327	67	0.00129	0.10383	0.1999	0.1803	2.133	0.08580	1069	1160	1334
B: M(2)	312	62	0.00105	0.10057	0.1944	0.1764	2.095	0.08589	1050	1147	1336
A: NM(−1)	440	98	0.00332	0.13292	0.2751	0.1735	2.066	0.08637	1031	1138	1347
B: M(3)	360	71	0.00159	0.10750	0.2118	0.1697	1.993	0.08518	1010	1113	1320
A: M(1)	368	74	0.00201	0.11409	0.2381	0.1674	1.984	0.08594	998	1110	1337
A: M(2)	495	81	0.00141	0.10461	0.1838	0.1439	1.684	0.08489	867	1003	1313

Regressions: A: U.I. = 1360 + 33 Ma, L.I. = 110 Ma; P = 0.00
B: U.I. = 1386 + 17 Ma, L.I. = 233 Ma; P = 0.00
(A+B): U.I. = 1371 + 14 Ma. L.I. = 151 Ma; P = 0.00

MO-RI-2 (Gulf Oil, PBW-2): Granite.

Fraction[1]	U (ppm)	Pb (ppm)	Pb-204/Pb-206	Pb-207/Pb-206	Pb-208/Pb-206	Pb-206/U-238	Pb-207/U-235	Pb-207/Pb-206	Pb-206/U-238	Pb-207/U-235	Pb-207/Pb-206
NM(−1)	188	51	0.00191	0.11832	0.1907	0.2376	3.008	0.09182	1374	1410	1464

Fraction[1]	Concentrations[2]		Pb Isotopic Composition[3]			Radiogenic Ratios[4]			Ages[5]		
	U (ppm)	Pb (ppm)	$\frac{Pb\text{-}204}{Pb\text{-}206}$	$\frac{Pb\text{-}207}{Pb\text{-}206}$	$\frac{Pb\text{-}208}{Pb\text{-}206}$	$\frac{Pb\text{-}206}{U\text{-}238}$	$\frac{Pb\text{-}207}{U\text{-}235}$	$\frac{Pb\text{-}207}{Pb\text{-}206}$	$\frac{Pb\text{-}206}{U\text{-}238}$	$\frac{Pb\text{-}207}{U\text{-}235}$	$\frac{Pb\text{-}207}{Pb\text{-}206}$
M(−1)	198	55	0.00225	0.12304	0.2031	0.2350	2.971	0.09170	1361	1400	1461
M(0)	429	111	0.00207	0.11946	0.2418	0.2140	2.673	0.09057	1250	1321	1438
M(1)	590	176	0.00404	0.14759	0.2983	0.2244	2.827	0.09138	1305	1363	1455
Regression: U.I. = 1482 + 9 Ma, L.I. = 321 Ma; P = 0.29											
MO-WE-1B (Union Carbide, M1J1): Granite.											
NM(−1)	454	112	0.00120	0.11092	0.1710	0.2194	2.854	0.09435	1279	1370	1515
M(−1)	634	143	0.00113	0.10915	0.1714	0.2008	2.590	0.09353	1180	1298	1498
M(0)	703	146	0.00114	0.10837	0.1829	0.1833	2.340	0.09258	1085	1225	1479
Regression: U.I. = 1549 + 10 Ma, L.I. = 237 Ma; P = 0.82. Probably false age; see text.											
NB-BF-1 (PC-103; Ohio Oil, #1 Pettett): Tonalite gneiss.											
NM(1)	362	100	0.00060	0.11681	0.0792	0.2673	4.003	0.10861	1527	1635	1776
M(1)	479	115	0.00095	0.12104	0.0946	0.2271	3.388	0.10817	1319	1502	1769
M(2)	406	87	0.00157	0.12881	0.1165	0.2023	2.995	0.10738	1188	1407	1756
Regression: U.I. = 1787 + 10 Ma, L.I. = 97 Ma; P = 0.05											
NB-CS-2 (PC-106; Ohio Oil, #1 Bremer): Augen Gneiss.											
NM(0)	201	47	0.00038	0.10557	0.1343	0.2199	3.044	0.10037	1282	1419	1631
M(1)	294	60	0.00041	0.10612	0.1453	0.1899	2.631	0.10051	1121	1309	1634
M(3)	397	65	0.00071	0.10958	0.1587	0.1484	2.042	0.09981	892	1130	1621
Regression: U.I. = 1639 + 12 Ma, L.I. = 32 Ma; P = 0.03											
NB-DA-2 (PC-105; Ohio Oil, #1 Dunse): Gneissic Granodiorite.											
NM(2)	470	119	0.00074	0.12035	0.0694	0.2466	3.748	0.11025	1421	1582	1804
M(2)	467	112	0.00105	0.12446	0.0796	0.2288	3.477	0.11022	1328	1522	1803
M(3)	561	131	0.00128	0.12739	0.0867	0.2206	3.346	0.11001	1285	1492	1800
M(4)	683	116	0.00190	0.13603	0.1096	0.1561	2.372	0.11021	935	1234	1803
Regression: U.I. = 1802 + 5 Ma, L.I. = 0 Ma; P = 0.25											

of the Lower Proterozoic terranes of Colorado eastward into western Kansas.

Several core or cuttings samples have been obtained which yield ages in the range 1.61 to 1.65 Ga in the southern part of the midcontinent region. Bickford et al. [1981a] reported on some from Missouri and northeastern Kansas, and additional samples are reported in this study (Figure 5D). One, MO-CM-2 (1.63 Ga), is granitic gneiss from central Missouri and is compatible with other samples in Missouri having similar ages. Another, NB-CS-2 (1.64 Ga), is from augen gneiss in southwestern Nebraska. Because of the small size of the original sample, we are not sure whether the age on this sample represents igneous crystallization of the protolith for the gneiss or whether it represents a metamorphic event. 1630 Ma old metamorphic nodes and anatectic granites have been found in the northern-most part of the Colorado Province [Premo, 1984], so either interpretation would fit the regional geology. A third sample, KS-NM-18 (1.61 Ga), from northeastern Kansas occurs in the immediate vicinity of several others that yield similar ages of 1.61 to 1.63 Ga [Bickford et al.,

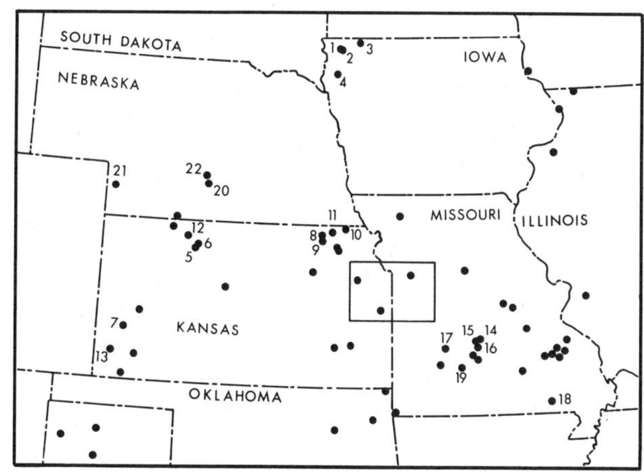

Fig. 4. Index map of the central United States showing locations of samples for which U-Pb ages on zircons have been obtained. Numbered localities refer to samples presented in this paper (Tables 1 and 2). Other localities are from reports cited in the text. Outlined area in northeastern Kansas/ eastern Missouri refers to Fig. 8.

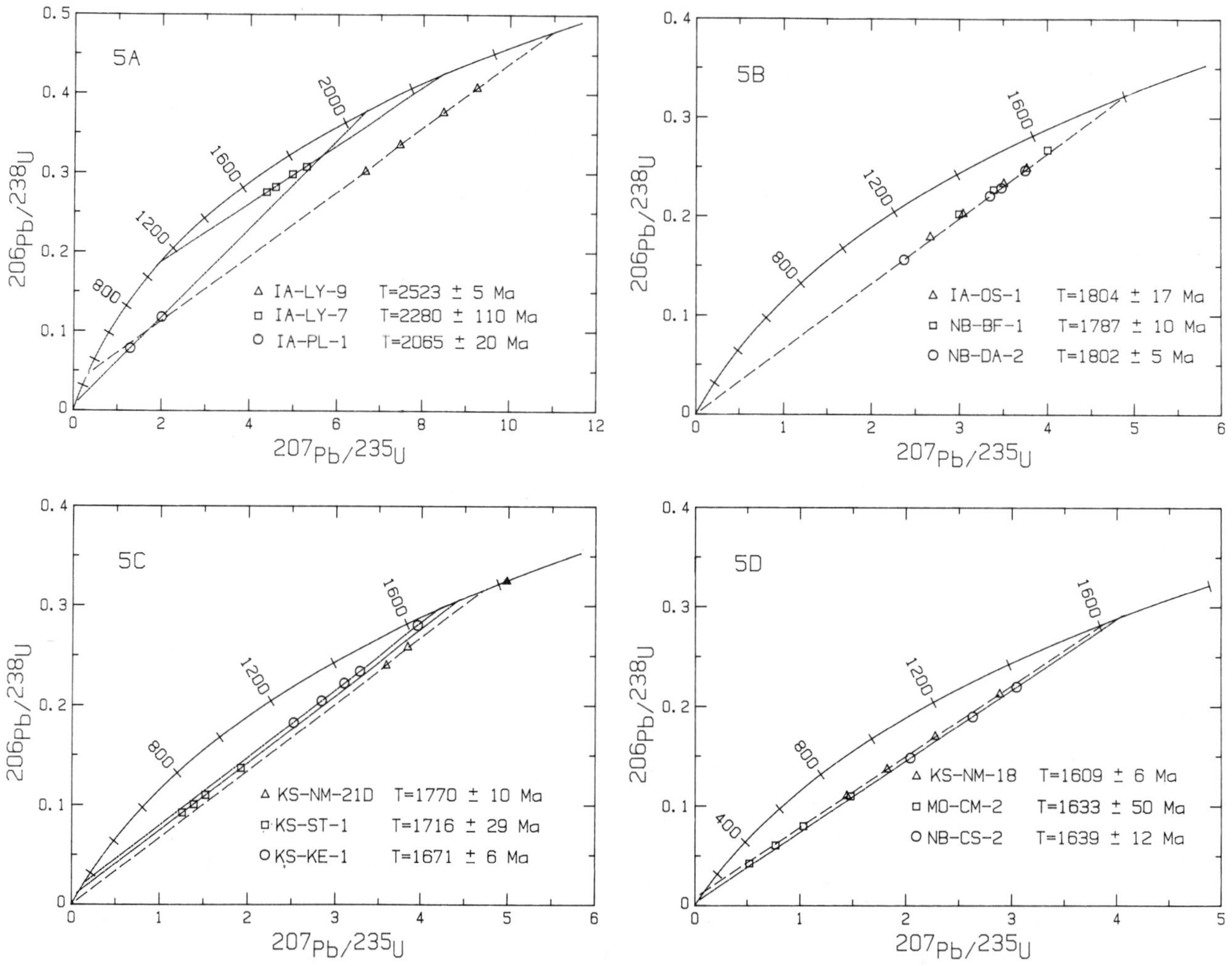

Fig. 5. Concordia diagram plots of U-Pb data for selected samples presented in this
Regression lines are not shown for all sets of data, but individual results can be
found in (A): Samples from northwestern Iowa that are apparently older than 2.0 Ga.
The age of the sample is well constrained; the two Early Proterozoic ages are still not
tightly constrained. See text for further details. (B): Samples from Iowa and
Nebraska that yield ages of about 1800 Ma. Regression line only shown for NB-DA-2.
(C): Samples from Kansas 1800 Ma. (D): Samples from Kansas, Nebraska, and Missouri
that yield ages between 1600 and 1650 Ma. Regression lines only shown for NB-CS-2 and
KS-NM-18.

1981a], and it is probably part of a suite of
plutons of that age.

Early Middle Proterozoic (1600 - 1300 Ma)

Several samples have yielded ages younger than
1600 Ma. The ones most easy to interpret are
those either about 1450 to 1500 Ma old (Figure
6A) or about 1350 to 1400 Ma old (Figure 6B).
Samples in these age ranges belong either to a
major period of anorogenic igneous activity about
1450 to 1500 Ma ago [Silver et al., 1977b;
Bickford et al., 1981a; Anderson, 1983; Hoppe et
al., 1983] that occurred throughout the early

Proterozoic orogenic crust south of the Archean
cratons, or to a similar period of igneous
activity that occurred in the south-central
United States 1340 to 1400 Ma ago [Thomas et al.,
1984; Bickford and Van Schmus, 1985]. The
samples yielding ages in this range are all
undeformed granites or intrusive dikes (Tables 1
and 2), and the zircon ages are interpreted as
primary crystallization ages.

Several other samples which have yielded
apparent ages in the range 1500 to 1600 Ma
(Tables 1 and 2) are more difficult to interpret.
Until this report, U-Pb zircon ages in that
interval were not known from the midcontinent

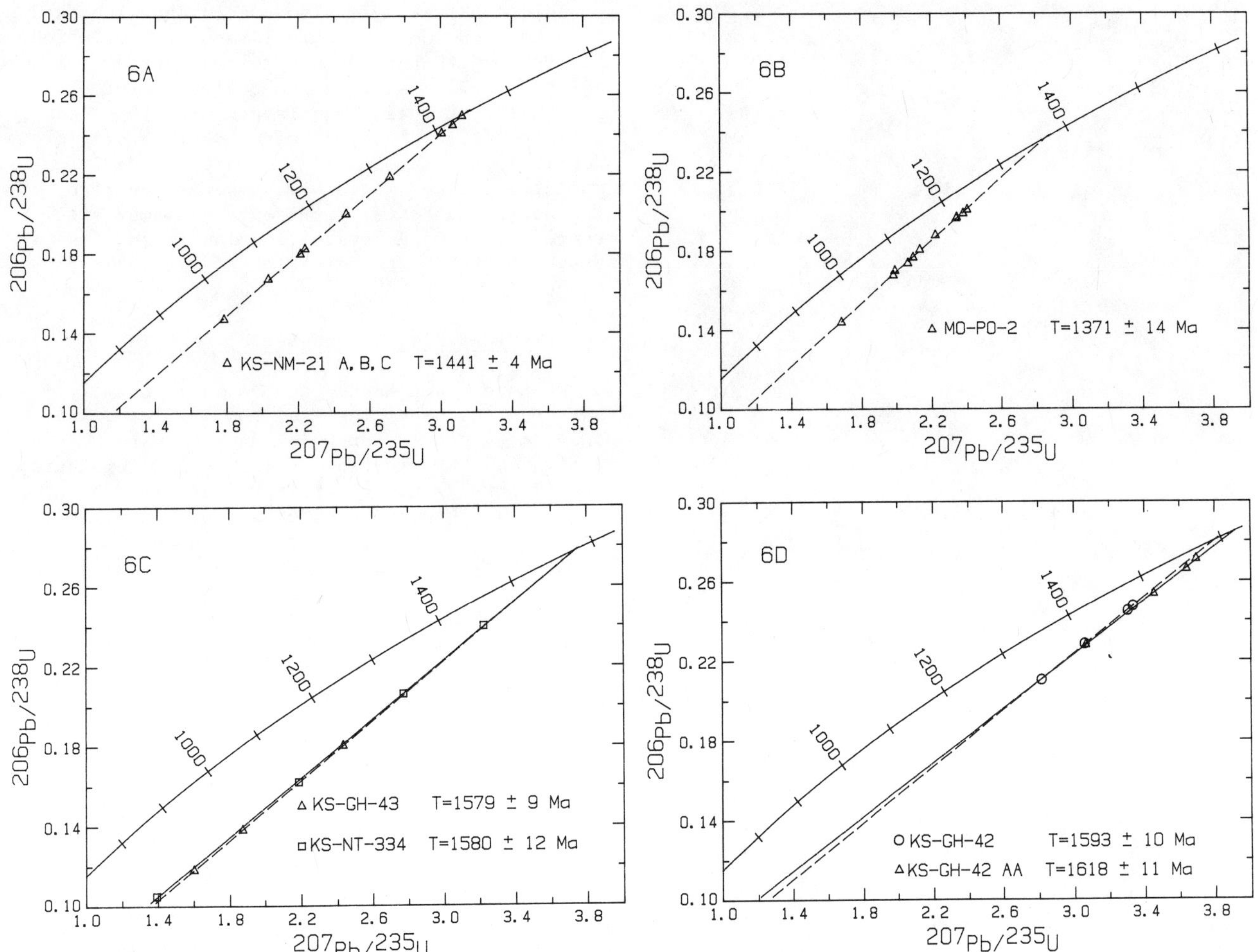

Fig. 6. Concordia diagram plots of U-Pb data for selected samples presented in this paper. Regression lines are not shown for all sets of data, but individual results can be found in Table 2; (A): Samples from three dikes from core KS-NM-21 (A,B,C), which are representative of units giving ages of 1450 to 1500 Ma. (B): Sample MO-PO-2, which is representative of samples that yield ages of 1350 to 1400 Ma. (C): Samples from Kansas that yield ages of 1550 to 1600 Ma. These ages are probably too low; see Fig. 6D and text. (D): Data for sample KS-GH-42 showing result of air abrasion (AA) on zircon population. The apparent age defined by unabraded zircons for this sample and those in Fig. 6C are biased to the low side due to metamorphic overgrowths. See also Fig. 7B.

region. Several rock units yielding these ages are gneissic and recrystallized (Tables 1 and 2), so that there is a significant probability that the zircons in them have been partially reset during high-grade metamorphism or consist of older cores and younger overgrowths. Unfortunately, in most cases the samples are too small to allow detailed separation of zircons to try to obtain specific ages of metamorphism or to try to determine if complex histories are present through analysis of several size splits, magnetic splits, and hand-picked fractions.

Three samples from northwestern Kansas (KS-NT-334, KS-GH-42, and KS-GH-43) have proven crucial to interpretation of these ages (Figures 6C and 6D). Thin sections of drill cuttings from these localities show that the basement units sampled are granites that have been cataclastically deformed and then statically recrystallized (Figure 7A). Zircons from KS-GH-42 (which was a relatively large sample) show definite indications of overgrowths on euhedral zircons. Because we obtained a large yield of zircons from KS-GH-42, we analysed two suite of fractions. In the first case we analysed the zircons without additional treatment, in a manner similar to the routine followed for the other, smaller samples, and obtained an apparent age of 1593 ± 10 Ma

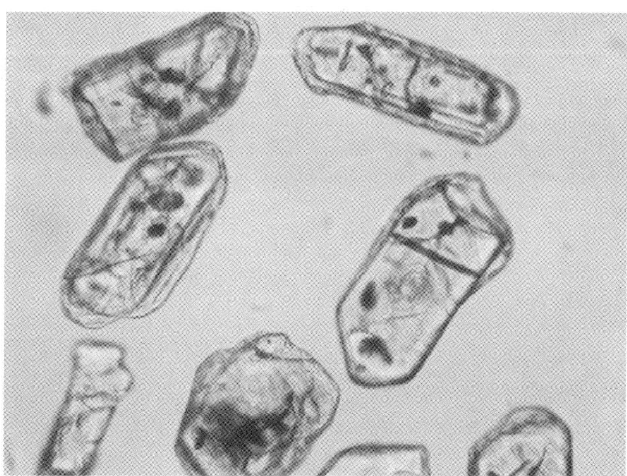

Fig. 7. (A): Photomicrograph of the recrystallized texture for KS-GH-43 that is typical of the
northwestern Kansas samples yielding ages of 1580
to 1590 Ma (Fig. 6C,D). Width of photo 3.5 mm.
(B): Photomicrographs of zircons separated from
KS-GH-42 showing metamorphic overgrowths before
removal by air abrasion. Larger zircons are about
150 microns long.

(Figure 6D). In the second case, we subjected
the zircons to air abrasion [Krogh, 1982] in
order to try to remove any metamorphic
overgrowths that may have been present. Those
zircons yielded an age of 1618 ± 11 Ma,
distinctly older than that of the first suite and
similar to other 1610 to 1650 Ma ages found in
the region. We believe that these results
demonstrate that the apparent ages derived from
untreated zircons for these three samples are too
low because of the presence of metamorphic
overgrowths, and that the correct ages are
probably about 1620 Ma; the lower ages may be due
to metamorphic effects from the 1450 to 1500 Ma
old plutonism.

This conclusion is also consistent with data
from core KS-NM-21, mentioned above, from which
all three ages have been obtained. Data from
core MO-LC-3, a sillimanite-bearing felsic
paragneiss, are also consistent with this model;
the zircons in this rock were probably originally
detrital, but have undergone partial resetting
during later metamorphism. However, we also
recognize that some of these results may still
represent primary crystallization ages; if so,
they are the first indication of events in the
midcontinent that have been unknown until now.

<u>Correlations With Aeromagnetic Anomalies</u>

We have examined the aeromagnetic anomaly
patterns in the vicinity of the samples discussed
above plus others for which data have been
published. Examination of both the aeromagnetic
map (Figure 2) and the filtered gravity map
(Figure 3) show a series of major NW-SE trending
linear anomalies extending through the Central
Plains Province, especially in Missouri and
Nebraska [Zietz, 1981] and a variety of other
"spot" anomalies or broader anomalies. In
general, we find no significant correlation of
magnetic anomalies with samples whose ages are
greater than 1600 Ma, but for samples with ages
less than 1500 Ma there is a distinct
correlation. Anderson [1983] has pointed out
that granites of the 1450 to 1500 Ma and 1350 to
1400 Ma old anorogenic suites may be either
ilmenite-bearing granites or magnetite-bearing
granites. In fact, most samples we have dated
in these age ranges appear to be magnetite series
granites, which generate distinct to strong
positive aeromagnetic anomalies. These range
from small, strong anomalies [Figure 8; Steeples
and Bickford, 1981] to somewhat broader anomalies
[Coates et al., 1983] or large regions of long
wavelength, generally positive aeromagnetic
character [Thomas et al., 1984; Bickford and Van
Schmus, 1985]. These relationships have been
used to help define the limits of the provinces
shown in Figure 1.

Discussion

Although data are still too fragmentary to
allow a definitive model for the Proterozoic
basement of the United States to be developed,
there have been enough additions in the past
several years to make a substantial improvement
on the model proposed earlier [Van Schmus and
Bickford, 1981]. Our current interpretation of
the geology is shown in Figure 1; two crustal
cross-sections are shown in Figure 9.

<u>Archean</u>

Portions of two major Archean cratons border
the Proterozoic terranes of the United States.
In the northeast the Superior Craton of the
Canadian Shield extends into Michigan, Minnesota,

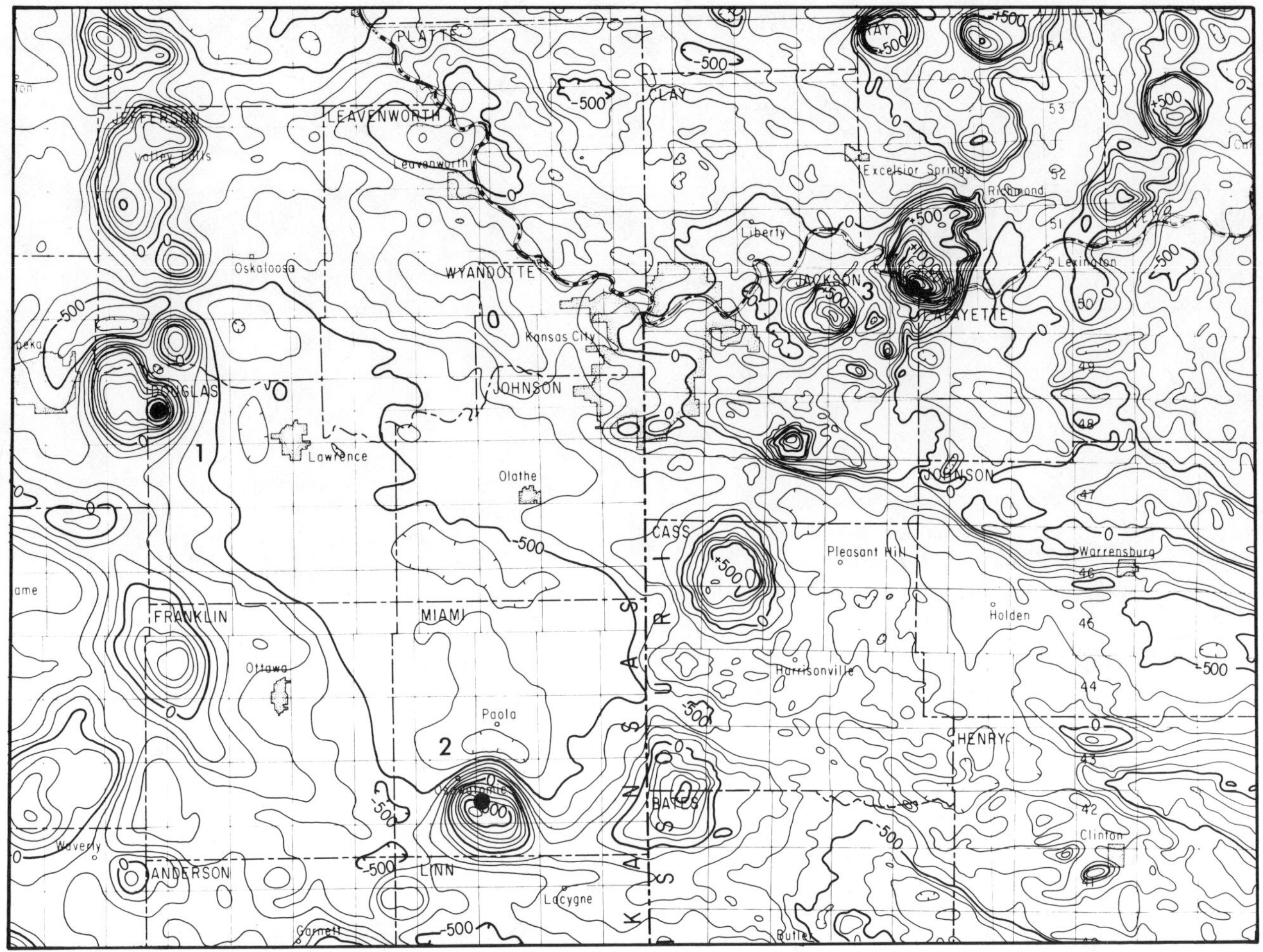

Fig. 8. Portion of the aeromagnetic map of eastern Kansas and western Missouri in the vicinity of Kansas City [Burchett et al., 1982]. Numbered localities refer to drill hole localities where samples have yielded ages of 1340 to 1370 Ma. Note that each occurs on a small, intense positive anomaly. Petrography of the samples shows that the anomalies are due to magnetite-bearing granites. Several similar anomalies in this area are thought to be due to 1340 to 1380 Ma old magnetite-bearing granites [Yarger, 1985] that are part of a major suite of such rocks [Thomas et al., 1984].

Wisconsin, and the Dakotas; in the northwest, the Wyoming Craton extends into Montana and the westernmost Dakotas [Condie, 1976; Sims and Peterman, 1981]. The principal questions pertinent to this report are those of the southern limit of each Archean craton and the extension of each into the structural belt of the western Dakotas.

The southern limit of the Wyoming Craton is well defined by the Cheyenne Belt [Houston et al., 1979], a major shear zone separating Proterozoic rocks of the Colorado Province from Archean rocks of the Wyoming Craton. Hills and Houston [1979] have described this boundary as a plate suture developed at a south-dipping subduction zone, but there may also be a significant component of transverse movement along the boundary. The southern limit of the Superior Craton is less well defined. South of Lake Superior the Archean craton is cut off abruptly by a major fault zone; to the south of this zone occur metavolcanic and plutonic rocks of the Penokean volcanic and plutonic belt of northern Wisconsin (Van Schmus, 1980). The absence of significant plutonic activity in the Archean craton to the north suggests that the boundary may be a south-dipping suture. An apparently isolated block of Archean crustal rocks, whose southern boundary is not exposed, occurs in central Wisconsin [Van Schmus and Anderson, 1977]; this block may represent a detached piece of the Superior province that was later re-accreted as part of the Penokean Province (see below). Archean rocks, from which

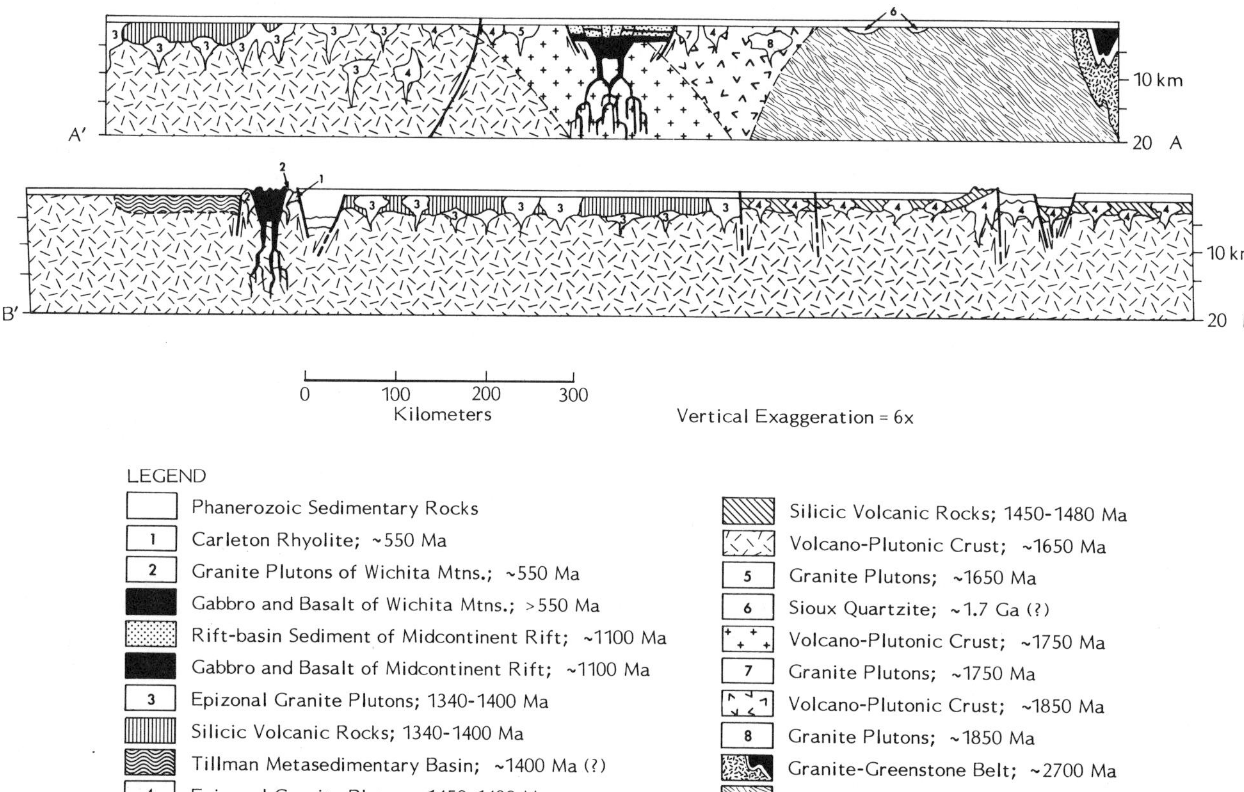

Fig. 9. Generalized cross section through the continental crust in the central United States, showing the relationships of the various lithologic and orogenic suites to each other. Section A-A' runs N-S, approximately along 96°W longitude (Fig. 1), but some units are projected laterally into the plane of section to show the inferred relationships. Section B-B' runs northeast-southwest (Fig. 1) and shows the inferred relationships of the granite-rhyolite terranes to the presumed underlying 1.6 to 1.8 Ga old deeper basement.

ages in excess of 3.5 Ga have been reported, are exposed in the Minnesota River Valley of southern Minnesota [Goldich and Wooden, 1980], but the southern boundary of those rocks in not exposed either.

The only control on the distribution of Archean rocks in the subsurface to the south is from northwestern Iowa, where monzonitic core (IA-LY-9) gives an age of 2523 ± 5 Ma. This age is less than that of most Archean rocks of Wyoming and Superior cratons, but it is comparable to the young end of the spectrum found for both. Van Schmus (unpublished) has found similar ages from parts of the Archean terrane in central Wisconsin. Zartman and Stern [1967] have also reported a similar age for the Little Elk Granite of the Black Hills in western South Dakota. There are no other radiometric ages or isotopic data to indicate how far south Archean crust extends from northwestern Iowa or central Wisconsin; the inferred southern limit shown in Figure 1 is based primarily on interpretation of aeromagnetic data (Figure 2), which shows a

sharp, linear gradient in northwestern Iowa just south of the dated sample [Storm Lake Trend, Anderson and Black, 1983]. This feature is approximately on strike with the southern limit of the Archean craton in northern Wisconsin and has been selected as a possible southern boundary of the Archean craton in northwestern Iowa. However, this is not required by available data, and there is no parallel feature in the gravity data for the same region (Figure 3); therefore, Archean basement could extend considerably farther south.

Early Proterozoic

Western Dakotas mobile belt. A major, north-south trending mobile belt underlies the western half of North Dakota and South Dakota. This belt has been interpreted as the southern extension of the Trans-Hudson Orogen of the Churchill Province of the Canadian Shield [Green et al., 1979, 1985]. Recent geochronologic studies on the Trans-Hudson Orogen in northern Saskatchewan [Van

Schmus et al., 1986] have shown that the main
Proterozoic orogenic activity occurred about 1830
to 1900 Ma ago; if the mobile belt of the western
Dakotas is an extension of that zone, similar
ages should occur there. Little is known in
detail about this belt. Gravity data [Arvidson
et al., 1984 and Figure 3] and aeromagnetic data
(Figure 2) show very clearly that the structural
trends are continuations of those in Canada
[Green et al., 1979, 1985], trending
predominantly north-south.

The western edge of the Superior Province is
well defined by both gravity and aeromagnetic
data [Dutch, 1983], but the eastern edge of the
Wyoming craton is not well defined by either
(Figures 2 and 3). Camfield and Gough [1977]
place the eastern edge of the Wyoming Craton
along the North American Central Plains (NACP)
conductive anomaly, but Peterman et al. [1983]
suggest that it may be slightly to the west.
There is little information available about the
geology and chronology of Proterozoic rocks
within the western Dakotas mobile belt; Peterman
et al. [1983] have reported that the rocks in the
belt consist at least in part of Archean basement
metamorphosed to granulite facies about 1.8 Ga
ago. Denison et al. [1984] reported that this
belt in South Dakota consists of a variety of
granitic and metamorphic rocks that are similar
to those in Nebraska [Treves and Low, 1984] and
consistent with components of an early
Proterozoic orogenic sequence. The major
uncertainty in the region is the southern limit
of this belt and its relationship with (or
transition into) Lower Proterozoic rocks in
Nebraska. There is a definite change in
aeromagnetic signature along the Nebraska–South
Dakota border that may represent this transition
(Figure 2).

<u>Penokean Province.</u>   To the east in Wisconsin,
the southern part of the Superior Craton is
overlain by cratonic margin sediments and bounded
by igneous and metamorphic rocks of the northern
Wisconsin magmatic terrane, which apparently
represents a suite of oceanic arcs that were
formed and accreted to the continental margin
about 1830 to 1890 Ma ago [Van Schmus, 1980;
LaBarge et al., 1984]. One might expect that
rocks of that type and age extend along the
southern margin of the Superior Craton westward
into western Iowa. However, the only rock whose
age and location support such an extrapolation is
a granite (IA-OS-1; 1.81 Ga) in northwestern
Iowa. This sample lies north of the Storm Lake
Trend, which defines the southern edge of the
Archean craton (see above), suggesting that the
granite is hosted by the Archean craton and is
not part of an accretionary assemblage south of a
major suture. Furthermore, the age is slightly
less than found for Penokean rocks in Wisconsin.
Therefore, interpretations showing the Penokean
Province extending into Iowa (Figure 1) may not
be valid. One possible alternative
interpretation is that the 1.81 Ga old pluton in

northwestern Iowa was formed by a period of
subduction along a more southerly north-dipping
suture a few tens of millions of years later than
the main Penokean activity, and that the Penokean
Province pinches out westward from Wisconsin.

A related question is how far south rocks of
the Penokean orogen extend. In southern
Wisconsin there is an extensive terrane of
rhyolite, epizonal granite, and overlying
quartzite [Smith, 1978; Dott and Dalziel, 1974].
The chemistry of the granites and rhyolites
suggests that they were derived from partial
melting of older crust [Anderson et al., 1980;
Smith, 1978], and zircon dating shows that they
are about 1760 Ma old [Van Schmus, 1980]. Thus,
we infer that these rocks overlie older basement,
probably in part or wholly represented by
southern extension of the Penokean Province.
Nelson and DePaolo [1985] have shown that a 1470
Ma old granite from the subsurface of
northwestern Illinois [Hoppe et al., 1983] has a
Nd isotopic signature indicating derivation from
older Proterozoic crust (approx. 1.9 Ga),
suggesting that the Penokean Province may extend
some distance to the south.

<u>Colorado Province.</u>   Lower Proterozoic units
south of the Wyoming Craton are referred to here
as the Colorado Province, since they are best
exposed in the State of Colorado. These units
are part of a northeasterly trending early
Proterozoic terrane recognized by Silver [1968]
in the southwestern United States, where they
occur in northern Arizona and northern New
Mexico; we believe that they extend eastward,
into the basement of eastern Colorado, western
Kansas and western Nebraska.

Rocks in the Colorado Province include
metasedimentary and metavolcanic rocks and
associated plutons. The metavolcanic rocks are
typically bimodal, whereas the metasedimentary
rocks range from metagraywacke where associated
with metavolcanics, to high grade metapelites
elsewhere [Tweto, 1980]. These rocks have been
interpreted as representing early Proterozoic
orogenic island arc assemblages that were
accreted to the North American continent [Condie,
1982; Bickford and Boardman, 1984]. Recent U-Pb
age studies on zircons from these rocks have
shown that the volcanic rocks and related plutons
formed 1700 to 1780 Ma ago; later granitic
plutons were emplaced about 1660 to 1680 Ma ago
[Premo, 1984; Bickford et al., 1984; Bickford and
Boardman, 1984; Condie and Bowring, 1984]. The
terrane is therefore significantly younger than
rocks of similar origin in the Penokean Province
[1830 to 1880 Ma, Van Schmus, 1980] and the
Trans-Hudson Orogen [1830 Ma to 1890 Ma, Van
Schmus et al., 1986]. DePaolo [1981] reported Nd
isotopic data that confirm earlier
interpretations [Zartman, 1974] that the early
Proterozoic crust of the Colorado Province
represents juvenile additions to the continent
and does not lie upon or primarily consist of
reworked Archean continental crust. Stacey and

Hedlund [1983] interpreted Pb isotopic data to indicate some involvement of older crustal material in the generation of the igneous rocks in the northern part of the Colorado Province. Hills and Houston [1979] have argued that the boundary (Cheyenne Belt) between rocks of the Colorado Province and the Wyoming Craton is a suture formed by a south-dipping subduction zone, since no igneous activity associated with the Colorado Province is found north of the boundary.

Southwestern Province. When Silver [1968] defined the age and distribution of the 1700 to 1780 Ma old northern basement terrane of the southwest United States (Colorado Province, above), he also recognized a terrane to the south of it that is about 100 Ma younger, having formed 1630 to 1700 Ma ago. This terrane, referred to here as the Southwestern Province, occurs in southeastern Arizona and southern New Mexico [Silver et al., 1977a], and we believe that it extends eastward through Texas, Oklahoma, and Kansas into Missouri and perhaps beyond [Van Schmus and Bickford, 1981]. Silver [1984] interpreted units of this terrane as parts of arc-basin systems that formed adjacent to and within the southern part of the Colorado Province; the rocks consist of basalt-rhyolite volcanic suites, relatively mature metasedimentary rocks, and syn- to post-tectonic granodioritic to granitic plutons. Stacey and Hedlund [1983] have reported Pb isotopic data from southwestern New Mexico that show that basement greater than 1.8 Ga old is lacking, similar to the case for the Colorado Province; these results are consistent with a model in which the 1630 to 1700 Ma old crust in this terrane represents new additions to North America during the early Proterozoic. Nd and Pb isotopic data are not yet sufficient to determine whether the primary crustal addition was about 1650 Ma ago, or whether the 1650 Ma rocks were developed from or upon 1750 Ma old crust. However, granitic plutons whose ages are 1650 to 1680 Ma occur as intrusions within volcanogenic rocks of the Colorado Province to the north [Bickford et al, 1984; Premo, 1984]. Therefore, it is very possible that the 1650 Ma old suite represents a younger belt adjacent to and partially overlapping the 1700 to 1780 Ma old Colorado Province [Silver, 1984]. The occurrence of extensive igneous and metamorphic activity about 1.65 Ga ago in the older belt indicates that the boundary between the two may be a north-dipping suture.

Central Plains Province. The Precambrian basement of the continental interior is bordered on the north and west by exposed Lower Proterozoic suites of the Penokean, Colorado, and Southwestern provinces, as well as the inferred southern extension of the Churchill Province in the western Dakotas mobile belt, as outlined above. It is bordered on the south and east by Middle Proterozoic terranes of the Grenville and Llano provinces (below). The continental interior can be divided into the Eastern Granite-Rhyolite Province, the Western Granite-Rhyolite Province, and the Central Plains Province (Figure 1). Sims and Peterman [1986] have summarized current petrographic and geophysical data for the northern part of this region, which they refer to as the Central Plains orogen. However, because there are probably several different orogenic events represented in this area, we prefer the more general designation Central Plains Province. The Precambrian subcrop in this province can be characterized by Lower Proterozoic host terranes with a variety of Middle Proterozoic intrusive bodies. Because of the lack of extensive subsurface control, however, it is not yet possible to determine the detailed crustal structure. Van Schmus and Bickford [1981] inferred that the region was probably underlain mostly by eastward extensions of the Colorado Province and the Southwestern Province. New data that are reported here allow some limits to be set and some further questions to be raised regarding correlation of the Central Plains Province to other terranes.

Perhaps the easiest results to interpret are the samples with ages of 1620 to 1650 Ma in northeastern Kansas and in Missouri (Figure 1). These ages indicate that the 1650 Ma terrane of southern New Mexico extends eastward across the Texas Panhandle and Oklahoma into southeastern Kansas and into Missouri. Most of the rocks from which such ages have been obtained are granitic, but one sample from Osage Co., Missouri is rhyolite [Bickford et al., 1981a]. The southern extent of this terrane is unknown because it is apparently buried beneath younger, 1300 to 1500 Ma old rhyolite and epizonal granite; however, petrographic studies of subsurface rocks in Texas [Denison et al., 1984] and Nd isotopic studies indicate it may extend far to the south, as indicated in Figure 9. We have, at present, insufficient data to define exactly the northern limit of 1650 Ma old rocks. As noted earlier, plutons of this age are known to intrude older rocks of the Colorado Province; Premo [1984], for example, has reported ca. 1630 Ma old ages for migmatitic rocks in southern Wyoming, where the dominant crustal ages are 1750 to 1780 Ma. Similar relationships occur in the Central Plains Province: sample NB-CS-2 from southwestern Nebraska is a migmatitic gneiss with an age of 1640 Ma, but occurs considerably north of granite from southwestern Kansas (KS-ST-1) that yields an age of 1716 Ma.

Interpretation of the northern, presumably older, portion of the Central Plains Province is more ambiguous, partly because of the lack of samples yielding ages in the 1700 to 1800 Ma range, and partly because some of the ages that have been obtained do not correlate well with either the Penokean Province (1830 to 1900 Ma) or the Colorado Province (1700 to 1780 Ma). For example, samples NB-BF-1 and NB-DA-2 from central Nebraska yield ages of 1787 ± 10 and 1802 ± 5 Ma, respectively, intermediate in age between the two possibly correlative regions. Sample IA-OS-1 from northern Iowa, mentioned above, also yields an intermediate age of 1804 ± 17 Ma. On the other hand, two samples from Kansas, KS-ST-1 (1716 ± 29 Ma) from the southwest and KS-NM-21D (ca. 1780 Ma) from the northeast, are consistent with correlation of these units with the Colorado Province. Sample KS-KE-1 (1663 ± 6 Ma), also from southwest Kansas, is similar in age to younger Lower Proterozoic plutons found in central Colorado [Bickford and Boardman, 1984] that may represent northerly intrusive activity from the Southwestern Province.

Although there is good evidence that the Colorado Province extends into western Kansas, we believe that it is best to avoid correlating the remainder of the northern part of the Central Plains Province with either the Colorado Province or the Penokean Province. It is generally 1700 to 1800 Ma old, but parts of it may correlate with one or another of the older terranes, or it may include substantial amounts of early Proterozoic crust formed independently of and at slightly different times than either of the other terranes. On the other hand, the entire older, northern portion of the early Proterozoic crust in the central United States may be part of a large, complex, time-transgressive orogen that extends from the Great Lakes area to the southwestern United States. If this were the case, there might be several smaller accretionary terranes included in the belt, by analogy with the western margin of North America [Coney et al., 1980; Ben-Avraham et al., 1981], but lack of exposure and data preclude their definition.

The boundary between the 1700 to 1800 Ma and the 1600 to 1700 Ma terranes in the Central Plains Province is still difficult to locate precisely. Because of the tendency for plutons and metamorphic effects of the younger orogen to occur to the north in the older suite, the best approach is probably to use the southernmost extent of units older than 1700 Ma for defining the boundary. There are relatively few samples with zircon ages in this range, and Sm-Nd data does not have the precision necessary to distinguish between ca. 1.75 Ga and 1.65 Ga source rocks for younger, crustally-derived plutons. Our best interpretation still follows that inferred earlier [Van Schmus and Bickford, 1981] and is consistent with the new data: an approximate boundary between a terrane dominated by 1650 Ma old units and one dominated by 1700 to 1800 Ma old units extends eastward from central New Mexico to the northeastern corner of Kansas, but the extension further to the northeast is poorly defined. The presence of extensive activity associated with the younger belt to the north in the older belt indicates that, if the boundary is a suture, it probably developed from

a north-dipping subduction zone. Because of these uncertainties, the geologic map in Figure 1 shows the Central Plains Province as an undivided early Proterozoic orogenic terrane; however, the southern and southeastern portions are younger (1630 to 1680 Ma) than the northern portions (1700 to 1800 Ma).

We have also examined the available geophysical data, such as shown in Figures 2 and 3, to try to resolve questions regarding correlation of the early Proterozoic terranes or structure within the terranes. There are many recognizeable geophysical anomalies, sets of anomalies, or lineaments present in the data. For example, Zietz [1981] described the major NW-trending fabric that can be seen in the aeromagnetic data of Missouri and Nebraska (Figure 2), and Yarger [1985] has seen similar trends in the aeromagnetic data for Kansas. Arvidson et al. [1984] have shown graphically that the gravity data reflect this same NW-SE trend of anomalies (see also Figure 3), and Kisvarsanyi [1984] has shown that there are several major structural features in Missouri with this trend. These are major features in geophysical maps of the continental interior [Dutch, 1983], but we have not yet been able to find any definitive correlation between the ages, lithologies, or structures of subsurface samples and the positions of these linear trends in northern Missouri, northern Kansas, or Nebraska.

There is one interesting trend in Figure 3 that might represent one of the major terrane boundaries. A series of small gravity lows extends from northeastern New Mexico to the northern part of lower Michigan, approximately along the southern limit of units older than 1700 Ma; it is conceivable that this trend represents a plutonic arc associated with a 1.65 Ga old subduction zone. Unfortunately, the aeromagnetic data do not confirm this trend, so that final definition of this boundary will have to await future studies.

There are many small, intense magnetic highs that are scattered throughout the Central Plains Province (Figure 8). Several of these have been penetrated by drill holes [Bickford et al., 1981a; Steeples and Bickford, 1981; Hoppe et al., 1983; Tables 1 and 2], and they have generally turned out to be magnetite-bearing granites that are part of the Middle Proterozoic anorogenic suite (below). In eastern Kansas and western Missouri many of them are about 1370 Ma old [Figure 8; Thomas et al, 1984; Yarger, 1985], but elsewhere they are typically about 1470 Ma old. We believe that such plutons are ubiquitous throughout the Central Plains Province, as they are in the Colorado Province, Penokean Province, and Southwestern Province [Silver et al., 1977b], and have not shown them individually. These plutons have also apparently induced significant local, perhaps regional, metamorphic effects as shown by the series of zircon samples that have

yielded ages in the range 1550 to 1600 Ma, as
discussed above in the "Results" section.

Middle Proterozoic

*Granite-rhyolite terranes of the southern
midcontinent.* The southern midcontinent is
mostly underlain by silicic volcanic rocks,
primarily rhyolite and dacite, and shallow
plutons of similar composition. Rocks of this
type are found in the basement subcrop from
western Ohio across Indiana, Illinois, and
southern Missouri into southern Kansas, most of
Oklahoma, and into the Panhandle region of Texas.
Most of the silicic volcanic rocks have features
suggesting pyroclastic origin, and the associated
plutons are typically fine-grained, pink to red
because of disseminated Fe oxides, and
characterized by granophyric to micrographic
texture. They are clearly subvolcanic plutons
related to the volcanic rocks which they
presumably intrude. Rocks of intermediate
composition, sedimentary rocks, and metamorphic
rocks are notably rare; mafic igneous rocks are
present, but volumetrically they are greatly
subordinate to felsic rocks. Although the
igneous rocks of this suite are commonly altered,
metamorphism beyond chlorite zone of greenschist
facies is not observed, and none of these rocks
has been penetratively deformed.

Rocks of this extensive basement region are
exposed only in three localities: (1) the St.
Francois Mountains of southeastern Missouri,
where about 900 sq km are underlain by granite,
rhyolite, and minor basaltic dikes and sills; (2)
the Spavinaw Creek locality in Mayes County,
northeastern Oklahoma, where only a few sq km of
granophyre are exposed; and (3) the Eastern
Arbuckle Mountains in southcentral Oklahoma,
where deformation during the Paleozoic Ouachita
Orogeny and subsequent erosion have exposed
Proterozoic basement in the core of the
Tishomingo-Belton Anticline [Denison, 1973].

The exposures in the St. Francois Mountains
have been studied by Bickford et al.[1981b],
Sides et al.[1981], and Cullers et al. [1981],
whereas the subsurface distribution of rocks in
the St. Francois Mountains area has been studied
by Kisvarsanyi [1981]. The region is
characterized by thick sequences of rhyolitic to
dacitic rocks, most of ash-flow origin , and
epizonal granitic plutons. Field relationships
suggest that many of the plutons are subvolcanic
and represent deeply eroded cauldrons; features
of the volcanic sequences suggest remnants of
calderas in several areas [Sides et al., 1981].
Kisvarsanyi [1981] has identified numerous ring-
like features involving plutonic rocks in the
subsurface and has interpreted these as ring
intrusions. The igneous rocks of the St. Francois
Mountains area are chemically evolved [Bickford
et al., 1981b, Cullers, et al., 1981].
Kisvarsanyi [1981], citing chemical data from
Viets et al. [1978], has referred to many of the
ring intrusives as "tin granites" and called
attention to similarities of these rocks with the
Mesozoic ring complexes of northern Nigeria
[Bowden and Turner, 1974]. As the only extensive
exposure of a major volcano-plutonic terrane, the
St. Francois Mountains must serve as the model
for the buried portion.

Geochronological study has been most
important in study of these mostly-buried rocks.
U-Pb ages of zircons from the St. Francois
Mountains [Bickford and Mose, 1975] are 1480 Ma
for most rocks studied, but one small pluton
(Munger Granite-Porphyry) yielded an age of 1380
Ma. So far, all of the samples we have dated
from the eastern midcontinent (central Missouri
Illinois, Kentucky, Indiana, Ohio) have yielded
ages within +/-30 Ma of 1480 Ma [Hoppe et al.,
1983]. Petrographic data [Denison et al., 1984]
suggest that most of the region east of the
Mississippi River is underlain by a nearly
continuous subcrop of such rocks, and we refer to
this region as the Eastern Granite-Rhyolite
Province (Figure 1). The limits of this terrane
in SE Missouri are well defined by subsurface
samples [Kisvarsanyi, 1979], but to the north
subsurface sample control is poor. However,
magnetic data show a broad band of positive
anomalies extending from SE Iowa across NW
Illinois and SE Wisconsin into Michigan (Figure
2). Yaghubpur [1979] interpreted these anomalies
in Iowa as a suite of granitic plutons, and
zircon ages from this region indicate that these
plutons are about 1450 to 1500 Ma old. Coates et
al. [1983] and Hoppe et al. [1983] placed the
northwestern limit of the Eastern Granite-
Rhyolite Province along the NW edge of this band
of anomalies, and we concur with this.

We have also studied a number of apparently
isolated plutons to the north and west, within
the older crust, that also yield zircon ages of
about 1480 Ma; these include the Wolf River
batholith of central Wisconsin [Van Schmus et
al., 1975], the subsurface Red Willow batholith
of SW Nebraska [Van Schmus, unpublished; Lidiak,
1972], and numerous bodies penetrated by drilling
in Missouri and Kansas [Bickford et al., 1981a].
These plutons are part of a broad terrane of so-
called "anorogenic" granitic rocks that extend
from Labrador to southern California and also
include anorthosite in the eastern part of the
continent [Silver et al., 1977b; Emslie, 1978;
Anderson, 1983].

The western extension of the granite-rhyolite
terrane (southwestern Missouri, southern Kansas,
most of Okhahoma, and panhandle Texas) consists
of petrographically similar rocks except for the
more coarse-grained to gneissoid plutons exposed
in the Eastern Arbuckle Mountains of south-
central Oklahoma. The rocks of this Western
Granite-Rhyolite Province (Figure 1), however,
were formed about 100 Ma later than the similar
rocks of the Eastern Granite-Rhyolite Province.
Zircon ages from this region [Bickford and Lewis,
1979; Bickford et al., 1981a; Thomas et al.,

1984] are consistently in the range 1340 to 1400 Ma. As noted earlier, the Sm-Nd data of Nelson and DePaolo [1985] indicate mantle separation ages of about 1.8 Ga for rocks throughout the granite-rhyolite terranes, thus supporting the idea that the granite and rhyolite occurs as a veneer, derived from and lying upon, older crust (Figure 9).

The northern boundary of the Western Granite-Rhyolite Province is well defined by aeromagentic data [Yarger, 1985; Zietz, 1982] as shown in Figure 2, but the western edge of it is not as well defined by either geophysical data or petrography of subsurface samples. The eastern boundary of the continuous granite-rhyolite veneer is defined by subsurface samples [Kisvarsanyi, 1979; Denison et al., 1984] and coincides with the magnetic boundary extending eastward from Kansas. The southern extent is not well defined, but this province is presumably terminated by the northern edge of the Llano Province (Figure 1).

Recently, a major pluton in the Wet Mountains of south-central Colorado, the San Isabel batholith, was dated at 1360 Ma [Thomas et al., 1984], indicating that the Western Granite-Rhyolite Province may have once extended at least that far west. The San Isabel batholith is presumably a deeper-seated magmatic body formed during this event and was exposed by Cretaceous and later uplift and erosion. Since it is demonstrably intrusive into crust of the Colorado Province that is at least 1.7 Ga old, it seems likely that similar, more deep-seated plutons, underlie the rhyolitic sheets and subvolcanic plutons of the continental interior, where they were probably emplaced within older crust of the eastward extension of the Colorado Province or, farther south, of the eastward extension of the 1650 Ma province (Figure 9). The presence of the 1380 Ma old Munger Granite Porphyry as an intrusion in the 1480 Ma old rocks of the St. Francois Mountains indicates that rocks of the younger suite extend at least as far east as southeastern Missouri. None are known to the east of the St. Francois Mountains, however.

Origin of the granite-rhyolite provinces. The presence of large volumes of high silica rocks with minor volumes of mafic rocks in a region underlain by older continental crust is strongly suggestive of either crustal anatexis and the formation of a bimodal basalt-rhyolite suite in a continental extensional environment (cratonic rifting or back-arc spreading), as in the Cenozoic Basin and Range Province of the western United States [Eaton, 1982], or formation of felsic magmas by fractional crystallization in a continental arc, such as in the Sierra Madre Occidental of western Mexico [Cameron et al., 1980]. Sm-Nd [Nelson and DePaolo] and trace element [Cullers et al., 1981] data are more consistent with models involving crustal anatexis, so that we consider fractional crystallization in a continental arc setting the

least attractive of the possibilities. One major problem in either case is the lack of intermediate to mafic rocks in the granite-rhyolite provinces. This problem can be circumvented in an extensional model by postulating that the mafic, mantle-derived magmas did not rise all the way to the surface, but ponded in the 1.6 to 1.8 Ga old lower crust and provided the heat source for partially melting that crust to form the felsic magmas.

The widespread granite-rhyolite terranes of the midcontinent thus likely indicate later extension over a very large region following the accretion of juvenile arc terranes between 1.8 and 1.6 Ga ago [Thomas et al., 1984]. The age difference between the eastern and western granite-rhyolite terranes shows that the rifting was not simultaneous, but occurred during two discrete periods, 1480-1440 Ma ago and 1400-1340 Ma ago. The older period apparently affected the entire Lower Proterozoic continental margin belt, but the later period was confined to the south-central U.S. and may overlap the older one locally. The southern limit of the pre-1600 Ma continent 1300 to 1500 Ma ago is not known. There could have been a continental margin farther to the south, or there could have been a much larger continent that was undergoing extension and which finally separated into two smaller masses, creating a new continental margin some time after 1300 Ma.

We do not consider continent-continent collision as a viable alternative model for formation of the granite-rhyolite provinces, since such an event would have produced extensive highlands or an extensive mountain chain, and subsequent erosion would have removed most, if not all, of the volcanic and epizonal plutonic rocks. The high degree of preservation over a broad, low elevation area is more consistent with an extensional environment than a collisional environment.

Midcontinent Rift System. One of the most prominent features on gravity and aeromagnetic maps of the United States is a series of major, generally linear anomalies that extend from central Kansas to Lake Superior [King and Zietz, 1971] and then turn southward into central Michigan (Figures 2 and 3). This system of anomalies clearly reflects geological features associated with a major event in the history of the North American continent. Although initially referred to descriptively as "midcontinent geophysical anomaly" or "midcontinent gravity high", this system is now commonly referred to as "Midcontinent Rift System" [Wold and Hinze, 1982; Van Schmus and Hinze, 1985].

Correlation of geophysical data with geology in the Lake Superior region has shown that the Midcontinent Rift System is associated with Keweenawan igneous activity 1100 to 1200 Ma ago. Several drill holes into geophysical anomalies in the western segment in Iowa, Nebraska, and Kansas have encountered mafic volcanic rocks, associated plutonic rocks, or clastic sedimentary rocks similar to those associated with the Keweenawan suite in the Lake Superior region. The system can be approximated as a long rift valley (or extensional basin) that was filled by basalts, sub-volcanic plutons, and post-volcanic sedimentary rocks. In the Lake Superior region the structure has been interpreted to show that the central basin fill has been subsequently uplifted to form a central horst complex, but Serpa et al. [1984] have interpreted seismic data in north-central Kansas to indicate that the structure simply consists of extensional normal faulting without a central uplift.

The bends and offsets in the rift system were probably influenced by pre-existing structures in the older Proterozoic basement [Klasner et al., 1982], but our understanding of the buried basement is insufficient to allow detailed correlations to be drawn. For example, it is not known whether the offset at the Kansas-Nebraksa border occurred along a pre-existing structure, was caused by later cross-faulting, or represents reactivation of an older structure.

Grenville-Llano Provinces. These terranes are mostly buried in the United States and include the southern portion of the Grenville Province and the rocks of the Llano Province that are exposed in the Llano Uplift of central Texas. Rocks of the Grenville Province make up the basement of North America east of the subsurface extension of the Grenville Front, a tectonic and metamorphic-grade boundary separating rocks of the Grenville Province from Archean rocks of the Superior Province where these rocks are exposed in Ontario and Quebec. Rocks in the Grenville Province of Canada and in the Adirondack Mountains region of New York are typically in granulite facies and yield ages ranging from about 1200 to 1000 Ma, although strongly metamorphosed rocks with greater primary ages are known, especially along the Grenville Front in Ontario [Krogh and Davis, 1969; 1970]. Rocks of Grenville type and age, high-grade metamorphics and related igneous rocks, occur in the subsurface as far west as eastern Ohio [Bass, 1960] and Kentucky [Denison et al., 1984] and in rare surface exposures in the Piedmont of the Appalachian Orogen [Denison et al., 1984], but they have not yet been observed to the west except in the Llano Uplift of central Texas, where similar rocks occur. Rocks in the Llano Uplift are mostly in amphibolite facies but yield ages in the same range.

The Grenville Province extends southwards into the subsurface near the west end of Lake Erie, but its exact western limit, the Grenville Front, is not well defined. The subsurface extension of the Grenville Province is marked by a series of prominent north-northeast trending gravity and magnetic anomalies (Figures 2 and 3) that transect northwesterly-trending, apparently older geophysical trends [Denison et al., 1984; Hinze et al., 1975; Zietz et al., 1966].

Subsurface samples in this zone include various schists and gneisses, marble and calc-silicate rocks, and two-feldspar granite. These rocks are similar to the granulite facies metasedimentary rocks and charnockitic granites typical of Grenville terranes to the north, but they are in amphibolite facies. Age determinations that are available are mostly Rb-Sr measurements on micas [Bass, 1960], and these yield ages in the 800 to 1000 Ma range. These ages are probably not indicative of the time of major metamorphism [about 1100 Ma; Silver, 1969] but they are similar to values obtained from micas separated from exposed rocks. In contrast to the situation to the north in Ontario, the subsurface Grenville Front separates the typical metamorphic-igneous suite from the silicic volcanic, epizonal granitic, and minor basaltic rocks of the 1450-1480 Ma-old Eastern Granite-Rhyolite Province. However, subsurface lithologic control is insufficient to locate the Grenville Front precisely. The location shown on Figure 1 is based on the age boundary defined by Bass [1960], but it can be seen from Figures 2 and 3 that the change in structural trends occurs somewhat farther east. Therefore, the actual crustal structural boundary may be east of that shown, an interpretation preferred by one of us (I.Z.).

Rocks of the Llano Province underlie a very large part of Texas and are exposed in the Llano Uplift in the central portion of the state. The rocks are both igneous and metamorphic and are distinctive, in both age and petrography, from the granite-rhyolite suites of the 1350-1400 Ma-old Western Granite-Rhyolite Province that underlies large areas to the north. Llano Province rocks are best known from the exposed area where three major units have been studied. These include the Valley Spring Gneiss, comprising more than 3.8 km of metarhyolite and meta-arkose; the Packsaddle Schist, more than 7 km thick and consisting of hornblende, biotite, actinolite, and muscovite schist, with marble and graphitic schists in its lower parts; and a series of massive granitic plutons. Garrison [1981] has suggested that the protolith of the Packsaddle Schist was an accumulation of shallow-water slope and shelf sediments, rich in organic matter, that were interbedded with mafic and felsic volcaniclastic rocks.

The Valley Spring Gneiss and the Packsaddle Schist have yielded ages of about 1170 Ma by Rb-Sr methods [Garrison et al., 1979]; these major metasedimentary and metavolcanic units were intruded by the massive granitic plutons, such as the Town Mountain Granite, about 1060 Ma ago [Zartman, 1964]. Garrison [1981] has also described a large serpentinite mass, the Coal Creek Serpentinite, within the uppermost part of the Packsaddle Schist, interpreting it as a portion of an ophiolite. If the serpentinite is indeed ophiolitic, its presence would be consistent with the suggestion of Garrison et al. [1979] that the Packsaddle schist, as well as various other meta-igneous rocks, is part of an island arc-plutonic complex.

The origin of the Grenville and Llano Provinces and the relationship of them to the approximately coeval Midcontinent Rift System have not been well explained. It appears likely that some form of continent-continent collisional event is involved, and this would imply that the cratonic margin of the pre-1200 Ma continent may have existed to the south and east of the present north and west limits of these provinces. However, we can not rule out the possibility that the pre-1200 Ma continental margin was formed by rifting of a much larger continental mass; if this occurred, it might have been the final manifestation of the extensional regime that produced the granite-rhyolite provinces.

Acknowledgments. We would like to give thanks to the personnel of the several petroleum and mineral exploration companies and state geological surveys who have given invaluable assistance in sampling existing cores and cuttings, who have donated drilling time for obtaining samples, or who have cooperated by allowing us to use their drill rigs and crews to obtain samples once they reached basement. Without their interest and assistance, this project could not have been done and could not continue. There are too many for all to be listed, but special recognition goes to John Deery and Ed Price (AMOCO Production Co.), Greg Williams (Petroleum Management Inc.), Webster Stickney (Houston International Minerals Co.), Tim Pierce (Hummon Oil Co.), Steve Hauk (Union Carbide), Ray Anderson (Iowa Geological Survey), Lynn Watney (Kansas Geological Survey), Eva Kisvarsanyi (Missouri Geological Survey), Marvin Carlson (Nebraska Geological Survey), and Sam Treves (Univ. Nebraska). We have been ably assisted in analytical work and sample processing by Lanny Latham, Rick Newill, Melisa Wardlaw, Wendel Hoppe, and several undergraduate assistants who have laboriously purified cuttings and extracted zircons. The work at Kansas is currently supported by NSF Grant EAR 82-19137. We also acknowledge past funding from the University of Kansas General Research Fund, NSF, Nuclear Regulatory Commission, and the Kansas Geological Survey. The use of the digital version of the aeromagnetic map was graciously permitted by The Phoenix Corp. T. Hildenbrand of the U.S. Geological Survey graciously had the black-and-white version of the gravity map in Figure 3 prepared for us.

References

Anderson, J. L., Proterozoic anorogenic granite plutonism of North America. Geol. Soc. Amer., Memoir, 161, 133-154, 1983.

Anderson, J. L., R. L. Cullers, and W. R. Van Schmus, Anorogenic metaluminous and peraluminous granite plutonism in the mid-Proterozoic of Wisconsin, USA. Contrib. Mineral. Petrol., 74, 311-328, 1980.

Anderson, R. R., and R. A. Black, Early Proterozoic development of the southern Archean boundary of the Superior Province in the Lake Superior region (abstract), Geol. Soc. Amer. Abstr. with Programs, 15, 515, 1983.

Arvidson, R. E., D. Bindschadler, S. Bowring, M. Eddy, E. Guinness, and C. Leff, Bouguer images of the North American craton and its structural evolution, Nature, 311, 241-243, 1984.

Bass, M. N., Grenville boundary in Ohio. J. Geol., 68, 673-677, 1960.

Ben-Avraham, Z., A. Nur, D. Jones, and A. Cox, Continental accretion: from oceanic plateaus to allocthonous terranes. Science, 213, 47-54, 1981.

Bickford, M. E., and S. J. Boardman, A Proterozoic volcano-plutonic terrane, Gunnison and Salida areas, Colorado. J. Geol., 92, 657-666, 1984.

Bickford, M. E., and R. D. Lewis, U-Pb geochronology of exposed basement rocks in Oklahoma, Geol. Soc. Amer. Bull., 90, 540-544, 1979.

Bickford, M. E., and D. G. Mose, Geochronology of Precambrian rocks in the St. Francois Mountains, southeastern Missouri. Geol. Soc. Amer. Spec. Pap. 165, 48 pp., 1975.

Bickford, M. E., and W. R. Van Schmus, Discovery of two Proterozoic granite-rhyolite terranes in the buried midcontinent basement: The case for shallow drill holes, Proceedings of the First International Symposium on Continental Drilling, Springer-Verlag, 1985 (in press).

Bickford, M. E., K. L. Harrower, W. J. Hoppe, B. K. Nelson, R. L. Nusbaum, and J. J. Thomas, Rb-Sr and U-Pb geochronology and distribution of rock types in the Precambrian basement of Missouri and Kansas. Geol. Soc. Amer. Bull., Part 1, 92, 323-341, 1981a.

Bickford, M. E., J. R. Sides, and R. L. Cullers, Chemical evolution of magmas in the Proterozoic terrane of the St. Francois Mountains, southeastern Missouri 1. Field, petrographic, and major element data. Jour. Geophys. Res., 86, 10365-10386, 1981b.

Bickford, M. E., R. L. Cullers, and W. R. Van Schmus, U-Pb zircon chronology of early and middle Proterozoic igneous events in the Gunnison, Salida, and Wet Mountains areas, Colorado (abstr.). Geol. Soc. Amer. Abstr. Prog., 16, 215, 1984.

Bowden, P. and D. C. Turner, Peralkaline and associated ring complexes in the Nigeria-Niger province, west Africa, in The alkaline rocks, edited by H. Sorensen, New York, John Wiley and Sons, 330-351, 1974.

Bowring, S. A., W. R. Van Schmus, and P. F. Hoffman, U-Pb zircon ages from Athapuscow aulacogen, East Arm of Great Slave Lake, N.W.T., Canada. Can. J. Earth Sci., 21, 1315-1324, 1984.

Burchett, R. R., Wilson, F. W., Anderson, R. R., and Satterfield, I.R., Magnetic map of the

Forest City Basin and adjacent regions of lava, Kansas, Missouri, and Nebraska. Neb. Geol. Surv., sale 1:500,000, 1983.

Cameron, K. L., M. Cameron, W. C. Bagby, E. J. Moll, R. E. Drake, Petrologic characteristics of mid-Tertiary volcanic suites, Chihuahua, Mexico, Geology, 8, 87-91, 1980.

Camfield, P. A., and D. I. Gough, A possible Proterozoic plate boundary in North America, Can. J. Earth Sci., 14, 1229-1238, 1977.

Coates, M. S., B. C. Haimson, W. J. Hinze, and W. R. Van Schmus, Introduction to the Illinois deep hole project. J. Geophys. Res., 88, 7267-7275, 1983.

Condie, K. C., The Wyoming Archean Province in the western United States, in The Early History of the Earth, edited by B. F. Windley, John Wiley and Sons, New York, 499-510, 1976.

Condie, K. C., Plate-tectonics model for Proterozoic continetal accretion in the southwestern United States. Geology, 10, 37-42, 1982.

Condie, K. C., and S. A. Bowring, Early Proterozoic supracrustal associations in the southwest: an update (abstract). Geol. Soc. Amer. Abstr. Prog., 16, 218, 1984.

Coney, P. J., D. L. Jones, and J. W. H. Monger, Cordilleran suspect terranes. Nature, 288, 329-333, 1980.

Cullers, R. L., R. J. Koch, and M. E. Bickford, Chemical evolution of magmas in the Proterozoic terrane of the St. Francois Mountains, southeastern Missouri 2. Trace element data. Jour. Geophys. Res., 86, 10388-10401, 1981.

Davis, D. W., Optimum linear regression and error estimation applied to U-Pb data. Can. J. Earth Sci., 19, 2141-2149, 1982.

Denison, R. E., Basement rocks in the Arbuckle Mountains, in Regional Geology of the Arbuckle Mountains, Oklahoma, Guidebook for Field Trip No. 5, Geol. Soc. Amer. Annual Mtng., Dallas, Tex., Oklahoma Geol. Sur., Norman, 43-49, 1973.

Denison, R. E., E. G. Lidiak, M. E. Bickford, and E. Kisvarsanyi, Geology and geochronology of Precambrian rocks in the central interior region of the United States. U.S. Geol. Survey, Prof. Paper 1241-C, 20pp, 1984.

DePaolo, D. J., Neodymium isotopes in the Colorado Front Range and crust-mantle evolution in the Proterozoic. Nature, 291, 193-196, 1981.

Dott, R. H., Jr., and I. W. D. Dalziel, Age and correlation of the Precambrian Baraboo quartzite of Wisconsin. J. Geol., 80, 552-568, 1974.

Dutch, S. I., Proterozoic structural provinces in the north-central United States. Geology, 8, 478-481, 1983.

Eaton, G. P., The Basin and Range province: origin and tectonic significance, in Ann. Rev. Earth Planet. Sci., edited by G. W. Wetherill, A. L. Albee, and F. G. Stehli, Annual Reviews, Inc., Palo Alto, 409-440, 1982.

Emslie, R. F., Anorthosite massifs, rapakivi granites, and late Proterozoic rifting of North America. Precamb. Res., 7, 61-98, 1978.

Garrison, J. R., L. E. Long, and D. L. Richmann, Rb-Sr and K-Ar geochronologic and isotopic studies, Llano Uplift, central Texas. Contr. Min. Pet., 69, 361-374, 1979.

Garrison, J. R., Jr., Coal Creek serpentinite, Llano Uplift, Texas:A fragment of an incomplete Precambrian ophiolite. Geology, 9, 225-230, 1981.

Goldich, S. S., and J. L. Wooden, Origin of the Morton Gneiss, southwestern Minnesota: Part 3. Geochronology. Geol. Soc. Amer., Spec. Paper 182, 77-94, 1980.

Green, A. G., Cumming, G. L., and D. Cedarwell, Extension of the Superior-Churchill boundary zone into southern Canada, Can. J. Earth Sci., 16, 1691-1701, 1979.

Green, A. G., W. Weber, and Z. Hajnal, Evolution of Proterozoic terrains beneath the Williston Basin. Geology, 13, 624-628, 1985.

Hildenbrand, T. G., R. W. Simpson, R. H. Godson, and M. F. Kane, Digital colored residual and regional Bouguer gravity maps of the conterminous United States with cut-off wavelengths of 250 km and 1000 km., Geophys. Inv. Map GP-953-A, U.S. Geol. Surv., 1:7,500,000, 2 sheets, 1982.

Hills, F. A., and R. S. Houston, Early Proterozoic tectonics of the central Rocky Mountains, North America. Contrib. Geology, University of Wyoming, 17, 89-109, 1979.

Hinze, W. J., R. L. Kellogg, and N. W. O'Hara, Geophysical studies of basement geology of southern peninsula of Michigan. Am. Assoc. Petrol. Geol. Bull., 59, p. 1562-1584, 1975.

Hoffman, P. F., and S. A. Bowring, A short-lived 1.9 Ga continental margin and its destruction, Wopmay Orogen, northwest Canada. Geology, 12, 68-72, 1984.

Hoppe, W. J., C. W. Montgomery, and W. R. Van Schmus, Age and significance of Precambrian basement basement samples from northern Illinois and adjacent states, J. Geophys. Res., 88, 7276-7286, 1983.

Houston, R. S., K. E. Karlstro, F. A. Hills, and S. B. Smithson, The Cheyenne Belt: a major Precambrian crustal boundary in the western United States (abstract). Geol. Soc. Amer. Abstr. Prog., 11, 446, 1979.

King, E. R., and I. Zietz, Aeromagnetic study of the midcontinent gravity high of central United States. Geol. Soc. Amer. Bull., 82, 2187-2208, 1971.

Kisvarsanyi, E. B., Geologic map of the Precambrian of Missouri. Contrib. Precamb. Geol. No. 7, Missouri Dept. Nat. Resources, Div. Geol. Land Surv., scale 1:1,000,000, 1979.

Kisvarsanyi, E. B., Geology of the Precambrian St. Francois terrane, southeastern Missouri. Contr. Precamb. Geol. No. 8, Rept. Inv. No. 64, Missouri Dept. Nat. Res., Div. Geol. Land Sur., 58 pp., 1981.

Kisvarsanyi, E. B., The Precambrian tectonic framework of Missouri as interpreted from the magnetic anomaly map. Contribution to Precambrian Geology, No. 14, Missouri Dept. Nat. Res., 19pp., 1984.

Klasner, J. S., W. F. Cannon, and W. R. Van Schmus, The pre-Keweenawan tectonic history of southern Canadian Shield and its influence in the formation of the Midcontinent Rift, in Geology and Tectonics of the Lake Superior Basin, edited by R. J. Wold and W. J. Hinze, Geol. Soc. Amer., Mem. 156, 27-46, 1982.

Krogh, T. E., A low contamination method for hydrothermal decomposition of zircon and extraction of U and Pb for isotopic age determinations. Geochim. Cosmochim. Acta, 37, 485-494, 1973.

Krogh, T. E., Improved accuracy of U-Pb zircon ages by the creation of more concordant systems using an air abrasion technique. Geochim. Cosmochim. Acta, 46, 637-649, 1982.

Krogh, T. E., and G. L. Davis, Geochronology of the Grenville Province. Carnegie Inst. Wash., Yrbk. 67, 224-230, 1969.

Krogh, T. E., and G. L. Davis, The age of metamorphism in the Grenville Province, and the age of the Grenville Front. Carnegie Inst. Wash., Yrbk. 68, 307-313, 1970.

LaBerge, G. L., P. E. Myers, and K. J. Schulz, Early Proterozoic plate tectonics: evidence from central Wisconsin (abstract). Geol. Soc. Amer. Abstr. Prog., 16, 567, 1984.

Lidiak, E. G., Precambrian rocks in the subsurface of Nebraska. Nebr. Geol. Surv. Bull., 26, 41pp., 1972.

Ludwig, K. R., Calculation of uncertainties of U-Pb isotopic age data. Earth Planet. Sci. Lett., 46, 212-220, 1980.

Ludwig, K. R., Plotting and regression programs for isotope geochemists, for use with HP-86/87 microcomputers. U. S. Geol. Surv., Open File Rept. 83-849, 94 pp., 1983.

Muhelberger, W. R., R. E. Denison, and E. G. Lidiak, Basement rocks in the continental interior of the United States, Am. Assoc. Petroleum Geol., 51, 2351-2380, 1967.

Nelson, B. K., and D. J. DePaolo, Rapid production of continental crust 1.7- 1.9 b.y. ago: Nd isotopic evidence from the basement of the North American midcontinent, Geol. Soc. Amer. Bull., 96, 746-754, 1985.

Peterman, Z. E., and R. E. Zartman, The early Proterozoic Trans-Hudson orogen in the northern Great Plains of the United States (abstract). Int. Basement Tect. Assoc. Abstr. with Prog., 6, 30, 1985.

Peterman, Z. E., S. S. Goldich, and R. E. Zartman, High-grade reworked Archean rocks in the basement of the Williston Basin, North Dakota (abstract). Geol. Soc. Amer., Abstracts with Programs, 15, 660, 1983.

Premo, W. R., U-Pb zircon geochronology of the Sierra Madre Range, Wyoming. M.S. Thesis, Univ. Kansas, Lawrence, 106 pp, 1984.

Serpa, L., T. Setzer, H. Farmer, L. Brown, J. Oliver, S. Kaufman, J. Sharp, and D. W. Steeples, Structure of the southern Keweenawan Rift from COCORP survey across the Midcontinent Geophysical Anomaly in northeastern Kansas, Tectonics, 3, 367-384, 1984.

Sides, J. R., M. E. Bickford, R. D. Shuster, and R. L. Nusbaum, Calderas in the Precambrian terrane of the St. Francois Mountains, southeastern Missouri. Jour. Geophys. Res. (Granite-Rhyolite Issue), 86, 10349-10364, 1981.

Silver, L. T., Precambrian batholiths of Arizona (abstract). Geol. Soc. Amer. Abstracts for 1968, Spec. Pap. 121, 558-559, 1968.

Silver, L. T., A geochronologic investigation of the anorthosite complex, Adirondack Mountains, New York, in Origin of anorthosite and related rocks, New York State Mus. and Sci. Ser., Mem. 18, 233-251, 1969.

Silver, L. T., Observations on the Precambrian evolution of northern New Mexico and adjacent regions (abstract). Geol. Soc. Amer. Abstr. Prog., 16, 256, 1984.

Silver, L. T., C. A . Anderson, M. Crittenden and J. M. Robertson, Chronostratigraphic elements of the Precambrian rocks of the southwestern and far western United States (abstract). Geol. Soc. Amer. Abstr. Prog., 9, 1176, 1977a.

Silver, L. T. M. E. Bickford, W. R. Van Schmus, J. L. Anderson, T. H. Anderson, and L. G. Medaris, The 1.4-1.5 b.y. transcontinental anorogenic perforation of North America (abstract). Geol. Soc. Amer. Abstr. Prog., 9, 1176, 1977b.

Sims, P. K., and Z. E. Peterman, Archaean rocks in the southern part of the Canadian Shield - A review. Spec. Pub. Geol. Soc. Australia, 7, 85-98, 1981.

Sims, P. K., and Z. E. Peterman, The early Proterozoic Central Plains orogen - a major buried structure in north-central United States. Submitted to Geology, 1986.

Smith, E. I., Precambrian rhyolites and granites in south-central Wisconsin. Geol. Soc. Amer. Bull., 89, 875-890, 1978.

Society of Exploration Geophysicists, Gravity anomaly map of the United States (exclusive of Alaska and Hawaii). Tulsa, Oklahoma, Soc. Explor. Geophysicists, 2 sheets, scale 1:2,500,000, 1982.

Stacey, J. S., and D. C. Hedlund, Lead-isotopic compositions of diverse igneous rocks and ore deposits from southwestern New Mexico and their implications for early Proterozoic crustal evolution in the western United States. Geol. Soc. Amer. Bull., 94, 43-57, 1983.

Stacey, J. S., and J. D. Kramers, Approximation of terrestrial lead isotope evolution by a two stage model. Earth Planet. Sci. Lett., 26, 207-221, 1975.

Steeples, D. W., and M. E. Bickford, Piggyback drilling in Kansas: an example for the continental scientific drilling program. Eos Trans. AGU, 62, 473-476, 1981.

Steiger, R. H., and E. Jager, Subcommission on geochronology: convention on the use of decay constants in geo- and cosmochronology. *Earth Planet. Sci. Lett., 28,* 359-362, 1977.

Thomas, J. J., R. D. Shuster, and M. E. Bickford, A terrane of 1350-1400 m.y. old silicic volcanic and plutonic rocks in the buried Proterozoic of the midcontinent and in the Wet Mountains, Colorado. *Geol. Soc. Amer. Bull., 95,* 1150-1157, 1984.

Treves, S. B., and D. J. Low, The Precambrian geology of Nebraska (abstract), *Geol. Assoc. Canada Program with Abstracts, 9,* 112, 1984.

Tvrdik, T. N., Petrology of the Precambrian basement rocks of the Matlock Project cores, northwestern Iowa, M.S. Thesis, Univ. Iowa, Iowa City, 113 pp., 1983.

Tweto, O., Precambrian geology of Colorado, in *Colorado Geology*, Rocky Mountain Assoc. of Geologists, 37-46, 1980.

Van Schmus, W. R., Chronology of igneous rocks associated with the Penokean orogeny in Wisconsin. *Geol. Soc. Amer. Spec. Paper 182,* 159-168, 1980.

Van Schmus, W. R., and J. L. Anderson, Gneiss and migmatite of Archean age in the Precambrian basement of central Wisconsin. *Geology, 5,* 45-48, 1977.

Van Schmus, W. R., and M. E. Bickford, Proterozoic chronology and evolution of the midcontinent region, North America, in *Precambrian Plate Tectonics,* edited by A. Kroner, Amsterdam, Elsevier Pub. Co., 261-296, 1981.

Van Schmus, W. R., and W. J. Hinze, The Midcontinent Rift System, *Ann. Rev. Earth Planet. Sci., 13,* 345-383, 1985.

Van Schmus, W. R., L. G. Medaris, and P. O. Banks, Geology and age of the Wolf River batholith, Wisconsin. *Geol. Soc. Amer. Bull., 86,* 907-914, 1975.

Van Schmus, W. R., M. E. Bickford, J. L. Lewry, and R. Macdonald, U-Pb geochronology of Precambrian rocks in the Trans-Hudson Orogen, northern Saskatchewan. *Can. J. Earth Sci.,* submitted, 1986.

Viets, J. G., E. L. Mosier, E. B. Kisvarsanyi, and S. K. McDanal, Spectrographic and chemical analyses of drill core from Precambrian igneous rocks of the St. Francois igneous province in southeast Missouri. *U.S. Geol. Sur. Open-File Report 78-402,* 12 pp., 1978.

Wold, R. J., and W. J. Hinze, Introduction, in *Geology and Tectonics of the Lake Superior Basin*, edited by R. J. Wold and W. J. Hinze, *Geol. Soc. Amer., Mem. 156,* 1-4, 1982.

Yaghubpur, A., Preliminary geological appraisal and economic aspects of the Precambrian basement of Iowa, PhD Thesis, Univ. Iowa, Iowa City, 294 pp., 1979.

Yarger, H. L., Kansas basement study using spectrally filtered aeromagnetic data, in The *Utility of Regional Gravity and Magnetic Anomaly Maps*, edited by W. J. Hinze, 213-323, Soc. Explor. Geoph., Tulsa, Oklahoma, 1985.

York, D., Least squares fitting of a straight line. *Can. J. Phys., 44,* 1079-1086, 1966.

York, D., Least squares fitting of a straight line with correlated errors. *Earth Planet. Sci. Lett., 5,* 320-324, 1969.

Zartman, R. E., A geochronologic study of the Lone Grove pluton from the Llano uplift, Texas. *J. Petrol., 5,* p. 359-408, 1964.

Zartman, R. E., Lead isotope provinces in the Cordillera of the western United States and their geologic significance. *Econ. Geol., 69,* 792-805, 1974.

Zartman, R. E., and T. W. Stern, Isotopic and geologic relationships of the Little Elk Granite, northern Black Hills, South Dakota. *U.S. Geol. Survey, Prof. Paper 575-D,* D157-D163, 1967.

Zietz, I., Aeromagnetic coverage of the midcontinent U.S.A. *Geol. Soc. Amer. Abstr. Prog., 13,* 588, 1981.

Zietz, I. (complier), Composite magnetic anomaly map of the United States, Part A — Conterminous United States, *Map GP954A,* U.S. Geol. Surv. 2 sheets, scale 1:2,500,000, 1982.

Zietz, I., E. R. King, W. Geddes, and E. G. Lidiak, Crustal study of a continental strip from the Atlantic Ocean to the Rocky Mountains. *Geol. Soc. Amer. Bull., 77,* 1427-1448, 1966.

M. E. Bickford and W. R. Van Schmus, Department of Geology, University of Kansas, Lawrence, Kansas 66045-2124.

I. Zietz, The Phoenix Corporation, 7921 Jones Branch Drive, McLean, Virginia 22102.

PROTEROZOIC TECTONIC ELEMENTS OF THE U.S. MAPPED BY COCORP DEEP SEISMIC PROFILING

Larry D. Brown

Institute for the Study of the Continents and Department of Geological Sciences
Cornell University, Ithaca, New York 14853

Abstract. Although seismic reflection profiling is not a traditional tool in Precambrian geology, COCORP and similar programs have obtained unique new information concerning the structure of Proterozoic crust using this technique. An extensive buried basin in the southern midcontinent, an asymmetric rift in northeastern Kansas, bimodal graben stratigraphy in central Michigan, a distinctly layered lower crustal wedge - perhaps highly conductive - beneath the Adirondacks, and an Archaean-Proterozoic suture in Wyoming are among the Proterozoic features delineated by COCORP surveys. Some of these features are quite prominent. Example: the Proterozoic basin in southern Oklahoma and northern Texas is inferred from a strong, persistent layering up to 10 km thick which underlies the Paleozoic cover. This sequence, which has been interpreted to correspond to clastic sedimentary and felsic volcanic rocks extending over an area of 2500 km^2 or more, is a striking new discovery from seismic profiling. Although the buried Proterozoic features in the U.S. often lack adequate outcrop or drill control, convincing correlations are possible. Example: the well-defined bimodal seismic stratigraphy found in buried Precambrian grabens of Michigan and Kansas suggests correspondence with the clastic Upper Keweenawan and volcanic Middle Keweenawan known from distant outcrops of these late Proterozoic sequences. Not only are distinct unconformities recognizable within the graben fill, the gross geometry of this major intracontinental rift is shown to be decidedly asymmetric. There is evidence from the seismic sections that some of these Proterozoic structures have been reactivated by Phanerozoic tectonics. Example: the deep layering of southern Oklahoma is abruptly disrupted along the southwest flank of the Wichita Uplift by apparent Precambrian normal faults that were active during the late Pennsylvanian with a reverse sense of motion. Interpretations of some of these seismic results are still largely speculative. COCORP profiles across the Grenville of the Adirondack dome reveal a lower crustal wedge of layered reflections which corresponds in depth with a high conductivity zone. Interpretations of this wedge range from its representing a layered igneous intrusion to the remnants of sediments underthrust during the Grenville orogeny. It has also been speculated that steeply-dipping reflections on COCORP profiles across a Proterozoic-Archaean terrane boundary in southeastern Wyoming mark a suture zone between crusts of differing reflection character and thicknesses as well as age. Although seismic surveys of Precambrian crust are still too few for confident generalizations, there is a growing perception that at least some Proterozoic crust exhibits a distinctive seismic reflection "signature", including a Moho that is less reflective than the crust-mantle boundary beneath peripheral Phanerozoic orogenic belts. These initial samplings suggest that the Proterozoic craton of the U.S., in spite of being masked by Phanerozoic cover, is a most exciting frontier for future seismic profiling.

Introduction

Multichannel seismic reflection profiling is usually associated with exploration for oil and gas in Phanerozoic sedimentary basins. However, the Consortium for Continental Reflection Profiling (COCORP) in the U.S. and similar programs in other countries (Oliver et al., 1986) have found reflection profiling to be extremely useful in probing crystalline and Precambrian terranes as well. Since most of the craton of the United States is Proterozoic in age, it is not surprising that some of COCORP's over 8000 km of reflection surveys (Figure 1) address late Precambrian structures and evolution.

COCORP's application of the multichannel reflection technique is very similar to that employed in the oil exploration industry (Brown, 1986). Data are collected by a professional contractor and subsequently processed by COCORP staff at Cornell University. Unlike oil explorationists, COCORP focusses on imaging structure in the continental basement. COCORP is particularly interested in deep crustal structure, and reflections from Moho depths are commonly recorded.

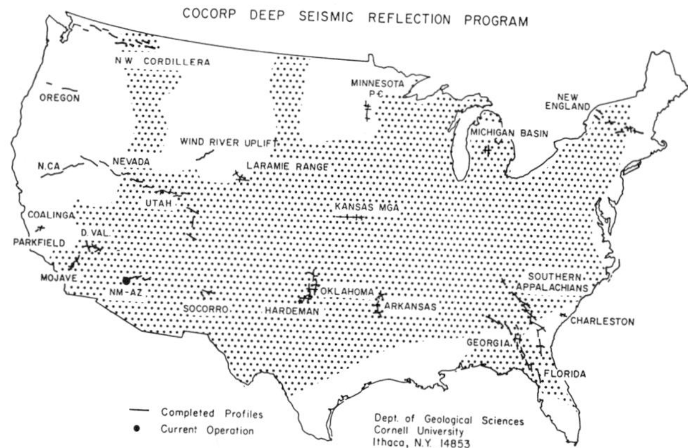

Fig. 1. Map showing location of COCORP deep seismic profiles. Stippled pattern represents inferred extent of Proterozoic basement (after Gibbs, 1986; Bickford et al., 1986).

Deep reflection profiling has proven especially effective in delineating crustal structure in Phanerozoic orogenic belts (e.g. Brown et al., 1986), especially when deep structures can be traced to outcrop or drillhole (Oliver, 1982). Its application to Precambrian problems has been more limited, partly because much of the Precambrian of the U.S. is buried beneath Paleozoic cover thus masking potential geologic control.

In spite of COCORP's past emphasis on younger structures, some of its most intriguing results have come from exposed and buried Precambrian terranes. Furthermore, it is sometimes difficult to differentiate reflections marking Phanerozoic structure from those generated by relict Precambrian features caught up in later deformations. This report briefly reviews some of COCORP's results which bear most directly on Proterozoic tectonics, including evidence of mid-crustal complexity in the central craton, a buried Proterozoic basin south of the Wichita Uplift, asymmetric grabens and bimodal stratigraphy of the Keweenawan rift, deep layering in the Grenville- age crust of the Adirondack Mountains, and a possible Proterozoic suture between disparate crustal blocks in southeastern Wyoming. These efforts are far from a complete sampling of the Proterozoic of the U.S., nor are they free from interpretational ambiguity. However, they clearly demonstrate how the seismic reflection technique offers novel and unique contributions to our understanding of variations in Proterozoic crust.

Complexity of the Proterozoic Basement

COCORP's first field surveys, in the Hardeman Basin of north Texas near the border with Oklahoma (Figure 2), traverse basement at least 1300 m.y. old (Denison et al., 1984). Although intended primarily as a test of the reflection technique as a deep exploration tool, these

relatively short seismic surveys remain a classic depiction of the heterogeneity of the deep continental crust (Oliver et al., 1976). Prominent, layered reflections in the upper crust cap a patchwork of short reflection segments, arcuate diffractions and zones of relative transparency that extend to the base of the sections (Figure 3). The heterogeneity implied by these patterns must be on the scale of hundreds of meters or less, and stands in contrast to the oversimplified "layer-cake" models of continental crust instilled by classical refraction surveys (Oliver and Kaufman, 1977).

Later and more extensive surveys in NE Kansas confirm a similar reflection character for the deep basement. Though the layered upper crustal sequence is missing further north, transparent zones and bands of linear and arcuate reflections are abundant (Brown et al., 1983a; Figure 4). In particular, the Kansas results show that beneath a thin, flat-lying Paleozoic cover is a shallow crust that is essentially transparent. Since reflections from greater depths are clear, this lack of reflections at shallow levels is not an artifact of inadequate signal penetration. The deeper reflections are frequently hyperbolic, suggesting diffraction from spatially compact sources (e.g. Schilt et al., 1981), and are distributed in a non-uniform manner. The deep reflection packages tend to cluster at mid-crustal depths, yet some rise to within a few km of the surface and reflections are common down to about 35 km depth.

Drilling indicates that the shallow basement in this area consists of 1.6 Ga old mesozonal granite with a 1.3 Ga metamorphic overprint (Bickford et al., 1981, 1986; van Schmus et al., this volume), lithologies whose gross homogeneity

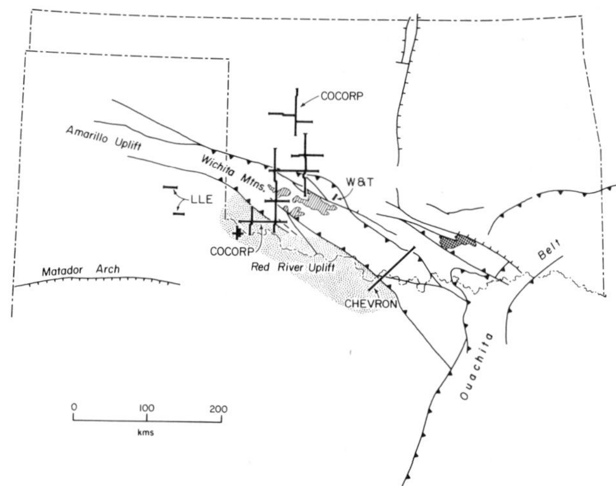

Fig 2. Location of COCORP profiles in southern Oklahoma and northern Texas. These surveys, together with evidence from oil industry lines in the area (as shown), help define the extent (stippled pattern) of a buried Proterozoic basin in this area. (Brewer et al., 1981).

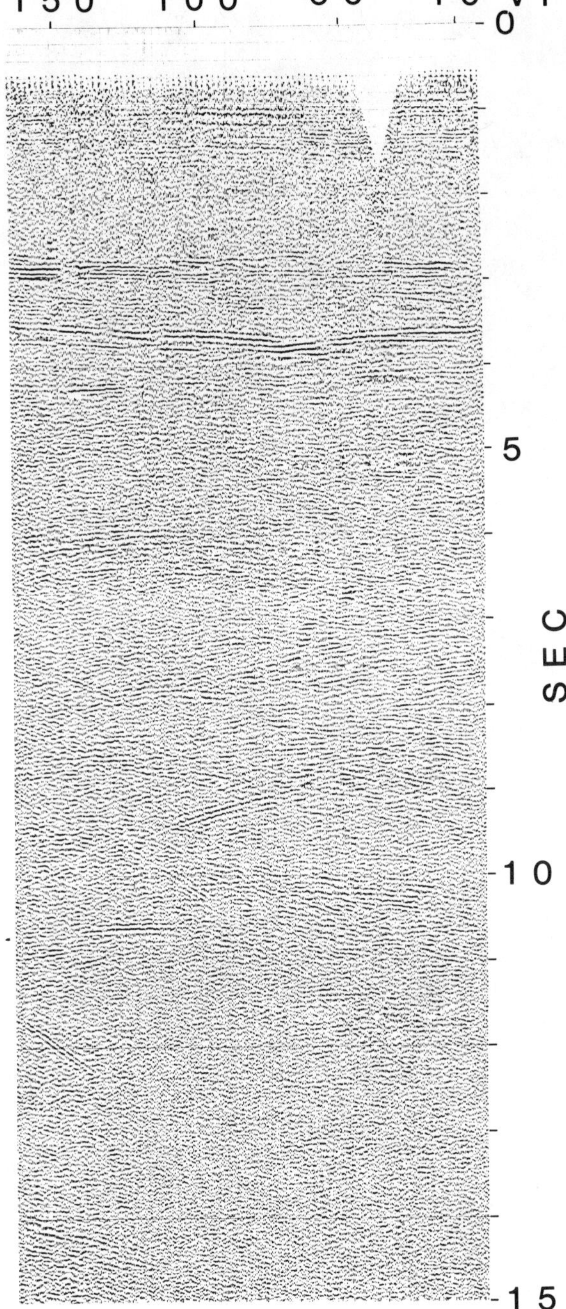

is easy to reconcile with the lack of reflections in the uppermost crust. The prominent reflections at mid-crustal depths are more problematic, since none extend to the surface for easy identification. Diffractions are ubiquitous, and most of the reflected and diffracted energy has a layered appearance. Seismic cross-lines confirm that these patterns vary in three dimensions (Brown et al., 1983a). Xenoliths found in kimberlite diatremes near the COCORP survey provide perhaps a few clues to their origin: in addition to granitic rocks these xenoliths include amphibolite, schist, diorite, gabbro and norite, and metagabbro and metanorite as well as granulite (Brookins and Meyer, 1974). Contacts between some of these lithologies, such as a gabbroic intrusion into granitic surroundings, might well give rise to diffractions of the type observed.

Specific lithologic identification of these reflectors must await deep drilling and/or further profiling to trace them to outcrop. Yet the reflections on these sections are also significant as representative of a "seismic signature" , i.e. a pattern of reflection geometry, abundance and amplitude that is sufficiently distinctive to be useful in the correlation of continental crust from one area to another. For example, a reflection character very similar to Kansas and the deeper portions of Hardeman County has been found for the relatively undisturbed block of Proterozoic crust making up the Colorado Plateau (Allmendinger et al., 1986). This character differs substantially from the reflection patterns in the adjacent Basin and Range, where penetrating faults and flat, layered lower crust dominates (Allmendinger et al., 1986). Whether these differences represent some basic contrast in the composition or structure of Proterozoic crust as opposed to crust that has been generated or strongly modified by latter tectonic episodes will continue to be debated until enough profiling is available from all types of crust to make such generalizations statistically valid.

One apparent contrast between Proterozoic and "later" crust that deserves special consideration is Moho reflectivity. The Kansas surveys lack distinctive Moho reflections, although similar surveys in many Phanerozoic terranes (e.g. Brown et al., 1980; Brewer et al., 1983b; Klemperer et al., 1986) show well-defined events that are correlatable, if not continuous, over considerable distances. Limited refraction measurements in Kansas suggest that the Moho should lie at about 40 km depth (13-14 sec; Stewart, 1968; Steeples, 1976). No prominent reflections are evident at that time, though there appears to be

Fig. 3 COCORP seismic section from its first field survey in Hardeman County Texas. The complex pattern of reflections at depth is clear evidence of crustal heterogeneity. The layered reflections above 1.6 sec correspond to Paleozoic sedimentary cover. The strong layered reflections between 3 and 5 seconds are inferred to be part of an extensive Proterozoic basin (see Figure 2). For this and subsequent seismic sections approximate depth = travel time X 3.

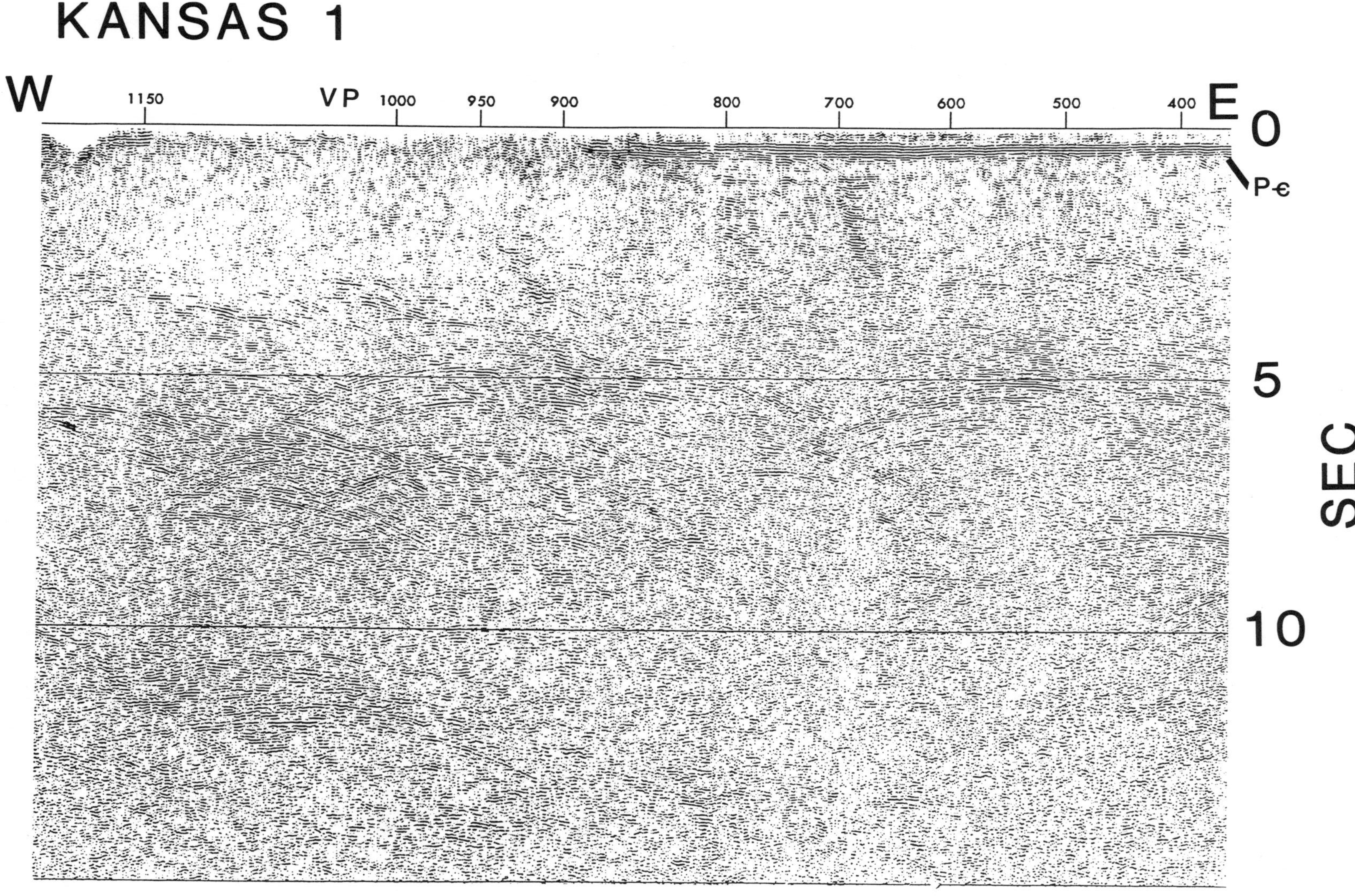

Fig. 4. Portion of COCORP seismic section from Kansas line 1 demonstrating complex reflection patterns in the middle crust This diffraction-dominated character resembles seismic sections from Hardeman County and the Colorado Plateau, perhaps representing a distinctive seismic signature for Proterozoic crust of this vintage.

Fig. 5. Portion of Oklahoma line 1. The layered reflections above 1.5 sec correspond to Paleozoic cover. The deeper reflections, extending down to 5 sec (ca. 15 km), correspond to the inferred Proterozoic sedimentary basin (Brewer et al., 1981). Note that the deep layering is disrupted at the south flank of the Wichita mountains.

a rapid decrease in the density of crustal reflections. This gradation suggests that the Moho in this area is more transitional than beneath crust subjected to Phanerozoic orogeny. One possible interpretation of this difference is that the Moho in the younger terranes represents relatively undisturbed underplating of igneous materials, perhaps still partially molten in some cases (e.g. Meissner, 1973; Klemperer et al., 1986). A study of the Colorado Plateau data suggests that the lack of sub-Moho reflections there is the result of acoustic homogeneity of the upper mantle rather than poor signal penetration (Mayer and Brown, 1986); similar arguments can be made for the Kansas results. Although sub-Moho reflectors are virtually absent from most COCORP surveys, including Kansas, they have been detected by deep seismic efforts in Great Britain (McGeary and Warner, 1985).

A New Proterozoic Basin

Stimulated by the Hardeman County results which show such strong layering in the shallow basement, COCORP collected additional seismic data in southern Oklahoma (Figure 2; Brewer et al., 1981). These surveys trace the layering northward to the southwestern flank of the Wichita Uplift (Figure 5), where they are abruptly disrupted by the Burch Fault. Although this fault exhibits clear SW-vergent reverse motion in the late Paleozoic, its earlier displacement history is unclear. Prominent layered reflections are absent N of the fault, perhaps indicating that the northern extensions of the layering were uplifted and then eroded by late Precambrian reverse faulting. However, weak, albeit layered, reflections are discernible on the seismic section beneath the Wichita Mountains

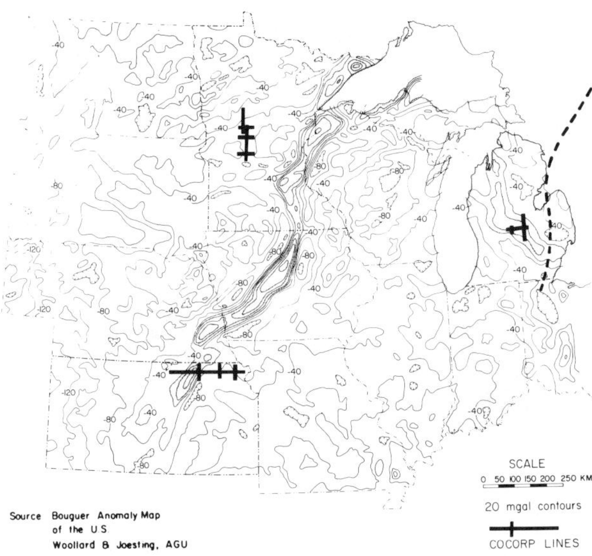

Fig. 6. Bouguer gravity map of the central
U.S. showing the Midcontinent Geophysical Anomaly
(MGA, also known as the Midcontinent Gravity
High) and mid-Michigan gravity high. COCORP
surveys in Michigan and Kansas (heavy lines)
delineate structure of the Keweenawan rift which
is inferred to be the source of the gravity
anomalies. Heavy lines in upper left are COCORP
surveys in the Archean terranes of Minnesota. The
heavy dashed line represents the inferred
position of the Grenville Front.

which appear to be offset downward relative to
the strong events to the south. If these weaker
events are correlative to the more prominent
layering, Precambrian normal faulting along the
Burch fault is indicated. It is also plausible
that substantial strike-slip motion, Precambrian
and/or Paleozoic, may have juxtaposed these
horizons, or that Cambrian plutonism in the
Wichita Uplift obliterated the northern extensions of the crustal layering. Subsequent COCORP
profiles north of the Wichita Uplift found no
evidence for similar basement layering, although
signal penetration through the thick Anadarko
Basin would probably have prevented their
detection if they existed (Brewer et al.,
1983a). Correlative reflectors are clearly absent
from the Kansas surveys much further north (Brown
et al., 1983a), where deeper crustal levels
subcrop and any comparable depositional sequence
may have been eroded away (Bickford et al.,
1981). Lynn et al. (1981) suggested, based on the
Hardeman County results, that the layering may be
igneous. However, Brewer et al. (1981) argue that
a sedimentary origin is more likely because of

its great extent, high amplitudes, and the
onlapping relations between individual reflections which suggest depositional unconformities. Together with oil industry surveys, the
COCORP lines indicate that the inferred basin
extends over an area of at least 2500 km^2 (Figure
2), varying in thickness from 7 to 10 km.
Volcano-sedimentary basins are well known in the
Proterozoic (e.g. Salop, 1977), with drilling in
the U.S. midcontinent encountering little-de-
formed anorogenic felsic volcanic rocks and
associated epizonal granites with ages of 1.2 -
1.5 Ga (Bickford et al., 1981, 1986; van Schmus
et al., this volume). A basin of felsic volcanics
and interlayered clastics is certainly consistent
with the character of the seismic layering on the
COCORP lines S of the Wichita Uplift, and it is
intriguing to speculate whether similar basins
lie hidden beneath Paleozoic cover in other parts
of the Proterozoic craton. Since Proterozoic
basins in other parts of the world have proven to
have significant hydrocarbon potential (Murray et
al., 1980), such basins could have economic as
well as scientific importance.

The Keweenawan Rift

Virtually bisecting the craton of the U.S. is
a major gravity and magnetic feature (Figure 6)
known as the Midcontinent Geophysical Anomaly
(MGA). This anomaly marks the course of the
Keweenwan rift system, an aborted attempt to
breakup the continent about 1 Ga ago (e.g. Chase
and Gilmer, 1973). Keweenawan rocks, including an
impressive sequence of basaltic extrusions, are
exposed along the MGA in the Lake Superior region
(Halls, 1966; Green, 1982; Ojakangas and Morey,
1982). The putative extensions of these rocks
toward Oklahoma and across Michigan are buried
beneath flat-lying Paleozoic strata.

COCORP's first attempt to image buried
Keweenawan structure was in central Michigan
(Brown et al., 1982). Seismic sections from those
surveys (Figure 7) show a remarkable bimodal
reflection sequence underlying the Cambrian
through Pennsylvanian sedimentary rocks of the
Michigan Basin proper. This deeper complex is
dominated by a relatively narrow trough of
strong, layered reflections between about 3 and 5
sec. Between the deep layered unit and the
overlying Paleozoic strata is a less reflective
zone. A deep borehole near the COCORP lines
encountered Precambrian arkoses and occasional
basalts just beneath the Paleozoic section (Sleep
and Sloss, 1978), though it did not penetrate to
the level of the deep layering. The drilling
results and the bimodal seismic appearance
strongly suggests a straightforward correlation
with Keweenawan stratigraphy (Brown et al.,
1982). If the upper, less reflective unit
corresponds to Upper Keweenawan clastics, as the
drillhole suggests, then the underlying reflec-

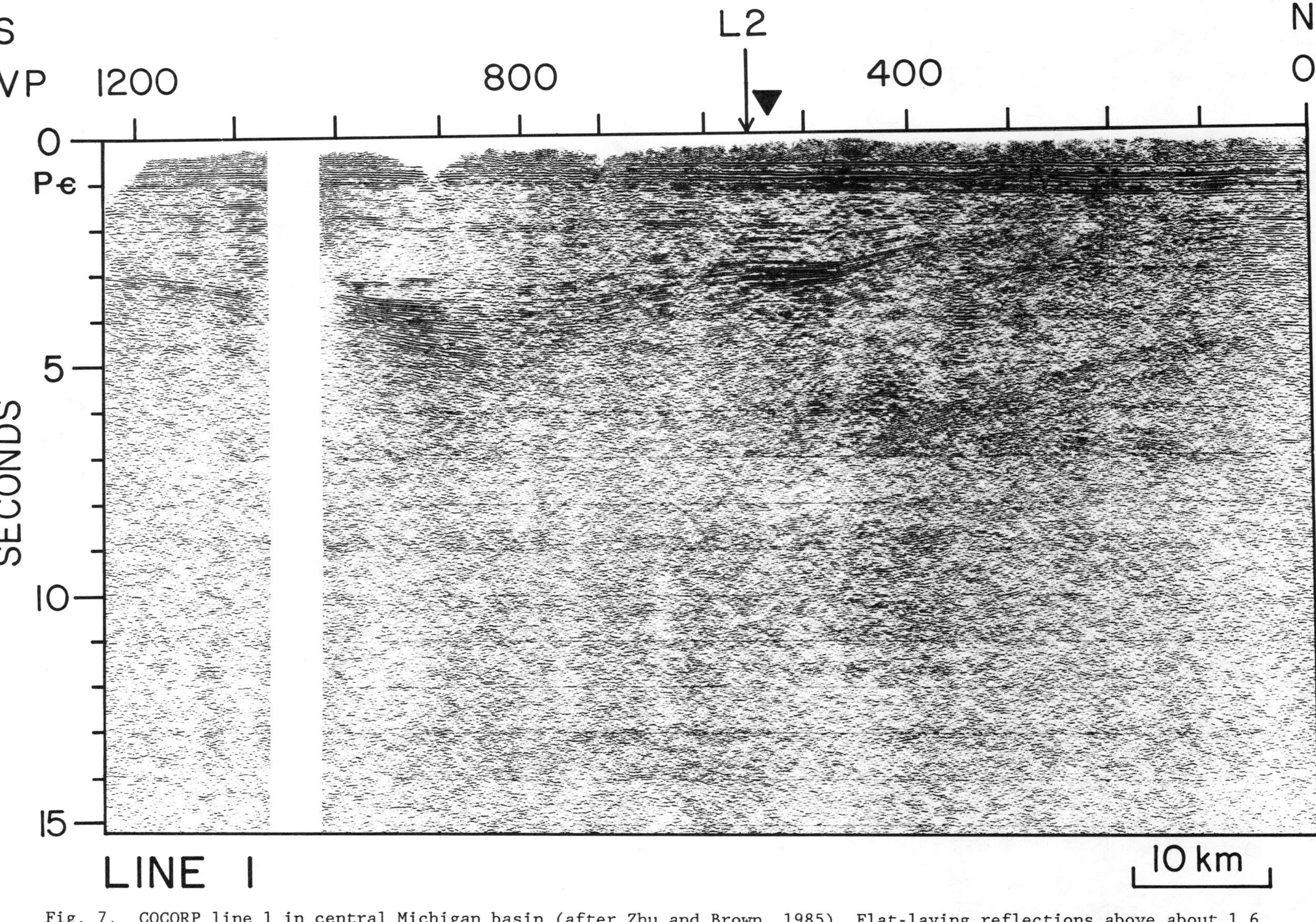

Fig. 7. COCORP line 1 in central Michigan basin (after Zhu and Brown, 1985). Flat-laying reflections above about 1.6 seconds correspond to the Paleozoic sedimentary rocks of the Michigan Basin proper. The deeper layered reflections, most prominent between 3 and 5 seconds in the center of the record, are interpreted to be Middle Keweenwan basaltic flows and intervening clastics. The intervening non-reflective zone is inferred to correspond to Upper Keweenawan clastics.

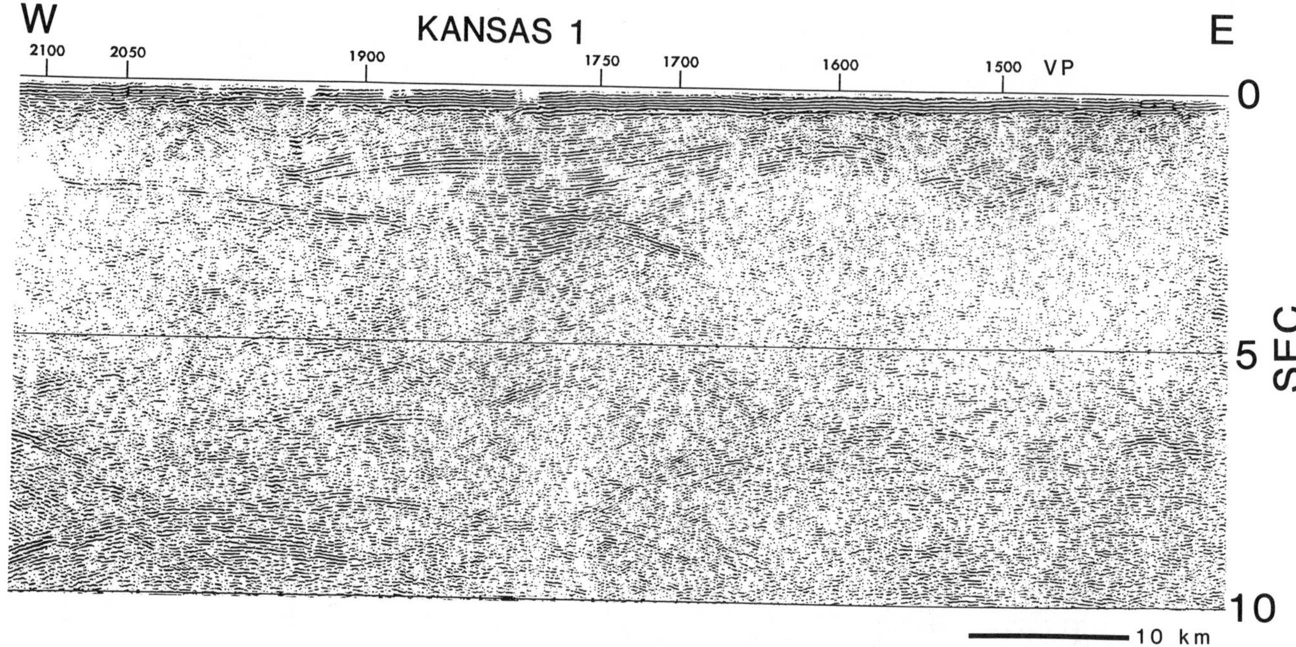

Fig. 8a.

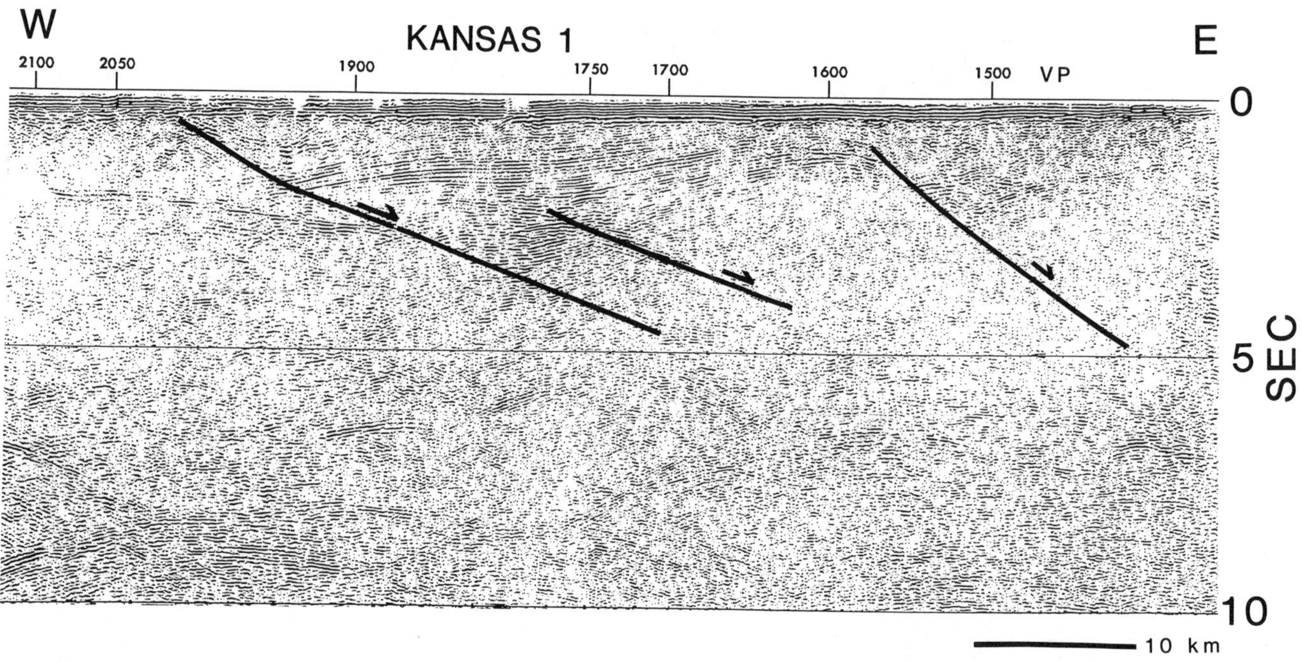

Fig. 8b.

Fig. 8. a). Portion of COCORP Kansas line 1, showing Keweenawan rift structure. b).
Interpretation of asymmetric rifting in the southern Keweenawan rift of Kansas (after
Serpa et al., 1984). As with the Michigan results, the rift fill is bimodal in seismic
appearance and similarly interpreted as layered basalts/clastics (Middle Keweenawan)
overlain by non-reflective clastics (Upper Keweenawan).

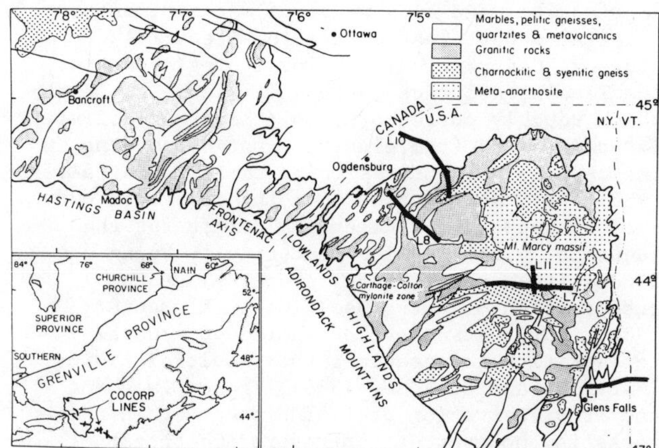

Fig. 9. Map of the Adirondacks of northeastern New York, showing the location of COCORP seismic lines across Grenville-age basement (Klemperer et al., 1983).

tive layering probably represents the interlayered basalt flows and clastic sedimentary rocks of the Middle Keweenawan. The cyclic juxtaposition of volcanics and clastics would be expected to provide stronger reflections than the immature clastics alone.

Recent reprocessing of the Michigan data by Zhu and Brown (1986) has greatly improved the seismic sections, leading to the recognition of an unconformity in the upper clastic sequence which may correspond to late Keweenawan compression (associated with the nearby Grenville orogeny?). In addition, reprocessing delineates major reflections beneath the deep layered sequence which may be Keweenawan or even pre-Keweenawan faults (Zhu and Brown, 1986).

Subsequent to the Michigan surveys, COCORP mounted an even larger effort to profile the southern end of the Keweenawan rift in Kansas (Figure 6; Serpa et al., 1984). The seismic section over this segment of the rift (Figure 8a) exhibits a moderately W-dipping sequence of strong, layered reflections lying near the apex of the MGA. As in Michigan, this reflective unit is overlain by a less reflective zone which is in turn buried beneath relatively undisturbed Paleozoic cover. The similarity in reflection character suggests a common interpretation: interbedded volcanics and clastics of the Middle Keweenawan overlain by Upper Keweenawan clastics.

In contrast to the Michigan results, the Kansas lines show clear evidence of extensional faults. Moderately E-dipping fault plane reflections and tilted basement blocks suggest an asymmetric rift geometry (Figure 8b) which Serpa et al. (1984) infer to result from domino-style block rotation along moderately-dipping faults. A similarly dipping, though weak, reflection can be

identified near the Nemaha Ridge, a major Paleozoic structure which subparallels the MGA to the east. This similarity lends credence to the speculation that the Nemaha Ridge represents Paleozoic reactivation of an outlying Keweenawan fault (Serpa et al., 1984).

The bimodal reflection character of the graben fill of the Keweenawan of Michigan and Kansas is quite prominent and resembles sequences from the younger intracontinental rifts, such as the Rio Grande rift of central New Mexico, as well as modern rifted ocean margins (Lillie, 1984). Lillie (1984) argues that this layering is in effect a reflection signature for early rift evolution, and may be recognizable even after being thrust beneath orogenic accretions, as has been inferred to have occurred in the Appalachians (Cook et al., 1981).

Deep Layering in Grenville Crust

A common drawback of the aforementioned surveys of Proterozoic structure is the lack of immediate surface exposures with which reflection patterns can be more directly correlated. In this respect, COCORP surveys in NE New York merit particular attention as they were conducted within and adjacent to the Adirondack Dome, a major exposure of Grenville basement (Figure 9). Since the granulite-facies surface rocks of the Adirondacks are generally inferred to have once been at mid-crustal depths (e.g. Bohlen et al., 1980), these surveys provide a opportunity for "calibrating" the seismic response of middle and lower crustal assemblages.

It has long been speculated that the layered lower crustal character evident on certain seismic reflection sections may be due to a ductile fabric in high-grade metamorphic rocks (e.g. Smithson, 1979; Phinney and Jurdy, 1979). Although the surface geology of the Adirondacks is one of considerable compositional and structural complexity with abundant evidence of ductile stretching (e.g. Wiener et al., 1984), the upper parts of the COCORP seismic sections (e.g. Figure 10) contain relatively few reflections and certainly nothing that looks like the pronounced deep layering in other areas (e.g. Brown et al., 1986; Matthews and Cheadle, 1986). Thus extremely heterogeneous basement may have a reflection signature as unremarkable as the granites of Kansas, at least at the seismic wavelengths used by COCORP. The layering in the lower crust must be due to factors other than ductile shear alone, perhaps the injection of sills (e.g. Meissner, 1973) or the presence of fluids (Matthews and Cheadle, 1986).

COCORP line 7 (Figure 10) traverses the southeastern skirt of the Mt. Marcy massif, the largest meta-anorthosite body in the Adirondacks (Isachsen, 1968). A reflection from the base of the meta-anorthosite cannot be identified,

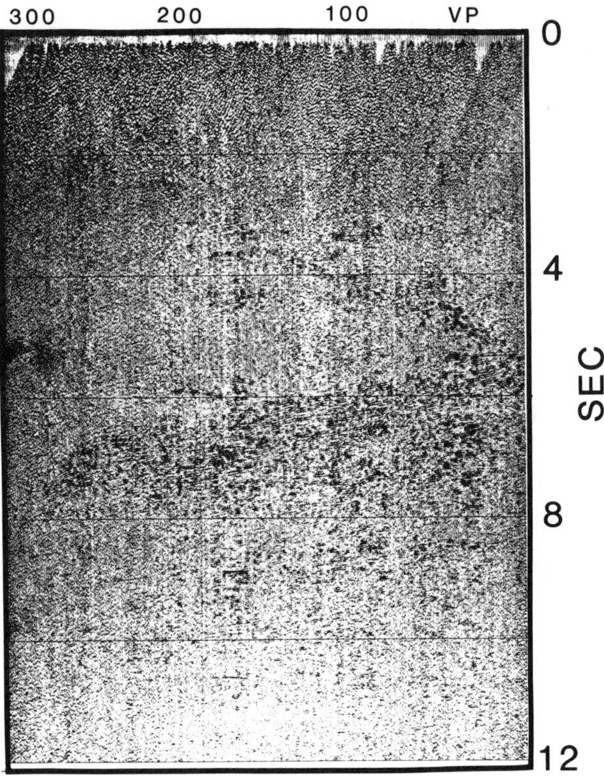

Fig. 10. Portion of COCORP line 7, which traverses the southern flank of the Mt. Marcy meta-anorthsite massif. The layered wedge at 5 to 8 sec (15-24 km) may be a layered igneous complex.

suggesting that this boundary is either non-reflective or too shallow for proper imaging. At the other extreme, a few scattered reflections can be identified at expected Moho travel times (about 11 sec), but a consistent arrival from the base of the crust is absent.

The most prominent feature of the Adirondack surveys is a wedge of short, layered reflections between 5 and 7 sec (15 and 21 km) beneath the Mt. Marcy massif (Figure 10). This deep layered series, referred to as the Tahawus sequence, has been the focus of most of the attention garnered by the Adirondack data (Brown et al., 1983b; Klemperer et al., 1985). Its NW-dipping upper surface, its multi-cyclic internal appearance and its location beneath the Marcy Massif and near the center of the Adirondack dome are perhaps clues to its origin, but thus far interpretation has been highly speculative. Klemperer et al. (1985) review some of the possible origins for the Tahawus sequence, including its being: a) a "modern" crustal intrusion, perhaps responsible for contemporary uplift and seismicity of the Adirondacks, b) a layered igneous complex, the parent or restite corresponding to the overlying anorthosite, and c) a wedge of metasedimentary material underthrust during the Grenville orogeny. One particularly intriguing aspect of the Tahawus sequence is that it lies at a depth which roughly corresponds to a high conductivity layer inferred from electromagnetic sounding in the central Adirondacks (e.g. Connerney et al., 1980). This correlation could be construed not only to support a sedimentary origin for the sequence, with the high conductivity being due to conductive minerals (e.g. graphite) or the preservation of entrapped fluids (Klemperer, 1985), but the suggestion that deep crustal reflectivity is sometimes the result of fluids (Matthews and Cheadle, 1986). The emplacement and/or preservation of fluids deep within ancient rocks is, however, a controversial concept.

In addition to the Adirondack profiles, COCORP's traverses of the Appalachian orogenic belt cross substantial tracts of Grenville crust that are only thinly covered by Paleozoic strata and largely unaffected by subsequent orogeny. A common result of deep seismic sections from these so-called "thin-skinned" forelands of the Appalachian system is the general lack of reflections from within the underlying Grenville basement (e.g. Cook et al., 1979; Lillie et al., 1983). There are exceptions, of course: in addition to the Tahawus sequence, deep reflections are evident beneath the Taconic thrust sheets of the New England Appalachians in Vermont and western New York (Brown et al., 1983b), and occasional reflections are scattered through other foreland surveys. Yet Grenville crust, at least that portion sampled to date, is substantially less reflective that the deep portions of the adjacent Phanerozoic accretions (e.g. Ando et al., 1983). If the eastern Grenville includes older accretions, why are they now non-reflective? Were they different in kind from Phanerozoic analogues? Have any original acoustic contrasts been "smeared" out by metamorphic processes over time? Or does this difference in reflectivity support the suggestions that the Grenville orogeny was essentially ensialic (e.g. Baer, 1981). Such issues cannot be fully addressed until related questions of technique are also resolved. For example, the lack of reflections beneath the Arkoma basin of the Ouachita foreland may be due to energy absorption by the thick (ca. 10 km) Paleozoic cover rather than Grenville "homogeneity" (Lillie et al., 1983). Studies of the variability of signal penetration in deep seismic profiling have only recently begun (e.g. Mayer and Brown, 1986).

A Proterozoic Suture in Southeast Wyoming

The prime targets in deep seismic profiling have typically been the major tectonic boundaries, whether large-scale faults which simply offset basement (e.g. the Wind River thrust; Smithson et al., 1978) or inferred "sutures" which juxtapose very dissimilar basements (e.g.

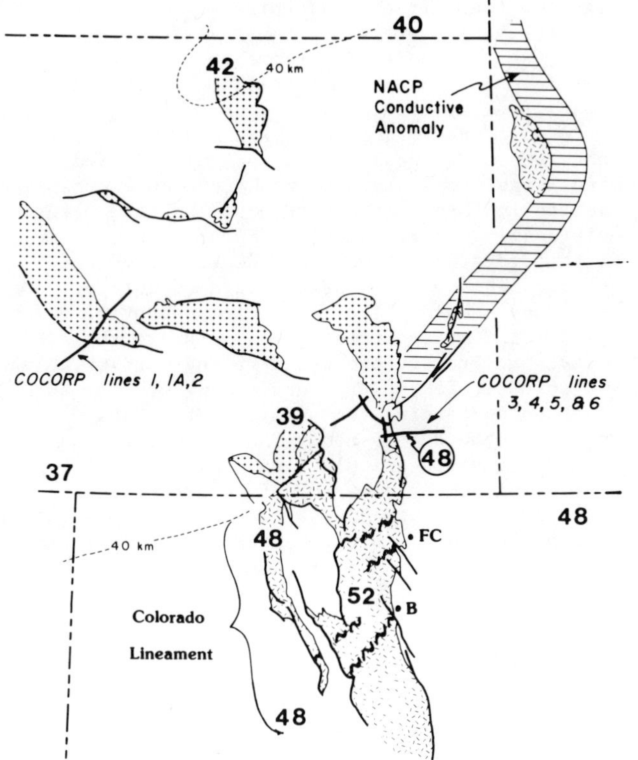

Fig. 11. Location of COCORP surveys in Wyoming. Numbers refer to estimates of Moho depth from refraction surveys in the area. From Allmendinger et al., 1982).

the Alleghenian suture of the southeastern Appalachians; Nelson et al., 1985) . Following COCORP's success in delineating the deep geometry of the Laramide Wind River overthrust in SW Wyoming, a survey was mounted to trace the Rocky Mountain Front in the Laramie Range of southeastern Wyoming (Figure 11). Although these surveys were primarily designed to understand the role of Laramide compression in producing Rocky Mountain uplifts, they also crossed a major Proterozoic tectonic feature, the inferred extension of the Mullen Creek-Nash Fork (MCNF) shear zone (Allmendinger et al., 1982).

The MCNF shear zone, exposed in the Medicine Bow Mountains, is part of the boundary between Archaean and Proterozoic crust known as the Cheyenne belt (Houston et al., 1979). North of the MCNF, early Proterozoic (1.7-2.5 Ga old) metasedimentary rocks, including quartzite, marble, and metavolcanics, overlie highly deformed and metamorphosed Archaean (at least 2.5 Ga old) gneiss. Younger Proterozoic basement lies south of this boundary, including the Sherman Granite batholith (1.41 Ga old) and the Laramie anorthosite (1.42-1.51 Ga old). A metamorphic event between 1.4 and 1.6 Ga affected both terranes, whose prior history is dissimilar. Hou-

ston et al. (1979) suggest that this feature marks a Proterozoic suture between colliding continental plates, although it has also been interpreted as a Proterozoic wrench fault system (the Colorado lineament of Warner, 1978). To the NE this feature may merge with the North American Central Plains conductive anomaly (Camfield and Gough, 1977).

The relevant seismic sections lack any striking reflection feature in the area where the extrapolated position of the MCNF shear zone is crossed (Figure 12). However, steeply dipping events on COCORP lines 4 and 5 are candidates for reflections from the MCNF structure. These events suggest a NE-striking plane which dips about 55° (migrated) to the SE. Allmendinger et al. (1982) suggest that these events mark a tectonic boundary which separates not only Archaean from Proterozoic basement, but differing seismic reflection characters and a change in crustal thickness. If this high-angle feature is the Proterozoic suture, it differs markedly from the low-angle overthrusting which characterizes COCORP results from the exposed Appalachians (e.g. Cook et al., 1981), though perhaps not so different from the moderately-dipping reflections interpreted to correspond to the Alleghenian suture buried beneath the Coastal Plain sediments of the southeasternmost Appalachians (Nelson et al., 1985).

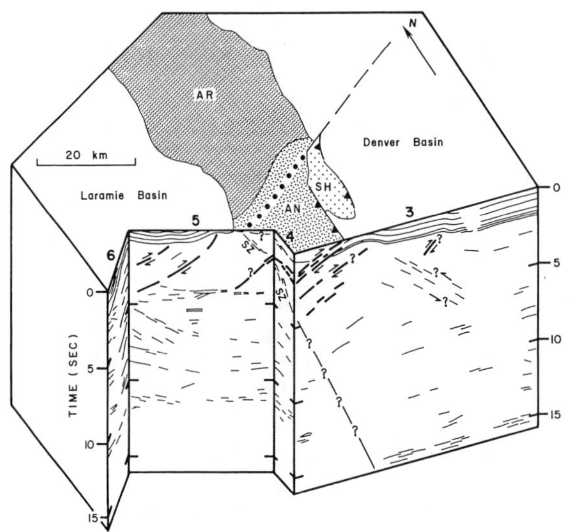

Fig. 12. Perspective line drawing of COCORP seismic sections from the Laramie range of southeastern Wyoming. A steeply dipping feature (SZ) may correspond to the extension of the Mullen Creek Nash Fork lineament (dotted- dashed line), a possible Proterozoic suture zone. This zone has been argued to juxtapose crust with differing reflection characters and thicknesses. AN- anorthosite; SH- Sherman granite; AR- Archean basement (Brewer et al., 1982).

Discussion

Although some of the more obvious manifestations of Proterozoic crustal structure as noted from COCORP deep seismic sections are reviewed here, it is important to recognize that some of the reflection patterns recorded by COCORP surveys in Phanerozoic orogenic belts may well correspond to relict Precambrian features. If such structures have been reactivated during later orogeny, it may be quite difficult to recognize their earlier history. Tectonic reactivation (though not necessarily involving Precambrian structure) has been argued from COCORP seismic sections from Oklahoma (see previous discussion of the Burch Fault), the southern Appalachians (Petersen et al., 1984) and west-central Utah (Allmendinger et al., 1986) as well as the Laramie Range (Allmendinger et al., 1982). Conversely, it may also be that some reflectors observed in Proterozoic crust may be younger features emplaced without corresponding surface manifestation. For example, it has been argued that the deep diffractions in Hardeman County, which lie well below known Proterozoic rocks, may represent Cambrian intrusions into older crust (Schilt et al., 1981).

Another issue in the interpretation of COCORP results from the craton is whether there is any evidence for Precambrian tectonics that are demonstrably different from Phanerozoic tectonics. Gibbs (1986) argues that COCORP results from the Archaean of Minnesota are consistent with Phanerozoic-style continental accretion, and further postulates that the same can be said for results across Proterozoic structures.

A related aspect is whether there is a distinct seismic reflection signature for older (e.g. Proterozoic crust) that distinguishes it from later continental accretions. As pointed out earlier, COCORP sections from northern Texas, Kansas and the Colorado Plateau, areas of Proterozoic crust relatively undeformed by subsequent tectonic activity, exhibit a gross character (transparent upper crust, diffraction dominated middle-crust, lack of distinct Moho reflections) that thus far appears to differ significantly from that in Phanerozoic orogenic belts (layered middle to lower crust, prominent Moho reflections), perhaps even from that of Grenville-age crust (relatively few deep crustal reflections). The Kansas and Colorado Plateau surveys may lie within the same Proterozoic orogenic belt (Sims and Peterman, 1986), and the deeper portions of the Texas/Oklahoma sections may correspond to that same orogen buried beneath a superficial, younger granite-rhyolite terrane (Bickford et al., 1986). Unfortunately, the quality of the Minnesota results appear to be too poor (Gibbs et al., 1985), and the Wyoming surveys too overprinted by Laramide events, to allow meaningful comparison to Archaean crust in this respect. However, deep seismic profiling, including efforts outside the U.S., has barely begun to identify and categorize the structural complexity of Phanerozoic, must less Precambrian crust, so that generalizations about crustal seismic "signatures" must still be treated as provisional at best.

In some cases, the interpretation of presumed Proterozoic features on deep seismic sections has been largely speculative, perhaps more so than results in Phanerozoic terranes. The reason for this seeming equivocation is simple: few, if any, of the Proterozoic features identified on the seismic sections have been traced to outcrop, nor have they been drilled. The circumstantial cases built for some interpretations, such as the postulated Proterozoic Basin in southern Oklahoma and northern Texas, **are** convincing. Others, such as those concerning the nature of the Tahawus sequence beneath the central Adirondacks, are merely plausible speculations. Some of these uncertainties may clear up if future profiling can trace events to the surface or drillable depths. However, to belabor these uncertainties is to largely miss the primary significance of reflection profiling of the cratons to date, which is that these older terranes contain major features at depth which may be detectable only by this technique.

Summary

The seismic reflection method is still relatively new in its application to problems of Precambrian geology. The most extensive efforts still lie in the U.S., but even these have sampled but a small portion of the Proterozoic crust which makes up North America. COCORP surveys in Kansas and Michigan define a distinct reflection stratigraphy associated with the late Proterozoic Keweenawan rift, an extensional feature which seems to have formed by asymmetric fault-block rotation. Profiles in northern Texas and southern Oklahoma reveal a major Proterozoic basin, 7 to 10 km thick and at least 2500 km^2 in extent, buried beneath the Paleozoic cover. A Proterozoic continental suture may correspond to a steeply-dipping feature on COCORP surveys from SE Wyoming, and metasedimentary rocks, still conductive, may lie deep within the Adirondack basement, perhaps overthrust during the Grenville orogeny. Although coverage is still too limited to justify confident generalization, there is at least the suggestion that some Proterozoic crust has a gross seismic reflection character that distinguishes it from more recently tectonized crust. In particular the Moho of the central U.S. craton appears to be less reflective than the Moho beneath the younger peripheral orogenic belts.

Some of the interpretations of reflection results from Proterozoic areas are admittedly speculative, due in part to the lack of surficial geologic control. Yet there is no doubt that major variations of crustal structure are being uniquely mapped by the seismic reflection technique. Other basins, suture zones and seismically distinctive crustal blocks may underlie the serene Paleozoic strata of the interior platforms. In spite of the paucity of geological control, the vast expanse of Proterozoic crust which lies hidden beneath this thin veneer is one of the most exciting frontiers for future deep seismic profiling.

Acknowledgements. The seismic data discussed here was collected by Petty-Ray Geophysical, a Division of Geosource, Inc. Some of the results were processed using the MEGASEIS seismic processing system, a product of Seiscom Delta. This review was supported by NSF Grant No. EAR-83-13378. INSTOC Contribution No. 56.

References

Allmendinger, R.W., J.A. Brewer, L.D. Brown, J.E. Oliver and S. Kaufman, COCORP profiling across the Rocky Mountain front in southern Wyoming, Part II: Precambrian basement structure and its influence on Laramide deformation, _Bull. Geol. Soc. Amer._, _93_, 1253-1263, 1982.

Allmendinger, R.W., H. Farmer, E. Hauser, J. Sharp, D. Von Tish, J. Oliver, and S. Kaufman, Phanerozoic tectonics of the Basin and Range-Colorado Plateau transition from COCORP data and geologic data: a review, in _Reflection Seismology and the Continental Crust: Continental Crust_, edited by M. Barazangi and L.D. Brown, pp. 257-268, Amer. Geophys. Union Geodynamic Series v. 14, 1986.

Ando, C.J., B. Czuchra, S. Klemperer, L.D. Brown, M. Cheadle, F.A. Cook, J.E. Oliver, S. Kaufman, T. Walsh, J.B. Thompson, Jr., J.B. Lyons, and J.L. Rosenfeld, Crustal profile of a mountain belt: COCORP deep seismic reflection profiling in New England Appalachians and implications for architecture of convergent mountain chains, _Amer. Assoc. Petrol. Geol. Bull._, _68_, 819-837, 1984.

Baer, A.J., A Grenvillian model of Proterozoic plate tectonics, in _Precambrian Plate Tectonics_, edited by A. Kröner, pp. 353-380, Elsevier, Amsterdam, 1981.

Bickford, M.E., K.L. Harrower, W.J. Hoppe, B.K. Nelson, and J.J. Thomas, Rb-Sr and U-Pb geochronology and distribution of rock types in the Precambrian basement of Missouri and Kansas, _Geol. Soc. Amer. Bull._, _92_, 323-341, 1981.

Bickford, M.E., W.R. Van Schmus, and I. Zietz, Proterozoic history of the midcontinent region of North America, _Geology_, _14_, 492-496, 1986.

Bohlen, S.R., E.J. Essene, and K.S. Hoffman, Update on feldspar and oxide thermometry in the Adirondack Mountains, New York, _Geol. Soc. Amer. Bull._, _Part I_, _91_, 110-113, 1980.

Brewer, J.A., L.D. Brown, D. Steiner, J.E. Oliver, S. Kaufman and R.E. Denison, Proterozoic basin in the southern midcontinent of the U.S. revealed by COCORP deep seismic reflection profiling, _Geology_, _9_, 569-575, 1981.

Brewer, J.A., R.W. Allmendinger, L.D. Brown, J.E. Oliver and S. Kaufman, COCORP profiling across the northern Rocky Mountain front in southern Wyoming, Part I: Laramide structure, _Bull. Geol. Soc. Amer._, _93_, 1242-1252, 1982.

Brewer, J.A., R. Good, J.E. Oliver, L.D. Brown, and S. Kaufman, COCORP profiling across the southern Oklahoma Aulacogen: Overthrusting of the Wichita Mountains and compression within the Anadarko basin, _Geology_, _11_, 109-114, 1983a.

Brewer, J.A., D.H. Matthews, M.R. Warner, J.R. Hall, D.K. Smythe and R.J. Whittington, BIRPS deep seismic reflection studies of the British Caledonides, _Nature_, _305_, 206-210, 1983b.

Brookins, D.G., and H.O.A. Meyer, Crustal and upper mantle stratigraphy beneath eastern Kansas, _Geophys. Res. Letters._, _1_, 269-272, 1974.

Brown, L.D., C.E. Chapin, A.R. Sanford, S. Kaufman and J.E. Oliver, Deep structure of the Rio Grande Rift from seismic reflection profiling, _J. Geophys. Res._, _85_, 4773-4800, 1980.

Brown, L.D., L. Jensen, J. Oliver, S. Kaufman and D. Steiner, Rift structure beneath the Michigan Basin from COCORP profiling, _Geology_, _10_, p. 645-649, 1982.

Brown, L.D., L. Serpa, T. Setzer, J. Oliver, S. Kaufman, R. Lillie, D. Steiner, and D. W. Steeples, Intracrustal complexity in the U.S. midcontinent: preliminary results from COCORP surveys in NE Kansas, _Geology_, _11_, 25-30, 1983a.

Brown, L.D., C. Ando, S. Klemperer, J. Oliver, S. Kaufman, B. Czuchra, T. Walsh, and Y.W. Isachsen, Adirondack-Appalachian crustal structure: The COCORP Northeast traverse, _Geol. Soc. Amer. Bull._, _94_, 1173-1184, 1983b.

Brown, L.D., Aspects of COCORP deep seismic profiling, in _Reflection Seismology and the Continental Crust: A Global Perspective_, edited by M. Barazangi and L.D. Brown, pp. 209-222, Amer. Geophys. Union Geodynamic Series v. 13, 1986.

Brown, L.D., M. Barazangi, S. Kaufman, and J. Oliver, The first decade of COCORP: 1974-1984, in _Reflection Seismology and the Continental Crust: A Global Perspective_, edited by M. Barazangi and L.D. Brown, pp. 107-120, Amer. Geophys. Union Geodynamic Series v. 13, 1986.

Camfield, P.A., and D.I. Gough, A possible Proterozoic plate boundary in North America, _Can. Jour. Earth Sci._, _14_, 1229-1238, 1977.

Chase, C.G., and T.H. Gilmer, Precambrian plate tectonics: the midcontinent gravity high, _Earth Planet. Sci. Letters_, _21_, 70-78, 1973.

Connerney, J.E.P., A. Nekut, and A.F. Kuckes, Deep crustal electrical conductivity in the Adirondacks, _J. Geophys. Res._, _85_, 2603-2614, 1980.

Cook, F.A., D.S. Albaugh, L.D. Brown, S. Kaufman, J.E. Oliver and R.D. Hatcher, Jr., Thin-skinned tectonics in the crystalline southern Appalachians: COCORP seismic reflection profiling of the Blue Ridge and Piedmont, Geology, 7, 563-567, 1979.

Cook, F.A., L.D. Brown, S. Kaufman, J.E. Oliver and T.A. Petersen, COCORP seismic profiling of the Appalachian orogen beneath the coastal plain of Georgia, Geol. Soc. Amer. Bull., Part I, 92, 738-748, 1981.

Denison, R.E., E.G. Lidiak, M.E. Bickford, and E.B. Kisvarsanyi, Geology and geochronology of Precambrian rocks in the central interior of the U.S., in Correlation of Precambrian Rocks of the United States and Mexico, edited by by J.E. Harrison, and Z.E. Peterman, p. C1-C20, U.S. Geol. Surv. Prof. Paper 1241-C, 1984.

Gibbs, A.K., B. Payne, T. Setzer, L.D. Brown, J.E. Oliver, and S. Kaufman, Seismic reflection study of the Precambrian crust of Central Minnesota, Bull. Geol. Soc. Amer., 95, 280-294, 1984.

Gibbs, A.K., Seismic reflection profiles of Precambrian crust: a qualitative assessment, in Reflection Seismology and the Continental Crust: The Continental Crust, edited by M. Barazangi and L.D. Brown, pp. 95-106, Amer. Geophys. Union Geodynamic Series v. 14, 1986.

Green, J.C., Geology of Keweenawan extrusive rocks, in Geology and Tectonics of the Lake Superior Basin, edited by R.J. Wold and W.J. Hinze, pp. 157-164, Geol. Soc. Amer. Mem. 156, 1982.

Halls, H.C., A review of the Keweenawan geology of the Lake Superior region, in The Earth Beneath the Continents, edited by J.S. Steinhart and T.J. Smith, pp. 3-27, Amer. Geophys. Union Mono. 10, 1966.

Houston, R.S., K.E. Karlstrom, F.A. Hills, and S.B. Smithson, The Cheyenne Belt: the major Precambrian crustal boundary in the western United States (abstract), Geol. Soc. Amer. Abstracts with Programs, 11, 446, 1979.

Isachsen, Y.W., The origin of anorthosites and related rocks: a summarization, in The Origin of Anorthosites and Related Rocks, edited by Y.W. Isachsen, pp. 435-445, N.Y. State Museum and Science Service Mem. 18, 1968.

Klemperer, S., L. Brown, J. Oliver, C. Ando and S. Kaufman, Crustal structure in the Adirondacks, in Seismic Expression of Structural Style, edited by A.W. Bally, pp. 1.5-12-16, Amer. Assoc. Petrol. Geol. Studies in Geology No. 15, 1983.

Klemperer, S.L., L.D. Brown, J.E. Oliver, C.J. Ando, B. Czuchra, and S. Kaufman, Some results of COCORP seismic reflection profiling in the Grenville - age Adirondack mountains, New York State, Can. J. Earth Sci., 22, 141-153, 1985.

Klemperer, S.L., T.A. Hauge, E.C. Hauser, J.E. Oliver, and C.J. Potter, The Moho in the northern Basin and Range province, Nevada, along the COCORP 40°N seismic reflection transect, Geol. Soc. Amer. Bull., 97, 603-618, 1986.

Lillie, R., K.D. Nelson, B. deVoogd, J. Brewer, J.E. Oliver, L.D. Brown, S. Kaufman, and G.W. Viele, Crustal structure of Ouachita Mountains, Arkansas: a model based on integration of COCORP reflection profiles and regional geophysical data, Amer. Assoc. Petrol. Geol. Bull., 67, 907-931, 1983.

Lillie, R.J., Tectonic implications of subthrust structures revealed by seismic profiling of Appalachian/Ouchita orogenic belt, Tectonics, 3, 619-646, 1984.

Lynn, H.B., L.D. Hale, and G.A. Thompson, Seismic reflections from the basal contacts of batholiths, J. Geophys. Res., 86, 10633-10638, 1981.

Matthews, D.H. and M.J. Cheadle, Deep reflections from the Caledonides and Variscides west of Britain and comparison with the Himalayas, in Reflection Seismology and the Continental Crust: A Global Perspective, edited by M. Barazangi and L.D. Brown, pp. 5-20, Amer. Geophys. Union Geodynamic Series v. 13, 1986.

Mayer, J.R., and L.D. Brown, Signal penetration in the COCORP Basin and Range-Colorado plateau survey, Geophysics, 51, 1050-1055, 1986.

McGeary, S. and M.R. Warner, Seismic profiling the continental lithosphere, Nature, 317, 795-797, 1985.

Meissner, R., The Moho as a transition zone, Geophysical Surveys, 1, 195-216, 1973.

Murray, G.E., M.J. Kaczor, and R.E. McArthur, Indigenous Precambrian petroleum revisited, Amer. Assoc. Petrol. Geol. Bull., 64, 1681-1700, 1980.

Nelson, K.D., J.A. Arnow, J.H. McBride, J.H. Willeman, J. Huang, L. Zheng, J.E. Oliver, L.D. Brown, and S. Kaufman, New COCORP profiling in the southeastern United States. Part I: Late Paleozoic suture and Mesozoic rift basin, Geology, 13, 714-718, 1985.

Ojakangas, R.W., and G.B. Morey, Keweenawan sedimentary rocks of the Lake Superior region: a summary, in Geology and Tectonics of the Lake Superior Basin, edited by R.J. Wold and W.J. Hinze, pp.157-164, Geol.Soc.Amer.Mem.156,1982.

Oliver, J.E., M. Dobrin, S. Kaufman, R. Meyer and R. Phinney, Continuous seismic reflection profiling of the deep basement, Hardeman County, Texas, Geol. Soc. Am. Bull., 87, 1537-1546, 1976.

Oliver, J.E., and S. Kaufman, Complexities of the deep basement from seismic reflection profiling, in The Earth's Crust, edited by J. Heacock, pp. 243-253, Amer. Geophys. Union Monograph 20, 1977.

Oliver, J., Tracing subsurface features to great depths: a powerful means for exploring the earth's crust, Tectonophysics, 81, 257-272, 1982.

Oliver, J., A global perspective on seismic reflection profiling of the continental crust, in _Reflection Seismology: A Global Perspective_, edited by M. Barazangi and L.D. Brown, pp. 1-3, American Geophysical Union Geodynamic Series, v. 13, 1986.

Petersen, T.A., L.D. Brown, F.A. Cook, S. Kaufman, and J.E. Oliver, Structure of the Riddleville Basin from COCORP seismic data, and implications for reactivation tectonics, _Jour. Geol._, 92, 261-271, 1984.

Phinney, R.A. and D.M. Jurdy, Seismic imaging of deep crust, _Geophysics_, 44, 1637-1660, 1979.

Salop, J.L., 1977, _Precambrian of the Northern Hemisphere and General Features of Geological Evolution_, Elsevier, New York, 378 pp, 1977.

Schilt, F.S., S. Kaufman, and G.H. Long, A 3-dimensional study of seismic diffraction patterns from deep basement sources, _Geophysics_, 46, 1673-1683, 1981.

Serpa, L., T. Setzer, H. Farmer, L. Brown, J. Oliver, S. Kaufman, J. Sharp and D.W. Steeples, Structure of the southern Keeweenawan Rift from COCORP surveys across the Midcontinent Geophysical Anomaly in northeastern Kansas, _Tectonics_, 3, 367-384, 1984.

Sims, P.K., and Z.E. Peterman, Early Proterozoic central plains orogen: a major buried structure in the north-central United States, _Geology_, 14, 488-491, 1986.

Sleep, N.H., and L.L. Sloss, A deep borehole in the Michigan Basin, _J. Geophys. Res._, 83, 5815-5819, 1978.

Smithson, S.B., Aspects of continental crustal structure and growth: targets for scientific deep drilling, _Contrib. to Geology_, _Univ. Wyoming_, 17, 65-75, 1979.

Smithson, S.B., J. Brewer, S. Kaufman, J. Oliver and C. Hurich, Nature of the Wind River thrust, Wyoming, from COCORP deep-reflection data and from gravity data, _Geology_, 6, 648-652, 1978.

Steeples, D.W., Preliminary crustal model for northwest Kansas (abstract.), _Eos Trans. AGU_, 57, 961, 1976.

Stewart, S.W., Crustal structure in Missouri by seismic refraction methods, _Seismol. Soc. Amer. Bull._, 58, 291-323, 1968.

Warner, L.A., The Colorado lineament: a middle Precambrian wrench fault system, _Geol. Soc. Amer. Bull._, 89, 161-171, 1978.

Wiener, R.W., J.M. McClelland, Y.W. Isachsen, and L.M. Hall, Stratigraphy and structural geology of the Adirondack Mountains, N.Y.: review and synthesis, in _The Grenville Event in the Appalachians and related topics_, edited by M.J. Bartholomew, pp. 1-55, Geol. Soc. Amer. Spec. Paper 194, 1-55, 1984.

Zhu, T., and L.D. Brown, COCORP Michigan surveys: reprocessing and results, _J. Geophys. Res._, in press, 1986.

EARLY PROTEROZOIC FOREDEEPS, FOREDEEP MAGMATISM, AND
SUPERIOR-TYPE IRON-FORMATIONS OF THE CANADIAN SHIELD

Paul F. Hoffman

Precambrian Geology Division, Geological Survey of Canada, Ottawa, Ontario K1A 0E4

Abstract. Foredeeps (foreland basins) are linear
asymmetric basins that migrate in front of, and become
incorporated within, foreland fold-and-thrust belts.
They develop as a flexural response to loading of the
continental lithosphere by thrust sheets. Many foredeeps evolve from oceanic trenches when rifted continental margins are drawn into subduction zones, but
they may also result from intracontinental thrusting.
Six 2.2 - 1.8 Ga foredeep sequences are described in
association with Proterozoic fold-and-thrust belts bor-
dering the >2.5 Ga Superior, Slave and North Atlantic
cratons of the Canadian Shield. The Proterozoic foredeeps differed from Phanerozoic examples in that mafic magmatism occurred in their axial zones and iron-
formations were deposited on their outer ramps. Foredeep magmatism, not ophiolite obduction nor initial
rifting, produced most of the volcanic rocks in the Proterozoic foreland belts described. Foredeep migration
produces diachronous deposition of facies. Consequently, axial-zone volcanic rocks occur stratigraphically
above outer-ramp iron-formations. Therefore, the
absence of volcanic rocks within the iron-formation
itself does not rule out contemporaneous down-dip volcanism. Such volcanism may have played an active role
in the origin of "Superior-type" early Proterozoic iron-
formations. For paleogeographic and geodynamic interpretations, it is critical to distinguish foredeep
sequences from underlying passive-margin and initial-
rift sequences.

Introduction

Foredeeps (also known as foreland basins and formerly as exogeosynclines) are linear asymmetric depressions located between actively prograding fold-and-
thrust belts and broad cratonic arches (Fig. 1). They
are the continental equivalents of oceanic trenches,
from which they commonly evolve when a passive continental margin underrides an accretionary prism (Fig. 2
and Stockmal et al., in press). However, they may also
develop as a consequence of thrusting initiated in an
intracontinental setting. Foredeeps are an inevitable
consequence of thrusting (Price, 1973) and they contain
a stratigraphic record of the transition, in space and
time, from shelf or cratonic conditions to orogeny.
Taiwan Strait, the Persian-Arabian Gulf and the Indo-
Gangetic Plain are well-known examples of active foredeeps.

Numerical simulations show that foredeeps can be
accounted for mechanically by the flexural loading of
continental lithosphere by overriding thrust sheets
(Beaumont, 1981; Jordan, 1981; Schedl and Wiltschko,
1984). The dimensions of foredeeps and outer arches
depend on the magnitude of the load and the tempera-
ture-dependent rheology of the flexed plate at the time
of loading. Relatively deep, narrow foredeeps develop
on relatively weak (hot) plates. The dimensions may be
subsequently altered due to viscous relaxation of the
plate and the effects of erosion. Relaxation causes
narrowing of the preserved foredeep as the arch migrates toward the load. Continental arches, unlike their
oceanic counterparts, may continue to grow because
they localize erosional unloading (Stephenson, 1984).
Continental margins typically experience repeated
tectonic loading events, producing superimposed
foredeeps and an evolving complexity of outer arches
(Quinlan and Beaumont, 1984). A significant and
neglected feature of many foredeeps is the growth of
normal faults on the outer ramp (Fig. 1), examples of
which are seen in seismic reflection profiles of the
Taconian (Middle Ordovician) foredeep of the Québec
Appalachians (Laroche, 1983) and the Carboniferous
Arkoma Basin of the Ouachitas in the south-central
United States (Buchanan and Johnson, 1968; Lillie,
1984).

The flexural subsidence mechanism of foredeeps contrasts with that of rift basins and rifted continental
margins (Beaumont et al., 1982). Although flexure
occurs at rifted margins due to sediment loading
(Dewey, 1982) it is not the primary subsidence mechanism. During active rifting, subsidence occurs as a
local isostatic consequence of crustal thinning (Vierbuchen et al., 1982). Following rifting, the primary or
"driving" subsidence results from thermal contraction
of the lithosphere as it cools and thickens (McKenzie,
1978; Royden and Keen, 1980; Beaumont et al, 1982). As
will be shown, Proterozoic foredeep sequences have
often been included with, or mistaken for, rifted margin
sequences.

Foredeeps are dynamic systems, in which the basin
migrates with respect to the plate receiving sediment
(Fig. 3). In accordance with Walther's Law, their ver-

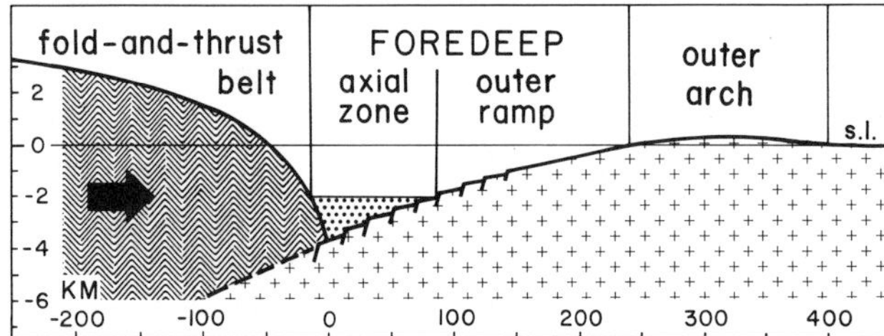

Fig. 1. Tectonic elements associated with foredeeps and nomenclature used in this paper. Note the step-like normal faults forming on the outer ramp. The geometric form of the fold-and-thrust belt is strictly schematic.

tical stratigraphic successions record the lateral range of depositional environments from the outer arch, expressed as a low-angle unconformity, to the front of tectonic accretion. A typical sequence upward from the unconformity is a transgressive coastal marine unit, a deepening-upward veneer of shale and chemical sediment deposited on the outer ramp, and axial zone turbidites (Fig. 4). The axial fill of most foredeeps is immature clastic detritus derived mainly from the fold-and-thrust belt but some are filled from the craton, as will the impending eastern Mediterranean foredeep be filled by the Nile Cone. Breccia beds are characteristic of foredeeps and may be composed of debris shed from normal fault scarps on the outer ramp or thrust sheets on the inner slope. Stratigraphic facies changes in the direction of migration reflect the evolution of the foredeep as a whole, normally changing from being relatively starved of sediment to overflowing (Fig. 3, 4). The older parts of a foredeep are inevitably incorporated within the fold-and-thrust belt. The "euxinic", "flysch" and "molasse" stages of the classic geosynclinal cycle (Pettijohn, 1957, p. 641) can be accounted for as foredeep deposits.

As few Proterozoic foredeeps have previously been recognized as such, it is the purpose of this paper to point out several possible early Proterozoic examples in

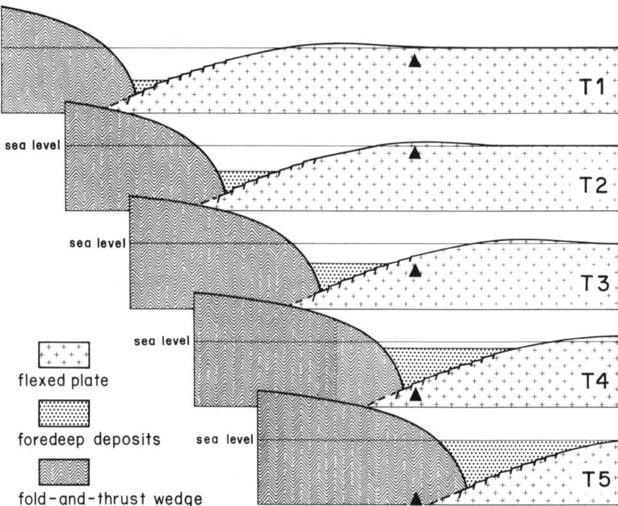

Fig. 3. Foredeep migration causes the site marked by a black triangle to experience uplift and erosion at time T2, deposition of outer-ramp sediment at T3, deposition of axial-zone sediment at T4, and incorporation into the actively deforming allochthonous wedge at T5. Note the progressive sedimentary infilling of the foredeep with time, a diachronous process also in a direction parallel to the foredeep axis.

the Canadian Shield (Fig. 5). In comparison with Phanerozoic foredeeps, two interesting and possibly interrelated differences emerge. First, Proterozoic foredeeps were much more active magmatically. This is surprising, as foredeeps are sites of underthrusting and therefore of below-normal geothermal gradients. Second, the deposition of Superior-type iron-formations occurred preferentially on the outer ramps of Proterozoic foredeeps. This explains the usual intermediate stratigraphic position of such iron-formations between

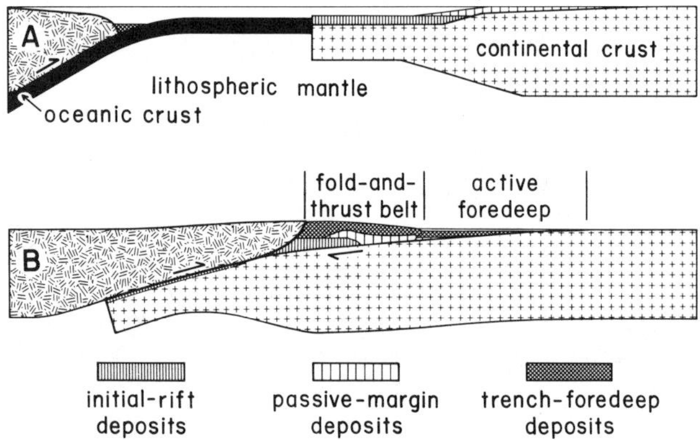

Fig. 2. Evolution of an oceanic trench (A) into a foredeep (B) by attempted subduction of a rifted continental margin.

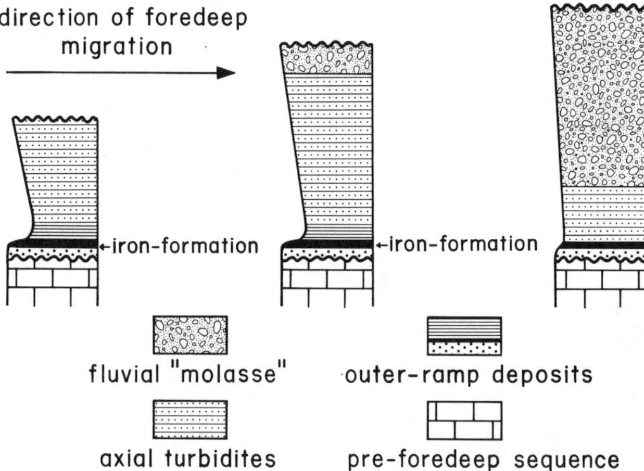

Fig. 4. Typical foredeep stratigraphic sequences. Middle column corresponds to the site of the black triangle in Figure 3. Paleocurrent directions are mainly axial.

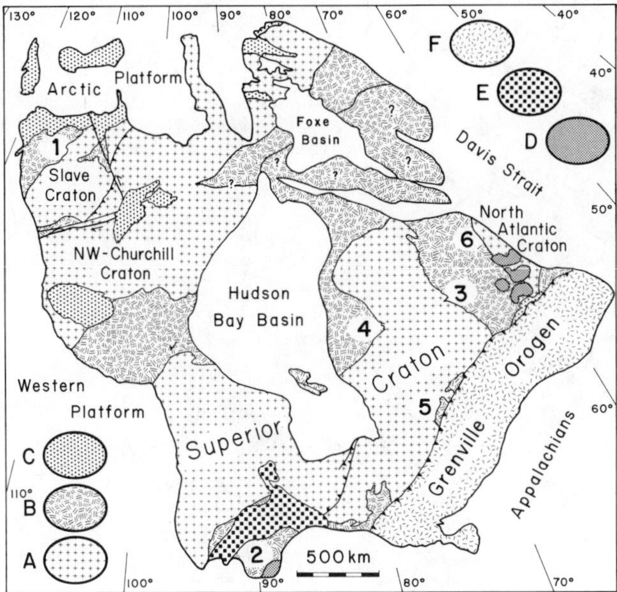

Fig. 5.   Major tectonic elements of the Canadian Shield and five cratonic margins preserving early Proterozoic foredeeps described in the text: 1 - Wopmay Orogen, 2 - Lake Superior region, 3 - Labrador "Trough", 4 - eastern Hudson Bay, 5 - Lake Mistassini, 6 - Ramah Group. Tectonic units: A - Archean cratons more-or-less structurally reactivated in the Proterozoic, B - early Proterozoic (2.5 - 1.6 Ga) epicratonic cover and orogens composed of juvenile early Proterozoic crust and strongly reactivated Archean crust, C - middle and upper Proterozoic (1.6 - 0.6 Ga) cratonic cover, D - middle Proterozoic (1.6 - 1.0 Ga) anorogenic granite-anorthosite intrusions, E - Lake Superior Rift (c 1.13 Ga) and related strata, F - mainly middle and early Proterozoic crust strongly reactivated by the Grenvillian Orogeny (1.2 - 1.0 Ga).

a platformal or shallow shelf sequence below, and a euxinic shale-turbidite sequence above.

Foredeep of Wopmay Orogen

Wopmay Orogen in northwest Canada (Fig. 5) stands out as a Proterozoic orogen about which a well integrated structural-stratigraphic account can be given. The eastern part of the orogen (Fig. 6) exposes a depositional prism, the Coronation Supergroup (formerly Coronation Geosyncline), that was deposited across the west-facing rifted margin of the Archean Slave Craton (Fig. 7). It was subsequently thrusted eastward (Fig. 8) during the Calderian Orogeny (1.89 - 1.87 Ga). The four main structural zones are, from the east: (1) a north-trending arch exposing Archean basement, (2) a west-dipping autochthonous homocline, (3) a thin-skinned, foreland fold-and-thrust belt (Fig. 8), and (4) an allochthonous metamorphic-plutonic hinterland that acted structurally as the rear half of the fold-and-thrust belt (Hoffman et al., in prep.). The foredeep deposits (Fig. 7) occur mainly in structural zones 2 and 3, where they

overlie shallow marine shelf sediments (Epworth Group) that culminate upward in a broad dolomite platform, the Rocknest Formation. The zone 3-4 boundary corresponds to the reefal margin of the Rocknest platform, which marks the allochthonous shelf-slope break (Fig. 6) and restores palinspastically to the western observed limit of autochthonous Archean crust (Fig. 7). During the Calderian Orogeny, the Epworth Group was diachronously depressed in front of the advancing fold-and-thrust belt, buried by deepwater foredeep sediments (Recluse Group) and then structurally detached from its basement, imbricated and translated eastward relative to the autochthon (Fig. 8).

The distribution of sedimentary facies in the foredeep prism is shown in a palinspastic cross-section (Fig. 7). The base of the foredeep is placed at a disconformity 0-10m below the top of the Rocknest Formation. This basal dolomite interval consists of mounded build-ups of branching columnar stromatolites that contrast strikingly with the underlying 100m of stratiform cryptalgal tufa sheets. The stromatolite build-ups are overlain by a sublittoral quartz siltstone unit (Tree River Fm) that contains thin beds of glauconitic ferrodolomite. On the autochthon, the siltstone is capped by a bed of granular hematite iron-formation, up to 10m thick, considerably replaced by spherulitic carbonate. Overlying the iron-formation is a euxinic shale with fine carbonaceous laminations (Fontano Fm). To this point, the vertical sequence records a rapid subsidence and increase in water depth resulting from passage down the migrating outer ramp of the foredeep (Fig. 3). The basal disconformity is poorly developed, reflecting a poorly developed outer arch. This suggests a lithosphere of low flexural rigidity due to the lingering thermal effects of rifting less than 10 Ma earlier (Hoffman and Bowring, 1984).

Evidence of normal faulting on the foredeep outer ramp is limited to the crust that was previously stretched at the time of initial rifting (Fig. 7). The autochthonous basement/cover contact is not offset by normal faults as far west as Exmouth Massif (Fig. 6). Megabreccias occur in the foredeep only west of the allochthonous shelf-slope break. They are mainly associated with the easternmost anticlinorium exposing initial-rift volcanics and capping carbonate reefs (Fig. 6). On the west side of the aniticlinorium, the euxinic shale (Fontano Fm) is in direct contact with the initial-rift assemblage and contains megabreccias derived from it. The slope facies of the passive-margin clastic unit (Odjick Fm) normally intervening stratigraphically between the initial-rift and foredeep sequences is absent, and also is not represented as clasts in the megabreccias. Different interpretations are possible. The contact may be a normal fault on the foredeep outer ramp if the absence of Odjick clasts in the megabreccia can be explained by the Odjick having been unlithified at the time. Alternatively, major block rotations during rifting, as in the Bay of Biscay (Montadert et al., 1979), could have produced ridges of volcanic rock capped by drowned reefs, which were never completely buried by Odjick slope-basin clastics (Fig. 7), and which may or may not have been structurally reactivated during foredeep migration.

The euxinic shale (Fontano Fm) is overlain by succes-

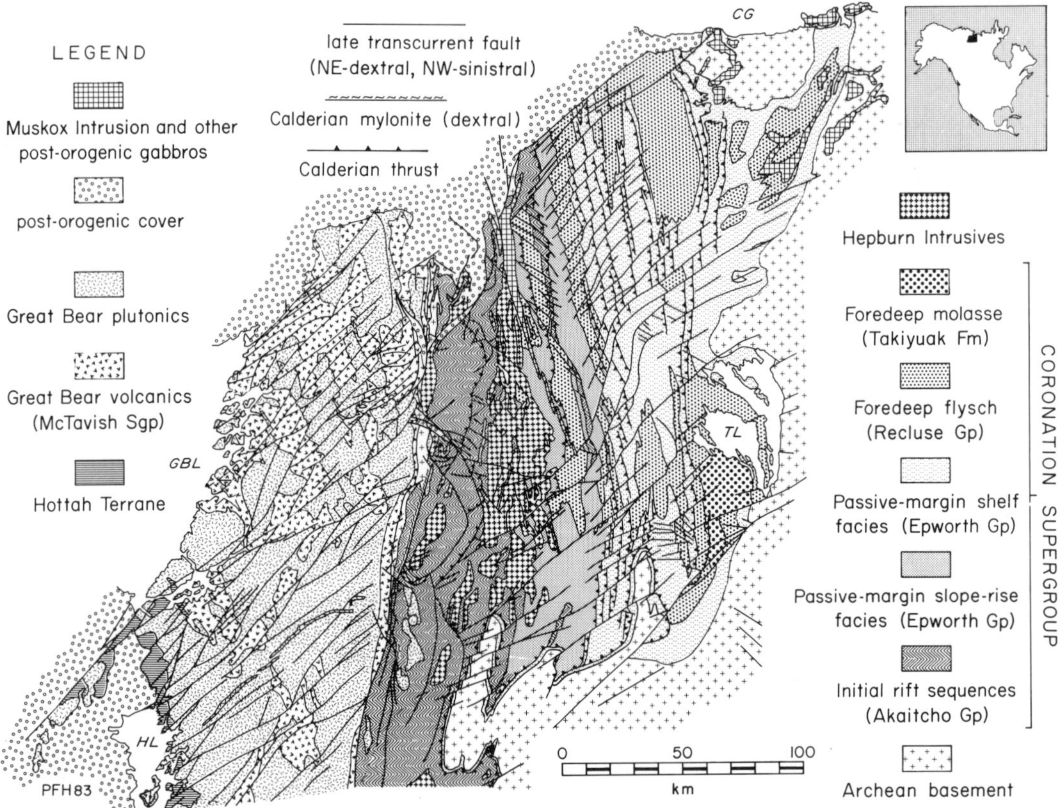

Fig. 6.   Simplified geological map of northern Wopmay Orogen. Prominent water bodies: CG - Coronation Gulf, GBL - Great Bear Lake, HL - Hottah Lake, TL - Takijuq Lake.

sive intervals of calcareous concretionary argillite (Kikerk Fm), laminated limestone-argillite rhythmite (Cowles Fm), and a regionally extensive dolomite-red-bed megabreccia (Fig. 7), which also occurs occurs in correlative intracratonic basins east and south of the basement arch (Campbell and Cecile, 1981; Hoffman, 1981). Several observations, including the indigenous nature of the clasts, the abundance of halite casts, and the gradational upper contact, point to this megabreccia as being the result of dissolution of basin-filling evaporites (Hoffman et al., 1977; Jackson, 1984). The overall vertical sequence above the euxinic shale may therefore reflect increasing salinity in the foredeep with time. The formation contacts would then approximate time-lines in a plane normal to the foredeep.

Axial turbidites (Asiak Fm) form the bulk of the foredeep deposits in the fold-and-thrust belt (Fig. 8). The turbidites are coarse grained, lithic-feldspathic wackes derived from a metamorphic-plutonic provenance like that exposed to the west. The wackes occur as parallel-sided, closely-spaced, graded beds forming mainly A-E type Bouma cycles and were deposited by southward flowing paleocurrents. Their sedimentological character indicates constricted flow in a relatively narrow trough, consistent with flexure of a weak lithospheric plate.

Eastward migration of the foredeep axis can be docu-

mented from the eastward stratigraphic "younging" of the onset of turbidite deposition (Fig. 7), assuming that the Fontano-Kikerk-Cowles-Takiyuak formation boundaries approximate time lines (see above). In the west half of the fold-and-thrust belt, the first turbidites appear in the Fontano Formation; in the east half they first appear in the Kikerk Formation. On the west side of the autochthonous Takijuq Syncline (Fig. 8), the first turbidites are in the Cowles Formation and, on the east side, there is no significant amount of sandstone until the fluvial Takiyuak molasse (Fig. 7).

Numerous gabbro sills (Morel Sills) intrude all foredeep units, including the axial turbidites and the Takiyuak molasse. They also intrude the underlying passive-margin sequence (Epworth Group). They have upper and lower chilled margins and contact metamorphic aureoles. Importantly, they predate all folding and thrusting (Fig. 9) and therefore the inescapable conclusion is drawn that the magmatism occurred in the foredeep axial zone. There is a major concentration of sills close to the allochthonous shelf-slope break (Fig. 6,7,9), suggesting that magma rose easily through the boundary zone between previously stretched and non-stretched Archean crust (Fig. 7). Morel Sills also occur as far east as the autochthonous Takijuq Syncline (Fig. 8), where they intrude the late fluvial foredeep sediments (Takiyuak Fm) and appear, here also, to predate folding.

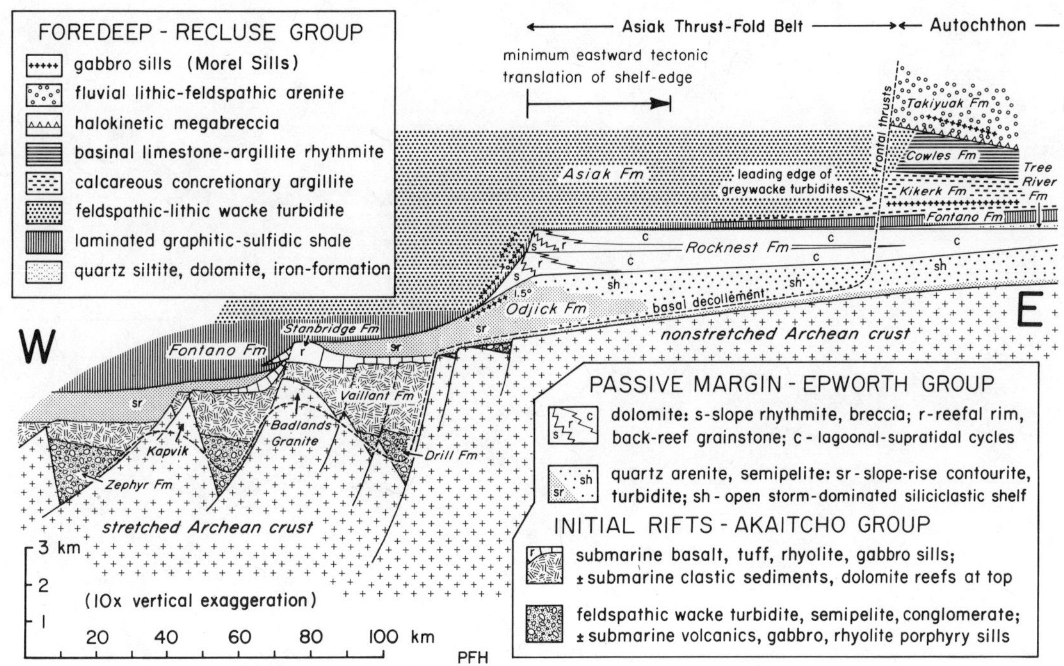

Fig. 7.   Stratigraphic reconstruction of the Coronation Supergroup, quantitatively accurate only east of the Rocknest shelf-edge. Note the eastward stratigraphic "younging" of the leading edge of Asiak Fm greywacke turbidites (axial-zone deposits).

The main areas of Morel Sills have no associated normal faulting.

Circum-Superior Foredeeps

The 2.7 Ga Superior Craton is nearly enclosed by belts of Early Proterozoic sedimentary and volcanic strata marginal to metamorphic-plutonic hinterlands (Fig. 5). Structurally, most of the belts have an autochthonous homocline dipping off the craton, flanked outboard by a foreland fold-and-thrust belt in which tectonic transport was directed toward the craton. Iron-formation is an important component of nearly all the belts and, as in Wopmay Orogen, overlies shallow-water shelf sediments and underlies euxinic shale and turbidites. The postulate of this paper is that the stratigraphically higher parts of these belts, beginning at low-angle unconformities beneath the iron-formation, were deposited in foredeeps. As with Wopmay Orogen, the proposed foredeeps were magmatically active.

Lake Superior Region

The Marquette Range Supergroup is the depositional prism preserved on the south margin of Superior Craton around western Lake Superior (Morey et al., 1982; Morey, 1983). It comprises an autochthonous south-dipping homocline (Mesabi and Gunflint ranges) and, to the south, an easterly trending belt of tight folds and a few thrusts having a general northward structural vergence (Morey, 1983 and references cited therein; Holst, 1984). Deposition began sometime after 2.1 Ga and

metamorphism of the prism associated with the Penokean Orogeny occurred at 1.85 - 1.82 Ga (Van Schmus, 1980). A currently prevalent view is that the orogeny resulted from accretion of volcanic island arcs and microcontinents, exposed in central Wisconsin, onto a rifted, south-facing, continental margin, to which the Marquette Range Supergroup is related (Cambray, 1978; Van Schmus and Bickford, 1981; Larue, 1983; Schulz et al., 1984; LaBerge et al., 1984).

The regional stratigraphy of the Marquette Range Supergroup is reviewed in Morey (1983) and a synopsis is shown in Figure 10. A major internal unconformity separates a continuous overlying sequence, the Animikie Group and its correlatives (Menominee, Baraga and Paint River groups), from an underlying sequence(s) that is preserved only in the south and even there is sporadically absent. The lower sequence(s) (Chocolay, Milles Lacs groups and their correlatives) may be related to rifting and shelf deposition on the south-facing, passive continental margin (Cambray, 1978; Larue, 1981a). The upper sequence (Animikie Group) is here interpreted as the fill of a northward migrating foredeep developed in front of the growing Penokean foldbelt.

The stratigraphic sequence of the Animikie Group is very similar to that of the Wopmay foredeep. A basal, transgressive, shallow-shelf quartzite passes gradationally upward into the main iron-formation (Ojakangas, 1983), which is overlain by euxinic shale passing upward into turbidites. Submarine volcanics, mainly mafic, occur in the south at various horizons, mostly above the iron-formation. Normal faulting associated with Ani-

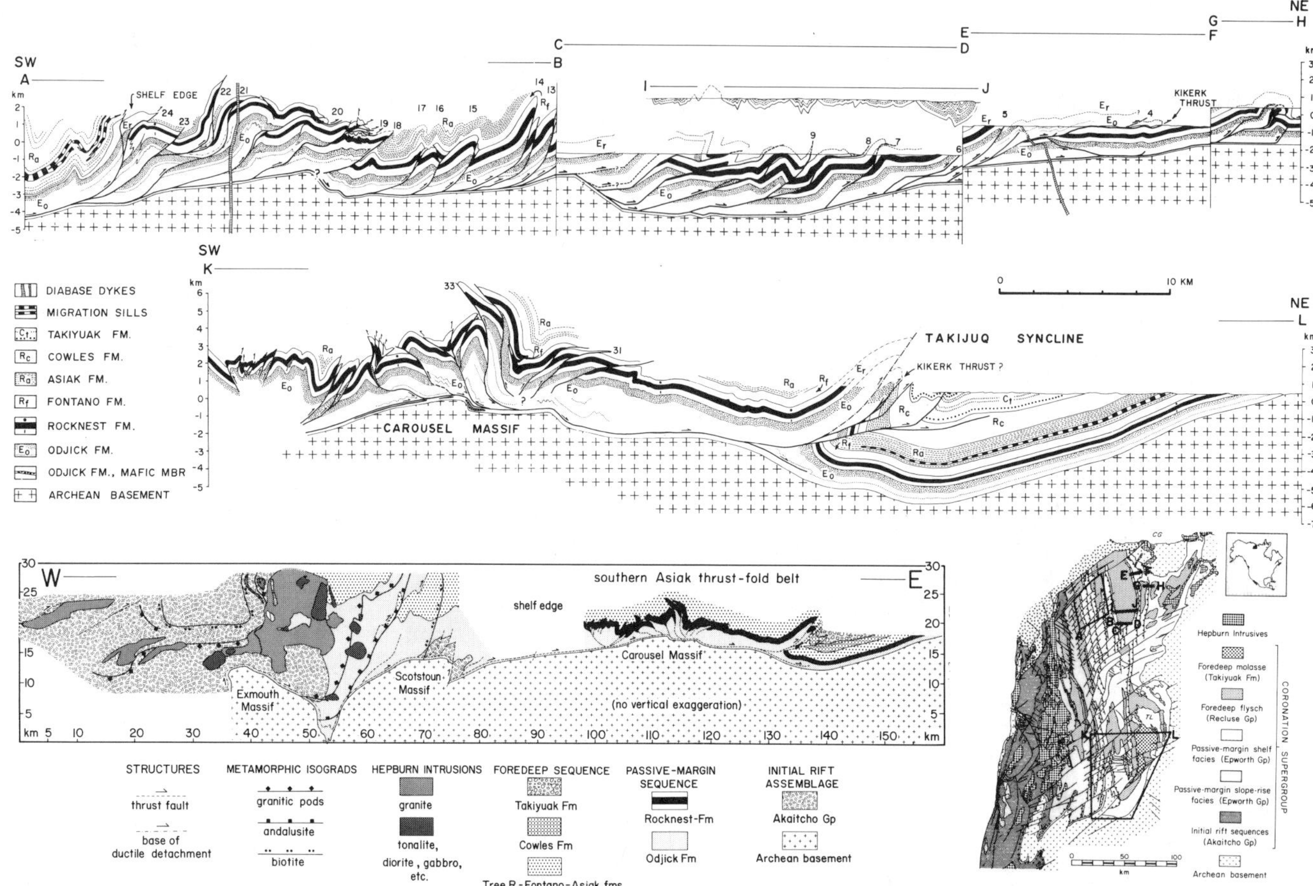

Fig. 8. Composite cross-sections of the foreland fold-and-thrust belt in Wopmay Orogen. See inset map for locations of sections A-L. The sections are reproduced with modifications from Tirrul (1983). For lithologic descriptions see Figure 7. Relatively closely-spaced, steeply-ramping thrusts splay from a basal décollement in pelite about 250 m above crystalline basement and accomplish a minimum 40-45% east-west shortening in the passive-margin sequence (Odjick and Rocknest formations). Comparable strain is achieved in the overlying foredeep turbidites (Asiak Fm) by tight, upright, chevron-type folds. The structural style indicates an active deformational wedge of relatively low taper (Tirrul, 1983), implying a mechanically weak basal décollement and/or relatively high rock-strength within the wedge (Davis et al., 1983). Following thrusting, further east-west shortening produced open folds in the basement and its autochthonous cover, as well as the overlying allochthon. Except for the late thick-skinned folding, the structure is very similar to that observed seismically in the Nankai Trough (Aoki et al., 1982), an active oceanic trench.

mikie sedimentation (Cambray, 1978; Larue, 1981b) may account for local unconformities, locally-derived conglomerates, and rapid thickness variations in the lower Animikie Group. Such normal faulting is here interpreted as occurring on the foredeep outer ramp, possibly reactivating earlier rift-related faults. This faulting may have facilitated localized erosion of the sub-Animikie sequence(s), but regional northward truncation of the older sequence(s) is more likely a consequence of erosion of a migrating outer arch.

Labrador Trough

The Kaniapiskau Supergroup is the Early Proterozoic depositional prism preserved on the northeast margin of Superior Craton (Dimroth et al., 1970; Dimroth, 1981; Wardle and Bailey, 1981; Wardle, 1982; LeGallais and Lavoie, 1982). It comprises an autochthonous east-dipping homocline and, to the east, a thin-skinned fold-and-thrust belt in which tectonic transport was directed toward the craton (Baragar, 1967; Seguin, 1969; Harrison et al., 1970; Wardle, 1982). The two structural zones are traditionally known as the "Labrador Trough", which passes eastward into a broad metamorphic-plutonic hinterland (Taylor, 1979) that separates the Archean Superior and North Atlantic cratons (Fig. 5).

The structure of the fold-and-thrust belt is complex and includes at least three distinct periods of deformation. The first involved brittle, thin-skinned, imbricate

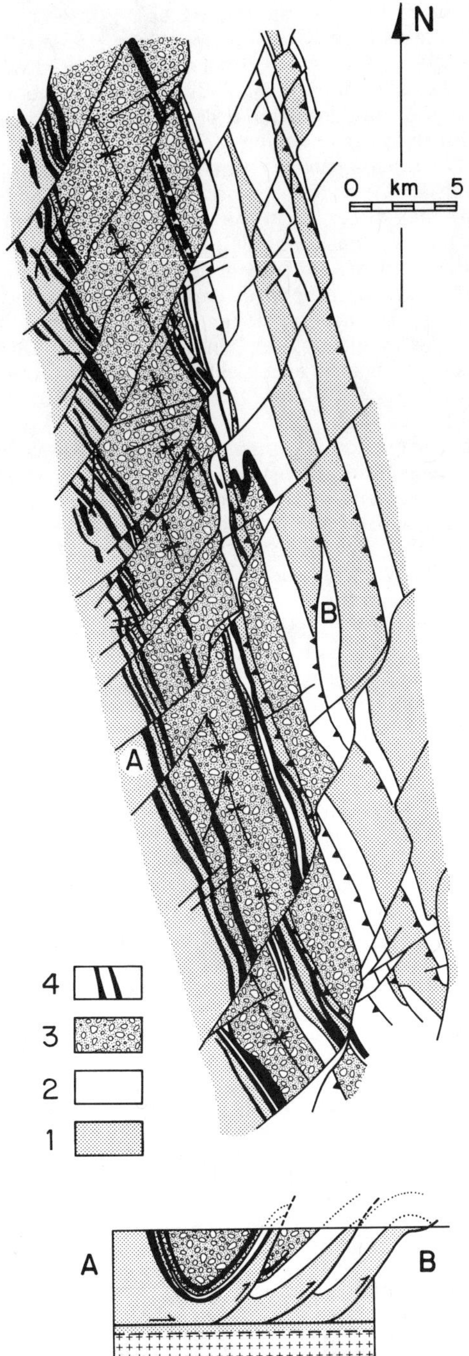

thrusting and related folding above a décollement in the Swampy Bay Group (Fig. 11) (Harrison et al., 1970; Wardle, 1982). The second is characterized by relatively ductile, basement-involved thrusting (Wardle, 1979), verging both toward and away from the craton. The last deformation involved brittle, high-angle(?) faulting of uncertain displacement type, possibly normal. New field observations and reinterpretation of existing geological maps suggest the presence of a relatively small number (3 or 4) of gently-ramping, far-travelled, thrust sheets, each more-or-less internally imbricated, that have been variably affected by the two younger deformations. The palinspastic implications of large-scale thrusting have not been fully dealt with in most attempts to reconstruct the stratigraphy, primary facies and paleogeography. In particular, although the "trough" has been thought of as an in situ basin of deposition, it is possible that most of the rocks preserved within it originated beyond its present eastern margin.

In the west, the Kaniapiskau Supergroup contains an important internal unconformity (Fig. 11), below which is a sequence of four major lithologic groups. Basal immature fluvial clastics (Seward Gp), containing alkaline mafic volcanics, grade upward into a marine, mixed carbonate-siliciclastic, shelf sequence (Pistolet Gp). Shelf drowning is indicated by the overlying euxinic shale and quartz-wacke turbidite sequence (Swampy Bay Gp and Attikamagen Fm in part), which in the east are host to major gabbro sill complexes and related subaqueous mafic volcanics (Dimroth, 1971). The euxinic shale-turbidite sequence shoals upward through westward-prograding stromatolite reef complexes to a supratidal dolomite platform (Denault Fm). The Seward and Pistolet groups may record rifting followed by subsidence of an east-facing, passive continental margin (Wardle and Bailey, 1981; LeGallais and Lavoie, 1981). The overlying deepwater sequence could represent an inner-shelf basin or, alternatively, a first foredeep. However, no fold-and-thrust is known to predate the internal unconformity, as would be expected if the Swampy Bay Group was deposited in a foredeep.

A more compelling case can be made that the sequence above the unconformity was deposited in a foredeep. The unconformity bevels progressively older strata to the west (Fig. 11), consistent with the deve-

Fig. 9. Geological map and cross-section of Morel Sills (black) concentrated at the passive-margin shelf-slope break. Stratigraphic units (see Fig. 7): 1 - Odjick Fm, 2 - Rocknest Fm shelf facies, 3 - foredeep deposits, mainly Asiak Fm turbidites, 4 - Morel Sills. The slope facies Rocknest Fm is too thin to be shown on the west limb of the major foredeep syncline. Note that the sills intrude both passive-margin and foredeep sediments, and that they are folded and thrusted. The same relations are observed for over 200km along strike.

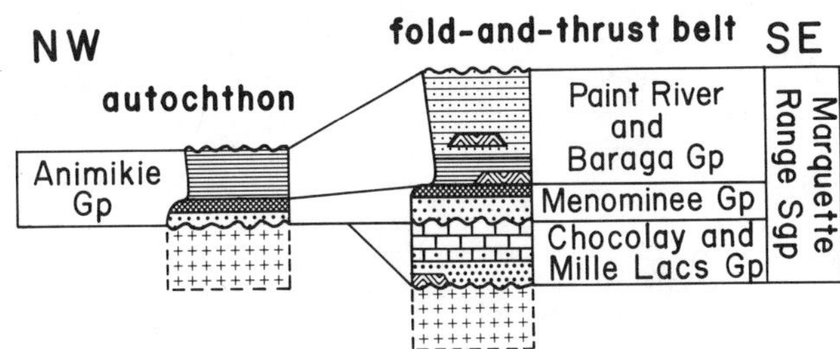

Fig. 10. Synopsis of stratigraphic relations in the Marquette Range Supergroup and its correlatives (Morey, 1983), Lake Superior region. For key to lithologies see caption to Figure 11. The Animikie Group and its correlatives are interpreted as foredeep deposits.

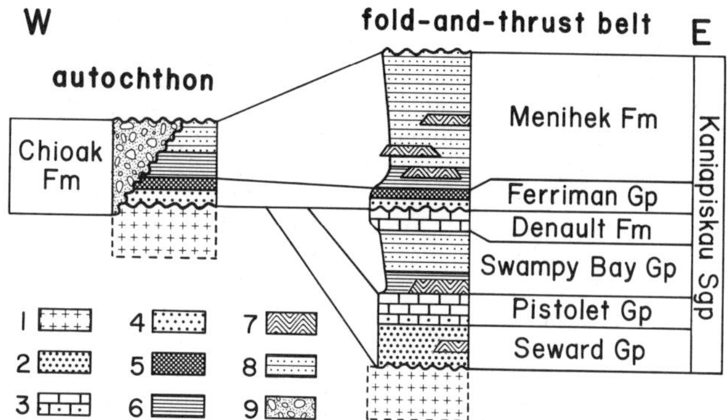

Fig. 11. Synopsis of stratigraphic relations in the Kaniapiskau Supergroup (LeGallais and Lavoie, 1982, modified based on observations by the author), Labrador "Trough". The Ferriman Group (Wishart quartzite and Sokoman iron-formation), Menihek and Chioak formations are interpreted as foredeep deposits. Key to lithologies: 1 - Archean basement, 2 - submature siliciclastic, fluvial to marine sediment, 3 - mixed carbonate-siliciclastic, marine-shelf sediment, 4 - supermature siliclastic, transgressive marine sediment, 5 - iron-formation, 6 - euxinic deepwater shale, 7 - mafic volcanic and subvolcanic rocks, mainly submarine, 8 - turbidites ("flysch"), 9 - immature fluvial conglomeratic sandstone ("molasse").

lopment of a flexural arch in the foreland prior to deposition of the overlying Wishart quartzite, which is a transgressive, high-energy, shelf deposit (Simonson, 1984). The vertical succession above the unconformity is essentially the same as in Wopmay Orogen and the Lake Superior region. The basal quartzite passes upward into the main iron-formation (Sokoman Fm), which is overlain by euxinic shale and a thick sequence of turbidites (Menihek Fm). These units are interpreted as outer ramp and axial zone deposits, respectively, of the foredeep. The structurally higher, more easterly derived, thrust sheets contain gabbro sill complexes and related subaqueous mafic volcanics hosted by the Menihek Formation and, locally, by the iron-formation (Evans, 1978; Gross and Zajac, 1983). The youngest sediments involved structurally in the fold-and-thrust belt are fluvial, polymictic conglomerate and lithic sandstone (Chioak Fm), which overlie an unconformity that cuts stratigraphically downward to the west (Fig. 11), indicating the continued existence of an outer arch.

Eastern Hudson Bay

Early Proterozoic strata on the northwest margin of Superior Craton (Fig. 5) are exposed in an autochthonous homocline around the southeastern periphery of Hudson Bay and, to the west, in a foldbelt exposed on the Belcher Islands and other island groups in eastern Hudson Bay. The structure of the belt is poorly known because most of it is below sea level, but there is a strong possibility that the Belcher Islands foldbelt developed above a shallow décollement, implying some eastward tectonic translation relative to the autochthon.

A synopsis of the regional stratigraphy, based on recent syntheses (Ricketts and Donaldson, 1981; Chandler, 1982, 1984), is shown in Figure 12. In the autochthon, a low-angle unconformity and associated normal faulting divide the depositional prism (Manitounik Supergroup) into lower and upper sequences (Richmond Gulf and Nastapoka groups respectively). In the Belcher Islands foldbelt, there is no angular discordance but karstic megabreccias in the upper Rowatt Formation (Fig. 12) may mark a correlative hiatus. The underlying sequence in the foldbelt comprises fluvial and tidal shelf clastics, carbonates deposited on shallow west-facing platforms and slopes, and subordinate units of flood basalt (Ricketts and Donaldson, 1981). This sequence is interpreted as a west-facing, intraplate, continental terrace (Bell and Hofmann, 1974; Ricketts and Donaldson, 1981).

The upper sequence is interpreted as a foredeep related to westward underthrusting, during collisional orogeny, of the cratonic margin beneath its own, detached, marginal prism (Bell and Hofmann, 1974; Ricketts and Donaldson, 1981; Mukhopadhyay and Gibb, 1981; Chandler, 1982). The vertical succession (Fig. 12) is virtually identical to that in the foredeeps already described. A thin basal carbonate gives way to a sheet of shallow-marine clastics (Mukpollo Fm), which grade upward into an iron-manganese-formation (Kipalu Fm), which is overlain by submarine mafic volcanics (Flaherty Fm). Above the volcanics is a euxinic shale that passes upward into a thick sequence of feldspathic-lithic turbidites (Omarolluk Fm), gradationally overlain by fluvial arkose (Loaf Fm) with southeast-directed paleocurrents.

In the autochthonous homocline, the foredeep sequence lies directly on Archean basement except in the Richmond Gulf Graben, an east-trending, pre-Nastapoka-Group structure (Chandler, 1982). Chandler and Schwarz (1980) interpret the graben as an aulacogen, relating the strongly southeastward-directed, fluvial paleocurrents within it to thermal doming in the **west** prior to continental separation. In this model, the Richmond Gulf Group should predate the passive-margin sequence (lower Belcher Group) of the Belcher

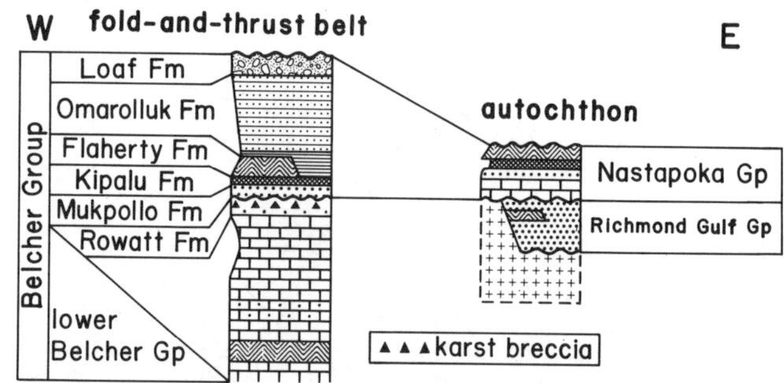

Fig. 12. Synopsis of stratigraphic relations in the early Proterozoic strata of eastern Hudson Bay (Ricketts and Donaldson, 1981; Chandler, 1984). For key to lithologies see caption to Figure 11. The Nastapoka Group and its correlatives are interpreted as foredeep deposits.

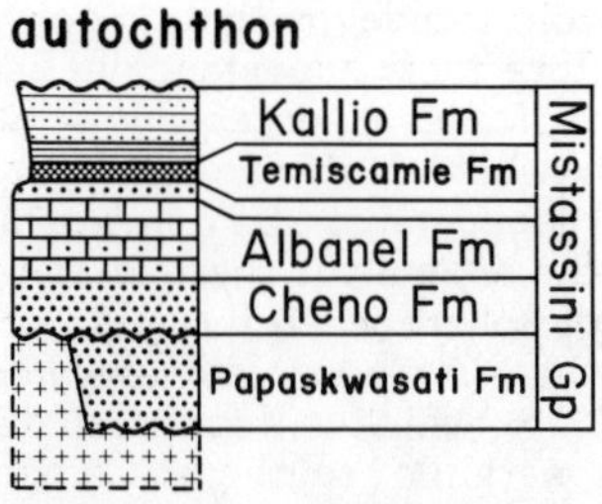

Fig. 13. Synopsis of stratigraphic relations in the Mistassini Group (Chown and Caty, 1973; Gross and Zajac, 1983). For key to lithologies see caption to Figure 11. Strata above the Albanel dolomite are interpreted as foredeep deposits.

Islands foldbelt (Fig. 12). This is consistent with the absence of detritus in the Richmond Gulf Group derived from the passive-margin sequence and the fact that direct correlation of the two is ruled out by the paleocurrent data (Chandler, 1984). Alernatively, the Richmond Gulf Group may be an impactogen, like the Rhine Graben (Sengor et al., 1978), developed contemporaneously with the outer arch and subsequently onlapped by the foredeep sequence as the arch migrated eastward. However, the absence of detritus derived from the passive-margin sequence is inconsistent with the Richmond Gulf Group being younger than the passive margin (Chandler, pers. comm., 1985).

<u>Lake Mistassini Region</u>

The southeast margin of Superior Craton was subjected to northwest-directed overthrusting, crustal thickening and deep erosion during the 1.2 - 1.0 Ga Grenvillian Orogeny (Rivers and Chown, in press). Only two autochthonous homoclinal outliers of Early Proterozoic strata are preserved north of the Grenville Front (Fig. 5). The southern outlier (Mistassini Group) contains the more complete sequence (Fig. 13) of the two (Chown and Caty, 1973; Gross and Zajac, 1983). A basal unit, not everywhere present, consists of conglomeratic fluvial arkose (Papaskwasati Fm). It is overlain by a more extensive clastic unit (Cheno Fm) which grades upward into shallow marine carbonate (Albanel Fm). Paleocurrents flowed southward (Chown and Caty, 1973) and the sequence could record the opening of a south-facing passive continental margin.

The sequence above the Albanel dolomite (Fig. 13) is a possible foredeep. A thin basal clastic sheet overlies the dolomite and passes upward into iron-formation (Temiscamie Fm), which is overlain by euxinic shale and turbiditic wackes (Kallio Fm). The sequence reflects increasing water depth, expected due to passage down the outer ramp of a northward or northwestward-migrating foredeep.

Ramah Group

The Ramah Group is the main early Proterozoic outlier on the west margin of the North Atlantic Craton (Fig. 5), exposed on the coast of northern Labrador. At its western limit, the Ramah Group has been overthrust by structurally reactivated Archean basement (Morgan, 1975a, 1975b, 1978). The thrusts and related folds in the Ramah Group form the eastern border of the early Proterozoic orogen separating the Superior and North Atlantic cratons.

Stratigraphically, the Ramah Group consists of two sequences of contrasting depositional environments, well described by Knight and Morgan (1981). The lower two formations (Fig. 14) are shallow-water, siliciclastic sediments deposited on a west-facing shelf under the influence of mainly west-directed bottom currents. The lower sequence is capped by a thin supratidal dolomite unit, which transgresses underlying deltaic and interdeltaic facies suggesting an abrupt change in depositional regimen. The supratidal dolomite is sharply overlain by the upper sequence, which consists of euxinic, deepwater sediments — black chert, bedded pyrite, and laminated, graphitic and pyritic shale. The shale becomes interbedded upward with pyritic dololutite and sandstone, into which are cut channels of dolomitic and intraformational breccia, interpreted as submarine fan deposits. Paleocurrents and facies changes between the channelized breccias and related outflow deposits indicate a regional west-dipping paleoslope, which would be the outer ramp in a foredeep model. The highest preserved sediments are "proximal" turbiditic sandstones of mixed basement and volcanic provenance, derived from the south. It is suggested here that the upper deepwater sequence was deposited in an eastward-migrating foredeep related to tectonic progradation of the thrust belt to the west.

The Ramah Group is intruded by gabbro sills. Although most numerous in the upper (foredeep) dololutite and shale, the sills also intrude the lower shelf sequence and the uppermost turbidites. They are folded and cut by west-dipping thrusts within the sediments and by the frontal basement thrust (Morgan, 1975b, 1978). If the upper sequence was deposited in a foredeep related to

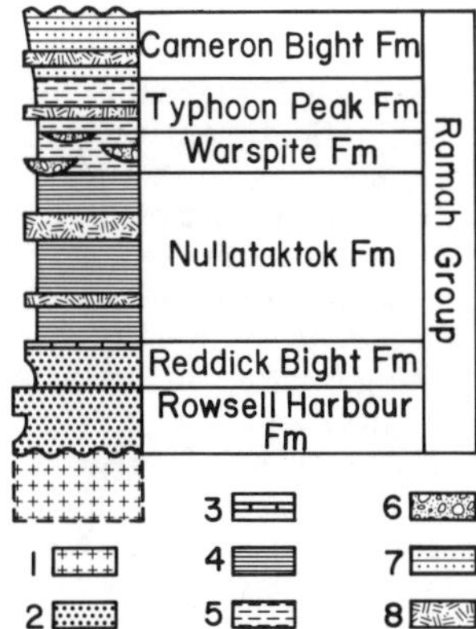

Fig. 14. Synopsis of stratigraphic relations in the Ramah Group (Knight and Morgan, 1981). Key to lithologies: 1 - Archean basement, 2 - shallow shelf siliciclastics, 3 - supratidal dolomite, 4 - euxinic shale, 5 - euxinic dololutite and shale, 6 - dolomitic debris-flow breccias, 7 - turbidites, 8 - altered gabbro sills. Strata above the Reddick Bight Fm are interpreted as foredeep deposits.

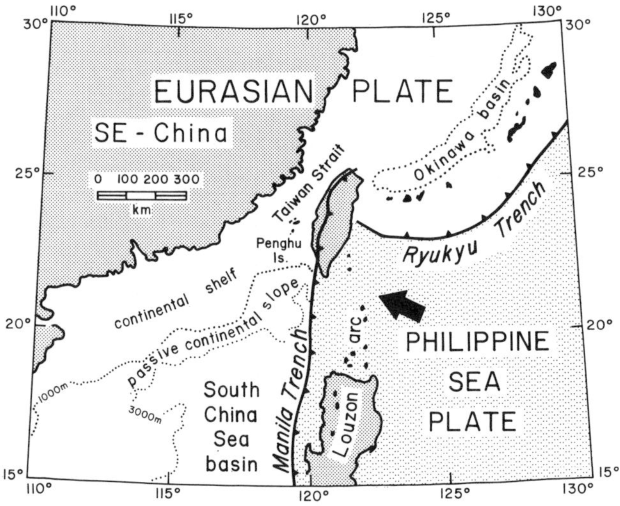

Fig. 15. Location of the active foredeep underlying Taiwan Strait, and its relation to convergence between the Eurasian and Philippine Sea plates, and oblique subduction of the rifted continental margin of southeast China at the Manila Trench. Teeth on the trench-lines are on the overriding plate. The island of Taiwan is an enlarged extension of the fore-arc accretionary prism bounding the Manila Trench. The arrow indicates the motion of the Philippine Sea Plate relative to the Eurasian Plate. The Penghu Islands in Taiwan Strait are underlain by Plio-Pleistocene foredeep basalt flows.

the observed thrusting, then the mafic magmatism must have occurred within the foredeep. Although most of the sills intrude outer ramp sediments, more than 0.5 km of synmagmatic cover would require that magmatism occurred beneath the axial zone turbidites (Cameron Brook Fm).

Foredeep Magmatism

All but one of the Early Proterozoic foredeeps described here was magmatically active. It is important to distinguish the foredeep magmatism from that associated with rifting. Rift-related assemblages, for example the Akaitcho Group of Wopmay Orogen (Fig. 7), normally underlie passive-margin sedimentary prisms because passive-margin subsidence follows as a thermally-driven consequence of rifting (McKenzie, 1978; Royden and Keen, 1980; Beaumont et al., 1982). Although the foredeep magmatism in the circum-Superior belts has often been attributed to rifting (eg. Wardle and Bailey, 1981; Baragar and Scoates, 1981; LeGallais and Lavoie, 1982), much of it post-dates the passive-margin sequence. According to stratigraphic and structural relations already given, the magmatism may have occurred in foredeep axial zones, where it was associated with turbidites and, to a lesser extent, on outer ramps, where it accompanied the deposition of iron-formation and euxinic shale. The magmatism is overwhelmingly mafic, indicating mantle derivation, the less than 1% rhyolite present suggesting a minimal contribution from melting of the continental crust and sediment.

Few Phanerozoic foredeeps contain coeval magmatic rocks and therefore their importance in Early Proterozoic foredeeps indicates a perceptible secular change in lithospheric behaviour. Hoffman and Bowring (1984) suggest that foredeep magmatism in Wopmay Orogen was related to the anomalous thermal state of the cratonic margin lithosphere when thrusting occurred only 10 Ma after rifting. A Cenozoic analogy occurs in Taiwan Strait (Fig. 15), which is a foredeep developed on the continental margin of southeast China where it is being obliquely overridden by the foreland fold-and-thrust belt of the island of Taiwan (Seno, 1977; Wu, 1979; Suppe, 1981; Davis et al., 1983). Foredeep magmatism is manifested by 320m of basaltic lava of early Pleistocene (Ho, 1975) or Pliocene (Taylor and Hayes, 1980) age that underlie the Penghu Islands, located in the strait about 80 km in front of the present leading edge of thrusting, which is advancing at a rate of about 70km/Ma (Seno, 1977). The age of opening of the continental margin of southeast China, as given by the oldest magnetic anomalies in the South China Sea (Taylor and Hayes, 1980) and a widespread "breakup" unconformity (Holloway, 1982), is mid-Oligocene (about 30 Ma). Therefore, Taiwan Strait and Wopmay Orogen have a common association of foredeep magmatism and the underthrusting of thermally immature, continental margin lithosphere. Why magmatism should be common in most Early Proterozoic foredeeps remains unclear however. It could reflect a thermal regime for Early Proterozoic lithosphere that was generally higher than later on, or alternatively a shorter average duration of passive continental margins (due to faster oceanic plate recycling), or some other factor.

The mechanisms of magma genesis and ascent beneath foredeeps are speculative. It would be helpful if the orientation of feeder dykes were known, as a clue to the state of stress in the plate through which the magma must have risen. Feeders oriented parallel to the foredeep axis might result from block subsidence of the plate (Freund et al., 1980). Those oriented perpendicular to the axis may be a consequence of sideways extension accomodating convergence of the plates as they lock in collision. An example of Cenozoic foredeep magmatism of the latter type occurs on the north margin of the Arabian Plate in southern Turkey (Lovelock, 1984), which is being overriden from the north by the Eurasian Plate as part of the Taurus-Bitlis-Zagros collision zone.

The areal distribution of magmatism in the Wopmay foredeep may be significant. It was strongly localized at the shelf-slope break in the underlying passive-margin sequence. Palinspastically, this restores to the transition zone between stretched and non-stretched, upper continental crust (Fig. 7). For some reason, this zone was magmatically activated with relative ease.

According to the interpretation advanced here, the most conspicuous mafic magmatic rocks in the forelands of early Proterozoic orogens in the Canadian shield are related neither to obduction of ophiolites nor to initial rifting of cratonic margins. This presents a problem for the determination of Proterozoic tectonic environments from direct comparisons of Proterozoic and Cenozoic rock chemistries. Cenozoic foredeep magmatism is rare and poorly characterized. The appa-

rent ubiquity of magmatism in early Proterozoic foredeeps makes one wonder about processes in oceanic
trenches of that age. It also raises questions regarding
the nature of Archean "greenstone belts".

Superior-type Iron-formations

The origin and significance of the giant 2.5 - 1.8 Ga
iron-formations are controversial subjects, having generated a vast literature (see Trendall and Morris, 1983,
and references therein). There are two aspects to the
problem, their confinement in geologic time and their
paleogeographic setting, of which the second is of more
concern here. Nearly all of the Superior-type iron-
formations overlie shallow shelf sequences and underlie
deepwater euxinic shale and turbidites. This strongly
suggests that their deposition was related to major
shelf-drowning events (Simonson, 1985). Sedimentary
facies of iron-formation indicate that depositional environments ranged from shallow marine shelves to deep
restricted basins (James, 1954) and the vertical succession of facies shows that the water deepened with time
(Ojakangas, 1983). These observations are well explained in the foredeep model.

The migratory nature of foredeeps may help to resolve some long-standing problems concerning the origin of Superior-type iron-formations. Several authors
(eg. Trendall, 1965; Gross, 1965; Cameron, 1983) have
favoured a genetic link with volcanism, but the associated volcanic rocks typically occur stratigraphically
above, rather than within, the iron-formation. This fits
the predicted sequence of a diachronous foredeep with
contemporaneous axial-zone magmatism and outer-
ramp iron-formation. Thus, the absence of volcanic
rocks directly associated with iron-formation does not
preclude contemporaneous down-dip volcanism. Axial-
zone magmatism (intrusive and extrusive) might drive
hydrothermal circulation, influencing the chemical and
thermal regimes of the foredeep and its interstitial
fluids.

Foredeep migration might also resolve the the para-
dox of iron-formations of great lateral extent being
deposited in highly restricted basins (James, 1954;
Trendall and Blockley, 1970; Becker and Clayton, 1972).
The instantaneous width of a foredeep places no limit
on the lateral extent of its migration.

In the foredeep model, iron-formations represent a
specific tectonosedimentary facies. Therefore, they
should not be the basis for interbasinal stratigraphic
correlation.

The foredeep model obviously does not solve the iron-
formation problem, nor does it deny the possible importance of an evolving atmosphere (Cloud, 1973). It does
provide a specific and dynamic paleogeographic setting
that has not previously been advocated for the deposition of giant early Proterozoic iron-formations.

Conclusions

Foredeep sequences deposited in front of, and incorporated partly within, coeval fold-and-thrust belts have
been recognized in the stratigraphically upper parts of
six early Proterozoic foreland belts in the Canadian
Shield. They closely resemble Phanerozoic foredeeps in
most ways but differ significantly in two respects.
First, they were much more active magmatically. The
geodynamic significance of this is uncertain, but the
most conspicuous mafic magmatism in the early Proterozoic forelands of the Canadian Shield originated
neither by the obduction of ophiolites nor the initial
rifting of cratonic margins. The second difference is
that Superior-type iron-formations characterize the
outer-ramp facies of early Proterozoic foredeeps. This
interpretation provides a more specific and dynamic
paleogeographic setting with which to reconsider the
problematic origin of such iron-formations. The prevalent deposition of iron-formations in early Proterozoic
foredeeps may be genetically related to foredeep magmatism.

The distinction between foredeep and pre-orogenic
sequences is essential for correct paleogeographic interpretations. For example, paleomagnetic data from
foredeep sequences deposited during collisional orogeny
cannot provide information regarding relative motions
prior to the collision.

In order to firmly document the evolution of foredeeps through geologic time, it would be desirable to
compare early and later Proterozoic foredeeps (eg.
Volta Basin, Damarides, Carpentarian, etc.) and attempt to recognize Archean foredeeps.

Acknowledgments. The opportunity to compare the
Wopmay foredeep with others in the Canadian Shield
arose through participation in the Decade of North
American Geology program (DNAG). The author is
grateful to Rein Tirrul for permission to reproduce
Figure 8 and to John Grotzinger for discussions and
ideas. Fred Chandler, Erich Dimroth, Rein Tirrul, Dick
Wardle and one anonymous reviewer made helpful critical comments on the manuscript. Ron Bell is thanked
for drafting most of the figures.

References

Aoki, Y., T. Tamano, and S. Kato, Detailed structure of
the Nankai Trough from migrated seismic sections, in
Studies in Continental Margin Geology, edited by J.S.
Watkins and C.L. Drake, pp. 309-324, Am. Assoc.
Petrol. Geol. Mem. 34, Tulsa, 1982.

Baragar, W.R.A., Wakuach Lake map-area, Quebec-
Labrador, Geol. Surv. Can. Mem. 344, pp. 1-174, 1967.

Baragar, W.R.A., and R.F.J. Scoates, The Circum-
Superior belt: a Proterozoic plate margin?, in
Precambrian Plate Tectonics, edited by A. Kröner,
pp. 297-330, Elsevier, Amsterdam, 1981.

Beaumont, C., Foreland basins, Geophys. J. Roy.
Astron. Soc., 65, 291-329, 1981.

Beaumont, C., C.E. Keen, and R. Boutilier, A comparison of foreland and rift margin sedimentary basins,
Phil. Trans. Roy. Soc. Lond. Ser. A, 305, 295-317,
1982.

Becker, R.H., and R.N. Clayton, Carbon isotopic evidence for the origin of banded iron-formation in
Western Australia, Geochim. Cosmochim. Acta, 36,
577-595, 1972.

Bell, R.T., and H.J. Hofmann, Investigations of the Belcher Group (Aphebian), Belcher Islands, N.W.T., Geol.
Assoc. Can. Program Abstr., p. 8, 1974.

Buchanan, R.S., and F.K. Johnson, Bonanza gas field: a model for Arkoma basin growth faulting, in Geology of the Western Arkoma Basin and Ouachita Mountains, Oklahoma, edited by L.M. Cline, pp. 75-85, Oklahoma City Geol. Soc. Guidebook, 1968.

Cambray, F.W., Plate tectonics as a model for the environment of deposition and deformation of the early Proterozoic of northern Michigan, Geol. Soc. Am. Abstr. Programs 10, p. 376, 1978.

Cameron, E.M., Genesis of Proterozoic iron-formation: sulphur isotope evidence, Geochim. Cosmochim. Acta, 47, 1069-1074, 1983.

Campbell, F.H.A., and M.P. Cecile, Evolution of the Early Proterozoic Kilohigok Basin, Bathurst Inlet-Victoria Island, Northwest Territories, in Proterozoic Basins of Canada, edited by F.H.A. Campbell, pp. 103-132, Geol. Soc. Can. Paper 81-10, 1981.

Chandler, F.W., The structure of the Richmond Gulf Graben and the geological environments of lead-zinc mineralization and of iron-manganese formation in the Nastapoka Group, Richmond Gulf area, New Quebec - Northwest Territories, in Current Res. Part A, Geol. Surv. Can., Paper 82-1A, 1982.

Chandler, F.W., Metallogenesis of an early Proterozoic foreland sequence, eastern Hudson Bay, Canada, J. Geol. Soc. Lond., 141, 299-313, 1984.

Chandler, F.W., and E.J. Schwarz, Tectonics of the Richmond Gulf area, Northern Quebec: a hypothesis, in Current Res. Part C, Geol. Surv. Can., Paper 80-1C, 1980.

Chown, E.H., and J.L. Caty, Stratigraphy, petrography and paleocurrent analysis of the Aphebian clastic formations of the Mistassini-Otish basin, in Huronian Stratigraphy and Sedimentation, edited by G.M. Young, pp. 49-71, Geol. Assoc. Can. Spec. Paper 12, 1973.

Cloud, P.E., Paleoecological significance of banded iron-formation, Econ. Geol., 68, 1135-1143, 1973.

Davis, D., J. Suppe, and F.A. Dahlen, Mechanics of fold-and-thrust belts and accretionary wedges, J. Geophys. Res., 88, 1153-1172, 1983.

Dewey, J.F., Plate tectonics and the evolution of the British Isles, J. Geol. Soc. Lond., 139, 371-412, 1982.

Dimroth, E., The evolution of the central segment of the Labrador Geosyncline, Part II: The ophiolitic suite, N. Jb. Geol. Paleont. Abh., 137, 209-248, 1971.

Dimroth, E., Labrador geosyncline: type example of early Proterozoic cratonic reactivation, in Precambrian Plate Tectonics, edited by A. Kröner, pp. 331-352, Elsevier, Amsterdam, 1981.

Dimroth, E., Baragar, W.R.A., Bergeron, R., and Jackson, G.D., The filling of the Circum-Ungava geosyncline, in Basins and Geosynclines of the Canadian Shield, edited by A.J. Baer, pp. 45-142, Geol. Surv. Can. Paper 70-40, 1970.

Evans, J.L., The geology and geochemistry of the Dyke Lake area, Labrador, Newfoundland Mineral Development Div. Rep. 78-4, St. John's, pp. 1-39, 1978.

Freund, R., D. Kosloff, and A. Matthews, A dynamic model of subduction zones, in Mechanisms of Continental Drift and Plate Tectonics, edited by P.A. Davies and S.K. Runcorn, pp. 17-40, Academic Press, London, 1980.

Gross, G.A., Geology of iron deposits in Canada: general geology and evaluation of iron deposits, Geol. Surv. Can. Econ. Geol. Rep. 22(1), pp. 1-181, 1965.

Gross, G.A., and I.S. Zajac, Iron-formation in fold belts marginal to the Ungava craton, in Iron-Formation: Facts and Problems, edited by A.F. Trendall and R.C. Morris, pp. 253-294, Elsevier, Amsterdam, 1983.

Harrison, J.M., J.E. Howell, and W.F. Fahrig, A geological cross-section of the Labrador miogeosyncline near Schefferville, Quebec, Geol. Surv. Can. Paper 70-37, pp. 1-34, 1970.

Ho, C.S., An introduction to the geology of Taiwan: explanatory text of the geological map of Taiwan, Ministry Econ. Affairs, Taipei, pp. 1-153, 1975.

Hoffman, P.F., Autopsy of Athapuscow Aulacogen: a failed rift arm affected by three collisions (abstr.), in Proterozoic Basins of Canada, edited by F.H.A. Campbell, pp. 97-102, Geol. Surv. Can. Paper 81-10, 1981.

Hoffman, P.F., and S.A. Bowring, Short-lived 1.9 Ga continental margin and its destruction, Wopmay orogen, northwest Canada, Geology, 12, 68-72, 1984.

Hoffman, P.F., I.R. Bell, R.S. Hildebrand, and L. Thorstad, Geology of the Athapuscow Aulacogen, east arm of Great Slave Lake, District of Mackenzie, in Current Res. Part A, Geol. Surv. Can. Paper 77-1A, pp. 117-129, 1977.

Hoffman, P.F., J.E. King, M.R. St-Onge, and R. Tirrul, Thin- and thick-skinned crustal shortening as portrayed in axial projections of eastern Wopmay orogen and the Cape Smith greenstone belt, northern Canadian shield, in Processes in Continental Lithospheric Deformation, edited by S.P. Clark, Jr., B.C. Burchfiel and J. Suppe, Geol. Soc. Am. Mem., in press.

Holloway, N.H., North Palawan block, Philippines: its relation to Asian mainland and role in evolution of South China Sea, Am. Assoc. Petrol. Geol. Bull., 66, 1355-1383, 1982.

Holst, T.B., Evidence for nappe development during the early Proterozoic Penokean orogeny, Minnesota, Geology, 12, 135-138, 1984.

Jackson, M.J., Sedimentology of an early Proterozoic calc-flysch to molasse transition in NW Canada: rhythmite to reef to river, Geol. Assoc. Can. Program Abstr. 9, p. 76, 1984.

James, J.L., Sedimentary facies of the Lake Superior iron-formations, Econ. Geol., 49, 253-293, 1954.

Jordan, T.E., Thrust loads and foreland basin evolution, Cretaceous, western United States, Am. Assoc. Petrol. Geol. Bull., 65, 2506-2520, 1981.

Knight, I, and W.C. Morgan, The Aphebian Ramah Group, northern Labrador, in Proterozoic Basins of Canada, edited by F.H.A. Campbell, pp. 313-330, Geol. Surv. Can. Paper 81-10, 1981.

LaBerge, G.L., P.E. Myers, and K.J. Schulz, Early Proterozoic plate tectonics: evidence from north-central Wisconsin, Geol. Soc. Am. Abstr. Programs 16, p. 567, 1984.

Laroche, P.J., Appalachians of southern Quebec seen through Seismic Line No. 2001, in Seismic Expression of Structural Styles 3, edited by A.W. Bally, pp. 3.2.1-7, Am. Assoc. Petrol. Geol., Tulsa, Oklahoma, 1983.

Larue, D.K., The Chocolay Group, Lake Superior region, U.S.A.: sedimentologic evidence for deposition in basinal and platform settings on an early Proterozoic craton, Geol. Soc. Am., 92, 417-435, 1981a.

Larue, D.K., The early Proterozoic pre-iron-formation Menominee Group siliciclastic sediments of the southern Lake Superior region: evidence for sedimentation in platform and basinal settings, J. Sed. Petrol., 51, 397-414, 1981b.

Larue, D.K., Early Proterozoic tectonics of the Lake Superior region: tectonostratigraphic terranes near the purported collision zone, in Early Proterozoic Geology of the Great Lake Region, edited by L.G. Medaris, Jr., pp. 33-48, Geol. Soc. Am. Mem. 160, Denver, 1983.

LeGallais, C.J., and S. Lavoie, Basin evolution of the lower Proterozoic Kaniapiskau Supergroup, central Labrador miogeocline (trough), Quebec, Bull. Can. Petrol. Geol., 30, 150-166, 1982.

Lillie, R.J., Tectonic implications of subthrust structures revealed by seismic profiling of Appalachian-Ouachita orogenic belt, Tectonics, 3, 619-646, 1984.

Lovelock, P.E.R., A review of the tectonics of the northern Middle East region, Geol. Mag., 121, 577-587, 1984.

McKenzie, D.P., Some remarks on the development of sedimentary basins, Earth Planet. Sci. Lett., 40, 25-32, 1978.

Montadert, L., O. de Charpal, D. Roberts, P. Guennoc, and J.-C. Sibuet, North-east Atlantic passive margins: rifting and subsidence processes, in Deep Drilling Results in the Atlantic Ocean, edited by M. Talwani et al., pp. 164-186, AGU Continental Margins and Paleoenvironment, series 3, Washington, 1979.

Morey, G.B., Animikie basin, Lake Superior region, U.S.A., in Iron-Formation: Facts and Problems, edited by A.F. Trendall and R.C. Morris, pp. 13-68, Elsevier, Amsterdam, 1983.

Morey, G.B., P.K. Sims, W.F. Cannon, M.G. Mudrey, Jr., and D.L. Southwick, Geologic map of the Lake Superior region, Minnesota, Wisconsin and northern Michigan, Minn. Geol. Surv. State Map Ser. S-13, scale 1:1 million, 1982.

Morgan, W.C., Geology of the Precambrian Ramah Group and basement rocks in the Nachvak Fiord-Saglek Fiord area, northern Labrador, Geol. Surv. Can. Paper 74-54, pp. 1-42, 1975a.

Morgan, W.C., Geology, Nachvak Fiord-Ramah Bay, Newfoundland-Quebec, Geol. Surv. Can. Map 1469A, scale 1:50,000, 1975b.

Morgan, W.C., Geology, Bears Gut-Saglek Fiord, Newfoundland, Geol. Surv. Ca. Map 1478A, scale 1:50,000, 1978.

Mukhopadhyay, M., and R.A. Gibb, Gravity anomalies and deep structure of eastern Hudson Bay, Tectonophysics, 72, 43-60, 1981.

Ojakangas, R.W., Tidal deposits in the early Proterozoic basin of the Lake Superior region: the Palms and the Pokegama formations: evidence for subtidal-shelf deposition of Superior-type banded iron-formation, in Early Proterozoic Geology of the Great Lakes Region, edited by L.G. Medaris, Jr., pp. 49-66, Geol. Soc. Am. Mem. 160, Denver, 1983.

Pettijohn, F.J., Sedimentary Rocks, 2nd edition, Harper, New York, pp. 1-718, 1957.

Price, R.A., Large-scale gravitational flow of supracrustal rocks, southern Canadian Rockies, in Gravity and Tectonics, edited by K.A. De Jong, and R. Scholten, pp. 491-502, Wiley, New York, 1973.

Quinlon, G.M., and C. Beaumont, Appalachian thrusting, lithospheric flexure, and the Paleozoic stratigraphy of the eastern interior of North America, Can. J. Earth Sci., 21, 973-996, 1984.

Ricketts, B.D., and J.A. Donaldson, Sedimentary history of the Belcher Group of Hudson Bay, in Proterozoic Basins of Canada, edited by F.H.A. Campbell, pp. 235-254, Geol. Surv. Can. Paper 81-10, 1981.

Rivers, T., and E.H. Chown, The Grenville Orogen in eastern Québec and western Labrador: definition, identification and tectonometamorphic relationships of autochthonous, parautochthonous and allochthonous terranes, in New Perspectives on the Grenville Problem, edited by A. Davidson, J.M. Moore and A. Baer, Geol. Assoc. Can. Spec. Paper, in press.

Royden, L., and C.E. Keen, Rifting process and thermal evolution of the continental margin of eastern Canada determined from subsidence curves, Earth Planet. Sci. Lett., 51, 343-361, 1980.

Schedl, A., and D.V. Wiltschko, Sedimentological effects of a moving terrain, J. Geol., 92, 273-287, 1984.

Schulz, K.J., G.L. LaBerge, P.K. Sims, Z.E. Peterman, and J. Klasner, The volcanic-plutonic terrane of northern Wisconsin: implications for early Proterozoic tectonism, Lake Superior region, Geol. Assoc. Can. Program Abstr. 9, p. 103, 1984.

Seguin, M., Configuration et nature du mode tectonique en bordure ouest de la partie centrale do la fosse du Labrador, Can. J. Earth Sci., 6, 1365-1379, 1969.

Sengor, A.M.C., K. Burke, and J.F. Dewey, Rifts at high angles to orogenic belts: tests for their origin and the upper Rhine graben as an example, Am. J. Sci., 278, 24-40, 1978.

Seno, T., The instantaneous rotation vector of the Philippine Sea plate relative to the Eurasian plate, Tectonophysics, 42, 209-226, 1977.

Simonson, B.M., High-energy shelf deposit: early Proterozoic Wishart Formation, northeastern Canada, in Siliciclastic Shelf Sediments, edited by R.W. Tillman, and C.T. Siemers, pp. 253-270, Soc. Econ. Paleont. Mineral. Spec. Publ. 34, Tulsa, 1984.

Simonson, B.M., Sedimentological constraints on the origins of Precambrian iron-formations, Geol. Soc. Am. Bull., 96, 244-252, 1985.

Stephenson, R., Flexural models of continental lithosphere based on the long-term erosional decay of topography, Geophys. J. Roy. Astron. Soc., 77, 385-413, 1984.

Stockmal, G., R. Boutilier, and C. Beaumont, Geodynamic models of convergent margin tectonics: the transition from rifted margin to overthrust belt, Geology, in press.

Suppe, J., Mechanics of mountain building and metamorphism in Taiwan, Geol. Soc. China Mem. 4, Taipei, pp. 67-89, 1981.

Taylor, B., and D.E. Hayes, The tectonic evolution of the South China basin, in The Tectonic and Geologic Evolution of Southeast Asian Seas and Islands, edited by D.E. Hayes, pp. 89-104, AGU Geophysical Monograph 23, Washington, D.C., 1980.

Taylor, F.C., Reconnaissance geology of a part of the Precambrian Shield, northeastern Quebec, northern Labrador and Northwest Territories, Geol. Surv. Can. Mem. 393, pp. 1-99, 1979.

Tirrul, R., Structure cross-sections across Asiak Foreland Fold-and-Thrust Belt, Wopmay Orogen, District of Mackenzie, in Current Research Part B, Geol. Surv. Can. Paper 83-1B, pp. 253-260, 1983.

Trendall, A.F., Origins of Precambrian iron-formations, Econ. Geol., 60, 1065-1070, 1965.

Trendall, A.F., and J.G. Blockley, The iron-formations of the Precambrian Hamersley Group, Western Australia, West. Aust. Geol. Surv. Bull. 119, pp. 1-366, 1970.

Trendall, A.F., and R.C. Morris, Iron-Formation: Facts and Problems, Elsevier, Amsterdam, pp. 1-558, 1983.

Van Schmus, W.R., Chronology of igneous rocks associated with the Penokean orogeny in Wisconsin, in Selected Studies of Archean Gneisses and Lower Proterozoic Rocks, Southern Canadian Shield, edited by G.B. Morey and G.N. Hanson, pp. 159-168, Geol. Soc. Am. Spec. Paper 182, Denver, 1980.

Van Schmus, W.R., and M.E. Bickford, Proterozoic chronology and evolution of the mid-continent region, North America, in Precambrian Plate Tectonics, edited by A. Kröner, pp. 261-296, Elsevier, Amsterdam, 1981.

Vierbuchen, R.C., R.P. George, and P.R. Vail, A thermal-mechanical model of rifting with implications for outer highs on passive continental margins, in Studies in Continental Margin Geology, edited by J.S. Watkins and C.L. Drake, pp. 765-780, Am. Assoc. Petrol. Geol. Mem. 34, Tulsa, Oklahoma, 1982.

Wardle, R.J., Geology of the eastern margin of the Labrador Trough, Newfoundland Mineral Development Div. Rep. 78-9, St. John's, pp. 1-22, 1979.

Wardle, R.J., Geology of the south-central Labrador Trough, Newfoundland Mineral Devlopment Div. Map 82-5, St. John's, scale 1:100,000, 1982.

Wardle, R.J., and D.G. Bailey, Early Proterozoic sequences in Labrador, in Proterozoic Basins of Canada, edited by F.H.A. Campbell, pp. 331-359, Geol. Surv. Can. Paper 81-10, 1981.

Wu, F.T., Recent tectonics of Taiwan, in Geodynamics of the Western Pacific, edited by S. Uyeda, R.W. Murphy and K. Kobayashi, pp. 265-299, Adv. Earth Planet. Sci. 6, Center Acad. Publ., Tokyo, 1979.

SEISMIC FEATURES OF PROTEROZOIC CRUST IN NORTHERN AUSTRALIA AND THEIR EVOLUTION

D. M. Finlayson

Bureau of Mineral Resources, Geology and Geophysics, P.O. Box 378, Canberra City ACT-2601, Australia

Abstract. The crust of the central part of the Proterozoic North Australian Craton is 51-54 km thick with an underlying mantle velocity of 8.16-8.20 km/s increasing to 8.29-8.38 km/s at 60-80 km. The crustal thickness decreases to 44-45 km at the craton margins to the north and northeast. Intra-crustal velocity increases are evident over a large depth range (19-40 km) with the increases being transitional both within the crust and at the crust/mantle boundary. Ubiquitous deep crustal reflections are characterised by their lack of continuity over any great distance. Geochemical and geochronological studies of the crustal evolution of Proterozoic Australia identify a major tectono-thermal event at 1850-1880 Ma extending over an area equivalent to one-third of the continent. The igneous rocks are almost exclusively I-type with a short prehistory. The style of metamorphism envisages rapid crustal thickening in the lower crust as a result of small-scale convection at about 2200-2000 Ma under a stationary or slowly-moving lithosphere. The event separates two major sedimentary cycles. Three younger anorogenic events with geochemical histories similar to the major event are also identified. It seems reasonable to infer that, in a variety of circumstances, a number of intrusive and underplating cycles near the base of the crust can produce a range of lower crustal velocity/depth distributions and thicknesses, and it is the intrusive nature of the process that we see in the character of many deep reflecting horizons.

Introduction

During the last ten years there have been considerable improvements in the quality and quantity of geochemical and geochronological data available from Precambrian Australia. Page et al. (1984) have emphasised the episodic formation of protolithic blocks during the Precambrian by the addition of primitive material from the mantle or juvenile lower crust. One highlight of this work has been the identification of a major tectono-thermal event across northern Australia within the time window of 1850-1880 Ma. The event can be identified across an area equivalent to about one-third of continental Australia. This widespread event contrasts with the ideas of progressive lateral accretion seen in some Phanerozoic terranes.

There have also been a number of explosion seismic investigations of the lithosphere in northern Australia, both at vertical incidence and at wide-angle incidence. These have led to an improved knowledge of the structural horizons within the crust and of the velocity structure of the lithosphere.

This report principally assesses the contribution of seismic survey results to our state of knowledge of the crust and upper mantle under Proterozoic northern Australia. These results are then viewed in the light of recent models for the evolution of Proterozoic Australia put forward by research scientists working in the fields of geochemistry and geochronology.

Crustal Evolution Model

The most recent evolutionary model for Proterozoic northern Australia is here summarised from the work of Page et al. (1984) and Etheridge et al. (1984). Page et al. (1984) recognised seven coherent episodes of crustal generation in Precambrian Australia; these were at 3.45-3.7 Ga, 3.2 Ga, 2.8 Ga, 2.5-2.7 Ga, 1.9-2.1 Ga, 1.5-1.8 Ga and 1.0-1.3 Ga. In particular they noted the lack of crustal events between 2.5 and 2.1 Ga. This was followed by the most important Proterozoic mantle/lower crustal event of continental proportions dated at 1.9-2.1 Ga for the age of primary crustal formation. It is recognised in the Halls Creek Inlier, the Arunta Block, the Tennant Creek, Pine Creek and Mount Isa Inliers and the Broken Hill Block, i.e. over about one-third of the area of continental Australia (Figure 1).

This event was followed 200-300 Ma later by final igneous crystallisation in the upper crust manifest as, and defined by, widespread metamorphic and/or felsic magmatism between 1.9 and 1.85 Ga. There is no geochemical evidence for the extensive involvement of a pre-existing Archaean crust. The event must therefore be regarded as the major Proterozoic crustal-forming event in Australia. However the event does not preclude the existence of an Archaean protolith.

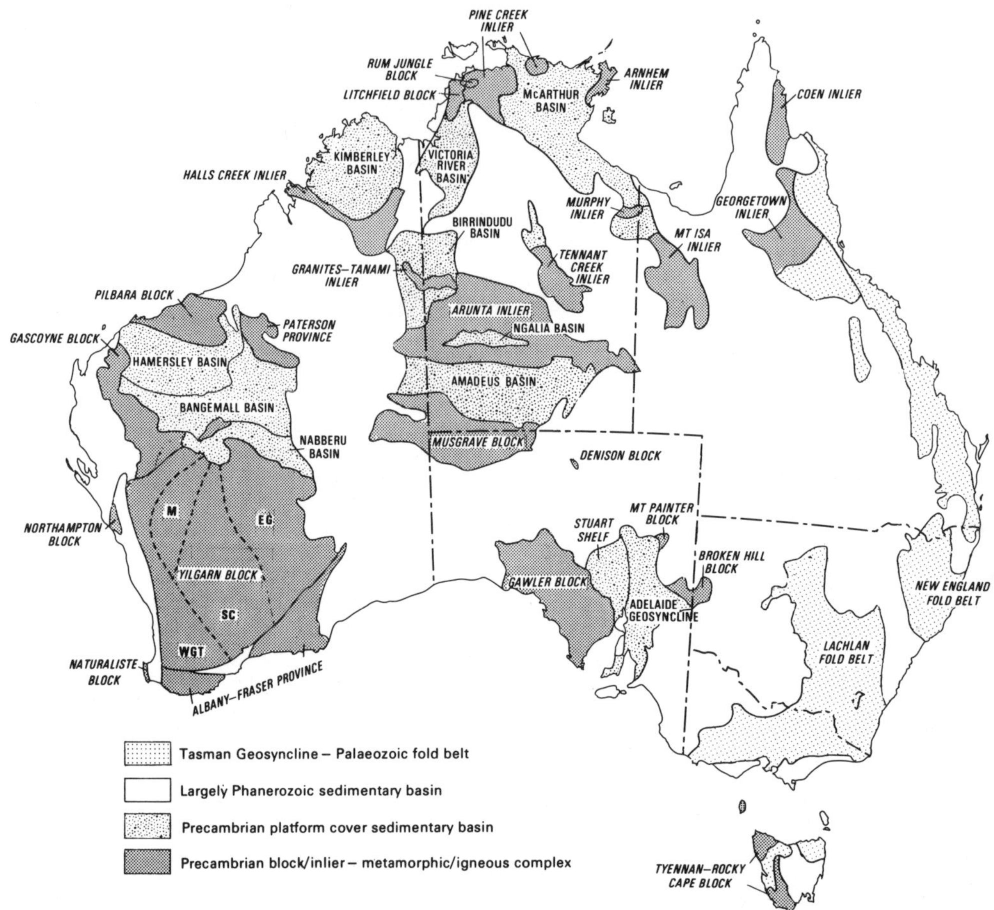

Fig. 1. Major Australian Precambrian orogenic provinces, sedimentary basins, and the Tasman fold belt (from Page et al.,1984).

The Pine Creek Inlier for instance, totally surrounds the Archaean Rum Jungle Block. It is quite possible that other small Archaean remnants exist under platform cover.

Etheridge et al. (1984) confine the time range of the major tectono-thermal event to 1850-1880 Ma, characterised by distinctive felsic igneous events of large magnitude and consistent chemistry. The igneous rocks are almost exclusively I-type, and isotopic modelling implies a short prehistory. Younger anorogenic felsic intrusives with ages of 1800-1780 Ma, 1760-1740 Ma and 1670-1640 Ma are geochemically distinct from the major orogenic rocks. The younger rocks are, however, also best modelled as being derived by partial melting of deep crustal mafic parents with short crustal residence times (Etheridge et al., 1984).

The major event at 1850-1880 Ma separates two sedimentary cycles which are seen by Etheridge et al. (1984) as being the result of stretching of a pre-existing continental lithosphere. The post-orogenic sedimentary sequence was subsequently deformed and metamorphosed at 1650-1550 Ma.

The style of metamorphism and the deduced P-T-t paths during the major orogenic event suggest rapid crustal thickening with ubiquitous low and (rarely) medium pressure metamorphic facies, somewhat different from modern collisional orogens with their trend towards high metamorphic facies and paired metamorphic belts.

The crust of the Early-Middle Proterozoic Australian provinces is envisaged as comprising sedimentary sequences from one or both sedimentary cycles, a thin upper-middle crustal layer of stretched Archaean protolith and a predominantly mafic lower crustal layer underplated about 2000 Ma. The mantle-derived magmatic event is seen as being the result of small-scale convection beginning about 2200-2000 Ma under a stationary or slow-moving lithospheric plate. The polygonal character of the Proterozoic terranes in Australia is seen as being analogous to the modern pattern of rifts and swells in the African continent (Etheridge et al., 1984).

Seismic Surveys

The locations of explosion seismic surveys in Proterozoic northern Australia from which

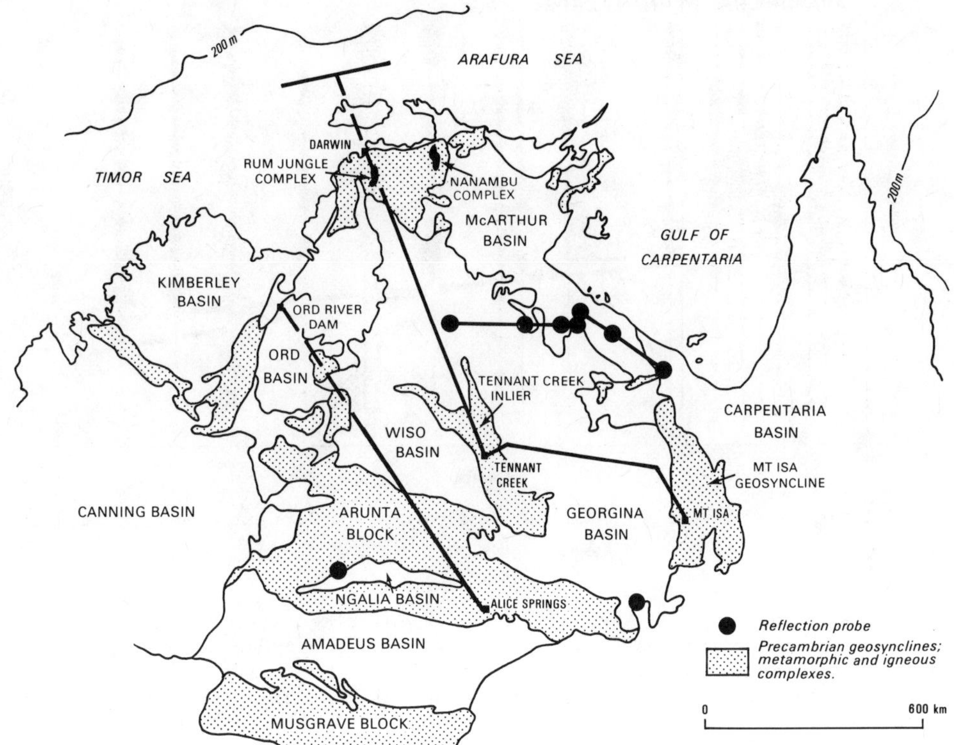

Fig. 2. Principal geological provinces of the North Australia Craton and central Australia, and locations of seismic investigations. Continuous line = seismic refraction recording stations; dashed line = no recorders.

lithospheric information can be obtained are shown in Figure 2. Most of these are wide-angle reflection/refraction surveys with recording distances of over 1000 km in places. Some surveys are, however over shorter distances and enable details of crustal velocity features to be interpreted. At a number of locations (Fig. 2) there are also data from deep reflection probes. These probes have not been conducted over large distances and are sufficient only to give an indication of the nature of deep reflecting horizons.

Western Arafura Sea-Pine Creek Inlier

Along a marine profile in the western Arafura Sea, Rynn and Reid (1983) interpreted a 34 km thick crust with an uppermost 2 km sedimentary section (Fig. 2). However shots from the same location were also recorded across the Pine Creek Inlier to the Tennant Creek Inlier. Hales and Rynn (1978) interpreted an upper mantle velocity of 8.18 km/s and a crustal thickness of 45 km. Below 76 km the P-wave velocity increases to 8.38 km/s. Figure 3 shows data recorded along the profile. Because there are few recordings at distances less than 150 km, very little detailed information can be interpreted about intra-crustal structure under the Pine Creek Inlier.

McArthur Basin

Two wide angle reflection/refraction profiles have been shot in the McArthur Basin region (Fig. 2). The western profile is almost wholly across the central part of the North Australian Craton; the eastern profile is probably more representative of the McArthur Basin sequence (Collins, 1983). Recording of vertical incidence waves was conducted at seven locations on the wide-angle reflection/ refraction profiles.

Under the eastern profile, the McArthur Basin sequence is about 3.3 km thick with velocities in the range 3.5-5.6 km/s. Under the western profile sediments up to 9 km thick with velocities of 4.1-5.9 km/s overly basement. There are no outstanding velocity discontinuities within the crust below basement and the crust/mantle boundary is interpreted as being gradational by Collins (1983) (Fig. 4). The upper mantle velocities are reached at a depth of 44 km under the eastern profile and 53 km under the western profile. Only the upper mantle velocity under the eastern profile (7.80 km/s) is considered to have been reliably determined.

The depth to the crust/mantle boundary under the McArthur Basin (44 km) (Fig. 4, profile 3) is close to the Hales and Rynn (1978) value for the Pine Creek Inlier (45 km) (Fig. 4, profile 2).

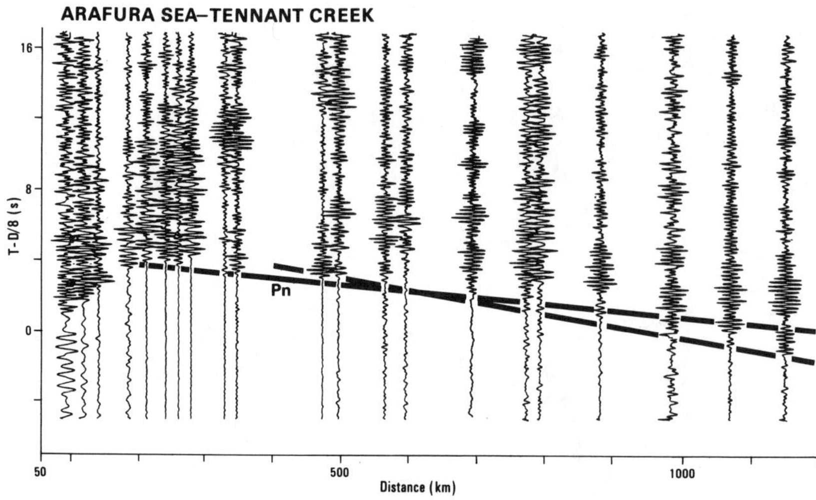

Fig. 3. Seismic record section from the Arafura Sea to Tennant Creek (from Hales and Rynn,1978).

However towards the central part of the North Australian Craton the Moho depth increases to about 53 km (Fig. 4, profiles 4 and 5).

Collins (1983) and Mathur (1983) have described the character of the reflections from the vertical-incidence recordings. In the upper crust at 2-way times of less than 7 s there are notable differences between the eastern and western profiles. The eastern profile has reflecting bands within the upper-crustal basement suggesting strongly layered basement rocks (Fig. 5). The western profile has no such reflections in the

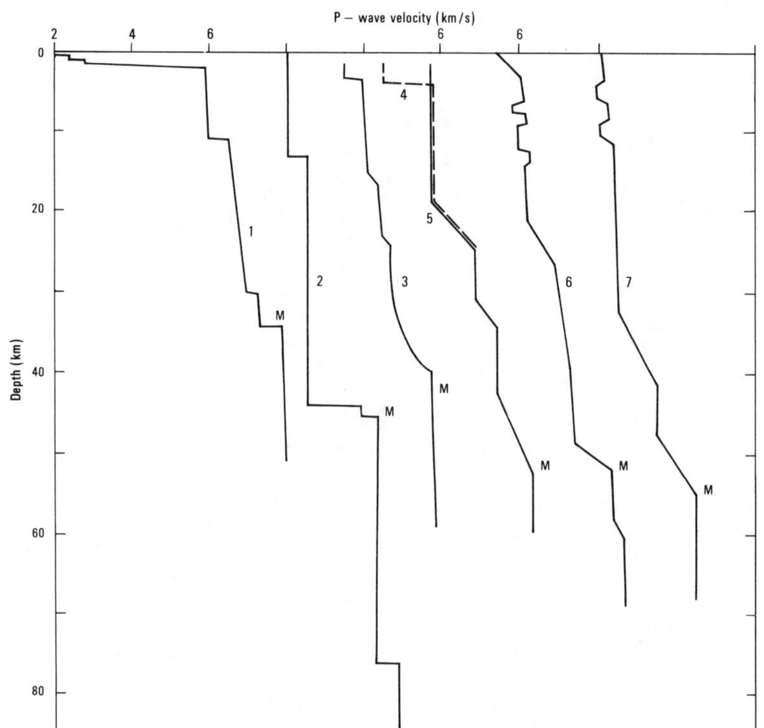

Fig. 4. Velocity/depth profiles for the North Australian Craton: 1) Arafura Sea; 2) Arafura Sea-Tennant Creek; 3) eastern McArthur Basin; 4 & 5) western McArthur Basin; 6) Tennant Creek-Mount Isa; 7) Mount Isa-Tennant Creek. M = Mohorovicic discontinuity.

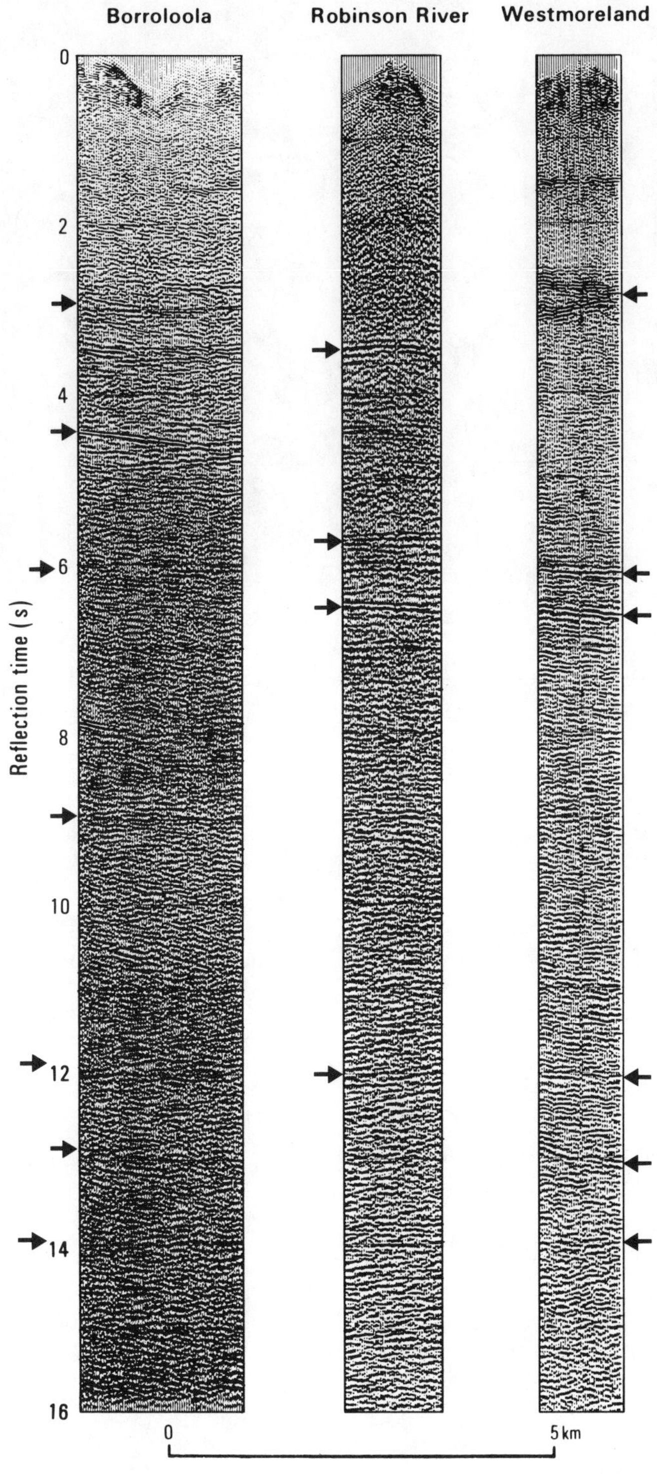

Fig. 5. Vertical incidence reflection sections from three sites in the eastern McArthur Basin (from Collins,1983), with some prominent reflectors indicated with arrows. Sites approximately 150 km apart. Recording summary:- Recorder, DFS-IV; source, explosives; spread length, 3 km; 36 traces at 83.3 m spacing; cross spread, 12 traces; 4 ms sampling to 16 s. Processing:- True amplitude recovery, time variant scaling, NMO and static corrections; 4-fold CDP stack.

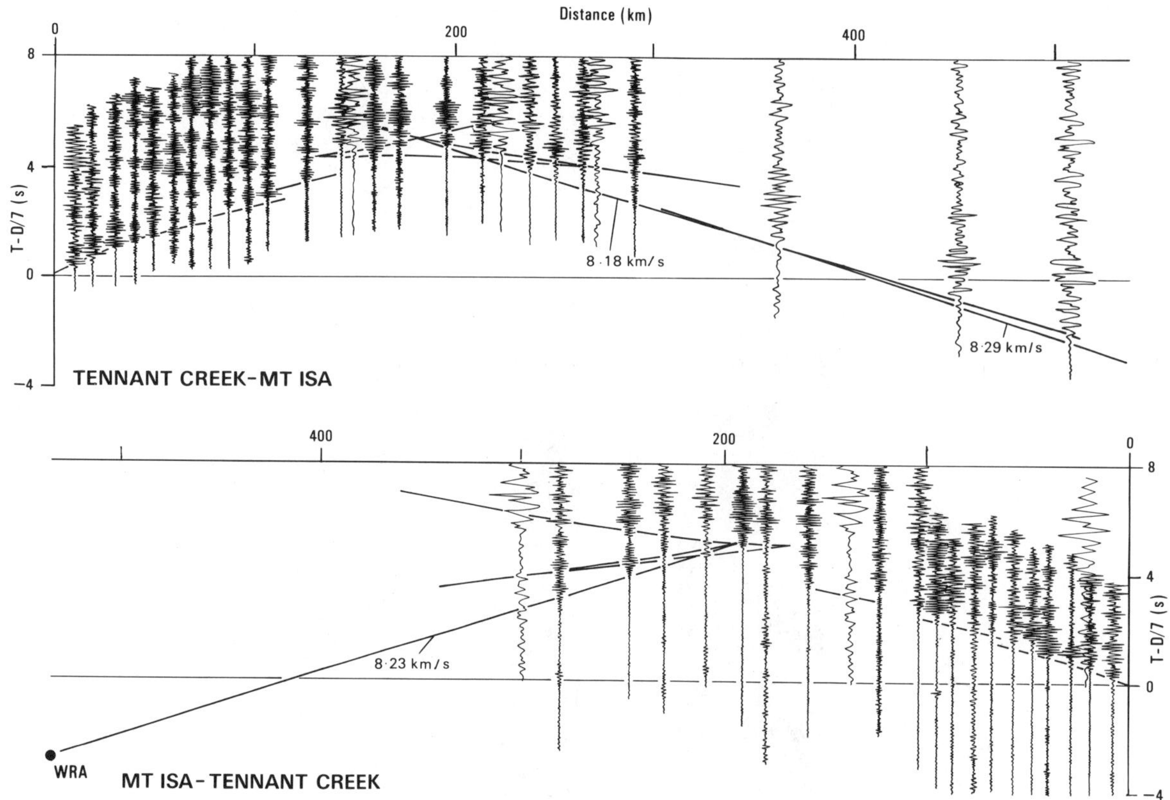

Fig. 6. Seismic record sections a) Tennant Creek - Mount Isa, and b) Mount Isa - Tennant Creek (from Finlayson, 1982).

upper crust, emphasising the difference between the upper crustal tectonics under the two profiles.

At 2-way reflection times greater than 7 s the character of the reflecting horizons is similar under both profiles, namely short (up to 1 km) weak segments down to the limit of recording (16 s). The reflections tend to increase in strength, continuity and coherency with depth.

Tennant Creek-Mount Isa

Between Tennant Creek and Mount Isa, seismic recordings have been made at distances up to 600 km (Finlayson, 1982). Basement rocks crop out at Tennant Creek and Mount Isa but the intervening basement is largely hidden by up to 10 km of late Proterozoic-Palaeozoic sediments in the Georgina Basin (Fig. 6).

Near Tennant Creek, Finlayson (1981, 1982) interpreted seismic data in terms of near-surface velocities of 5.5-6.2 km/s being representative of the Warramunga Group low-grade metamorphics and weathered granites ovelying amphibolite facies rocks of older metamorphic domains with velocities greater than 6.2 km/s. These latter rocks are possibly interspersed with lower velocity rocks (5.7-5.9 km/s). Near Mount Isa, the upper crustal

velocities of 6.0-6.2 km/s correspond to the Leichhardt Metamorphics with interspersed granites. Between Tennant Creek and Mount Isa an upper mantle velocity of 8.16-8.20 km/s is reached at depths of 51-54 km. The Moho depth is in accord with the value for the western profile in the McArthur Basin. The upper mantle velocity agrees with that of Hales and Rynn (1978) for their north-south traverse. A velocity increase at 61 km depth in the sub-crustal lithosphere to 8.29 km/s is determined by Finlayson (1982) whereas the Hales and Rynn (1978) data indicate an increase to 8.38 km/s at 76 km depth (Fig. 4).

There is no seismic reflection profiling near the Tennant Creek-Mount Isa traverse but, farther south in the Georgina Basin, Mathur (1983a,b) has described profiling data at the edge of the Precambrian craton. The character of the reflections is similar to that under the western McArthur Basin, namely, a general lack of reflections in the upper crustal basement but short discontinuous reflecting segments in the lower crust showing increasing strength, continuity and coherency with depth (Fig. 7).

Ord Dam-Alice Springs

Some reconnaissance data are available from a survey conducted in 1970-71 when recordings were

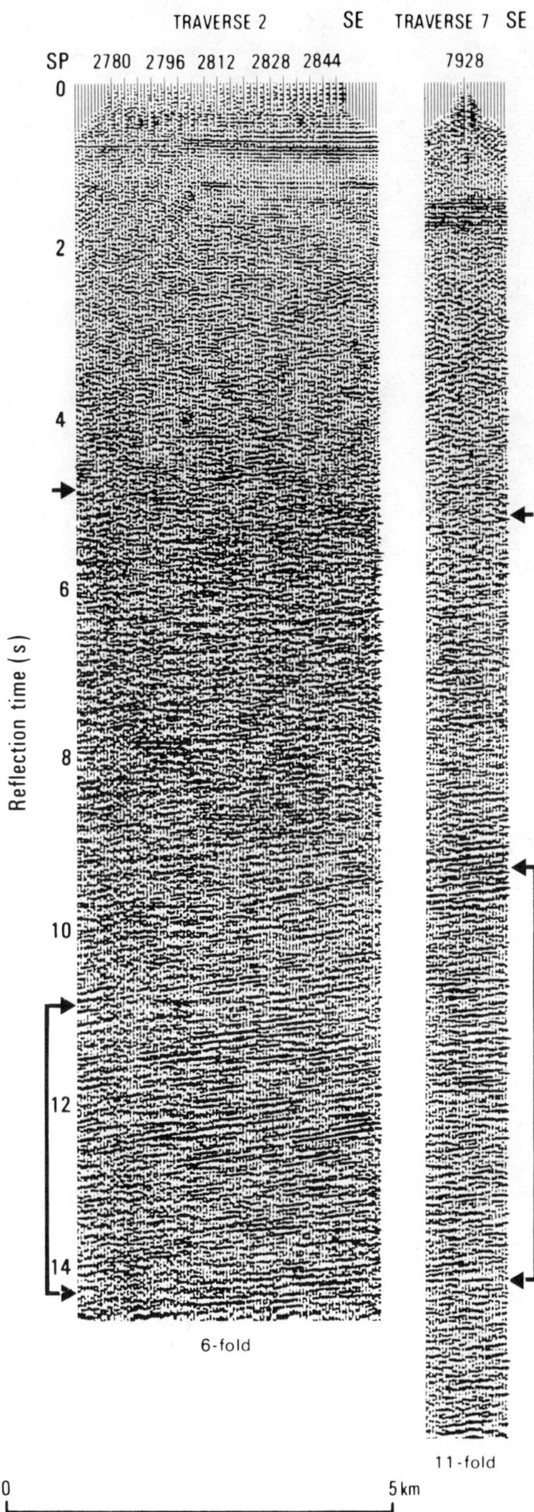

Fig. 7. Vertical incidence reflection sections
from the southern Georgina Basin (from
Mathur,1983a,b), with the start of prominent
lower crustal reflections and reflection bands

made, at widely spaced stations, of shots fired at
the Ord River damsite near the northern end of the
Halls Creek Inlier. The survey was designed as an
upper mantle investigation and no useful
information can be expected about even the gross
features of crustal structure.

Denham et al. (1972) indicate that, at
distances less than 864 km across the
Granites-Tanami and Arunta Blocks, an upper mantle
velocity of 8.20 km/s was determined, in broad
agreement with the 8.16-8.20 km/s determined from
the more detailed later work of Finlayson (1982)
and Hales and Rynn (1978). A value of 8.17 km/s
was estimated from the Ord dam shots in the
direction of Tennant Creek. However the early
travel-times (by 1.5-2.0 s) for sites towards
Tennant Creek, compared with sites towards Alice
Springs, points to there being marked lateral
differences in velocity structure along these two
azimuths in the western part of the North
Australian Craton.

Seismic Character of the Crust

The seismic data from Proterozoic terranes in
northern Australia must be regarded as being of a
reconnaissance nature in terms of providing detail
from within the crust. It is therefore premature
to draw a well demonstrated model for the
intra-crustal seismic features over the whole
region. However, it is still worthwhile
summarising those features which are evident from
the data available.

Finlayson (1982) has indicated that the crust
of the central part of the North Australian Craton
is 51-54 km thick and that the upper mantle
velocity is 8.16-8.20 km/s. In the middle/lower
crust P-wave velocities of 6.85 km/s occur at
depths of about 26 km near Tennant Creek whereas
such velocities are not reached until about 37 km
near Mount Isa. There is therefore a trend for
midcrustal velocities to increase from east to
west. The lower crust is characterised by
velocities of 7.3-7.5 km/s.

Collins' (1983) data from his western traverse
across the McArthur Basin re-emphasises the
crustal thickness of about 53 km in the central
part of the North Australian Craton and that the
velocity changes are gradational in nature rather
than characterised by sharp increases. Finlayson
(1982) models the crust/mantle boundary as
occurring over a range of 3-7 km and Collins
(1983) models it over a range of 2-10 km. Thus
the crust/mantle boundary seems to be transitional
over an average depth range of 5-6 km.

indicated by arrows. Recording summary:-
Recorder, DFS-IV; source,explosives; spread
length, 1-4 km; 24 traces at 41.7-83.3 m spacing;
2 ms sampling to 16 s. Processing:- True
amplitude recovery; time variant scaling; NMO and
static corrections; 6-11 fold stack.

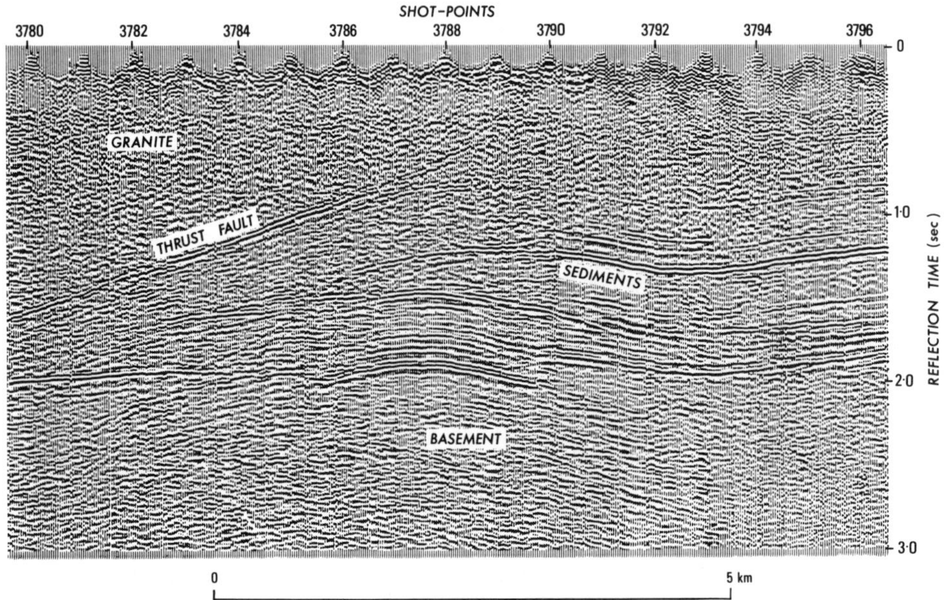

Fig. 8. Seismic reflection profile across the northern margin of the Ngalia Basin onto the Arunta Block, showing overthrusting of granitic rocks onto basinal sediments (from Wells et al.,1972). Recording summary:- Recorder, analogue SIE PMR-20; source, explosives; 24 traces at 46 m intervals; record length, 5 s; single fold coverage. Processing:- A-D conversion; static and NMO corrections; time variant deconvolution.

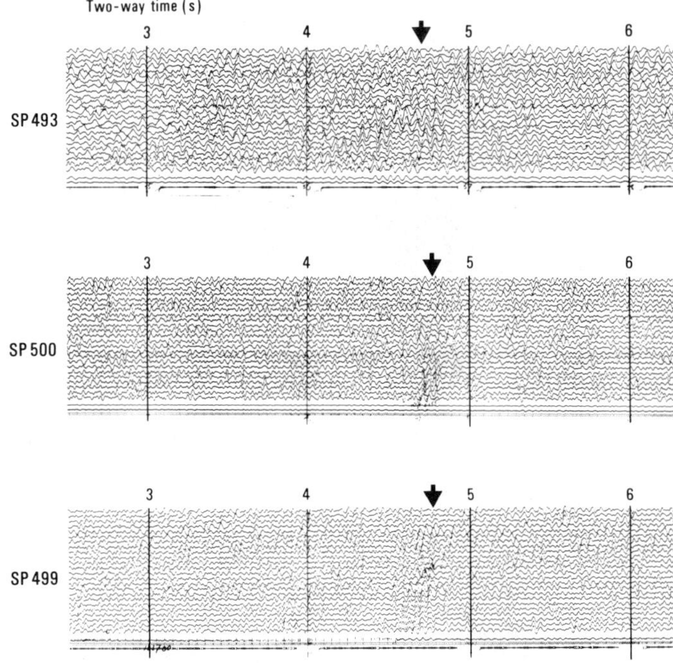

Fig. 9. Vertial incidence seismic monitor records from the Broken Hill Block showing variation in detecting a horizon at 4.5-5.0 s from adjacent shots (from Branson et al.,1976). Recording summary:- Recorder, Electro-Tech DS7-700; source, explosives; 48 traces at 90 m intervals (2 rows of 24, 6 m apart); 20 s records.

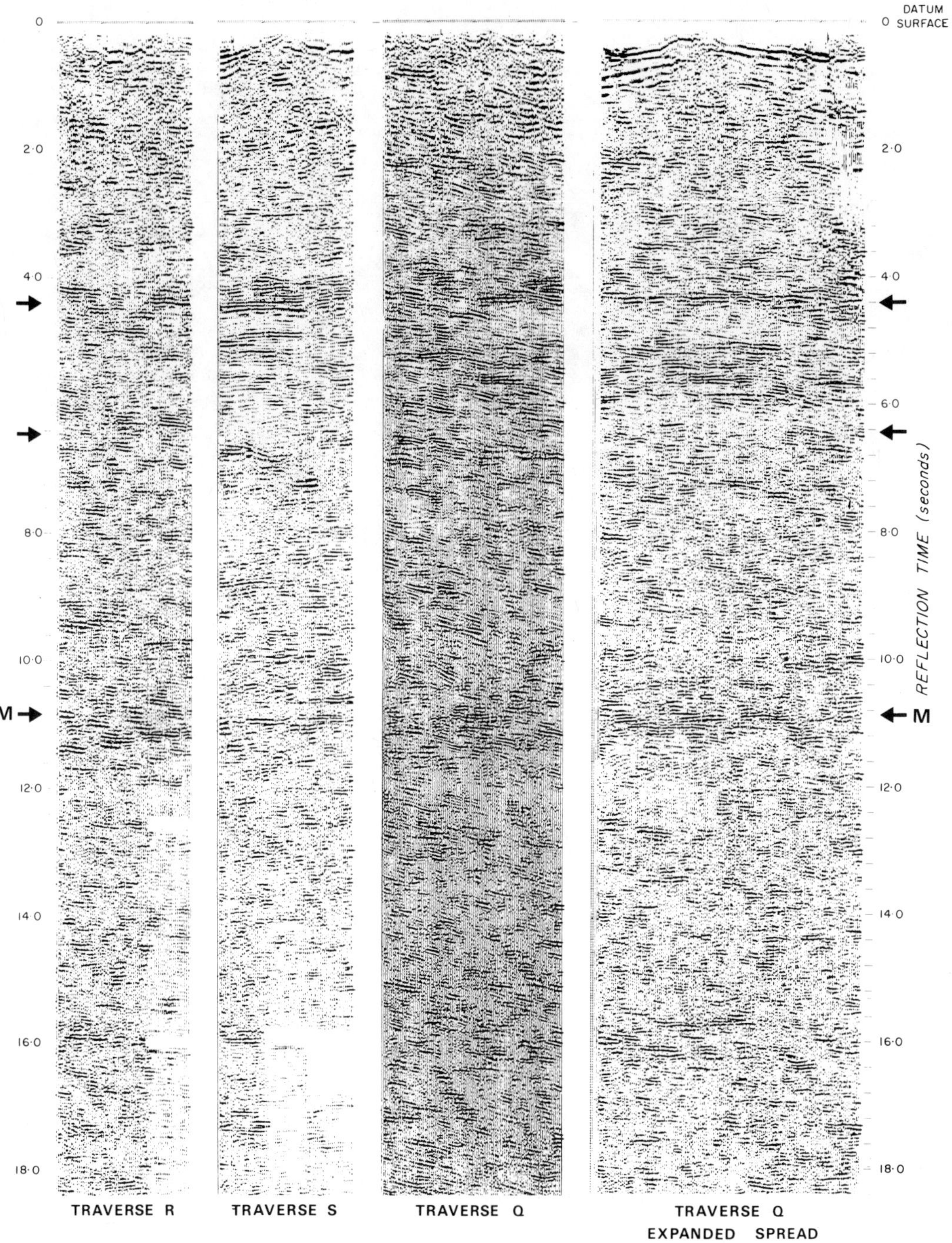

Fig. 10. Seismic reflection recordings from Hines Hill, Yilgarn Block, Western Australia (from Mathur,1974), with prominent velocity horizons determined from wide-angle reflection/refraction surveys indicated by arrows. Recording summary:- Recorder, analogue Electro-Tech DS7-700; 48 traces at 90 m intervals (2 rows of 24, 6 m apart); 20 s records; source, explosives. Processing:- A-D conversion, 8 ms sampling; static and NMO corrections, time variant filter, single fold.

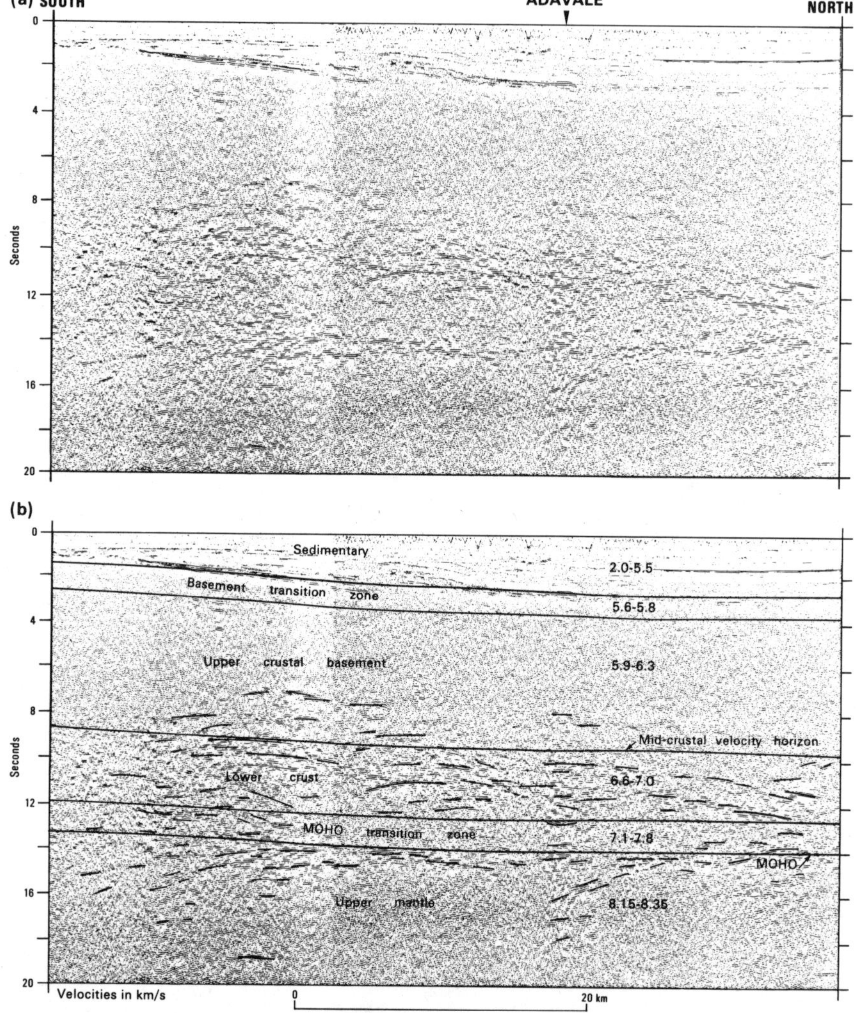

Fig. 11. Deep reflection record section from the Eromanga Basin (near Adavale) (6-fold CDP with time variant filter after stack and coherency scaling), together with velocity features determined from coincident wide-angle reflection / refraction data (from Finlayson & Collins, 1986).

Both Finlayson (1982) and Collins (1983) interpreted velocity increases between the upper and lower crust over a large depth range (19-40 km). Such a range indicates the variability of intra-crustal velocity features and, consequently, the complexity of composition/metamorphic effects throughout the crust.

The crust under the eastern profile of the McArthur Basin is 44 km thick (Collins, 1983), i.e., about 9 km thinner than under the central part of the North Australian Craton. The traverse is wholly within the McArthur Basin and is perhaps more representative of the basin setting. The crustal thickness is similar to that found from less detailed work by Hales and Rynn (1978) for the Pine Creek Inlier to the northeast (45 km). This thinner crust may be associated with the attenuation of the lithosphere in the process of basin formation at the margins of the craton.

The upper crustal basement is, under most of the seismic profiles, overlain by sedimentary cover varying in thickness up to 10 km. Basement rocks are exposed at Tennant Creek and Mount Isa. Finlayson (1982) interprets low velocity zones down to depths of 15 km within the upper crustal basement at these two locations to account for offset travel-times.

Such low velocity zones, determined from data recorded at the shorter distance ranges, are features in the upper crust of the Tennant Creek Block and the Mount Isa Geosyncline, and the nature of the rocks in such zones remain speculative. There are examples of imbrication of igneous and sedimentary sequences and these may be

TABLE 1. Continental Regions With Mean Crustal Velocity of 6.5 km/s or Greater (From Prodehl, 1984)

Region	Mean vel. km/s	Depth to M km	Pn vel. km/s
Ukranian Shield	6.5-6.6	43-45	8.1-8.2
Voronish Massif	6.5-6.6	45-50	7.8-8.0
Baltic Shield	6.5-6.8	43	8.1-8.3
Caladonian (Europe)	6.5-6.6	30-41	8.0-8.2
Fore-Sudetic region (Silesia)	6.5	30	8.2
Outer Carpathians	6.5-6.6	40-57	8.0-8.1
S. Ukraine (Crimea)	6.6	47-52	8.0
Arabian Shield	6.5-6.6	40-45	8.1
Kalahari Craton	6.5-6.6	45-49	7.9-8.2
Indian Shield	6.5	42	8.0
Tien Shan	6.6-6.7	58	7.8
Urals	6.7-6.8	47	7.8
Grenville-Superior	6.5-6.6	36-50	8.0-8.5
Appalachians	6.7-6.8	50-55	7.9-8.0
Great Plains (Arkansas)	6.6-6.7	45-47	8.1-8.2
Great Plains (Oklahoma- E.Colorado- New Mexico)	6.5-6.6	46-51	8.2
Great Plains (E.Montana)	6.5-6.6	44-52	8.0-8.4
Great Plains (Canada)	6.5	49	8.0
Rocky Mts. Belt	6.5	50	7.9
Cent. Sierra Nev.	6.5	51	7.9
Brazil Precambrian	6.5	42	8.2
Lachlan Fold Belt	6.5-6.6	42-52	8.0
N. Austrian Craton	6.5	52	8.1

not uncommon; Figure 8 shows an example from the Arunta Block (Wells et al., 1972). Also, the role of fluids in the upper crust to depths of 12 km has been highlighted by the results of deep drilling in Precambrian terrains of the USSR (Kozlovsky, 1984) and these will have an influence on seismic velocities.

The characteristics of vertical incidence reflection records have been described earlier. In the McArthur Basin, Collins (1983) found no exact correspondence between reflecting horizons and velocity increases throughout the crust. However, there is an association of the reflections between 5 and 7 s two-way-time on Figure 5 with mid-crustal velocity increases determined by Collins (1983), and those at 12-13 s with the velocity increase at the crust/mantle boundary. The reflection profiling data were only recorded on short traverses and hence the limitations of generalising from the data sets should be recognised. Figure 9 illustrates the difficulties of even reproducing the same intra-crustal events at 4.5-5.0 s two-way-time

from adjacent shots on the Broken Hill Block (Branson et al., 1976).

The profusion of seismic reflections in the lower crust entitles one to doubt their interpretation and significance without other related evidence. The seismic reflection profiling method, of necessity, samples only horizons sub-normal to the raypaths and will, therefore, only provide negative evidence of steeply dipping structures in the form of lack of reflections, offsets in sub-horizontal reflectors and diffractions. The reflection data from Proterozoic Australia is limited to 16 s at most, and it is possible that the crust/mantle boudary was not sampled in the southern Georgina Basin where the crustal thickness may approach 50 km.

There is, however, optimism that zonation boundaries can be seen in Precambrian terrains with BMR techniques. Figure 10 shows data from Hines Hill on the Yilgarn Block of Western Australia where reflections at 4-5 s, 6-7 s and at 11 s were identified with velocity changes determined from refraction data (Mathur, 1974). In Phanerozoic Australian provinces the boundaries of reflection zones can be identified with velocity changes determined from refraction data; Figure 11 illustrates data from the Eromanga Basin (Finlayson & Collins, 1986).

Thus, although detailed structures may be difficult to define in the lower crust, the general character of these structures may be defined. The observations made in Proterozoic Australia are not greatly dissimilar from those made on other continents. Such matters were the subject of discussion at the recent meeting on the deep structure of the continents from reflection profiling data (Klemperer, 1984) and comments on possible mechanisms applying in the lower crust are discussed later in this paper.

Comment

The seismic data from Proterozoic Australia, as indicated earlier, can only be regarded as providing some basic constraints on the structure of the crust and uppermost mantle. However, taken together with data from Phanerozoic Australia, they begin to constrain models of crustal evolution. Future surveys will undoubtedly examine more closely the specific consequences of tectonic models now that they have been formulated more precisely. In the meantime it is worth commenting on the contributions so far made by seismic data to the concepts envisaged for continental evolution.

Prodehl's (1984) world-wide review of crustal velocities indicates that many stable continental provinces are characterised by a crustal thickness in excess of 40 km and a mean crustal velocity in excess of 6.5 km/s. Some of his data are summarised in Table 1. In central/eastern Australia, Finlayson and Mathur (1984) have drawn attention to the observation that the depth to the Moho under outcropping igneous, metamorphic and

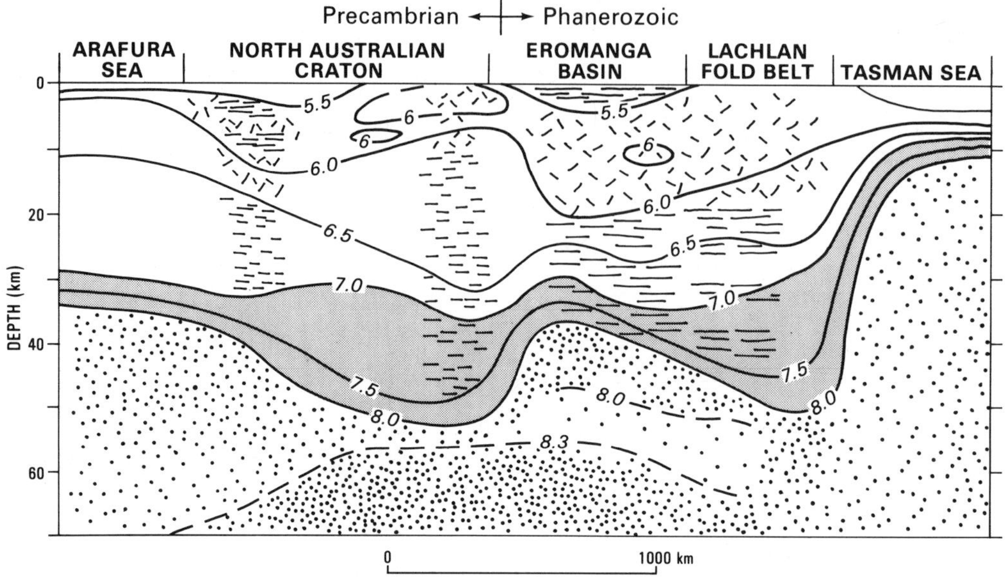

Fig. 12. Diagramatic lithospheric section between the Arafura Sea and the Lachlan Fold Belt. Iso-velocity lines are shown in km/s. Note the increased lower crustal section with velocity greater than 7 km/s under the outcropping igneous, metamorphic and fold belt provinces (from Finlayson & Mathur, 1984).

fold belt provinces is 10-15 km greater than under intracratonic basins and that this is largely the result of a greater thickness of lower crustal material with a velocity over 7 km/s (Fig. 12).

Rutland (1982) has reviewed some of the Australian geological and geophysical data contributing towards an evolutionary model for the continental crust. He proposes development by a multistage process involving the reworking of older crustal material. Using velocity/depth models derived from seismic surveys in major provinces of Australia, Rutland postulates that "normal" velocity lower crustal rock at depths of greater than 20 km has been reworked and underplated by "high" velocity rocks of basaltic and more mafic composition in the granulite facies.

Wass and Hollis (1983) put forward a mechanism of fluctuating geothermal flux from the mantle to explain the geochemistry and thermal history evident in crustal eclogitic and granulitic xenoliths in southeast Australia (Fig. 13). The episodic nature of geochemical differentiation from the mantle that Wass and Hollis (1983) envisage may be the process seen in the geochemical and geochronological data from the Proterozoic provinces of northern Australia. This process would apply throughout geological time and account for the thickness of high-velocity material in the lower crust and the character of many lower crustal reflections.

In eastern Queensland, Ewart et al. (1980) have proposed a model of crustal underplating based on the analysis of megacryst composition in Tertiary mafic lavas. The results are interpreted in terms of intrusion and fractionation within the crust/mantle interface region, and their model envisages the crustal thickening amounting to about 40% of the depth to the upper mantle (Fig. 14).

The seismic velocity gradients in the crust are consistent with a heterogeneous crust being vertically differentiated according to the density changes resulting from compositional and metamorphic variations. Ferguson et al. (1979), using the analysis of kimberlitic xenoliths, indicate that a basaltic lower crust in the granulite facies produces a multiplicity of mineralogical phase changes that are consistent with gradational P-wave velocities in the lower crust and near the crust/mantle boundary. They emphasise the heterogeneous nature of the lower crust which is contributing to the bulk velocities determined by seismic methods.

In the Phanerozoic provinces of eastern Australia Finlayson and Mathur (1984) emphasise the different tectonic environments predominating in the upper and lower crust during crustal formation as seen now in the different characteristics of reflecting horizons in the upper and lower crust. The temperature regime is undoubtedly a major factor influencing the seismic character, with brittle fracture and other structures being evident in the seismic character of the cool upper crust and more diffuse structures being evident in the more ductile, hotter, lower crust.

Hale and Thompson (1982) have examined the

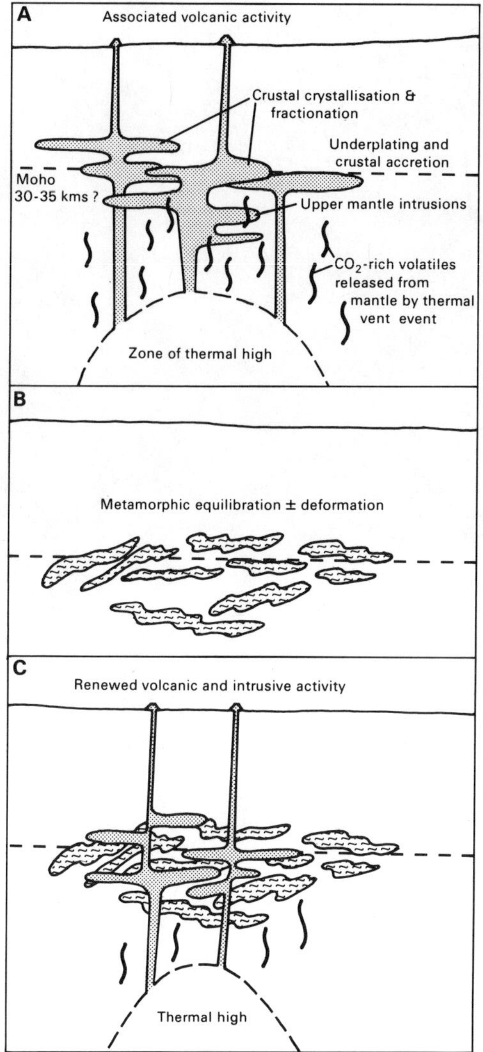

Fig. 13. Crustal accretion model of Wass and Hollis (1983). A, B, and C represent successive stages which may be repeated throughout geological time.

reflection character of the Moho under conterminous USA and discussed the possible structures/rock-types leading to multiple reflections in the lower crust. These include granulite facies rocks of sedimentary and volcanic origin, but they also emphasise that cumulate layering is a more reasonable explanation for reflection zones in magmatically active areas where extensive crustal melting would be expected. Smithson (1978) has emphasised the highly variable composition of the lower crust, being a region where igneous and metamorphic conditions overlap.

It therefore seems quite reasonable to infer that, under a variety of different circumstances, successive underplating episodes at the base of the crust and intusions into the crust can produce a deep crustal sill complex with a range of lower

crustal velocity/depth distributions, and it is the intrusive nature of the process that we see in the character of some deep reflecting horizons. Geochemical modelling (Ewart et al., 1980; Wass & Hollis, 1983; Etheridge et al., 1984) envisages episodes of crustal magma intrusion, crystallisation and fractionation accompanied by crustal underplating and accretion. These processes would be repeated at times of high thermal flux from the mantle throughout geological history. It seems likely that such episodes of lower crustal intrusion and reworking play an important part in determining the present-day seismic character of the lower crust.

With the intrusive/underplating model envisaged, it would be unreasonable to expect seismic reflecting horizons to be coherent over more than a few kilometers in the lower crust. The long time span over which such processes have applied in Proterozoic Australia would tend to lead to further degradation of the reflecting horizons by later thermal and tectonic events, but not necessarily to their extinction altogether.

Although the emphasis in this paper has been towards attributing the seismic characteristics of the crust to the major crustal generation episodes in Proterozoic Australia involving mantle-derived material, other reasons for these characteristics must also be considered. Lateral crustal movements have been shown to produce major seismic reflection features associated with overthrusts, e.g., the Wind River Thrust in western USA (Smithson et al., 1978), and the Outer Isles and Flannan Thrusts off northern Scotland (Brewer et al., 1983). Allegre and Hirn (1984), among others, have emphasised the horizontal interfingering of crustal segments during periods of crustal shortening, as well as vertical interpenetration of mantle material. Mylonite layering has also been shown to produce prominent reflections in regions of crustal shortening (Smithson et al., 1984). A sill-like complex in the lower crust may be a factor enabling crustal shortening and mylonite zonation in the lower crust. Blundell (1984) has emphasised, with 3-D ray-tracing, the complexity of possible reflections from even relatively simple structures in the lower crust.

It would be well worth investigating in detail some of the problems associated with reflection profiling across, say, the Musgrave Block or the Fraser Range in Australia where substancial cross-sections of continental crust are exposed. An important aspect of such work would be the seismic holography of structural detail using offset recording as well as normal incidence recording.

There is also a need for greater detail in defining the seismic velocities of particular features as an aid to determining densities and bulk compositions. Drummond and Collins (pers. comm.), from studies in terrains of all ages and taking into account temperature and pressure effects, found a positive correlation between average velocities in the crystalline crust and

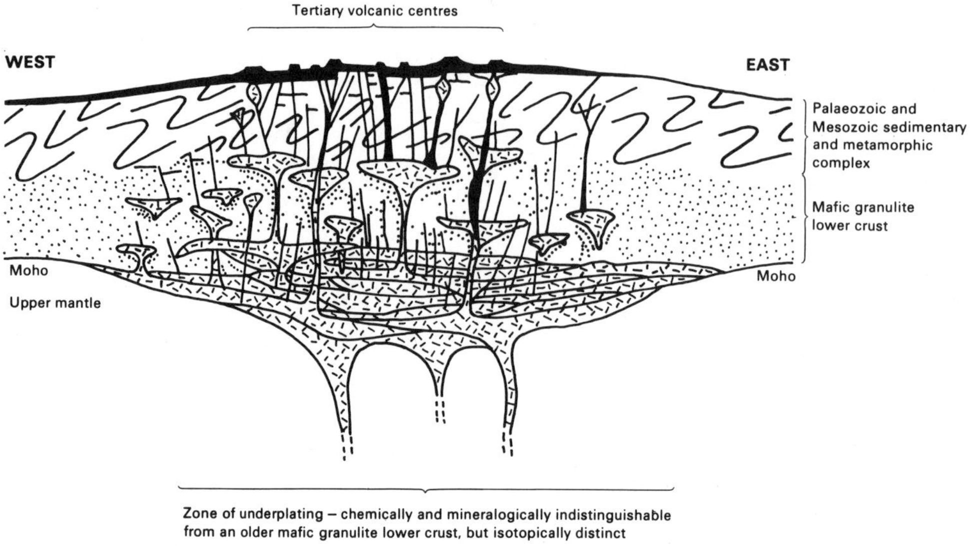

Fig. 14. Schematic cross-section of the crust under southeastern Queensland (from Ewart et al., 1980) illustrating the concept of crustal underplating by multiple basaltic intrusion and fractionation.

crustal thickness. This is attributed to the addition of high velocity material to the lower crust, with the upper crustal velocities remaining largely unchanged. The increased velocity with crustal thickness could not be achieved by rearrangement of crustal rocks.

Finally, the questions regarding structure and composition in the lower crust must, at this stage, still remain open. Specific detailed seismic investigations in both the upper and lower crust should be an objective of future seismic surveys using a variety of techniques.

Acknowledgements. This paper is published with the approval of the Director, Bureau of Mineral Resources, Geology and Geophysics, Canberra.

References

Allegre, C.J. & Hirn, A., The crustal overthrusting, major feature in mountain belt. The horizontal and vertical mozaic model. Abstracts of the International Symposium on Deep Structure of the Continental Crust: Results from Reflection Seismology, Cornell University, June 1984, 1, 1984.

Blundell, D.J., Modelling the lower crust. Abstracts of the International Symposium on Deep Structure of the Continental Crust: Results from Reflection Seismology, Cornell University, June 1984, 6, 1984.

Branson, J.C., Moss, F.J. & Taylor, F.J., Deep crustal reflection seismic test survey, Mildura, Victoria and Broken Hill, NSW 1968. Aust. Bur. Miner. Resour. Report 183, 1976.

Brewer, J.A., Matthews, D.H., Warner, M.R., Hall, J., Smythe, D.K. & Whittington, R.J., BIRPS deep seismic reflection studies of the British Caledonides. Nature, 305, 206-210, 1983.

Collins, C.D.N., Crustal structure of the southern McArthur Basin, northern Australia, from deep seismic sounding. BMR J. Aust. Geol. & Geophys., 8, 19-34, 1983.

Denham, D., Simpson, D.W., Gregson, P.J. & Sutton, D.J., Travel times and amplitudes from explosions in northern Australia. Geophys. J. Roy. Astron. Soc., 28, 225-235, 1972.

Etheridge, M.A., Wyborn, L.A., Rutland, R.W.R., Page, R.W., Blake, D.H. & Drummond, B.J., Workshop on Early to Middle Proterozoic of northern Australia. Aust. Bur. Miner. Resour. Record 1984/31.

Ewart, A, Baxter, K. & Ross, J.A., The petrology and petrogenesis of the Tertiary anorogenic mafic lavas of southern and central Queensland, Australia - possible implications for crustal thickening. Contrib. Mineral. Petrol., 75, 129-152, 1980.

Ferguson, J., Arculus, R.J. & Joyce, J., Kimberlite and kimberlitic intrusives of southeastern Australia: a review. BMR J. Aust. Geol. & Geophys., 4, 227-241, 1979.

Finlayson, D.M., Reconnaissance of upper crustal seismic velocities within the Tennant Creek Block. BMR J. Aust. Geol. & Geophys., 6, 245-252, 1981.

Finlayson, D.M., Seismic crustal structure of the Proterozoic North Australian Craton between Tennant Creek and Mount Isa. J. Geophys. Res., 87, 10569-10578, 1982.

Finlayson, D.M. & Mathur, S.P., Seismic refraction and reflection features of the lithosphere in northern and eastern Australia, and continental growth. Annales Geophysicae, 6, 711-722, 1984.

Finlayson, D.M. & Collins, C.D.N., Lithospheric velocities beneath the Adavale Basin, Queensland, and the character of deep crustal reflections. BMR J. Aust. Geol. & Geophys., 10, 23-37, 1985.

Hale, L.D. & Thompson, G.A., The seismic character of the continental Mohorovicic discontinuity. J. Geophys. Res., 87, 4625-4635, 1982.

Hales, A.L. & Rynn, J.M.W., A long-range controlled source seismic profile in northern Australia. Geophys. J. Roy. Astron. Soc., 55, 663-644, 1978.

Klemperer, S.L., Seismic reflections of the continental crust. Nature, 311, 409, 1984.

Kozlovsky, Ye.A., The world's deepest well. Scientific American, 251(6), 106-112, 1984.

Mathur, S.P., Crustal structure in southwestern Australia from seismic and gravity data. Tectonphysics, 24, 151-182, 1974.

Mathur, S.P., Deep reflection probes in eastern Australia reveal differences in the nature of the crust. First Break, 1, 9-16 1983a.

Mathur, S.P., Deep reflection experiments in northeastern Australia, 1976-78. Geophysics, 48, 1588-1597, 1983b.

Page, R.W., McCulloch, M.T. & Black, L.P., Isotopic record of major Precambrian events in Australia. Proc. 27th Inter. Geol. Cong., Moscow, 4-14 Aug., 1984, Vol. 5, 25-72, 1984.

Prodehl, C., Structure of the earth's crust and upper mantle. In, Geophysics of the Solid Earth, the Moon and the Planets, (eds. K.Fuchs and H.Soffel), Springer-Verlag, Berlin, 96-206, 1984.

Rutland, R.W.R., On the growth and evolution of continental crust: a comparative tectonic approach. J. & Proc. Roy. Soc. NSW, 115, 33-60, 1982.

Rynn, J.M.W. & Reid, I.D., Crustal structure of the western Arafura Sea from ocean bottom seismograph data. J. Geol. Soc. Aust., 30, 59-74, 1983.

Smithson, S.B., Modeling continental crust: structural and chemical constraints. Geophys. Res. Letts., 5, 749-752, 1978.

Smithson, S.B., Brewer, J., Kaufman, S., Oliver, J. & Hurich, C., Nature of the Wind River thrust, Wyoming, from COCORP deep reflection data. Geology, 6, 648-652, 1978.

Smithson, S.B., Johnson, R.A., Hurich, C.A & Fountain, D.M., Crustal reflections and crustal structure. Abstracts of the International Symposium on Deep Structure of the Continents: Results from Reflection Seismology, Cornell University, June, 1984, 80, 1984.

Wass, S.Y. & Hollis, J.D., Crustal growth in southeastern Australia - evidence from lower crustal eclogitic and granulitic xenoliths. J. Metamor. Geol., 1, 25-45, 1983.

Wells, A.T., Moss, F.J. & Sabitay, A., The Ngalia Basin, Northern Territory - recent geological and geophysical information upgrades prospects. Austr. Petrol. Expl. Ass. J., 12(1), 144-151, 1972.

Sm-Nd ISOTOPIC CONSTRAINTS ON THE EVOLUTION OF PRECAMBRIAN CRUST IN THE AUSTRALIAN CONTINENT

M.T. McCulloch

Research School of Earth Sciences, Australian National University, G.P.O. Box 4, Canberra, A.C.T. 260, Australia

Abstract.   In this study, results from an extensive survey of the Sm-Nd isotopic systematics in the major Precambrian belts of the Australian continent are presented. Major periods of primary crustal formation are identified and an attempt is made to distinguish between formation of new crust during the Early Proterozoic and the reworking of Archean crust. The Pine Creek, Halls Creek, Granites-Tanami, Arunta, Mt Isa, Georgetown, Broken Hill and Olary terrains have a narrow range of Sm-Nd model ages (T_{DM}) of from 2130 Ma to 2290 Ma. Samples from the Mt Isa and Pine Creek Inliers with U-Pb zircon crystallisation ages of from 1783 Ma to 1886 Ma define a correspondingly narrow range in εNd values of -2.5 ± 1.0. Whilst contamination by Archean crust cannot be completely discounted these magmas have extremely coherent geochemistry and in general do not possess inherited Archean U-Pb zircon systematics. This together with the small range in negative εNd values is interpreted as reflecting an ~400 Ma protolith prehistory rather than large scale assimilation of Archean crust. Evidence for direct involvement of Archean crust is only found in those Proterozoic terrains immediately adjacent to Archean cratons (e.g. Gascoyne Block of northwestern Western Australia). Younger Mid-Proterozoic crust is present in the Musgrave Block of central Australia and parts of the Albany Province of southwestern Australia where T_{DM} ages of 1770 Ma to 1890 Ma occur. These results indicate that at least one-third of the present-day exposed area of the Australian upper crust consists of material added during the Proterozoic. A crustal age distribution pattern derived from the data imply a relatively short period (~200 Ma) of high crustal growth (~10 km³/yr) and only minimal volumes of whole scale mantle recycling of continental crust.

Introduction

Although there is general consensus on the importance of crustal growth during the Late Archean [Moorbath, 1977; McCulloch and Wasserburg, 1978; Taylor and McLennan, 1981] the role and relative volumes of younger crustal additions is much less certain [Reymer and Schubert,1984]. This is mainly due to the difficulty of identifying what are fundamentally new additions to the crust as opposed to reworking or remelting of pre-existing crust. In the Australian continent this problem is exacerbated, as the exposed basement, particularly in the central and northern portion, consists predominantly of complex poly-metamorphosed inliers.

In the Proterozoic the role of Archean crust is enigmatic. In some nonuniformitarian models, it is the dominant component with Proterozoic mobile belts representing intracratonic features such as rift or collision zones and hence consisting of predominantly reworked Archean crust [e.g. Rutland, 1976; Katz, 1985; Etheridge et al., 1986]. Alternate models view Proterozoic crustal development in essentially contemporary plate-tectonic terms [e.g. Burke et al., 1976; Hoffman, 1980; Windley, 1981] with the Proterozoic mobile belts representing laterally accreted island arc terrains and hence Archean crust is an incidental component. A model compatible with the concept of horizontal plate tectonics but differing from the Wilson cycle in that anhydrous subcrustal lithosphere is partially subducted, has been proposed by Kroner [1981, 1983]. In this paper I present the results of an extensive survey of the Sm-Nd isotopic systematics in the Proterozoic and Archean orogenic belts of the Australian continent (Figure 1). Although of a reconnaissance nature, an attempt is made to identify the major periods of primary crustal formation [McCulloch and Wasserburg, 1978] and in particular to distinguish between Proterozoic crustal formation events and Proterozoic reworking of dominantly Archean crust. The quantification of rates of crustal formation has important implications, not only for determining overall rates of continental growth and recycling, but also for constraining general models of continental crustal development. This study uses the strategy and procedures of McCulloch and Wasserburg [1978]. Some of the preliminary results were reported by McCulloch and Hensel [1984] and Page et al. [1984].

Sm-Nd Isotopic Systematics

One of the main aims of this study is the delineation of the major periods of crustal formation in the Australian continent. For this purpose the now well established Sm-Nd model age technique [DePaolo and Wasserburg, 1976a; McCulloch and Wasserburg, 1978; Nelson and DePaolo, 1985] has been applied. Since the initial application of Sm-Nd model ages the main refinement has been the recognition of a more complex depleted mantle evolution [e.g. DePaolo, 1981; McCulloch and Compston, 1981], rather than a chondritic Sm/Nd reservoir (CHUR) as initially inferred [DePaolo and Wasserburg, 1976, a,b]. This is illustrated in Figure 2 where εNd versus time trajectories are shown. The CHUR evolution corresponds to εNd = o, whereas depleted mantle evolves to increasingly more positive εNd values. For the purposes of

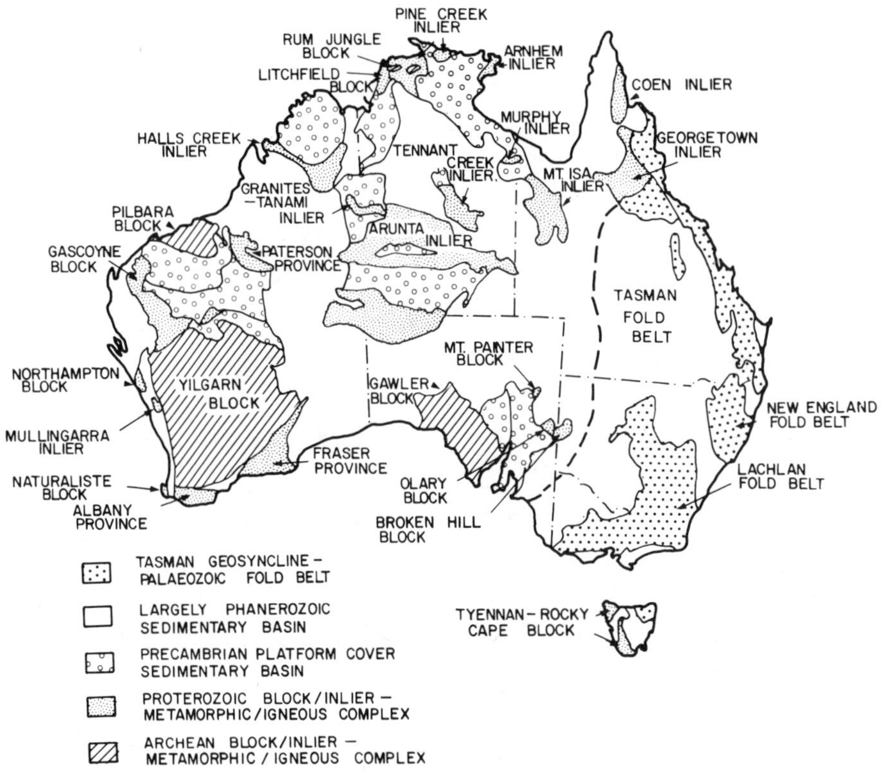

Fig. 1. Map showing the distribution of the major Australian Precambrian orogenic provinces, sedimentary basins and the Phanerozoic Tasman Fold Belt. The largest segments of Archean crust are in the Yilgarn and Pilbara Blocks of Western Australia and poorly exposed remnants are present in the Gawler Block of South Australia, and the Pine Creek and Rum Jungle complexes of the Northern Territory. Proterozoic crust is abundant throughout the Australian continent.

calculating T^{Nd} model ages, a relatively simple model has been used in the present study. In this model it is assumed that a mantle depletion event occurred at 2750 Ma and the present-day depleted mantle is equivalent to MORB (εNd $\approx$ 10). A more complex and possibly more plausible model involving continuously depleting mantle has been proposed by DePaolo [1981] (see dashed line in Figure 2). However in view of the obvious spatial and temporal heterogeneities in the mantle it is now clear that no single mantle evolutionary curve can be justified and for this reason the simpler model is adopted in this study [see also Liew and McCulloch, 1985].

Also shown in Figure 2 is the evolution of Proterozoic and Archean LREE enriched crust with the latter evolving rapidly to negative εNd values. The T^{Nd} model age is given by the intersection of the sample and depleted mantle evolutions. In general the T^{Nd} model ages are not equivalent to crystallization ages with the latter defined for example by U-Pb zircon ages (Table 2). There are a number of possible explanations for this commonly observed discrepancy. Firstly, assimilation of older crustal materials may occur and will result in model ages which reflect the contribution of both older and younger components [see equation 6 of McCulloch and Wasserburg, 1978]. This is illustrated in Figure 2 where mixtures of juvenile Proterozoic and older Archean crust are shown.

Another possible reason for apparent discrepancies

between T^{Nd} model ages and crystallization ages is an extended protolith prehistory. It is now generally appreciated that the derivation of continental crustal materials from the mantle requires multi-stage differentiation and partial melting processes. This is represented in Figure 2 by an extended evolution of a protolith with an Sm/Nd fractionation factor f_1 where $f_1 = [(Sm/Nd)_1/(Sm/Nd)_{CHUR} -1]$. A dominantly mafic protolith could be expected to have a LREE enrichment intermediate between the final differentiated crust and the mantle source i.e. $f_1 < f_0$ and $f_1 > f_2$ where f_0, f_1 and f_2 are the Sm/Nd fractionation factors of the mantle source, protolith and crust respectively (see Figure 2). In this case the T^{Nd} model age would also be intermediate between the protolith age and the final crystallization age. In most previous studies it has been assumed that the protolith prehistory has been negligible (i.e. ≤ 100 Ma.). Although this may be a reasonable assumption in the Early Archean, in this study it will be shown that protolith prehistory is of paramount importance, particularly in the Early Proterozoic.

A third possibility is that the mantle source reservoir may have a different composition to that assumed for the model reservoir. Although it is clear that the upper mantle is heterogeneous on both small and large scales [e.g. White and Hoffman, 1982] and that there are enriched reservoirs [McCulloch et al., 1983c], as already pointed out, there is now widespread evidence (ops. cited) that the bulk of the

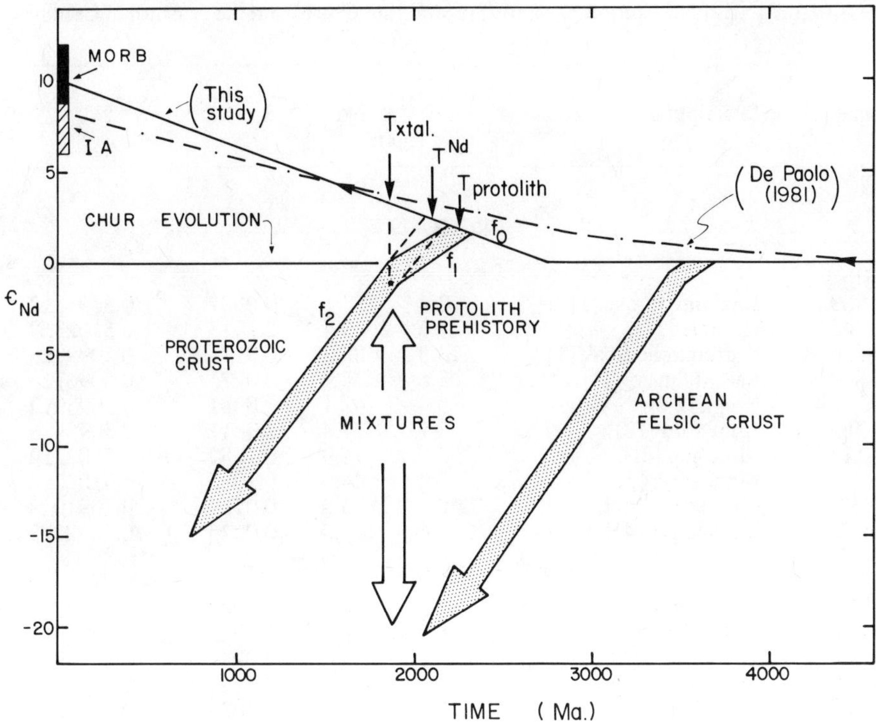

Fig. 2. Schematic diagram showing the evolution of ^{143}Nd/^{144}Nd as fractional deviation in parts in 10^4 from the evolution in a reservoir with chondritic Sm/Nd. Solid line shows the depleted mantle evolution used in this study to calculate T^{Nd} model ages. Dashed curve is an alternate evolutionary model of DePaolo [1981]. Also shown are the systematics expected if large scale assimilation of Archean crust occurs during production of younger Proterozoic crust. In this case mixtures of Archean and Proterozoic crust would produce a large range of εNd values. Effects of an extended protolith prehistory on εNd values and T^{Nd} ages is illustrated with the Sm/Nd fractionation factors shown schematically for a mantle reservoir (f_0), crustal protolith (f_1) and crust (f_2) respectively with $f_0>f_1>f_2$. For a significant protolith prehistory, εNd values may be significantly lower than those anticipated for primary mantle derived magmas.

convecting upper mantle is being progressively depleted. Thus assuming that this is due to progressive extraction of continental of crustal materials, then it is unlikely that significant volumes of crust could have been extracted from enriched mantle sources.

Geological Setting and Samples

The Australian continent can be divided into three broad domains (Figure 1). The large Archean cratons of the Yilgarn and Pilbara Blocks of Western Australia with their encircling Proterozoic mobile belts, the central and northern portions of Australia which contain the main Proterozoic inliers and the Phanerozoic Tasman fold belt of eastern Australia. In this paper a compilation of new and published (indicated by an asterisk in Table 1) isotopic data is presented for the Proterozoic inliers of central and northern Australia, the western mobile belts of western Australia and the Archean Yilgarn and Pilbara Blocks. Proterozoic crust is also an important component in the Paleozoic Tasman fold belt, particularly in southeastern Australia [McCulloch and Chappell, 1982], but this region will be discussed in detail elsewhere.

Archean

To provide a basis for comparison with the Proterozoic terrains limited Sm-Nd data are presented for the Archean Pilbara and Yilgarn Blocks. More extensive isotopic studies of the Yilgarn Block and its margins are reported by McCulloch and Compston [1981], McCulloch et al. [1983 a,b], and Fletcher et al. [1983 a,b, 1985] and the Pilbara Block by Hamilton et al. [1981], Jahn et al. [1981] and Collerson and McCulloch [1983]. For a recent review see Page et al. [1984].

The Yilgarn Block is one of the largest intact segments of Archean crust and contains rocks ranging from 3700 Ma [deLaeter et al., 1981; Williams et al., 1983,] to ~2600 Ma in age. [Turek and Compston, 1971; Oversby, 1975 and Bickle et al., 1983]. In Table 1 Sm-Nd analyses are presented for the youngest phase, the post-tectonic granites that outcrop extensively in the eastern portion but are also an important component throughout the Yilgarn Block. These samples give T^{Nd} ages of 2600 Ma to 2750 Ma. The T^{Nd} ages are essentially identical to the Pb-Pb whole-rock and mineral ages [Oversby, 1975] as well as the Rb-Sr whole-rock ages [Turek and Compston, 1971 and Bickle et al., 1983]. This is

Sample	Description	Sm	Nd	$^{147}Sm/^{144}Nd$	$^{143}Nd/^{144}Nd§$	T^{Nd}
		(ppm)				(Ma)

ARCHEAN

Yilgarn Block

Sample	Description	Sm	Nd	$^{147}Sm/^{144}Nd$	$^{143}Nd/^{144}Nd§$	T^{Nd} (Ma)
1. 71-909	Buldania Rocks [1]	2.4	18.1	0.0807	0.50983±2	2620
2. 71-742	Karonie [1]	1.6	9.4	0.1013	0.51023±2	2600
3. 71-736	Karramindie Soak [1]	13.3	88.2	0.01910	0.50998±3	2680
4. 72-864	Lake Johnston [1]	3.6	28.0	0.0773	0.50967±3	2750
5. GA 4137	Ringa [2]*	5.4	42.1	0.0781	0.50930±2	3240
6. 76-265	Jimperding [2]*	3.7	27.4	0.0811	0.50953±3	3020
7. 77-1130	Mortigup [2]*	3.0	16.9	0.1082	0.51022±2	2770
8. 499	Mundaring [3]*	4.4	24.0	0.1085	0.51027±3	2710
9. 60734	Mt Narryer [4]*	0.7	5.4	0.0789	0.50910±3	3510
10. 60735	Mt Narryer [4]*	8.0	55.0	0.0875	0.50922±2	3630

Pilbara Block

Sample	Description	Sm	Nd	$^{147}Sm/^{144}Nd$	$^{143}Nd/^{144}Nd§$	T^{Nd} (Ma)
11. 76324	Duffer Fm. [5]	3.2	18.6	0.1049	0.50977±2	3400
12. 49205	Duffer Fm. [5]	5.2	28.9	0.1083	0.50983±2	3430
13. 15255	Shaw Bath. [6]	5.9	37.8	0.0949	0.50952±2	3440
14. 26218	Shaw Bath. [6]	4.8	33.9	0.0855	0.50935±2	3380
15. 15207	Shaw Bath. [6]	5.1	29.6	0.1039	0.50988±2	3190
16. 12584	Mt Edgar Bath. [7]	3.1	21.4	0.0877	0.50946±2	3300
17. 26240	Mt Edgar Bath. [7]	5.4	23.5	0.1382	0.51061±2	(3170)
18. 77-713	Boobina Porph. [8]	2.5	13.4	0.1131	0.51003±3	3270
19. 77-712	Spinaway Porph. [8]	9.6	55.2	0.1069	0.50996±2	3160

Rum Jungle Block

Sample	Description	Sm	Nd	$^{147}Sm/^{144}Nd$	$^{143}Nd/^{144}Nd§$	T^{Nd} (Ma)
20. RJ1	paragneiss [9]	3.1	21.0	0.0892	0.50986±2	2790
21. GA 693	orthogneiss [9]	7.5	38.0	0.1192	0.51046±2	2710
22. GA 662	orthogneiss [9]	2.3	14.8	0.0929	0.50957±2	3300
23. GA 697	metadolerite [9]	2.5	8.1	0.1829	0.51179±2	(2020)
24. GA 676	granitoid [9]	3.6	23.2	0.0937	0.50987±2	2890

Pine Creek Inlier

Sample	Description	Sm	Nd	$^{147}Sm/^{144}Nd$	$^{143}Nd/^{144}Nd§$	T^{Nd} (Ma)
25. 7212-4048A	Nanambu Complex [10]	6.8	42.3	0.0977	0.51015±2	2620

Gawler Block

Sample	Description	Sm	Nd	$^{147}Sm/^{144}Nd$	$^{143}Nd/^{144}Nd§$	T^{Nd} (Ma)
26. 516-431	Eyre Pen. [11]	4.9	28.9	0.1024	0.51017±2	2690
27. 516-427	Eyre Pen. [11]	14.3	91.6	0.0947	0.51017±2	2540
28. P1455	Tarcoola [12]	4.3	28.6	0.0909	0.51004±2	2610
29. P1456	Tarcoola [12]	7.5	43.2	0.1044	0.51028±2	2600
30. P1748	Tarcoola [12]	5.3	36.0	0.0891	0.50994±2	2690

EARLY PROTEROZOIC

Pine Creek Inlier

Sample	Description	Sm	Nd	$^{147}Sm/^{144}Nd$	$^{143}Nd/^{144}Nd§$	T^{Nd} (Ma)
31. 7212-4081C	Nimbuwah Complex [10]	7.0	40.0	0.1055	0.51059±2	2240
32. 7212-4069D	Nimbuwah Complex [10]	5.6	44.3	0.0771	0.51018±3	2230

Halls Creek Inlier

Sample	Description	Sm	Nd	$^{147}Sm/^{144}Nd$	$^{143}Nd/^{144}Nd§$	T^{Nd} (Ma)
33. 51706	Sally Downs [13]	3.9	21.7	0.1083	0.51071±2	2130
34. 30806	Sophie Downs [13]	5.7	32.1	0.1075	0.51058±2	2290
35. GA 1241	Hart dolerite [14]	4.0	17.3	0.1396	0.51110±2	2220
36. GA 5132	Hart dolerite [14]	5.1	22.5	0.1377	0.51111±2	(2160)

Table 1. Continued

Sample	Description	Sm	Nd	^{147}Sm/^{144}Nd	^{143}Nd/^{144}Nd§	T^{Nd}
		(ppm)				(Ma)
Granites-Tanami Inlier						
37. 7249-5024	Winnecke Granophyre [15]	14.1	15.0	0.1596	0.51129±3	(2460)
38. 7249.5023	Winnecke Granophyre [15]	6.7	32.1	0.1256	0.51090±2	2210
39. 7249.5029	Mt Winnecke Fm. [15]	8.9	44.3	0.1212	0.51086±3	2180
Arunta Inlier						
40. 72902009	Kanandra [16]*	4.3	25.9	0.0998	0.51055±2	2180
41. 73913007	Ongeva [16]*	7.5	41.8	0.1079	0.51071±2	2120
42. 73933007	Mt Hay [16]*	1.3	7.9	0.974	0.51051±2	2190
Tennant Creek Inlier						
43. 78063318	Tennant Creek granite [17]	9.8	51.1	0.1167	0.51068±2	2340
44. 75063317	Bernborough volcanics [18]	12.0	60.0	0.1205	0.51074±2	2340
45. 73063284A	orthogneiss; basement [16]*	8.6	43.9	0.1181	0.51059±3	2500
46. 73063284B	orthogneiss; basement [16]*	8.6	56.3	0.0930	0.51038±2	2270
47. 73063284C	orthogneiss; basement [16]*	6.0	30.8	0.1172	0.51061±3	2450
Mt. Isa Inlier						
49. 74-368A	Leichhardt Met. [19;20]	8.4	50.5	0.1007	0.51054±2	2210
50. 7920.5312	Leichhardt Met. [19;20]	8.5	47.5	0.1080	0.51060±2	2300
51. 7920.5309	Kalkadoon Granite [19;20]	8.3	56.5	0.0888	0.51043±2	2140
52. 74-360	Argylla Fm. [19;20]	13.0	65.8	0.1197	0.51084±2	2180
53. 7720.5009	Yaringa Met. [21]	5.7	33.2	0.1044	0.51058±2	2230
Georgetown Inlier						
54. 73303024	Einasleigh Met. [16]*	7.5	42.8	0.1063	0.51056±3	2290
55. 82303068	Cobbold [16]*	4.7	27.4	0.1045	0.51054±3	2280
56. 80303065	Einasleigh Met. [16]*	2.5	7.6	0.1971	0.51205±3	(1640)
57. 3009	granitoid [22]	3.9	20.8	0.1119	0.51073±2	2170
58. G63	granitoid [22]	2.3	11.7	0.1184	0.51081±2	2190
Coen Inlier						
59. 70-933	Coen Met. I [23]	8.9	45.2	0.1145	0.51080±3	2130
60. 70-952	Coen Met. II [23]	5.5	33.8	0.0985	0.51073±2	1940
61. 70-883	Dargalong	7.6	44.5	0.1029	0.51065±2	2110
Broken Hill Block						
62. S5	S1 [24]	5.7	32.4	0.1071	0.51061±2	2240
63. BHLO	S2 [24]	16.2	73.5	0.1347	0.51095±2	(2350)
64. 82-428	Potosi S4 [24]	11.0	45.9	0.1443	0.51115±2	(2260)
65. 82-429	Alma S3 [24]	8.7	46.8	0.1121	0.51067±2	2260
Mt Painter Block						
66. 570	orthogneiss [25]	33.5	164.7	0.1229	0.51100±2	2010
67. 1265	orthogneiss [25]	11.1	43.2	0.1549	0.51141±2	(2040)
Olary Block						
68. OL-1	granitoid [26]	4.3	21.0	0.1252	0.51085±2	2280
69. OL-2	granitoid [26]	3.4	21.0	0.0971	0.51045±2	2250
Houghton Inlier						
70. 69-153	granitic gneiss [27]	0.9	5.8	0.0992	0.51044±3	2300
71. 69-149	granitic gneiss [27]	0.9	5.3	0.1033	0.51041±2	2420
Gawler Volc.						
72. GV976	Gawler Range [28]	39.9	174.9	0.1381	0.51092±2	(2430)

Table 1. Continued

Sample	Description	Sm	Nd	$^{147}Sm/^{144}Nd$	$^{143}Nd/^{144}Nd$§	T^{Nd}
		(ppm)				(Ma)
Paterson Province						
73. 48929B	Rudall Complex [29]	10.3	63.4	0.0989	0.51048±2	2250
74. 48929F	Rudall Complex [29]	8.8	53.8	0.0993	0.51049±2	2250
75. 48925E	Rudall Complex [29]	5.0	30.8	0.0989	0.51035±3	2410
Gascoyne Block						
76. 69103	Minnie Creek [30]*	7.6	46.0	0.1002	0.51054±2	2200
77. 60706	Jackson Well [30]*	13.5	83.0	0.0990	0.51040±2	2350
78. 60711	Dunawah Well [30]*	6.0	34.0	0.1089	0.51034±2	2620
79. 281E	Errabiddy [30]*	4.9	35.0	0.0852	0.50924±2	3520
Northampton Block						
80. 1579	Mary Springs [3]*	13.0	68.0	0.1124	0.51081±3	2080
81. 1760	Riverside [3]*	4.3	23.0	0.1124	0.51085±3	2020
Mullingarra Inlier						
82. 1673	Ikewah Range [3]*	6.9	37.0	0.1135	0.51086±2	2030
83. 1797	Yandanooka Hills [3]*	10.4	55.0	0.1144	0.51085±2	2060
Perth Basin Basement						
84. 1758	N. Yardarino [3]*	8.1	41.0	0.1189	0.51093±	2030
85. 1757	Jurien No. 1 [3]*	1.8	9.5	0.1139	0.51074±3	2200
86. 1672	Sue No. 1 [3]*	7.6	47.0	0.0974	0.51066±2	2010
Fraser Province						
87. GA 4065	granulitic gneiss [31]	3.0	11.7	0.1539	0.51144±3	(1940)
88. Ga 4061	granulitic gneiss [31]	6.6	26.7	0.1501	0.51134±4	(2050)
Albany Province						
89. 56425	Rooneys Bridge [32]*	7.0	36.0	0.1193	0.51073±3	2330
90. 55811C	Northcliffe [32]*	8.0	38.0	0.1243	0.51083±3	2290
91. 55812B	Chudalup [32]*	11.1	64.0	0.1048	0.51061±3	2200
92. 55813A	Crystal Springs [32]*	14.1	81.0	0.0982	0.51067±3	2010

MID PROTEROZOIC

Sample	Description	Sm	Nd	$^{147}Sm/^{144}Nd$	$^{143}Nd/^{144}Nd$§	T^{Nd}
Albany Province						
93. FAGR	Albany [33]	9.4	58.7	0.0968	0.51079±3	1850
94. AA	Albany [33]	22.9	136.8	0.1014	0.51079±2	1920
95. END-G	Albany [33]	10.3	66.8	0.934	0.51069±2	1920
Musgrave Block						
96. GA 3092	granulitic gneiss [31]	22.7	139.6	0.0993	0.51086±3	1800
97. GA 3093	granulitic gneiss [31]	6.9	38.1	0.1.94	0.51093±3	1870
98. GA 3084	granulitic gneiss [31]	11.8	54.9	0.1302	0.51117±2	1890
99. GA 2425	Ernabella Adamellite [31]	15.2	71.6	0.1284	0.51120±4	1810
100. 69-1443	granulitic gneiss [34]*	15.7	70.0	0.1359	0.51127±3	(1840)
101. 70-203	granitic gneiss [34]*	4.6	24.3	0.1149	0.51107±3	1770

Sample	Description	Sm	Nd	$^{147}Sm/^{144}Nd$	$^{143}Nd/^{144}Nd$[§]	T^{Nd}
		(ppm)				(Ma)

LATE PROTEROZOIC

Naturaliste Block

Sample	Description	Sm	Nd	$^{147}Sm/^{144}Nd$	$^{143}Nd/^{144}Nd$	T^{Nd}
102. 82-427a	Sugar Loaf [35]	4.3	23.7	0.1104	0.51150±2	1130
103. 82-426	Sugar Loaf [35]	8.7	37.2	0.1409	0.51153±3	(1480)
104. IF-1	Augusta [36]*	20.5	145.0	0.0860	0.51129±2	1160

*Sm-Nd data from reference cited. §$^{143}Nd/^{144}Nd$ normalised to $^{146}Nd/^{142}Nd = 0.636151$, errors are 2σ mean.

$$T^{Nd} = \frac{1}{\lambda} \ln \left\{ 1 + \frac{(^{143}Nd/^{144}Nd)_{meas} - (^{143}Nd/^{144}Nd)_{ref}}{(^{147}Sm/^{144}Nd)_{meas} - {}^{147}Sm/^{144}Nd)_{ref}} \right\} \quad \text{where } \lambda = 6.54 \times 10^{-12} \text{ yr}$$

$(^{143}Nd/^{144}Nd)_{ref} = 0.51235$ and $(^{147}Sm/^{144}Nd)_{ref} = 0.225$. For $T^{Nd} > 2750$ Ma
$(^{143}Nd/^{144}Nd)_{ref} = 0.511836$ and $(^{147}Sm/^{144}Nd)_{ref} = 0.1967$.

References: [1] Oversby (1975). [2] McCulloch et al. (1983b). [3] Fletcher et al. (1985). [4] de Laeter et al. (1981). [5] Pidgeon (1978). [6] de Laeter et al. (1975). [7] de Laeter and Blockley (1972). [8] Pidgeon (1984). [9] Richards et al. (1966). [10] Page et al. (1980). [11] Fanning et al. (1980). [12] Hensel and Fanning, pers. comm. (1986). [13] Ogasawara (1984). [14] Bofinger (1967). [15] Page et al. (1976). [16] Black and McCulloch (1984). [17] Black (1984). [18] Black (1977). [19] Page (1978). [20] Page (1983). [21] Page (1976). [22] Black et al. (1979). [23] Cooper et al. (1975). [24] Willis et al. (1983). [25] Teale, pers. comm. [26] Teale, pers. comm. (1984). [27] Cooper and Compston (1971). [28] Hensel, pers. comm. (1984). [29] Chin and de Laeter (1981). [30] Fletcher et al. (1983a). [31] Arriens and Lambert (1969). [32] Fletcher et al. (1983b). [33] Stephenson (1980). [34] Gray et al. (1981). [35] Compston and Arriens (1968). [36] Fletcher, pers. comm. (1986).

indicative of minimal involvement of significantly older crustal protoliths and is not consistent with previous suggestions [Oversby, 1975] that these granites represent remobilised ~3300 Ma crust. Although these granites do not contain any evidence for crust older than about 2800 Ma it should be noted that recent U-Pb zircon studies [Compston et al., 1986] of the Kambalda volcanics have found xenocrystic zircons of age up to ~3400 Ma. Thus, there is some evidence for the presence of Early Archean crust in the east Yilgarn Block but its extent is currently unknown.

The plutonic and extrusive felsic volcanics of the Pilbara Block define a relatively limited range of Early Archean ages of from 3170 Ma to 3440 Ma (Table 1). Felsic volcanics in the Duffer Formation (samples 11 and 12 in Table 1) of the Warrawoona Group [Hickman, 1981], give model ages of 3400 Ma and 3430 Ma. Sample 11 has a U-Pb zircon age of 3452 ± 16 Ma [Pidgeon, 1978] giving an $^{\varepsilon}Nd$ value at extrusion of 0.6. A similar $^{\varepsilon}Nd$ value of 0.4 is also obtained from the slightly younger Boobina Porphyry which has a U-Pb zircon age of 3307 ± 19 Ma [Pidgeon, 1984]. These slightly positive $^{\varepsilon}Nd$ values imply a limited crustal prehistory for the volcanic protoliths (<100 Ma). The gneisses (samples 13 to 17 in Table 1) give T^{Nd} ages of 3440 Ma to 3190 Ma. These model ages are not however resolvable from those of the Warrawoona Group, suggesting that the gneiss protoliths were formed essentially contemporaneously and from similar protoliths as those of the volcanics. The youngest samples examined in the Pilbara Block are the post-tectonic Moolyella and Cooglegong granites from the Mt. Edgar and Shaw batholiths respectively. These samples have Rb-Sr whole-rock ages of 2600 Ma [deLaeter et al., 1974] but yield significantly older T^{Nd} ages of 3120 Ma and 3200 Ma. The T^{Nd} ages partially overlap with those from the surrounding granitic gneisses. Thus in contrast to the Yilgarn Block the post-tectonic granites of the Pilbara are essentially wholely derived by intracrustal melting of the pre-existing Early Archean crust. This is also the case for the Spinaway Porphyry (Fortescue Group in the Pilbara Block) which has a U-Pb zircon age of 2768 ± 16 Ma [Pidgeon, 1984] but a T^{Nd} age of 3160 Ma. This result is also indicative of remelting of the older crust.

One of the initial aims of this study was to document the extent of Archean crust in localities other than those of the Yilgarn and Pilbara Blocks. Rocks of unequivocal Archean age are present in only two other localities, the Pine Creek Inlier of northern Australia and the Gawler Block of southern Australia. In the Rum Jungle complex of the Pine Creek Inlier a previously unrecognised Early Archean (~3300 Ma) gneiss terrain has been identified. This terrain may possibly be correlative with gneisses of similar age from the Pilbara Block. In the central and eastern portion of the Pine Creek Inlier Late Archean gneisses (the Nanambu complex) are

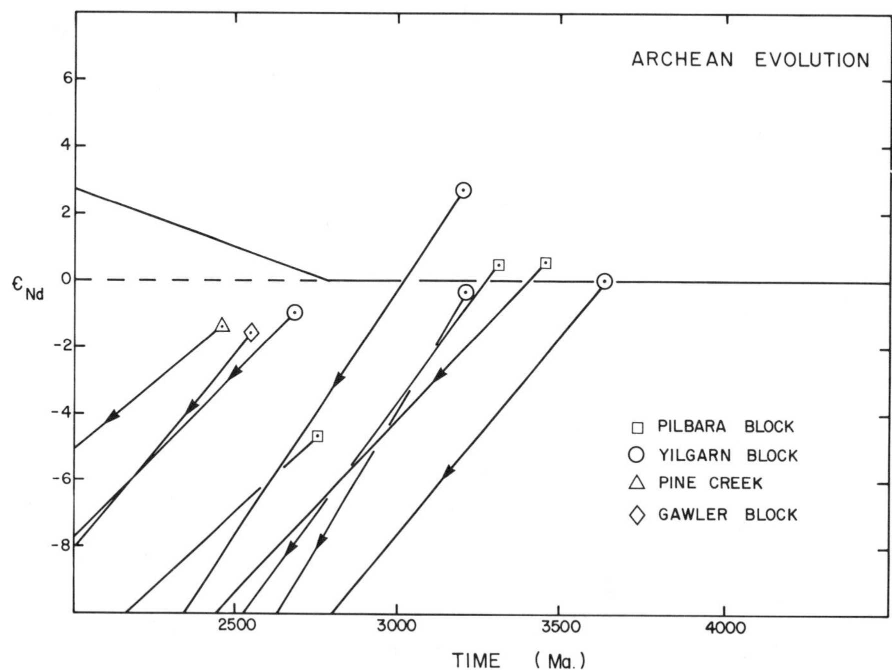

Fig. 3. Evolution of initial ^{143}Nd/^{144}Nd (as εNd values) versus crystallisation ages in Australian Archean felsic rocks. Ages are from U-Pb zircon studies (Table 2). Early Archean rocks (solid symbols) generally have more positive εNd values compared to late Archean samples (open symbols).

widespread [Needham et. al., 1980]. The sample analysed in this study (sample 25 in Table 1) has a T^{Nd} age of 2620 Ma and a U-Pb zircon age of 2470 ± 26 Ma [Page et al., 1980]. The εNd value of -1.4 for the Nanambu complex indicates the presence of an older component, possibly similar to that found in the Rum Jungle complex of the Pine Creek Inlier.

In the Gawler Block, Archean ages have been obtained from two localities in the southern Eyre Peninsular and the Tarcolar Inlier in the northern section of the Block. In the Eyre Peninsula the oldest T^{Nd} age of 2690 Ma has been obtained for a granulitic augen-gneiss (sample 30) for which Cooper et al., (1975) and Fanning et al., (1980) report a Rb-Sr age of 2590 ± 130 Ma and a U-Pb zircon age of 2550 ± 13 Ma [Fanning et al., 1980].   A sample of the cordierite-garnet gneiss from this area has both a younger T^{Nd} age of 2540 Ma as well as Rb-Sr and U-Pb ages of 2412 ± 72 Ma and 2305 ± 10 Ma respectively [Fanning et al., 1980]. The Tarcolar Inlier in the northwest portion of the Gawler Block, has T^{Nd} ages of 2600 Ma to 2690 Ma, which are similar to those from the eastern portion of the Yilgarn block.

In Figure 3 the εNd versus time trajectories are shown for Archean samples for which U-Pb zircon ages are available. All the samples have low ^{147}Sm/^{144}Nd ratios (i.e. they are LREE enrichments) and therefore have well defined intersections with the mantle growth curve. Hence the T^{Nd} model ages are relatively well defined. In addition the samples evolved rapidly to negative εNd values and for example by 1850 Ma had acquired pronounced negative εNd values of from -8 to -24.   As noted previously, mafic Archean rocks, for example the basaltic or ultramafic portions of greenstone belts, do not generally have LREE enrichments

and therefore have essentially constant εNd values and hence their formation ages are not well defined with the Sm-Nd system (e.g. sample 23 in Table 1).

A notable feature of Figure 3 is the difference between Early and Late Archean εNd values (calculated using U-Pb zircon ages from Table 2). The Early Archean samples (i.e. >3000 Ma) generally have more positive εNd values (+2.7 to -0.4) whereas those from the late Archean all have negative εNd values (-1.0 to -4.6). The simplest explanation for this feature is that the source region of the Late Archean magmas has incorporated a significant proportion of older Archean crust whereas the Early Archean materials are derived predominantly from depleted mantle sources (see Figure 2). In the Pilbara, Yilgarn and Pine Creek blocks this explanation is consistent with the juxtaposition of both Early and Late Archean terrains. However an alternative explanation is a more extended protolith prehistory in the late Archean. This will be shown to be especially relevant to the Early Proterozoic terrains, and is discussed in more detail in the following section.

<u>Early Proterozoic</u>

In this study Early Proterozoic crust is loosely defined as Proterozoic rocks having U-Pb zircon ages >1800 Ma and T^{Nd} ages >2000 Ma. The majority of the samples have T^{Nd} ages in the exceedingly narrow range of 2130 Ma to 2290 Ma. This includes samples from the Pine Creek Inlier, Halls Creek Inlier, Granites-Tanami Inlier, Arunta Inlier, Mt. Isa Inlier, Georgetown Inlier, Broken Hill Block and Olary Inlier. Considering that the analytical uncertainty in the determination of T^{Nd} ages is typically ± 40 Ma (2 sigma mean) the limited

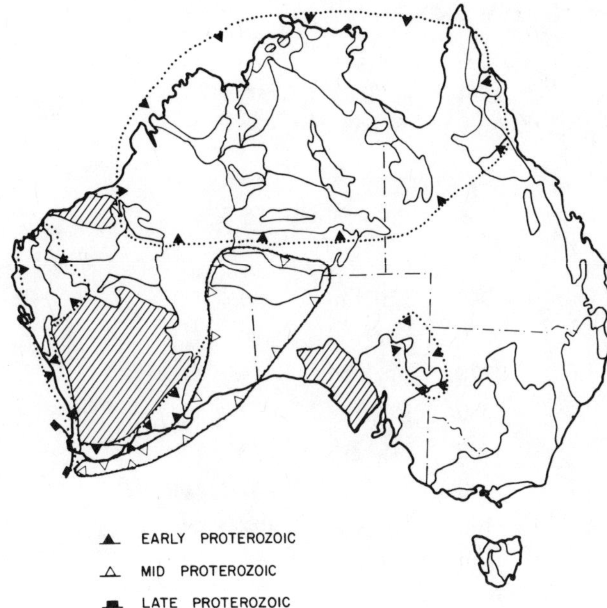

Fig. 4. Map showing the generalised distribution of Early Proterozoic (TNd ages from 2000 Ma to 2300 Ma), mid-Proterozoic (TNd ages from 1500 Ma to 2000 Ma) and late Proterozoic (TNd ages less than 1500 Ma) crust. Proterozoic crust (not shown) is also present in the Lachlan Fold Belt [McCulloch and Chappell, 1982].

range in TNd ages ($\pm$ 80 Ma) over such a large geographic area is remarkable and is interpreted as having geologic significance. Slightly younger model ages of from 1940 Ma to 2130 Ma are present in the Coen and Mt Painter Inliers and in the Albany/Fraser Province.

Only two provinces have a range of significantly older TNd ages. These are the Gascoyne Block and Tennant Creek Inlier with TNd ages of 2200 Ma to 3650 Ma [Fletcher et al., 1983] and 2270 Ma to 2500 Ma (Table 1) respectively. The study by Fletcher et al., [1983] of the margin of the Proterozoic Gascoyne Province and Archean Yilgarn Block clearly illustrates that incorporation of Archean crust into a Proterozoic orogenic terrain has dramatic and obvious effects on the Sm-Nd isotopic systematics. This is in part due to the large contrast in crustal formation ages between the Gascoyne Province (~2200 Ma) and the Archean crust (~3600 Ma), and that the boundary between the terrains was defined using structural criteria [Gee, 1979]. By analogy the range of model ages in the Tennant Creek Inlier can also be interpreted as indicating the involvement of late Archean crust (of age ~2500 Ma to 2600 Ma) although Archean crust has not previously been identified in this area. Relatively minor Archean components are also probably present in the Paterson Province (sample 75) and the Houghton Inlier and Gawler volcanics, which are either alongside or within Archean provinces. However with the notable exception of the Tennant Creek Inlier it appears that contamination by Archean crustal components is restricted to those Proterozoic terrains which are adjacent to known exposures of Archean crust. Evidence for Archean crust in the Tennant Creek Inlier is still equivocal [Black and McCulloch, 1984] and requires further confirmation.

The Albany Province appears to represent a unique terrain in that a wide range of TNd ages of from 1850 Ma to 2330 Ma are present. This area borders on the southwestern margin of the Yilgarn Block and the older model age (sample 89) [Fletcher et al., 1983a] is probably due to contamination by either 2700 Ma or 3100 Ma crust at the margin of the Yilgarn Block. The Early Proterozoic ages (2200 Ma and 2290 Ma) and Mid-Proterozoic ages (1850 Ma and 1920 Ma) may indicate at least two Proterozoic crustal formation events or alternatively mixing between Mid-Proterozoic and Late Archean crust.

<u>Mid-Proterozoic</u>

As already noted, Mid-Proterozoic crust is present only to a limited extent in the Albany Province. The largest occurrence of Mid-Proterozoic crust is in the Musgrave Block of central Australia (Figure 1) where TNd ages of 1770 Ma to 1890 Ma are present (Table 1). The samples analysed in this study include both the granulite-facies gneisses of the southern portion and the Ernabella Adamellite in the east which have previously yielded Rb-Sr whole rock ages of 1530 $\pm$ 250 Ma and 1100 $\pm$ 27 Ma respectively [Arriens and Lambert, 1969]. Two of the samples (101 and 102) are from the western portion of the Musgrave and near the Giles complex [Gray et al., 1981]. The TNd ages of 1770 Ma and 1840 Ma are not distinguishable from those elsewhere in the Musgraves. From these results it is evident that the Musgrave Block is approximately 400 Ma younger than the Early Proterozoic Inliers. Furthermore Mid-Proterozoic age crust is also probably present in the Albany and perhaps Fraser provinces of southwestern Australia but not, as previously inferred by Mathur and Shaw [1982] and Plumb, [1979], in the northwestern part of Australia (i.e. in the Paterson Province).

<u>Late Proterozoic</u>

With the exception of the Tasman fold belt Late Proterozoic (ie <1200 Ma) crust has so far only been identified in the Naturaliste Block of southwestern Western Australia. A relatively felsic and LREE enriched granulite sample from the northern portion of the Naturaliste Block (Sugar Loaf) gives a depleted mantle model age of 1130 Ma. A less fractionated mafic granulite gives a similar result within analytical uncertainties. These results are confirmed by a TNd age of 1180 Ma [Fletcher pers. comm., 1986] for a garnet gneiss (sample 105 in Table 1) from Augusta, the southern-most portion of the Naturaliste Block. These results indicate an unexpectedly young age for the granulite protolith of the Naturaliste Block and insofar as this terrain appears to be of limited extent it may provide a means of refining the reconstruction of the once contiguous southern Australia and the east Antarctic Shields.

Evolution of Early Proterozoic Crust

The generalised distribution of Proterozoic crust is shown in Figure 4. It is apparent from this figure that large volumes of Early Proterozoic crust are present in the northern and central portions of Australia as well as on the margins of the Archean Yilgarn Block. The key question is to what extent does this Early Proterozoic crust represent new crustal additions as opposed to reworking or remelting of pre-existing

Sample	^{147}Sm/^{144}Nd	^{143}Nd/^{144}Nd	$\varepsilon^*_{Nd}(T)$	T^{Nd} (m.y.)	U-Pb age (m.y.)	$\Delta T^\dagger$ (m.y.)	Ref.
ARCHEAN							
5.[††] GA 4137	0.0781	0.50930±2	-0.4	3240	3210±40	30	[1]
6. 76-265	0.0811	0.50953±3	2.7	3020	3200±50	-180	[1]
7. 77-1130	0.1082	0.51022±2	-1.0	2770	2680±50	-110	[1]
10. 60735	0.0875	0.50922±2	0.0	3630	3630±4	0	[2]
11. 76324	0.1049	0.50977±2	0.6	3400	3452±16	-52	[3]
18. 77-713	0.1131	0.51003±3	0.4	3270	3307±19	-37	[4]
19. 77-712	0.1069	0.50996±2	-4.6	3160	2768±16	392	[4]
25. 7212.4048A	0.0977	0.51015±2	-1.4	2620	2470±45	150	[5]
26. 516-431	0.1024	0.51017±2	-1.6	2690	2550±13	140	[6]
EARLY PROTEROZOIC							
31. 7212.4081C	0.1055	0.51059±2	-2.5	2240	1866±8	374	[5]
43. 78063318	0.1167	0.51068±2	-3.4	2340	1870±20	470	[7]
44. 75063317	0.1205	0.51074±2	-3.1	2340	1870±15	470	[7]
49. 74-368A	0.1007	0.51054±2	-2.3	2210	1865±3	345	[8]
50. 7920.5312	0.1080	0.51060±2	-2.7	2300	1886±35	414	[9]
51. 7920.5309	0.0888	0.51043±2	-1.6	2140	1862±25	278	[8]
52. 74-360	0.1197	0.51084±2	-1.8	2180	1783±5	397	[9]

$* \; \varepsilon_{Nd}(T) = 10^4 \; ((^{143}Nd/^{144}Nd)^T_{sample} - (^{143}Nd/^{144}Nd)^T_{CHUR}) \; / (^{143}Nd/^{144}Nd)^T_{CHUR}$
where T is U-Pb zircon age.

$\dagger \; \Delta T = T^{Nd} - (U\text{-}Pb) \; age$ $\dagger\dagger$ Sample No. from Table 1

References:[1] Nieuwland and Compston (1981). [2] Williams et al. (1983). [3] Pidgeon (1978). [4] Pidgeon (1984). [5] Page et al. (1980). [6] Fanning et al. (1980). [7] Black (1984). [8] Page (1978). [9] Page (1983).

Archean crust? This question is of obvious importance for terrains that are adjacent to the Archean blocks, but is also of general relevance, as T^{Nd} model ages are commonly older than the U-Pb zircon ages (e.g. in the late-Archean). This is evident from Table 2 where the time differences (ΔT) between U-Pb zircon ages and T^{Nd} model ages are tabulated. The Early Proterozoic samples have U-Pb zircon ages of from 1783 ± 5 Ma to 1886 ± 35 Ma with the T^{Nd} ages being 278 Ma to 470 Ma older. An equivalent representation of this data is given in Figure 5 where initial ^{143}Nd/^{144}Nd ratios (expressed as εNd values) are plotted against U-Pb zircon ages. The mean εNd value is -2.5 with a range of ± 1 epsilon units. The typical analytical uncertainty is ± 0.5 epsilon units (2 sigma mean) and thus the observed variation is only a factor of two larger than the analytical uncertainty.

Understanding the mechanisms responsible for the extremely limited range of negative εNd values is of crucial importance to Early Proterozoic crustal evolution. As described previously, a common explanation for the deviation of εNd values below the model mantle curve is contamination by Archean crust [e.g. Nelson and DePaolo, 1985, Patchett and Arndt, 1986]. If this is the case in the Early Proterozoic of Australia, then the range of εNd values implies assimilation of up to 40% to 60% of late Archean crust (assuming similar Nd concentration in the mantle derived and assimilated components). This is an unrealistically large amount of crustal assimilation. The amount of Archean crustal involvement can be minimised if the assimilated crustal components have higher Nd concentrations. If for example the crustal component had a Nd concentration x2 greater than

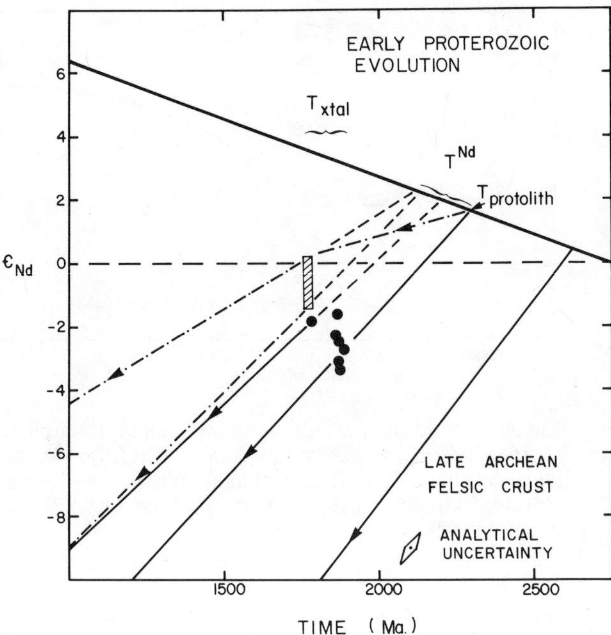

Fig. 5. Diagram showing initial $^{143}Nd/^{144}Nd$ (as ε_{Nd} values) at the time of magma crystallisation. Crystallization ages are from U-Pb zircon ages (Table 2). The shaded field is for mafic volcanics from the Mt Isa Inlier [Sun and McCulloch, 1985]. Early Proterozoic felsic samples have an exceedingly narrow range of ε_{Nd} values of from -1.6 to -3.4 (solid symbols). The preferred explanation for the negative and limited range in ε_{Nd} values is that the protoliths for both the felsic and mafic volcanics have had extended prehistories. For the felsic samples a protolith prehistory of ~400 Ma is inferred if the Sm/Nd ratios are not significantly fractionated during magma production at 1840 Ma to 1880 Ma.

the proposed mantle component, then the amount of assimilation that would be required is reduced to 20% to 30%. Alternatively, the proportions of assimilation would be reduced if it is assumed that older (i.e. > 3000 Ma) crust is involved instead. However with the exception of the Rum-Jungle complex, Early Archean crust appears to be restricted to the Pilbara and Yilgarn Blocks and therefore crust of this age is unlikely to be widely involved in the genesis of Early Proterozoic crust.

In addition to the large proportions of Archean crust that is required to account for the negative ε_{Nd} values, another major difficulty with large scale assimilation of Archean crust is the lack of any other supporting evidence. For example the detailed and extensive U-Pb zircon studies of Page [1978, 1983], Page et al. [1980,1986] and Black [1984] have not as yet revealed any evidence for the widespread presence of inherited Archean zircons. In addition, geochemical analyses of Early Proterozoic felsic crust [e.g. Bultitude and Wyborn, 1982; Wyborn and Page, 1983] shows distinctively high levels of K_2O, Rb, Zr, La, Ce, Th, U and low MgO and CaO at relatively low SiO_2 levels of 60-65%. These chemical features like the ε_{Nd} values (or T^{Nd} model ages) are highly coherent [Wyborn, 1985] throughout the entire central portion of Australia which is equivalent to an area $>4 \times 10^6$ Km^2.

Thus an Archean contaminant would be required to have an extremely specific and uniform composition as well as being involved in constant proportions (40% ± 10%).

Due to these difficulties, an alternative two stage model is proposed for Early Proterozoic crustal development. This model assumes an extended (~400 Ma) protolith prehistory for the Early Proterozoic crust. As shown in Figure 2 an extended protolith prehistory will produce negative ε_{Nd} values at the time of crystallisation, if the Sm/Nd fractionations are such that $f_1 < f_0$. That is, the protolith is LREE enriched relative to the mantle source reservoir. The protolith prehistory is not strictly constrained by the T^{Nd} model ages as fractionation of Sm/Nd may occur during partial melting of the protolith at T_{xtal} (see Figures 2 and 5). However, for $f_2 < f_1$ the oldest model age (~2300 Ma) provides a minimum estimate of ~400 Ma for the protolith prehistory. The small range in T^{Nd} values observed in the felsic samples may therefore be accounted for by either a range of protolith prehistories (see range of ΔT values in Table 2) or more probably variable Sm/Nd fractionation at T_{xtal}. Using equation 4 of McCulloch and Black [1984] relative fractionations of ~10% (i.e. $f_2/f_1 \sim 1.1$) will account for the limited range of T^{Nd} model ages of the felsic samples.

Also shown in Figure 5, are the ε_{Nd} values of mafic volcanics from the Mt Isa Inlier [data from Sun and McCulloch, 1985]. These are distinctly less negative (ε_{Nd} values of from 0 to -1.5) than the felsic materials but are still significantly below the model depleted mantle reservoir. This may be explained by similar models, that is contamination by Archean crust or an extended protolith prehistory. If the latter is invoked, the less negative ε_{Nd} values of the mafic volcanics simply implies a lower degree of LREE enrichment in the mafic protolith (ie f_1[mafic] > f_1[felsic]) or alteratively a shorter protolith prehistory. A variation of this model is contamination of juvenile (i.e. ~1800 Ma) mantle derived mafic magmas by the pre-existing Early Proterozoic felsic protolith. This is plausible as the ~1800 Ma volcanism is widespread and would require a large heat source which may be represented by the mafic volcanism. A difficulty is that at least in the Mt Isa Inlier, the mafic and felsic volcanism are not strictly consanguinous, with the mafic volcanics being ~100 Ma younger [Page, 1983]. Thus they are unlikely to be representative of the heat source responsible for the large scale lithospheric melting at ~1800 Ma. Contamination of the mafic volcanics by Archean crust, also cannot be excluded, but as discussed for the felsic volcanics, there is as yet no direct evidence for for widespread involvement of Archean crust. Further detailed studies are in progress to resolve these possibilities.

Implications for Crustal Growth

Regardless of the detailed interpretation of the isotopic data it is clear that large volumes of continental crust were formed during the Early Proterozoic. This is apparent in Figure 6 which shows that approximately 1/2 of all the samples analysed, have T^{Nd} model ages of 2200 ± 100 Ma. However this is not an accurate estimate of the abundance of crust of this age, as the sampling was biased towards Early Proterozoic terrains. A more accurate representation of the age distribution of Australian continental crust is shown in Figure 7. In this figure an attempt has been made to estimate the relative proportions of various age terrains. This assumes uniform crustal thickness and an areal distribution of age

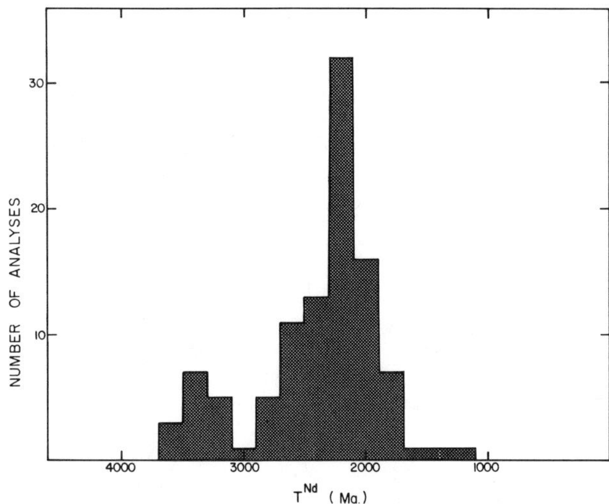

Fig. 6. Histogram showing number of analyses versus T^{Nd} model ages (Table 1). Peaks in the Early Archean (~3600 Ma) and Early Proterozoic (~2200 Ma) are clearly evident but in part reflect biased sampling.

domains as shown in Figure 4. The largest uncertainty is the unknown age of crust beneath the Phanerozoic and Precambrian platform cover sedimentary basins. These basins, cover over one-half of the Australian continent. Isotopic analyses of drill core that has intersected basement beneath these basins would clearly be of great value. However, assuming continuity of the exposed basement that is present in the inliers, then a reasonably accurate quantitative estimate of the age distribution pattern can be made. This

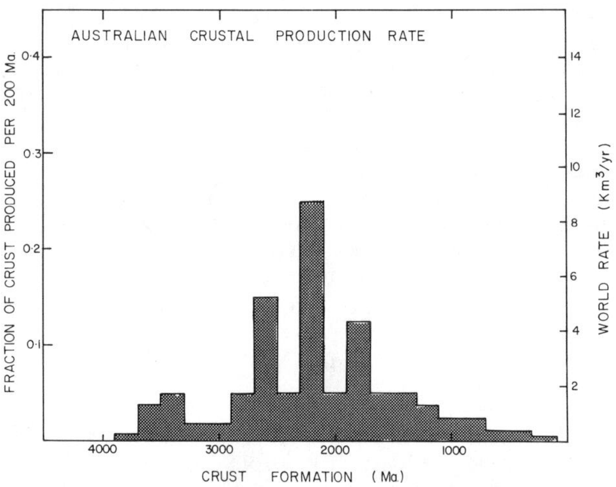

Fig. 7. Fraction of Australian continental crust produced per 200 Ma interval. Crustal formation ages are obtained from T^{Nd} model ages. Approximate areal distribution of Archean and Proterozoic crust is shown in Figure 4. Four main periods of high crustal production are apparent at, ~3600 Ma, ~2600 Ma, ~2200 Ma and ~1800 Ma. Also included in the overall crustal production are the Proterozoic and younger crustal components in the Phanerozoic Tasman Fold Belt.

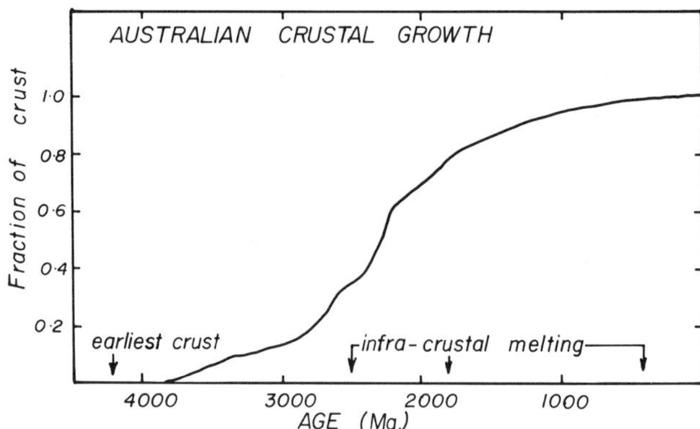

Fig. 8. Generalised diagram showing cumulative fraction of crustal growth for the Australian continent. Infra-crustal melting events have been documented at 2600 Ma (Pilbara), 1840 Ma to 1880 Ma (central Australia) and at ~400 Ma (Tasman Fold Belt).

latter assumption is reasonable as there is no evidence from seismic or gravity data [Finlayson, 1982] for large scale crustal discontinuities between the basins and the inliers. It is noted however that modelling of relatively large gravity anomalies produced by the Ngalia and Amadeus Basins of central Australia, suggests large variations (~20 km) in crustal thickness [Lambeck, 1984]. This has obvious implications for estimates of overall crustal volumes even if the rather tenuous assumption can be made that the age of the lower crust and exposed upper crust are similar.

Despite the uncertainties inherent in obtaining a quantitative age distribution for continental crust, it is nevertheless informative to examine the robust features of such a distribution. In particular the peaks in crustal production at ~3600 Ma, ~2600 Ma., ~2200 Ma and ~1800 Ma (Figure 7). Although further work will almost certainly refine the relative sizes of the peaks, the present data requires that they remain as important periods of significantly higher than average crustal production. For example, even if evidence is found for wide-scale assimilation of large proportions (i.e. ~ 40%) of late Archean crust, this will only have the effect of enhancing the ~2600 Ma peak while reducing the size and age of the ~2200 Ma peak. The Early Proterozoic will still nevertheless remain as an important period of crustal production.

From Figure 7 it is clear that from the presently available data for the Australian continent, that the Early Proterozoic is the single most important period of crustal production. Assuming no recycling of continental crust into the mantle through geologic time, Figure 8 indicates that over one third of the present Australian continent was produced during the Early Proterozoic. If Australia is typical of continental growth in other cratons, then this implies a world-wide crustal production rate during the Early Proterozoic of ~10 km^3/yr. This is approximately a factor of ten larger than present-day production rates [McLennan and Taylor, 1983].

Evidence for an important world-wide crustal production event during the Early Proterzoic is also found in North America [DePaolo, 1981; Hoffman and Bowring, 1984; Nelson and DePaolo, 1985; Patchett and Arndt, 1986], Sweden [Wilson et. al., 1985], Finland [Patchett and

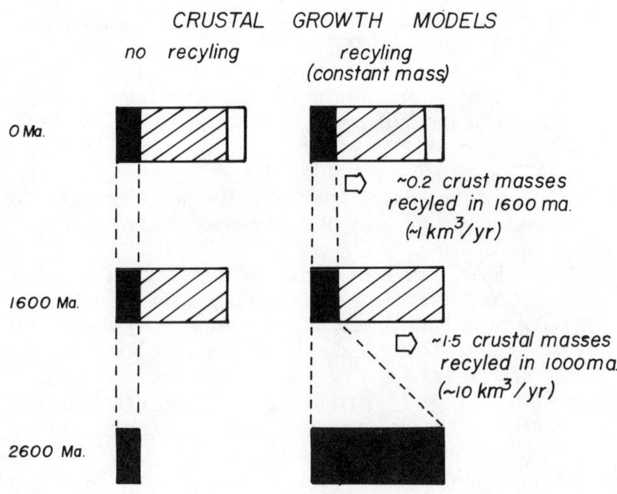

Fig. 9. Schematic diagram illustrating two endmember crustal growth models. For the case of no mantle recycling of continental crust, the age distribution pattern directly indicates the rate of growth of continental crust. In this model the peaks in the age distribution at ~2600 Ma, 2200 Ma and to a lesser extent 1800 Ma (Figure 7) can be accounted for by limited periods (~200 Ma) of more rapid crustal growth. To account for the observed age distribution pattern, the alternate model of a near steady-state constant continental mass with crustal recyling [e.g. Armstrong, 1981] requires extremely rapid recycling of crust from ~1600 Ma to 2600 Ma equivalent to ~10 km³/yr or ~1.5 crustal masses recycled in ~1000 Ma.

Kouvo,1986] and Greenland [Patchett and Bridgwater, 1984]. In the mid-continental craton of North America, U-Pb zircon geochronological studies [Van Schumus and Bickford, 1981] have recognised a number distinct terrains with major periods of magmatism at ~1820 Ma to 1860 Ma, ~1350 Ma to 1480 Ma and at ~1100 Ma .

An alternate view which has been discussed by Gurnis and Davies [1985], is that the observed peaks on crustal growth reflect instead variations in the rates of recycling of continental crust into the mantle. This is illustrated qualitatively in Figure 9 where models of no recycling versus recycling with constant continental mass [e.g. Armstrong, 1981] are compared. To account for the observed crustal age distribution pattern (Figure 7) an extreme model involving continental recycling with constant mass, requires the recycling of at least 1.5 crustal masses in the interval from 1600 Ma to 2600 Ma .This also implies high crustal production rates (~10 km³/yr) for a period of 1000 Ma (to maintain constant continental mass) whereas the no recycling model requires only short periods (<200 Ma) of high crustal production. This suggests that massive recycling of continental crust such that constant continental masses are retained is an unlikely scenerio. These arguments do not of course eliminate the possibility of recycling of minor volumes of continental crust into the earth's mantle.

Conclusions

Although our understanding of the tectonic setting of the protoliths responsible for producing the large volume of Early Proterozoic crust is limited, it is nevertheless worthwhile considering the available constraints and models for crustal genesis. Firstly, the available geological evidence, as summarised in this volume [Etheridge et al., 1986], is not easily reconcilable with the operation of contemporary style tectonics in the Early Proterozoic of northern Australia. In particular the occurrence of bimodal volcanic suites and the lack of subduction related components such as andesites and ophiolitic and blue schist terrains, argues against subduction induced magmatism.

An alternative model for growth of Early Proterozoic crust via underplating and rifting of Archean crust has been considered by Etheridge et al., [1986]. This model is appealing as it can in principle satisfy both the isotopic and geochemical requirements of a protolith having an extended prehistory as well as a relatively uniform chemical composition. In this scenerio, relatively mafic material (but capable of producing felsic melts upon partial melting) is underplated onto pre-existing Archean crust at ~2300 Ma. At ~1850 Ma this material is partially melted to produce the widespread felsic magmatism. The commonly associated mafic volcanics may be produced by higher degrees of partial melting of the mafic underplate or as discussed previously may represent direct melts of the heat source (asthenospheric?) responsible for the ~1850 Ma magmatism. However there are a number of unsatisfactory aspects to this model. Firstly, an essential element is underplating onto Archean crust, for which there is little or no evidence. In addition it would require large volumes of Archean crust to be buried beneath younger basin cover and only the Proterozoic rift zones to be exposed today. Furthermore the Sm-Nd isotopic data requires a large time difference (~400 Ma) between underplating (2200 Ma to 2300 Ma) and magmatism (1780 Ma to 1850 Ma) and hence it is difficult to appeal to a single mantle convection system to account for both the underplating and subsequent partial melting [Etheridge et al., 1986].

These difficulties may in part be resolved if it is assumed that the underplating occurred onto pre-existing basaltic Archean crust or alternatvely highly attenuated Archean felsic crust. Light-REE depleted Archean basaltic crust would not have distinguishable Nd isotopic characteristics from upper mantle material. It is also conceivable that the Archean-Proterozoic boundary (i.e. at 2400 Ma to 2500 Ma) represents the onset of a 'quiescent' period of underplating resulting from a secular change in the upper mantle convection system. The term quiescent is suggested as there is a notable absence of igneous rocks of age 2400 Ma to 2000 Ma. The onset of underplating may have been dependent on proximity to the large Archean cratons and in part represented thickening of oceanic lithosphere. At ~1850 Ma partial melting of the underplated protoliths occurred as part of a worldwide thermal event probably reflecting a major perturbation in the upper mantle convection system at that time.

Although the isotopic data presented here do not characterise the geologic environment, they do nevertheless provide some basic constraints. These constraints consist of firstly a well defined chronological framework. As described by Page et al., [1984] and summarised in Table 2, U-Pb zircon ages define the main period of Early Proterozoic magmatism as occurring from 1840 Ma to 1880 Ma. In addition the Sm-Nd isotopic data discussed in the previous section indicates formation of a protolith at ~2300 Ma. The self consistency of the isotopic data argues against significant involvement of Archean felsic crust. This inference is also supported by limited Sm-Nd studies of older cycle metasediments (eg sample 53) which gives the same T^{Nd} model ages (i.e. 2100 Ma to 2300 Ma) as the main period of felsic magmatism. The absence of any obvious Archean signatures is important as it is not readily compatible with an intra-cratonic setting. The final constraint is that isotopic data not only from Australia but also from North America [Nelson and DePaolo, 1985] and Scandinavia [Wilson et al., 1985, Patchett and Kouvo, 1986] indicates that the Early Proterozoic is an important period for continental crustal production.

Acknowledgements. This study would not have been possible without ready access to well documented samples from throughout the Australian continent. I thank Lance Black, John Cooper, John deLaeter, Mark Fanning, Ian Fletcher, Andrew Glikson, Hans Hensel, Masa Ogasawara, Virginia Oversby, Rod Page, Bob Pidgeon, John Richards, Barney Stevens, Shen-Su Sun, Graham Teale, Alec Trendal and Lesley Wyborn for the provision of samples, Hans Hensel for assistance with the isotopic analyses and Joan Cowley for drafting diagrams. Discussions with Rod Page, Lesley Wyborn and Shen-Su Sun on the geochemistry and isotopic aspects of Mt Isa were particularly informative. Shen-Su Sun and Ian Fletcher kindly made available unpublished analyses. Jon Patchett, Alfred Kroner and Ian Fletcher are thanked for providing helpful reviews and Shen-Su Sun and Lesley Wyborn for comments on an early version of this paper. Rod Page provided both scientific and moral encouragement when my efforts were flagging. Finally, I am grateful to Vic Wall and Bruce Hobbs for pointing out the complexity of both the Broken Hill vintage reds and metamorphic basement.

References

Armstrong, R.L., Radiogenic isotopes: the case for crustal recyling on a near-steady-state no-continental-growth earth, _Phil. Trans. Roy. Soc. Lond. A, 301_, 433-472, 1981.

Arriens, P.A. and Lambert, I.B., On the age and strontium isotopic geochemistry of granulite-facies rocks from the Fraser Range, Western Australia and the Musgrave Ranges, central Australia, _Spec. Publ. Geol. Soc. Aust., 2_, 377-388, 1969.

Bickle, J.J., Champan, H.J., Bettenay, L.F., Groves, D.I. and deLaeter, J.R., Lead ages, reset rubidium-strontium ages and implications for the Archean crustal evolution of the Diemals area, central Yilgarn Block, Western Australia, _Geochim. Cosmochim. Acta, 47_, 907-914, 1983.

Black, L.P., A Rb-Sr geochronological study on the Proterozoic Tennant Creek Block, central Australia, _BMR J. Aust. Geol. Geophys., 2_, 111-122, 1977.

Black, L.P., Bell, T.H., Rubenach, M.J. and Withnall, I.W., Geochronology of discrete structural-metamorphic events in a multiply deformed Precambrian terrain, _Tectonphys., 54_, 103-137, 1979.

Black, L.P., U-Pb zircon ages and a revised chronology for the Tennant Creek Inlier, Northern Territory, _Aust. J. Earth Sci., 31_, 123-131, 1984.

Black, L.P. and McCulloch, M.T., Sm-Nd ages of the Arunta, Tennant Creek and Georgetown Inliers of northern Australia, _Aust. J. Earth Sci., 31_, 49-60, 1984.

Bofinger, V.M., Geochronology in the east Kimberley area of Western Australia, Unpubl. _Ph.D. Thesis, Australian National University,_ 1967.

Burke, L., Dewey, J.F. and Kidd, W.S.F., Precambrian palaeomagnetic results compatible with contemporary operation of the Wilson cycle, _Tectonics, 33_, 287-299, 1976.

Chin, R.J. and deLaeter, J.R., The relationship of new Rb-Sr isotopic dates from the Rudall Metamorphic Complex to the geology of the Paterson Province, _West. Aust. Geol. Surv. Ann. Rep.,_ 80-87, 1980.

Collerson, K.D. and McCulloch, M.T., Field and Sr-Nd isotopic constraints on Archean crust and mantle evolution in the east Pilbara Block, Western Australia (Abstr.), 6th Aust. Geol. Conv., Canberra, _Geol. Soc. Aust.,_ 167-168, 1983.

Compston, W. and Arriens, P.A., The Precambrian geochronology of Australia, _Can. J. Earth Sci., 5_, 561-583, 1968.

Compston, W., Williams, I.S., Campell, I.H., and Gresham, J.J., Zircon xenocrysts from the Kambalda volcanics: age constraints and direct evidence for older continental crust below below the Kambalda-Norseman greenstones, _Earth Planet. Sci. Lett., 76_, 299-311, 1986.

Cooper, J.A. and Compston, W., Rb-Sr dating within the Houghton Inlier, South Australia, _J. Geol. Soc. Aust., 17_, 213-219, 1971.

Cooper, J.A., Webb, A.W. and Whitaker, W.G., Isotopic measurements in the Cape York Peninsula area, north Queensland, _J. Geol. Soc. Aust., 22_, 285-310, 1975.

DePaolo, D.J. and Wasserburg, G.J., Nd isotopic variations and petrogenetic models, _Geophys. Res. Lett., 3_, 249-252, 1976a.

DePaolo, D.J. and Wasserburg G.J., Inferences about magma sources and mantle structure from variations of $^{143}Nd/^{144}Nd$, _Geophys. Res. Lett., 3_, 743-746, 1976b.

DePaolo, D.J., Neodymium isotopes in the Colorado Front Range and crust-mantle evolution in the Proterzoic, _Nature, 291_, 193-196, 1981.

deLaeter, J.R. and Blockley, J.G., Granite ages within the Pilbara Block, Western Australia, _J. Geol. Soc. Aust., 19_, 363-370, 1972.

deLaeter, J.R., Lewis, J.D. and Blockley, J.G., Granite ages within the Shaw Batholith of the Pilbara Block, West. Australia, _Geol. Surv. Ann. Rep.,_ 82-91, 1974.

deLaeter, J.R., Fletcher, I.R., Rosman, K.J.R., Williams, I.R., Gee, R.D. and Libby, W.G., Early Archaean gneisses from the Yilgarn Block Western Australia, _Nature, 292_, 322-324, 1981.

Etheridge, M.A., Rutland, R.W.R. and Wyborn, L.A.I., Orogenesis and tectonic process in the Early to Middle Proterozoic of northern Australia, _this volume,_ 1986.

Fanning, C.M., Oliver, R.L. and Cooper, J.A., The Carnot Gneisses, a metamorphosed Archaean supracrustal sequence in southern Eyre Peninsula. In: Symposium on the Gawler Craton, (Abstr.) _J. Geol. Soc. Aust., 27,_ 47, 1980.

Finlayson, D.M., Seismic crustal structure of the Proterozoic north Australian craton between Tennant Creek and Mt Isa, _J. Geophys. Res., 87_, 10569-10578, 1982.

Fletcher, I.R., Williams, S.J., Gee, R.D. and Rosman, K.J.R. Sm-Nd model ages across the margins of the Archaean Yilgarn Block, Western Australia; northwest transect into the Proterozoic Gascoyne Province, _J. Geol. Soc. Aust., 30_, 167-174, 1983a.

Fletcher, I.R., Wilds, S.A., Libby, W.G. and Rosman, K.F.R., Sm-Nd model ages across the margins of the Archaean Yilgarn Block, Western Australia-II; southwest transect into the Albany-Fraser Province, *J. Geol. Soc. Aust., 30,* 333-340, 1983b.

Fletcher, I.R., Wilde, S.A. and Rosman, K.J.R., Sm-Nd model ages across the margins of the Archaean Yilgarn Block, Western Australia-III; the western margin, *Aust. J. Earth Sci., 32,* 73-82, 1985.

Gee, R.D., Structure and tectonic style of the Western Australian Shield, *Tectonophysics, 58,* 327-369, 1979.

Gray, C.M., Cliff, R.A. and Goode, A.D.T., Neodymium-strontium isotopic evidence for extreme contamination in a layered basic intrusion, *Earth Planet. Sci. Lett., 56,* 189-198, 1981.

Gurnis, M., and Davies, G.F., Simple parametric models of crustal growth, *J. Geodynamics, 3,* 105-135, 1985.

Hamilton, P.J., Evensen, N.M., O'Nions, R.K., Glikson, A.Y. and Hickman, A.H. Sm-Nd dating of the North Star Basalt, Warrawoona Group, Pilbara Block, Western Australia, *Spec. Publ. Geol. Soc. Aust., 7,* 187-192, 1981.

Hickman, A.H., Crustal evolution of the Pilbara Block, Western Australia, *Geol. Soc. Aust., Spec. Publ., 7,* 57-69, 1981.

Hoffman, P.F., Wopmay orogen: a Wilson cycle of Early Proterozoic age in the northwest of the Canadian Shield, in Strangway, D.W., ed., *The continental crust and its mineral deposits, Geol. Assoc. Canada Spec. Paper, 20,* 523-549, 1980.

Hoffman, P.F. and Bowring, S.A., Shortlived 1.9 Ga continental margin and its destruction, Wopman orogen, northwest Canada, *Geology, 12,* 68-72, 1984.

Jahn, B.M., Glikson, A.Y., Paucat, J.J. and Hickman, A.H., REE geochemistry and isotopic data of Archaean silicic volcanics and granitoids from the Pilbara Block, Western Australia: implication for the Early crustal evolution, *Geochim. Cosmochim. Acta, 45,* 1633-1652, 1981.

Katz, M.B., The tectonics of Precambrian craton-mobile belts: progressive deformation of polygonal miniplates, *Precambrian Res., 27,* 307-319, 1985.

Kroner, A., Precambrian plate tectonics, *in* Kroner A.,ed., *Precambrian plate tectonics,* Amsterdam , Elsevier, 57-90, 1981.

Kroner, A., Proterozoic mobile belts compatible with the plate tectonic concept, *Geol. Soc. Am. Memoir, 161,* 59-74, 1983.

Lambeck, K., Structure and evolution of the Amadeus, Officer and Ngalia basins of central Australia, *Aust. J. Earth Sci., 31,* 25-48, 1984.

Liew, T.C. and McCulloch M.T., Genesis of granitoid batholiths of Peninsular Malaysia and implications for models of crustal evolution: evidence from a Nd-Sr isotopic and U-Pb zircon study, *Geochim. Cosmochim Acta, 49,* 587-600, 1985.

Mathur, S.P. and Shaw, R.D., Australian orogenic belts: evidence for evolving plate tectonics? *Earth Evol. Sci., 4,* 281-308, 1982.

McCulloch, M.T. and Wasserburg, G.J., Sm-Nd and Rb-Sr chronology of continetal crust formation, *Science, 200,* 1003-1011, 1978.

McCulloch, M.T. and Compston, W., Sm-Nd age of Kambalda and Kanowna greenstones and heterogeneity in the Archean mantle, *Nature,* 294, 322-327, 1981.

McCulloch, M.T. and Chappel, B.W., Nd isotopic characteristics of S- and I-type granites, *Earth Planet. Sci. Lett., 58,* 51-64, 1982.

McCulloch, M.T., Collerson, K.D. and Compston, W., Growth of Archaean crust within the Western Gneiss Terrain, Yilgarn Block, Western Australia, *J. Geol. Soc. Aust., 30,* 155-160, 1983a.

McCulloch, M.T., Compston, W. and Froude D.O., Sm-Nd and Rb-Sr dating of Archean gneisses, eastern Yilgarn Block, Western Australia, *J. Geol. Soc. Aust., 30,* 149-153, 1983b.

McCulloch, M.T., Jaques, L., Nelson, D.R. and Lewis, J.D., Nd and Sr isotopes in kimberlites and lamproites from Western Australia: an enriched mantle origin, *Nature, 302,* 400-403, 1983c.

McCulloch, M.T. and Hensel, H.D., Sm-Nd isotopic evidence for a major Early to Mid-Proterozoic episode of crustal growth in the Australian continent, *XXVII Internat. Geol. Conv. Moscow, Vol. II ,* 351-352, 1984.

McLennan, S.M. and Taylor, S.R., Continental freeboard, sedimentation rates and the growth of continental crust, *Nature, 306,* 169-172, 1983.

Moorbath, S., Ages, isotopes and evolution of Precambrian continental crust, *Chem. Geol., 20,* 151-187, 1977.

Needham, R., Crick, I.H., and Stuart-Smith, P.G., Regional geology of the Pine Creek Geosyncline, in *Uranium in the Pine Creek Geosyncline,* edited by J. Ferguson and A.B. Goleby, International Atomic Energy Agency, Vienna, 1-22, 1980.

Nelson, B.K. and DePaolo, D.J., Rapid production of continental crust 1.7 to 1.9 b.y. ago: Nd isotopic evidence from the basement of the North American mid-continent, *Geol. Soc. America Bull., 96,* 746-754, 1985.

Ogasawara, M., Petrogenesis of Proterozoic granotoids from Halls Creek mobile zone, Western Australia, *XXVII Internat. Geol. Conv. Moscow, Vol. IV,* 413, 1984.

Oversby, V.M., Lead isotopic systematics and ages of Archaean acid intrusives in the Kalgoorlie-Norseman area, Western Australia, *Geochim. Cosmochim. Acta, 39,* 1107-1125, 1975.

Page, R.W., Blake, D.H. and Mahon, M.W., Geochronology and related aspects of acid volcanics, associated granites, and other Proterozoic rocks in the Granites-Tanami region, northwestern Australia, *BMR Jour. Aust. Geol. Geophys., 1,* 1-13, 1976.

Page, R.W., Response of U-Pb zircon and Rb-Sr total-rock and mineral systems to low-grade regional metamorphism in Proterozoic igneous rocks, Mount Isa, Australia, *J. Geol. Soc. Aust., 25,* 141-164, 1978.

Page, R.W., Compston, W. and Needham, R.S., Geochronology and evolution of the late-Archaean basement and Proterozoic rocks in the Alligator Rivers Uranium Field, Northern Territory, Austrtalia. In: *Uranium in the Pine Creek Geosyncline,* J. Ferguson and A.B. Goleby (eds), Int. Atomic Energy Agency, Vienna, 39-68, 1980.

Page, R.W., Timing of superposed volcanism in the Proterozoic Mount Isa Inlier, Australia, *Precambrian Res., 21,* 223-245, 1983.

Page, R.W., McCulloch, M.T. and Black, L.P., Isotopic record of major precambrian events in Australia, *Proc. 27th Int. Geol. Congress, 5,* 25-72, 1984.

Patchett, P.J. and Bridgwater D., Origin of continental crust of 1.9-1.7 Ga age defined by Nd isotopes in the Ketilidian terrain of South Greenland, *Contrib. Mineral. Petrol., 87,* 311-318, 1984.

Patchett, P.J. and Kouvo O., Origin of continental crust of age 1.9-1.7 Ga age: Nd isotopes and U-Pb zircon ages in the Svecokarelian terrain of South Finland, *Contrib. Mineral. Petrol., 92,* 1-12, 1986.

Patchett, P.J. and Arndt, N., Continental crust genesis on earth: need for systematic Nd isotopic study of terrains 3.8 Ga to present, *Lunar Planet. Sci. (abstr.) XVII,* 650-651, 1986.

Pidgeon, R.T., 3450 m.y. -old volcanics in the Archaean layered greenstone succession of the Pilbara Block, Western Australia, *Earth Planet. Sci. Lett., 37,* 421-428, 1978.

Pidgeon, R.T., Geochronological constraints on volcanic evolution of the Pilbara Block, Western Australia, *Aust. J. Earth Sci., 31,* 237-242, 1984.

Plumb, K.A., The tectonic evolution of Australia, *Earth Sci. Rev. 14,* 205-249, 1979.

Reymer, A. and Schubert G., Phanerozoic addition rates to the continental crust and crustal growth, *Tectonics, 3,* 63-77, 1984.

Richards, J.R., Berry, H. and Rhodes, J.M., Isotopic and Lead-alpha ages of some Australian zircons, *J. Geol. Soc. Aust., 13,* 69-96, 1966.

Rutland, R.W.R., Orogenic Evolution of Australia *Earth-Science Reviews, 12,* 161-196, 1976.

Sun, S.S. and McCulloch, M.T., Chemical and Nd isotope study of Late Archaean to Middle Proterozoic mafic volcanic rocks in northern Australia, Abstr., Tectonic and Geochemistry of Early to Middle Proterozoic Fold Belts, *BMR record 1985/28,* 1985.

Stephenson, N.C.N., Precambrian amphibolites and basic granulites of south coast of Western Australia, *J. Geol. Soc. Aust., 27,* 91-104, 1980.

Taylor, S.R., and McLennan, S.M., The composition and evolution of the continental crust: rare earth element evidence from sedimentary rocks, *Phil. Trans. R.Soc. Lond. A 301,* 381-399, 1981.

Turek, A. and Compston, W., Rubidium-strontium geochronology in the Kalgoorlie region, *Geol. Soc. Aust. Spec. Publ., 3,* 72, 1971.

White, W.M. and Hoffman, A.W., Sr and Nd isotope geochemistry of oceanic basalts and manfle evolution, *Nature, 296,* 821-825, 1982.

Williams, I.S., Page, R.W., Froude, D., Foster, J.J. and Compston, W., Early crustal components in the Western Australian Archaean: zircon U-Pb ages by ion microprobe analysis from the Shaw Batholith and Narryer metamorphic belt, (Abstr.) 6th Aust. Geol. Conv., Canberra, *Geol. Soc. Aust. Abstr., 9,* 109-171, 1983.

Willis, I.L., Brown, R.E., Stevens, B.P.J. and Stroud, W.J., Stratigraphy of the Early Proterozoic Willyama Supergroup and implications for evolution of the Broken Hill Block, (Abstr) 6th Aust. Geol. Conv., Canberra, *Geol. Soc. Aust. Abstr., 9,* 92-94, 1983.

Wilson, M.R., Hamilton, P.J., Fallick, A.E., Aftalion, M. and Michard, A., Sm-Nd, U-Pb and O isotope systematics of granites and Proterozoic crustal evolution in Sweden, *Earth Planet. Sci. Lett., 72,* 376-388, 1985.

Windley, B.F., Precambrian rocks in the light of the plate-tectonic conncept , in Kroner, A., ed., *Precambrian plate tectonics,* Amsterdam, Elsevier, 1-20, 1981.

Wyborn, L.A.I. and Page, R.W., The Proterozoic Kalkadoon and Ewen Batholiths, Mount Isa Inlier, Queensland: source, chemistry, age and metamorphism, *BMR Journal of Australian Geology and Geophysics, 8,* 53-69, 1983.

Wyborn, L.A.I., Geochemistry and origin of a major Early Proterozoic felsic igneous event of northern Australia and evidence for substantial vertical accretion of crust, Abstr. Tectonic and Geochemistry of to Middle Proterozoic Fold Belts, *BMR record 1985/28,* 1985.

 OROGENESIS AND TECTONIC PROCESS IN THE EARLY TO MIDDLE PROTEROZOIC OF NORTHERN AUSTRALIA

M.A. Etheridge, R.W.R. Rutland and L.A.I. Wyborn

Bureau of Mineral Resources, Geology and Geophysics, GPO Box 378, Canberra, ACT 2601, Australia

Abstract. The early Proterozoic terranes of northern Australia were affected by an essentially isochronous orogenic event between about 1850 and 1880 Ma ago. We have termed this event the Barramundi orogeny. The Barramundi orogeny was immediately preceded by a basin-forming episode that resulted from local extension of the pre-existing Archean continental crust. These basins contain three separate sequences. A lower rift sequence is overlain by a laterally extensive, shallow water succession ascribed to post-rift thermal subsidence. A rapid increase in subsidence rate gave rise to a thick flysch sequence, which we interpret to have heralded the main orogenic phase. The Barramundi orogeny is characterised by an extensive and uniform suite of I-type granitoids and comagmatic volcanics, which show significant petrological and chemical differences from modern subduction related magmatism. We interpret this Barramundi igneous association to have been derived by partial melting of a lower crustal mafic layer which was underplated at around 2000 Ma ago. The metamorphic and deformational style of the Barramundi orogeny suggests that abnormally thick crust was not produced. Low pressure (andalusite-sillimanite) metamorphic facies series predominate, and the limited P-T path data suggest that near isobaric cooling was more common than the adiabatic decompression typical of modern orogens. The character of the Barramundi orogeny, together with the absence of ophiolites, paired metamorphic belts and other diagnostic features of modern orogeny, suggest that it was essentially ensialic. We propose a preliminary model in which contemporaneous onset of small-scale mantle convection triggered mantle melting, underplating and continental extension above a polygonal array of upwelling zones. Heat loss led to the termination of convection, thermal subsidence of the stretched regions and cooling of the lithosphere. Cooling of enriched mantle below the early Proterozoic basins is suggested to have triggered crust-mantle delamination and a A-subduction. Delamination provided the driving force for the compressional orogeny and the enhanced heat flow to produce the Barramundi association from the previously underplated layer.

Introduction

Two major schools of thought exist concerning the tectonic setting of orogeny in the early to middle Proterozoic. The first, typified by Burke et al. [1976], Windley [1981] and Hoffman [1980], is based on the application of actualistic concepts of modern orogeny, including the various components of the Wilson cycle to Precambrian tectonic processes. The second has proposed that Precambrian tectonics is typified by secular changes within a plate tectonic framework, and in particular that basin formation and orogeny are commonly ensialic, with sea-floor spreading and subduction being relatively unimportant in the orogenic process [Rutland 1973, 1976, 1982; Kröner, 1977, 1983].

The actualistic school has been strongly influenced by such factors as 1.) the similarity between some Precambrian sequences and those of modern passive continental margins, 2.) the predominance of so-called calcalkaline orogenic igneous activity, 3.) the occurrence of mafic rocks with MORB-type chemical signatures, and 4.) the importance of thrusting as a macroscopic deformation process. The alternative school has emphasised 1.) the virtual absence of ophiolitic rocks and paired metamorphic belts, 2.) the presence of an ensialic basement to parts of terranes, 3.) the paucity of andesitic volcanism, 4.) the presence of broad basins with dominantly shallow water sequences, 5.) the lack of significant vertical uplift, and 6.) the absence of paleomagnetic evidence for significant relative motion between cratons bord-ering Proterozoic orogens.

In this paper we analyse the character of a spatially extensive, essentially isochronous orogenic event that affected all of the early Proterozoic domains of northern Australia [which are separated by extensive areas of younger platform cover; Geological Society of Australia, 1971]. Limited evidence for the same event also exists in early Proterozoic domains of southern and western Australia. The orogeny terminated a widespread extensional basin-forming event and has a consistent style from terrane to terrane. The igneous, structural, and metamorphic character of the oro-

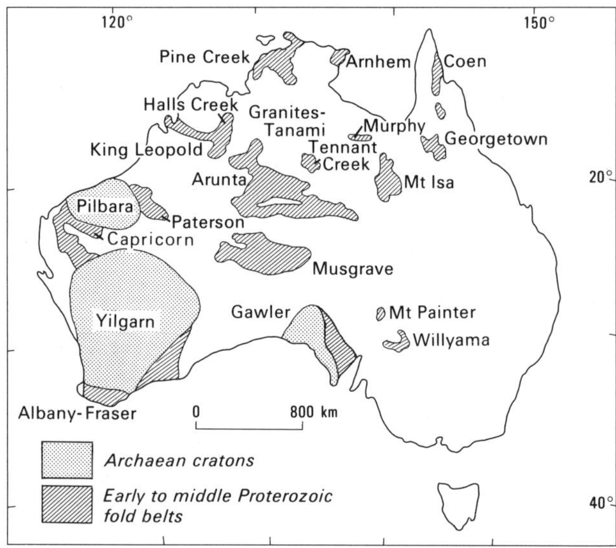

Fig. 1. Distribution of early to middle Proterozoic fold belts in Australia. The two major
Archean cratons are also shown, and smaller Archean remnants occur within the Pine Creek and
Gawler domains.

geny place important constraints on its tectonic
setting and support the concept of a spatially
extensive, isochronous, ensialic orogeny that did
not involve significant lateral continental
accretion. It therefore contrasts with the actualistic plate tectonic model of diachronous orogeny
related to lateral accretion at migrating continental margins.

Chronotectonic Framework

A major tectonothermal event between about 1900
and 1600 Ma ago is known from virtually all Proterozoic provinces [Gastil, 1960; Moorbath, 1978].
For example, it corresponds in Canada to the Hudsonian orogeny, in Scandinavia to the Svecofennian, in Africa to the Eburuian, and in China to
the Luliangian [Dahzong and Songnian, 1985]. In
Australian provinces (Figure 1) there is a fairly
wide spread of isotopic ages which purport to date
this event [Compston and Arriens, 1968; Page et
al., 1984; McCulloch, this volume]. However, the
chronological framework for this time period has
been substantially refined in the last 10 years by
the application of the zircon U/Pb dating technique [e.g., Page, 1978] and by closer collaboration
between geochronologists, petrologists, and structural geologists in the application of Rb/Sr techniques to rocks with complex deformational and
metamorphic histories [Black et al., 1979; Wyborn
and Page, 1983a].

Page et al. [1984] have critically assessed the
data for the Australian Precambrian on the basis
of this experience, with the result that a widespread orogenic episode is well bracketed between

1820 and 1920 Ma over the whole of northern Australia (Table 1). By rejecting those Rb/Sr dates
that are not well controlled by petrological,
structural and metamorphic studies, we suggest
that this orogenic event can be further constrained to between 1850 and 1880 Ma. It is
particularly well defined in the Mount Isa and
Pine Creek orogenic domains. In the Pine Creek
domain (Figure 1), U/Pb zircon data on volca,cla-
stic units predating and postdating the orogenic
event bracket it between 1870 Ma and 1880 Ma
[Page, 1985]. Similar data from pre-orogenic
metavolcanics and post-orogenic granitoids southeast of Mt Isa (Figure 1) constrain the deformation to between 1850 and 1870 Ma [Wyborn and Page,
1983a; Page, 1985].

The orogeny was characterized, especially at
its close, by a distinctive felsic igneous event
of large magnitude (Table 1)and rocks of remarkably consistent chemistry, which have been recognised in the Halls Creek, Pine Creek, Arunta,
Tennant Creek, Gawler, and Mount Isa domains.
Therefore, a major orogenic/magmatic episode affected virtually all of the early Proterozoic orogenic domains of Australia within a period of 30 to
60 Ma, with little evidence of the lateral migration of orogeny that characterises Phanerozoic
orogeny and produces assemblages of contrasting
tectonic zones. The orogeny terminated a widespread cycle of sedimentary basin formation [broa-
dly corresponding to the Nullaginian of Dunn et
al., 1966], and was followed in some domains,
after an interval of up to 150 Ma, by one or more
basin-forming episodes [Carpentarian of Dunn et
al., 1966]. We propose the name Barramundi orogeny for this key event in the Precambrian tectonic
evolution of northern Australia. Preliminary data
from the Gawler Province (Figure 1) suggests that
this event is Australia wide [Fanning et al.,
1986].

Tectonostratigraphic History

This paper is concerned mainly with the style
and timing of the Barramundi orogeny, rather than
with the volcano-sedimentary sequences that pre-
and postdate it. However, we will briefly summarise the character of those sequences, with particular emphasis on their implications for basin-
forming processes. Previously for the Australian
Proterozoic, the Tectonic Map of Australia
[Geological Society of Australia, 1971] distinguished precratonic, transitional, and cratonic domains within these broad sequences, and this classification does have general significance in
relation to the stabilisation of the Proterozoic
continental crust. However, we consider that a
more useful subdivision can be made on the basis
of age and the unit's position in the basin-
forming process. In particular, each sequence is
variably deformed and metamorphosed, so a classification scheme based largely on their degree of
deformation takes insufficient account of the
tectonic setting of the basin-forming processes.

TABLE 1. Summary of Information Available for Australian Early Proterozoic Domains

Domain	pre-Barramundi Sediments as % of Domain area	Tectonostrat Sequences Documented[1]	Age of Barramundi Orogeny	Outcrop Area Barramundi Ign Assocn(km^2)	Archean Basement	Deformational Style	Tectonic Transport Direction[6]	Metamorphic Facies Series
Gascoyne	>80%	R,S,T[1]	~1850??[3]	unknown	present	D1/D2	north-south[6]	low pressure
King Leopold	~80%	?S,T	1840-1910[2]	4000	inferred[4]	D1/D2	N-->S+rt lat[6]	low pressure
Halls Creek	~80%	R,S,T	1850-1910[2]	9500	inferred[4]	D1+/D2[5]	NW-->SE+left lat	low pressure
Pine Creek	~80%	R,S,T	1860-1880	11000	present	D1/D2	ENE-->WSW	low-med press
Granites/ Tanami	60-80%	?S,T	1820+[3]	3400	unknown	complex	unknown	?low pressure
Arunta	?50-90%	?S,?T	1820+[3]	1500+	unknown	D1/D2+[5]	north-south	low pressure
Tennant Ck	~100%	T	1860-1870	1200	unknown	?D1/D2	unknown	low pressure
Mount Isa	<10%	?S,T	1860-1870	6000	inferred[4]	D1/D2	?east-west	low pressure
Gawler	?50-80%	R,S,T	1840-1950[2]	????	present	D1/D2	?east-west	low pressure

[1] The sources of information are either quoted in the text, or are original to this paper.
[2] R-rift phase sedimentation, S - thermal subsidence phase, T - turbidites (see text).
[3] The upper limits of these age ranges are based on poorly constrained Rb/Sr data.
[4] These ages are based on Rb/Sr model ages of metamorphosed granites, and are therefore minimum ages.
[5] Archean basement inferred from Sm/Nd model ages [Sun and McCulloch, 1985] or U-Pb zircon ion probe data [Page, 1985].
[6] Refer to Fig. 6, D1+ and D2+ indicate more than one period of deformation of that style.
[6] Interpreted from both D1 and D2 events, 'north-south' indicates polarity is unknown, 'N-->S' implies polarity.

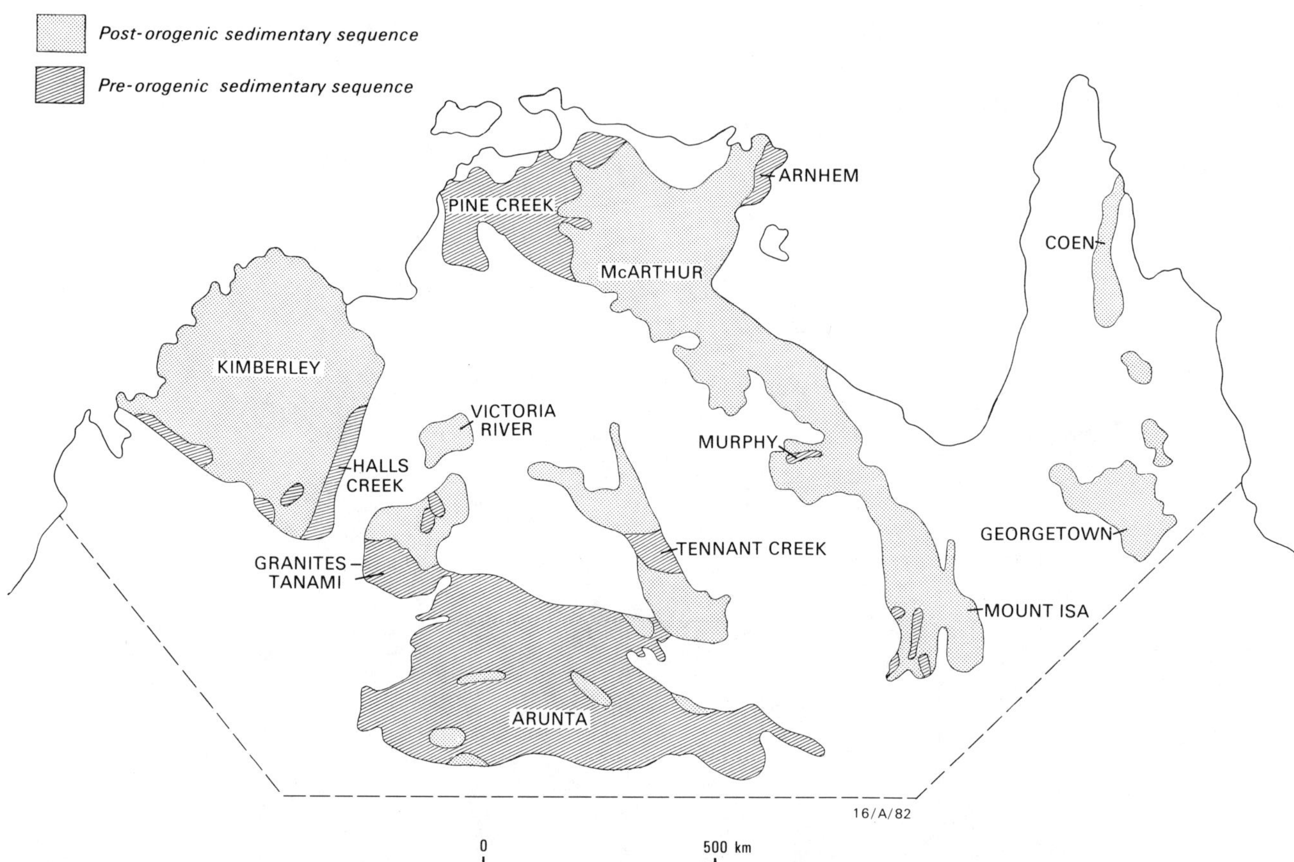

Fig. 2. Subdivision of northern Australian early to middle Proterozoic sequences into those which predate and those which postdate the Barramundi orogeny. There is significant uncertainty about this subdivision in the Arunta domain.

The sedimentary cycle that preceded the Barramundi orogeny [Nullaginian of Dunn et al., 1966] is best represented in the Halls Creek and Pine Creek domains (Figure 2), where its stratigraphy and structure are reasonably well known. It is also present or inferred in the Arunta, Mount Isa, Tennant Creek, [Page et al., 1984], Capricorn [Thorne, 1986] and Gawler [Fanning et al., 1986] domains (Table 1). This sedimentary cycle can be divided into three sequences of differing tectono-stratigraphic character. The three sequences are:

1. A lower clastic sequence, quartz-rich and commonly demonstrably fluviatile, containing predominantly mafic volcanic rocks. This sequence is interpreted to represent the rift phase of crustal extension which initiated basin formation in this cycle. It is represented by the Ding Dong Downs Volcanics and Saunders Creek Formation in the Halls Creek domain [Hancock and Rutland, 1984] and the Namoona, Kakadu, and at least part of the Mount Partridge Groups in the Pine Creek domain [Needham and Stuart-Smith,

1985]. Locally this sequence rests unconformably on Archean rocks.

2. A middle, finer grained, dominantly clastic sequence that is commonly carbonaceous and includes carbonates and iron formations. Units within this sequence are generally laterally uniform and extensive and are interpreted to represent the post-stretching thermal subsidence (sag) phase of basin evolution. The Biscay Formation [Halls Creek domain, Hancock and Rutland, 1984] and the South Alligator Group [Pine Creek domain, Needham et al., 1980] are considered to belong to this sequence.

3. An upper turbidite facies that may be synorogenic in part, represented by the Olympio Formation of the Halls Creek domain and the Finniss River Group of the Pine Creek domain [Hancock and Rutland, 1984; Needham et al., 1980].

The orogeny was followed by further sedimentation in most provinces, but the style and timing of the younger episode(s) of basin formation var-

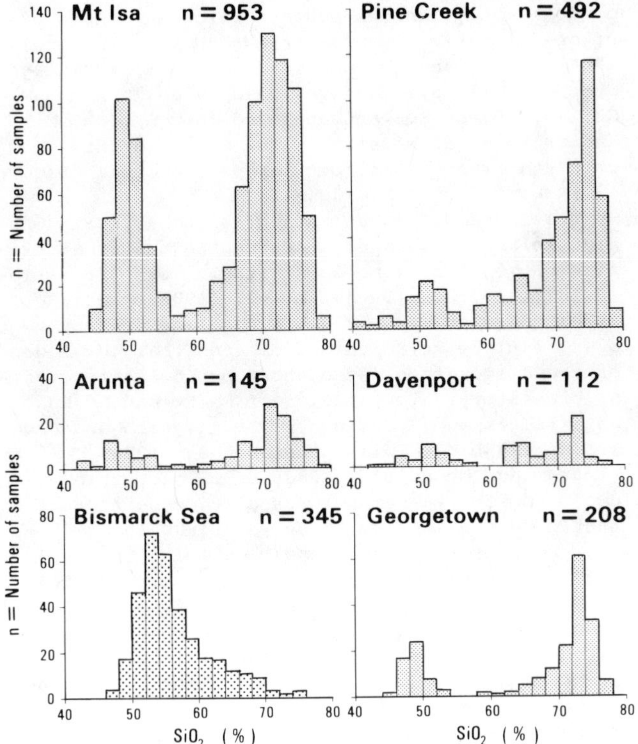

Fig. 3. Comparison between the SiO_2 contents of five Australian Proterozoic terranes and that of a modern arc suite (Bismark Sea, lower left). Data are from Wyborn and Page [1983a], Rossiter and Ferguson [1980], BMR [unpublished] and Johnson [1977].

ied somewhat from province to province. Associated with the orogeny or immediately postdating it there was a major magmatic episode represented by extensive batholithic intrusives and comagmatic, largely ignimbritic extrusives.

The extrusives are unconformable on rocks of the first sedimentary cycle, though they may be broadly folded by the late upright event of the major orogeny [Needham and Stuart-Smith, 1985]. They correspond in part to the transitional tectonism of the Tectonic Map of Australia. There is some evidence that these volcanics and their associated sediments were largely related to rifts formed during limited crustal extension that effectively terminated the Barramundi orogeny [Needham and Stuart-Smith, 1984]. Widespread sedimentation took place over much of Northern Australia in the period 1800 to 1600 Ma ago. Plumb [1979] has referred to this sedimentation as transitional or platform, depending largely on whether or not the rocks are significantly deformed. In the Mt Isa domain (Figure 1), for example, post-Barramundi sedimentation took place in two discrete episodes, both apparently initiated by

continental extension, at about 1800 Ma and 1680 Ma ago respectively. However, we are not concerned with the details of these younger events in this paper.

Magmatic History and Geochemistry

Major igneous rock associations occur within the first cycle basin sequence, associated with the Barramundi orogeny, and throughout a 300 to 400 Ma post-orogenic period. Individual suites within this complex array have remarkably similar ages and geochemical distribution patterns from domain to domain [Wyborn and Page 1983a, 1983b; Wyborn, 1985]. Perhaps the most remarkably uniform of these igneous rock associations is that related to the Barramundi orogeny. Following a brief general description of the range of igneous rock types throughout the early to middle Proterozoic, we will describe the Barramundi association in some detail because it has important implications for tectonic style.

In the first cycle basin sequence the igneous rocks are predominantly mafic and include sills, lopoliths and large layered intrusions [Hamlyn, 1977, 1980]. Felsic intrusives are very rare, and volcanics are not widespread through the sedimentary sequence. The mafic rocks include those with both MORB and continental affinities [Rossiter and Ferguson, 1980; Hancock and Rutland, 1984; Hynes and Gee, 1986], and some, such as the Woodward (Halls Creek) and Zamu (Pine Creek) Dolerites, have virtually identical chemistry and setting in different domains.

The orogenic phase is almost entirely felsic with large I-type granitic batholiths and extensive ignimbrite sheets. Throughout post-Barramundi basin development, a succession of bimodal and felsic volcano-plutonic suites were emplaced, virtually all of them anorogenic. For example, the Mt Isa domain contains major felsic suites of either I or A-type affinity emplaced at 1800 Ma, 1740 Ma, 1670 Ma and 1550 Ma ago [Bultitude and Wyborn, 1982; Wyborn and Page, 1985].

The composition of the felsic magmas emplaced throughout this post-orogenic sedimentary cycle is quite distinctive from that associated with the orogenic event. The post-orogenic granites contain higher abundances of K_2O, TiO_2, P_2O_5, Th, and U, as well as of the high field strength cations such as Zr, Nb, and Y. They also have a more restricted silica range than the orogenic granites (from 68-78 wt %) and generally contain less Al_2O_3 and Sr.

Over the period of about 500 Ma from the earliest first cycle volcanism to the youngest post-orogenic granites, the igneous record in northern Australia is dominated by mafic, felsic or distinctly bimodal suites. A compilation of over 1900 analyses shows that intermediate compositions with between 56 and 63% SiO_2 are rare (Figure 3), in contrast to many modern oceanic and continental

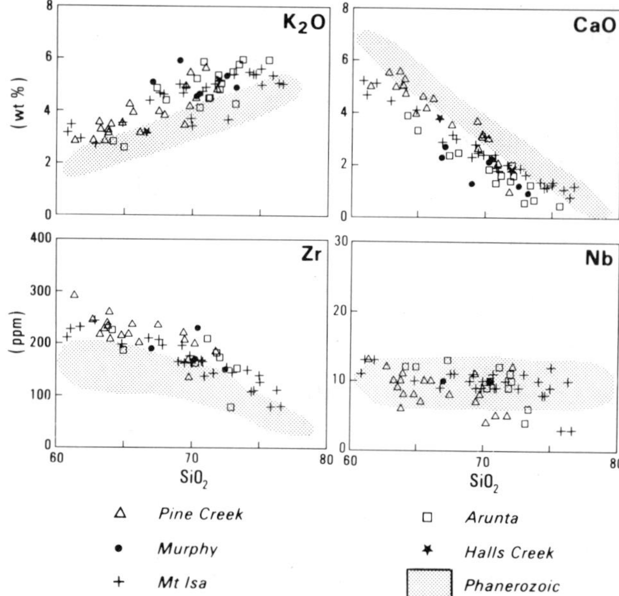

Fig. 4. Selection of Harker variation diagrams for the Barramundi association, with the field for Phanerozoic I-type granite suites for comparison. Proterozoic data from Wyborn & Page [1983a], BMR [unpublished], Dow & Gemuts [1969], Ferguson et al. [1980], Sweet et al. [1981]; Phanerozoic data from Hine et al. [1978], Bateman & Chappell [1979], Whalen et al. [1982], Griffin et al. [1978].

arcs [e.g. Johnson, 1974; Lipman et al., 1972; Pitcher et al., 1986].

The Barramundi Igneous Association. The Barramundi igneous association is dominated by felsic igneous rocks that range from 60 to 78% SiO_2, with compositions between 68 and 75% SiO_2 predominant. The granites and their comagmatic volcanics have been identified in the Halls Creek, Pine Creek, Tennant Creek, Arunta and Mt Isa domains (Figure 1), with a total outcrop area in excess of 36,000 km^2 (Table 1). Where dated by U/Pb zircon techniques, the emplacement ages of the suite are largely confined to between 1840 and 1870 Ma. The majority of the volcanics are subaerial, and the granites commonly intrude their volcanic ejecta.

Despite their wide geographical distribution, these orogenic granites and volcanics form a remarkably uniform petrographic and chemical association. The volcanics range in composition from crystal-rich dacites to rhyodacites and rhyolites. Where unmetamorphosed, the volcanics contain phenocrysts of quartz, plagioclase, clinopyroxene, hornblende, and biotite. The granites range from hornblende-bearing tonalites through granodiorites to monzogranites and syenogranites [I.U.G.S. nomenclature, Streckeisen, 1973]. The petrographic characteristics of the Barramundi association place it in the I-type category of Chappell and White [1974]. I-type magmas are derived from predominantly igneous sources which have not undergone significant surficial weathering [Chappell, 1984; White and Chappell 1983]. The association is chemically distinctive, being high in K_2O, La, Ce, Rb, Th, and U, and depleted in MgO, CaO, Ni, and Cr (even at relatively low SiO_2 contents), compared with Phanerozoic I-type analogues (Figure 4) [Wyborn and Page, 1983a].

The Barramundi association is characterised by high Rb/Sr ratios and, where unaffected by alteration or metamorphism, low initial $^{87}Sr/^{86}Sr$ ratios (0.703 to 0.706) [Bennet et al., 1975; Page, 1978; Page et al., 1980]. Modelling of source ages from the Rb/Sr data indicates that the source of the association had a short but definite crustal prehistory. Combination of Rb/Sr and Sm/Nd data [Black and McCulloch, 1984; Wyborn and Page, 1983a, 1983b; McCulloch and Hensel, 1984; Windrum and McCulloch, 1983; McCulloch, this volume] from the Barramundi association point to a mantle derivation age of between 1900 and 2200 Ma.

Page et al. [1984] and McCulloch [this volume] have recognized this as a "primary crustal formation" event. It is separated from the Barramundi orogeny by a major period of continental extension and basin formation. We interpret this event as a major period of underplating of Archean crust, and relate it to the initiation of the Early Proterozoic basins. We therefore regard the major crustal growth event documented isotopically by Page et al [1984], McCulloch [this vol ume], and Reymer & Schubert [1986] as essentially one of vertical accretion rather than of lateral accretion by subduction-related processes. A lateral accretion model implies that extensive subduction was taking place, and that the commonly elongate volcano-plutonic complexes of the Barramundi association are analogous to modern arc or Cordilleran complexes. The general description of the association as calcalkaline has served to support this model, but we would like to point out some difficulties in applying it to the early Proterozoic of northern Australia.

The term calcalkaline was originally coined as a mineralogical classification [Holmes, 1920] and was later consolidated as one of the major chemical groups of igneous rocks [Peacock, 1931; Tilley, 1950; Irvine and Barragar, 1971]. More recently, magma composition has been related to source chemistry and process, and thence to tectonic environment. Thus the term calcalkaline has become synonomous with oceanic and continental arcs, because the classical examples of calcalkaline suites are those from the circum-Pacific region [Irvine and Baragar, 1971; Christiansen and Lipman, 1972].

There is no doubt that the Barramundi association resembles the more felsic endmembers of typical calcalkaline suites, but it must be emphasized that the distinction between calcalkaline and tholeiitic, alkaline and other suites is based largely on their chemistry and mineralogy at intermediate to basic compositions. This can be illustrated on an AFM diagram (Figure 5), which shows that the members of the Barramundi associa-

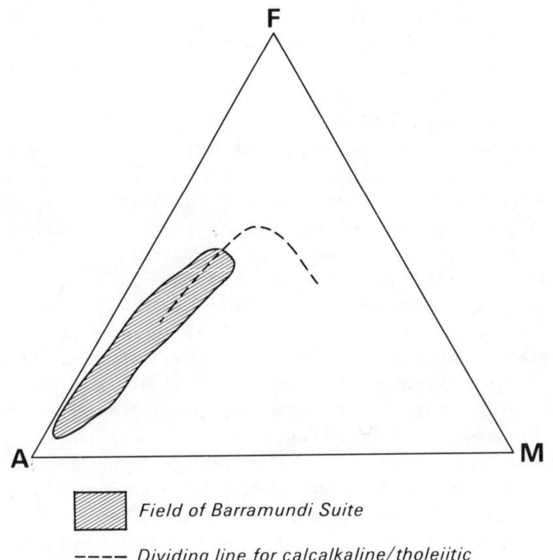

Fig. 5. The field for Barramundi association geochemistry shown on a conventional AFM diagram, illustrating the ambiguity of classification of the suite. Dividing line is that of Irvine and Baragar [1971].

tion are too felsic to define the nature of the iron-enrichment trend at intermediate compositions that helps to distinguish the calcalkaline group. The Barramundi association is dominated by rhyolitic to dacitic compositions, which, as Irvine and Baragar [1971] pointed out, are notoriously difficult to classify, and which are commonly called calcalkaline without sufficient justification. We therefore conclude that, because it lacks the intermediate to basic end members necessary for classification, there are no grounds for describing the Barramundi association as calcalkaline, with the consequent tectonic implications.

What then does the chemistry of this association indicate about its tectonic environment? The most striking feature of the association is its chemical homogeneity from domain to domain. Some minor variation does occur (e.g., Arunta and Halls Creek are slightly lower in MgO), but samples of the Barramundi association from all domains give rise to well-defined linear trends for all major and trace elements on Harker diagrams (Figure 4). The association is also isotopically uniform (after accounting for the effects of metamorphism) and constrained within a narrow range of ages. This isotopic and geochemical uniformity over such a large area requires a homogeneous and extensive source.

Subduction of young oceanic lithosphere with a small component of Archean-derived sediment has been cited to explain the isotopic character of this and other similar early to middle Proterozoic suites [Patchett, 1983; Patchett and Bridgwater, 1984]. However, the Barramundi association does not resemble the modern subduction-related igneous

suites in a number of respects. First, as discussed above, the association lacks intermediate to mafic compositions typical of modern island and continental arcs (Figure 3). Second, the range of chemical variation within and between domains is small compared to that within and between individual modern arcs or even segments of arcs. The chemical data from the Barramundi association come from 7 separate domains, most of which seem to have been discrete fold belts, albeit on a smaller scale than modern igneous belts associated with plate margins. Subduction models for the association would therefore have to explain how these discrete subduction zones gave rise to identical mixtures of primitive crustal-derived melts with an Archean component to produce the observed uniformity of isotopic and geochemical character. Moreover, the simple geological setting of the Proterozoic volcanics and the lack of any chemical or temporal polarity seems incompatible with the subduction model.

The isotopic and geochemical character of the Barramundi association is, however, consistent with derivation from a uniform, mafic lower crustal layer accreted by underplating at about 2000 Ma ago. There is little evidence of reworking of pre-existing Archean crust into either the underplated source or the Barramundi association magmas, and we address this problem below. The underplated layer would have been derived from a mantle enriched in K_2O, Rb, Ca, Ce, Th, and U to provide a source for the highly enriched I-type magmas of the association. Mafic magmas in the age range 2.4 to 2.0 Ga enriched in these same incompatible elements have been documented from Scotland, Antarctica, Greenland, and Wyoming [e.g. Weaver & Tarney, 1981; Sheraton & Black, 1981] suggesting that parts of the early Proterozoic mantle were also enriched in these elements.

Deformational and Metamorphic Style

There are few detailed structural and metamorphic studies of what we now recognize as the Barramundi orogeny. However, there appears to be a consistency of style from domain to domain. This deformational and metamorphic style is typified by the somewhat enigmatic combination of widespread nappe tectonics and low, or less commonly medium pressure metamorphism. Some aspects of the orogenic style are described here, mainly to elucidate their significance for the tectonic setting of the Barramundi event.

One or more early nappe-style deformational events have been documented in the Halls Creek [Hancock and Rutland, 1984] and Pine Creek [Johnston, 1984] domains and are strongly implied by the available evidence (mainly extensive bedding-parallel foliations and refolded intrafolial folds) in the Arunta, Mount Isa and Tennant Creek domains. Large scale thrusts have been identified, but extensive inverted sequences resulting from recumbent folding are not common. At low metamorphic grade, discrete thrusts and thrust

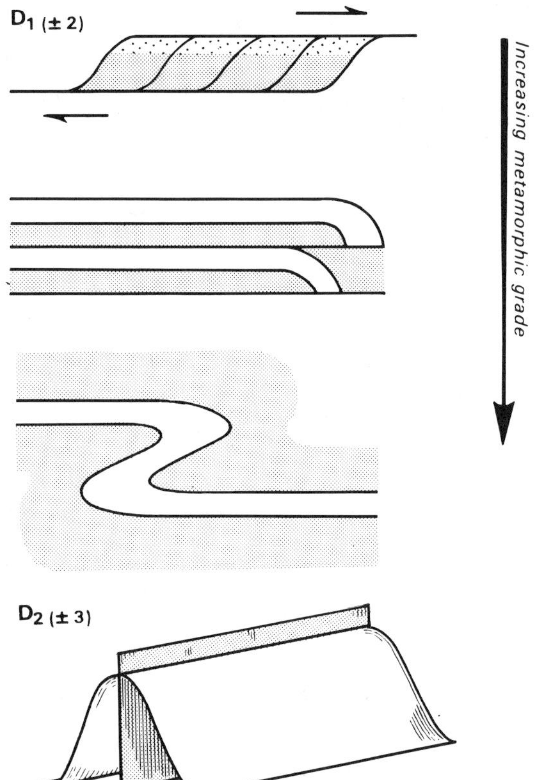

Fig. 6. Summary sketch of the typical structural styles of the Barramundi orogeny, comprising one (or rarely more) nappe-style event overprinted by one or more upright folding episodes.

duplexes of typical foreland style occur. With increasing grade, layer-parallel foliations and tight to isoclinal mesoscopic intrafolial folds are common (Figure 6).

Unlike many modern collisional orogens, however, moderate to steep bedding dips abound, because the sub-horizontal structures are overprinted by tight upright folds of one or more episodes (Figure 6). These generally have gentle to moderate plunges, occur at all scales, and dominate the map patterns at 1:100 000 to 1:500 000 scales. The upright folds occur at all metamorphic grades and stratigraphic levels and commonly imply 30 to 50% regional shortening throughout much of the early Proterozoic sequence. Where the nappe movement direction has been determined, it is generally at a high angle to the axial plane traces of the upright folds, inferring a consistency of gross tectonic transport direction within each domain.

There has been only limited modern study of tectonic transport directions in the various domains. However, our assessment of the published data (Table 1) demonstrates that there is little consistency of movement direction from domain to

domain. In general, the structural grain seems to parallel interpreted trends of the early Proterozoic basins, although Hancock and Rutland [1984] have proposed oblique convergence across the Halls Creek and King Leopold domains.

Taken at face value, the early thrusting followed by regional shortening implies substantial crustal thickening during the Barramundi orogeny, seemingly consistent with a typical collisional orogen. However, there are a number of lines of evidence that suggest that crustal thickness did not greatly exceed its current value of 40-45 km [Collins, 1983; Finlayson, 1982] during the orogeny. First, high stratigraphic levels at low metamorphic grade are preserved in many of the early Proterozoic terranes, placing strict limits on the amount of post-orogenic isostatic rebound and erosion. Second, the relatively undeformed volcanic sequences that followed immediately after or even overlapped the orogeny are preserved over wide areas, on both low and high grade metamorphics. Finally, the metamorphic facies and limited P-T-t path data are inconsistent with crustal overthrusting models, as discussed in more detail below.

The Barramundi Orogeny was typified by low to rarely medium pressure metamorphic conditions in all terranes from which reliable metamorphic data are available (Table 1). Andalusite-sillimanite facies series predominate, with only sparse evidence of assemblages indicative of the kyanite-sillimanite facies series. In the Halls Creek province, low pressure assemblages are virtually ubiquitous [Gemuts, 1971; Gellatly et al., 1974; Hancock and Rutland, 1984; Plumb et al., 1985]. Gemuts [1971] has reported andalusite-K-feldspar-bearing assemblages without muscovite, which represent very low pressures. He has also reported sillimanite replacing andalusite, and kyanite replacing andalusite and sillimanite. These temporal relationships among the aluminosilicates imply that the P-T-t path of these rocks involved increasing temperature (± pressure) at low pressure, followed by essentially isobaric (or weakly decompressive) cooling (Figure 7). Anticlockwise P-T-t paths of this type have also been reported from the Arunta [Warren, 1983] and Willyama [Philipps and Wall, 1981; Hobbs et al., 1984] complexes in the sequences that immediately postdated the Barramundi orogeny. In the Pine Creek domain Ferguson [1980] has reported the only known assemblages of the kyanite-sillimanite facies series. However, andalusite has also been described from medium grade schists within this domain [Roberts, 1960]. Limited data from the Capricorn province in Western Australia suggest that the metamorphic conditions of what is inferred to have been the Barramundi orogeny were within the low pressure facies series [Williams et al., 1983; Blight and Barley, 1983].

The predominance of low pressure facies series and the indication of anticlockwise P-T-t paths in the Barramundi orogeny distinguishes it from classical modern collisional orogens [Oxburgh and

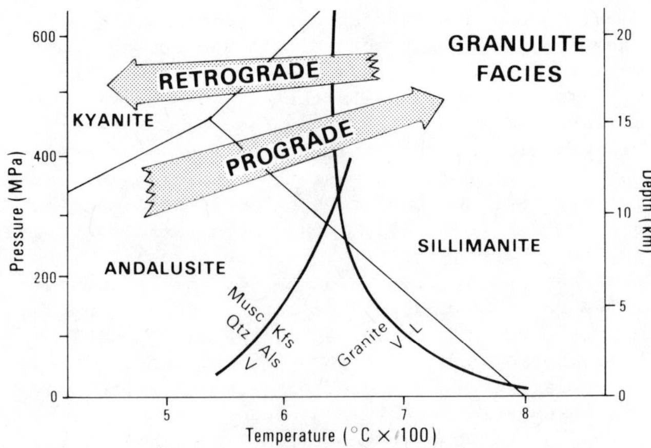

Fig. 7. Apparent P-T-t path for the Barramundi orogeny and susequent middle Proterozoic metamorphic events. Only limited data are available for the Barramundi orogeny.

Turcotte, 1974; Thompson, 1981]. The problem of the very high geothermal gradients implied by the andalusite-sillimanite facies series has been highlighted by Thompson [1981], Hobbs et al. [1984] and Wickam and Oxburgh [1985]. It is generally explained by crustal thinning and/or by addition of mantle melts to the lower crust. According to simple conductive models, low pressure metamorphic conditions cannot be produced by simple crustal thickening, unless large amounts of heat are also added from the mantle [England and Richardson, 1977; Oxburgh and Turcotte, 1974].

In summary, the deformational and metamorphic style of the Barramundi orogeny, although not thoroughly documented, differs substantially from that of modern collisional orogens. Despite the widespread occurrence of thrusts there is little evidence of abnormal crustal thickening, which suggests that the orogeny developed largely on previously extended and thinned crust, with the shortening simply restoring normal crustal thicknesses.

Discussion

The Barramundi orogeny and its preceding basinal phase have much in common with early Proterozoic orogens of almost precisely the same age worldwide. The Wopmay Orogen, so well documented by Hoffman and his colleagues [Hoffman, 1980; Hoffman and Bowring, 1984; Hoffman et al., 1986], is widely regarded as a model for the application of plate tectonic concepts to the early Proterozoic, and is analogous in many ways to the Australian terranes. There is almost universal agreement among workers in the early Proterozoic of North America that the tectonic setting at that time was virtually the same as today, with plate tectonic processes dominating basin formation and orogeny. In particular, the early basinal phase is likened to that of a modern passive margin,

generally with the inference that continental extension proceeded to the stage of sea-floor spreading [Hoffman, 1973, 1980; Easton, 1981, 1983; Stauffer, 1984]. The orogenic phase (Hudsonian) is generally interpreted as an ocean continent collison of Andean type [Hoffman, 1980; Gibb, 1975; Hildebrand and Bowring, 1984; Hoffman and Bowring, 1984; Hoffman et al., 1986]. The widespread linear magmatic belts in many of the terranes are almost universally described as calcalkaline, and interpreted as continental arcs [Hoffman, 1980; Hoffman and McGlynn, 1977; Fumerton et al., 1984; Stauffer, 1984; Hoffman and Bowring, 1984; Hildebrand and Bowring, 1984].

Given this overwhelming support for the application of actualistic tectonics to the North American early Proterozoic, and the gross similarity of the Australian and North American terranes what is the basis for proposing an alternative model for the Australian early Proterozoic? First, it should be emphasized that the difference between our model and modern plate tectonic analogues is one of degree rather than kind. Horizontal motions of crustal (and probably lithospheric) segments are involved, but the motions are considered to have been limited. Basin formation is considered to have resulted from continental extension, as on modern passive margins, and the Barramundi orogeny is essentially a compressional orogeny with much in common with collisional tectonic process. The key distinction between our model and that proposed for the North American (and to a lesser extent for the Scandinavian and African) terranes is that continental extension did not proceed to the stage of sea-floor spreading, and that the subsequent compression did not involve subduction of oceanic lithosphere [Kröner, 1983, 1984]. The key pieces of evidence which have led us to question actualistic models for the Barramundi orogeny and its preceding basinal phase are as follows:

1.) There is a remarkable contemporaneity of process in widely scattered domains over much of the Australian continent, and indeed on the global scale. The increasing body of U/Pb zircon dating also brackets the equivalents of the Barramundi I-type intrusive/volcanic suite between 1840 and 1880 Ma in North America and Scandinavia. Even allowing for the likelihood that plate tectonic activity has been episodic throughout earth history, the restriction of global convergent orogenic activity to a few tens of millions of years is difficult to relate to a major period of sea-floor spreading. The limited reliable dating on volcanic rocks in the pre-orogenic basin phase suggests that it was short-lived, and predated the orogeny by a few tens of millions of years at most [Page, 1985; Hoffman and Bowring, 1984]. Therefore, the period of active horizontal "plate" motion is restricted to less than 80 Ma, and possibly to as little as 50 Ma, including both extensional and compressional phases for all domains studied. Certainly, there is no evidence in northern Australia of progressive lateral accretion around one

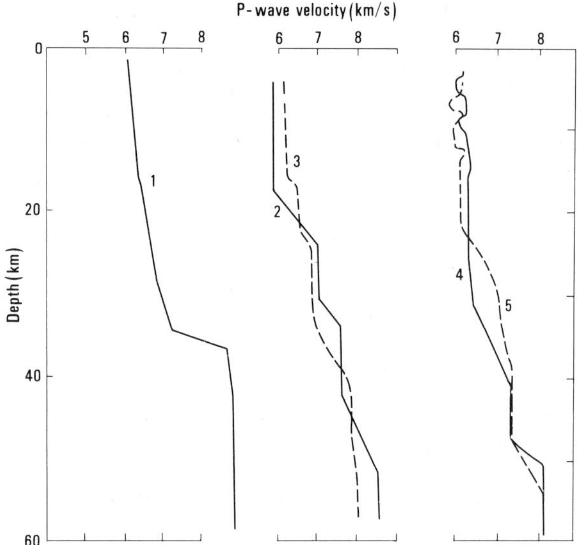

Fig. 8. Velocity-depth curves for 1) Archean Pilbara craton; 2,3) Proterozoic MacArthur Basin; 4,5) Tennant Creek-Mt Isa traverse [data from Finlayson, 1982; Drummond, 1982; Collins, 1983]. Note the thick high velocity lower crust beneath the Proterozoic terranes, in contrast to the thinner, lower velocity Archean crust.

or more Archean nuclei over a period of 100 Ma or more, as might be expected from a number of modern analogues.

2.) There is no evidence that any of the early Proterozoic sediments of the pre-Barramundi basinal phase were deposited on oceanic crust. Ophiolites are unknown, and, wherever basement to the basin sequences is observed, it is Archean or earliest Proterozoic and continental in character. The early Proterozoic basins appear to have developed by stretching of pre-existing continental crust, in similar fashion to modern passive margins [McKenzie, 1978; Le Pichon and Sibuet, 1981]. However, continental extension can clearly be terminated prior to sea-floor spreading in the modern plate tectonic environment, resulting in the development of thick sedimentary basins on a continental basement [e.g., North Sea, Barton and Wood, 1984; Bass Strait, SE Australia, Etheridge et al., 1984, 1985].

3.) The most consistent and characteristic feature of the Barramundi orogeny is the extensive and remarkably uniform I-type magmatism. The most common expression of this major igneous event is as narrow, elongate volcano-plutonic complexes of broadly calcalkaline character, and it has consequently been widely compared with modern continental arc magmatism [e.g.Wilson, 1978]. However, as we pointed out above, the geochemical character of the Barramundi association in northern Australia displays a number of important differences from modern arc magmatism. In particular, the association is dominated by rocks of acid compositions

($>60\%$ SiO_2), and it displays a remarkable geochemical uniformity both within and between domains. The isotopic and geochemical character of the association suggests that it was derived from a fractionated, mafic igneous source with a short crustal prehistory. This source must have been extremely uniform, as must the conditions of partial melting. Neither modern continental nor oceanic arc suites display such compositional uniformity from terrane to terrane on this scale. We regard the most likely source for the Barramundi suite to be an underplated layer generated during mantle upwelling accompanying the extensional basin formations. Evidence for this underplated layer is seen in a number of seismic refraction traverses across early Proterozoic terranes in northern and western Australia. A high velocity (6.8-7.2 km/s) lower crustal layer up to 15 km thick has been interpreted beneath a number of terranes (Figure 8) [Finlayson, 1982; Drummond 1981]. In addition, Drummond [1981] has shown that this high velocity layer is not present beneath Archean nuclei which border the Proterozoic terranes.

4.) Since no oceanic remnants remained at the end of the Barramundi orogeny, any plate tectonic model for the event must terminate with a continent-continent collision. Modern continent-continent collision results in substantial crustal thickening, uplift and extensive erosion. Despite the evidence for extensive thrusting, major duplication of sequences, especially involving basement is unusual. Sequences which have been thickened significantly by both thrusting and tight, upright folding display little evidence of isostatic rebound, in either their present erosion levels or their metamorphic facies. The metamorphic facies are virtually everywhere low pressure, with more evidence for isobaric cooling than for adiabatic decompression. One can certainly find some of these characteristics in a few modern terranes, but we consider it significant that they are ubiquitous in Australian early Proterozoic provinces and are common to many contemporaneous provinces in other continents.

Tectonic Setting of Australian Early Proterozoic Orogeny - A Working Hypothesis

Our understanding of the extensive Australian early Proterozoic terranes is still fairly primitive, and it would be presumptuous to erect a specific and detailed tectonic model for their evolution. However, we present here an outline of a working hypothesis that is currently undergoing testing and evaluation. The hypothesis is based on the premise that widespread, contemporaneous underplating and continental extension in linear belts were initiated by the onset of small-scale mantle convection. Using the models of Anderson [1984] combined with that of McKenzie and Weiss [1975], small-scale mantle convection is considered to have been triggered by the blanketing effect of a large, slow moving or stationary con-

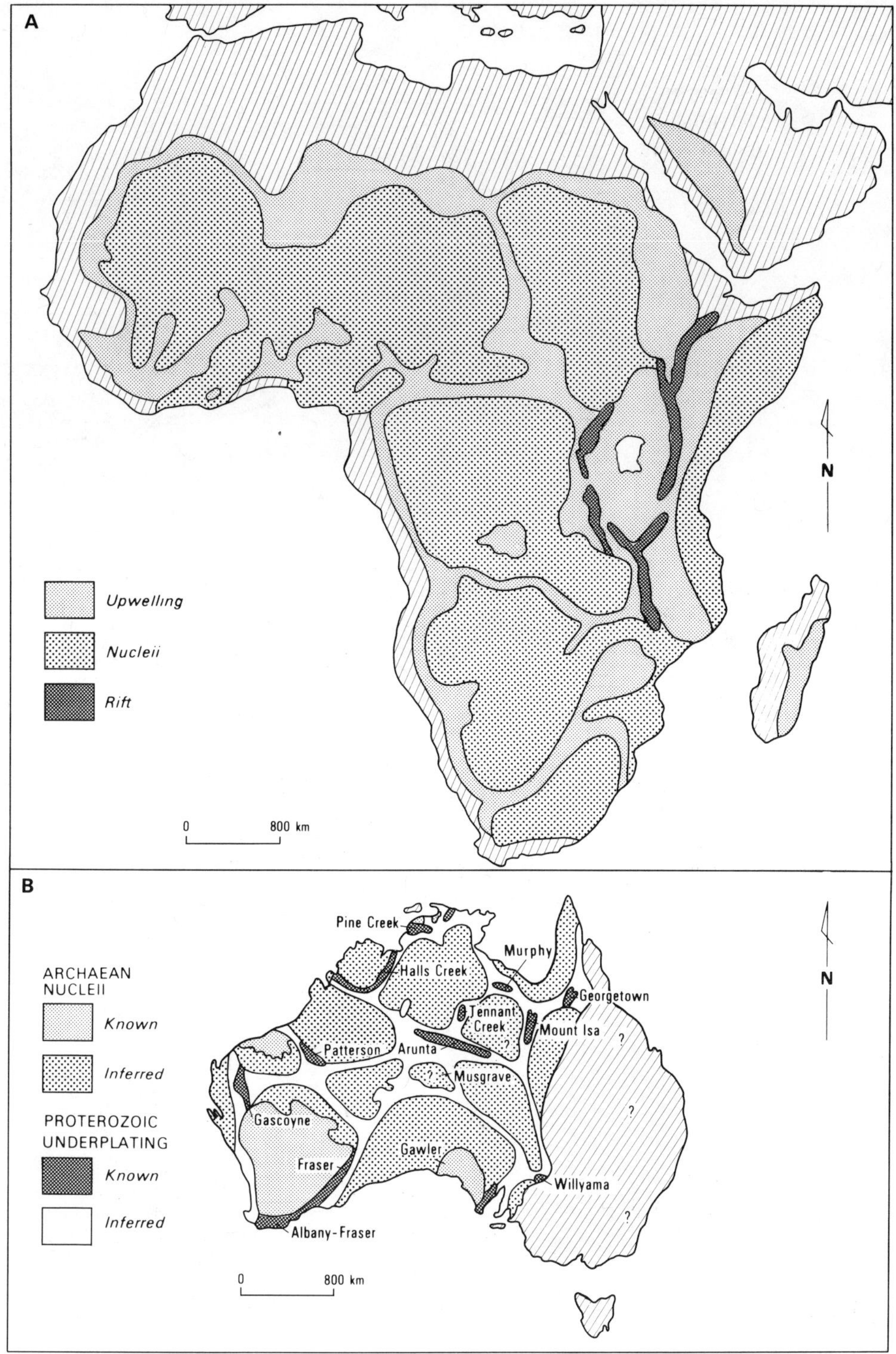

Fig. 9. (a) Polygonal array of modern plateaus, swells and ridges (with or without rifts) surrounding broad basins in Africa [after Holmes, 1965; McKenzie and Weiss, 1975]. (b) Possible extent of Archean nuclei during development of early Proterozoic fold belts, based on the African analogy and a small-scale convection model. The extent to which the unexposed Archean nuclei on this figure were involved in the early Proterozoic basin-forming and orogenic processes is poorly understood.

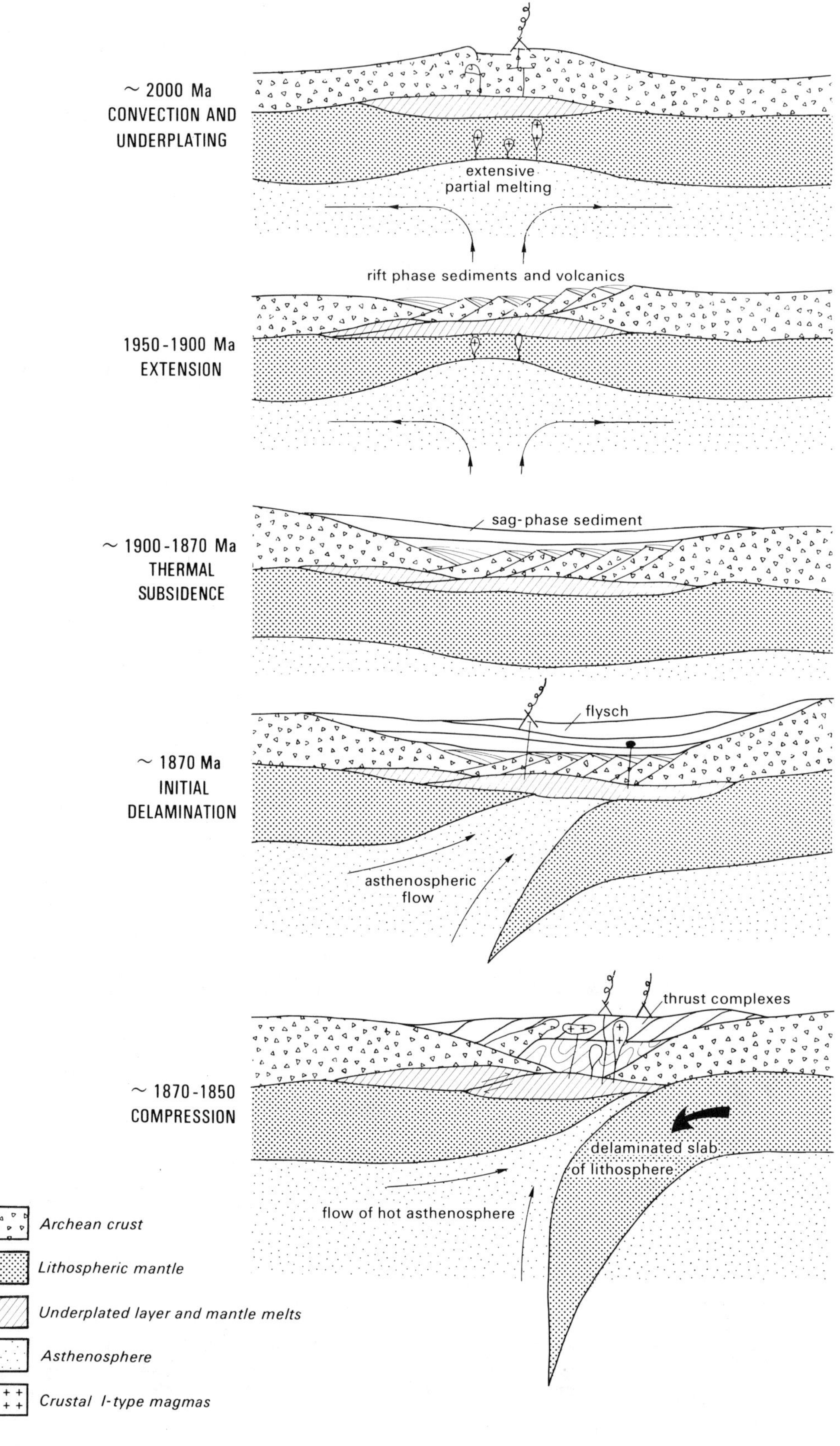
~ 2000 Ma
CONVECTION AND
UNDERPLATING
extensive
partial melting
rift phase sediments and volcanics
1950-1900 Ma
EXTENSION
sag-phase sediment
~ 1900-1870 Ma
THERMAL
SUBSIDENCE
flysch
~ 1870 Ma
INITIAL
DELAMINATION
asthenospheric
flow
thrust complexes
~ 1870-1850
COMPRESSION
delaminated slab
of lithosphere
flow of hot asthenosphere
Archean crust
Lithospheric mantle
Underplated layer and mantle melts
Asthenosphere
Crustal I-type magmas

tinental mass. The extent of that mass is unknown, and the precision of existing paleomagnetic data is insufficient to assess continental motion during this period.

The notion of relatively rapid, small-scale mantle convection was developed by McKenzie and Weiss [1975] to explain the polygonal array of swells, plateaus, and rift valleys surrounding broad shallow basins in modern Africa [Holmes, 1965]. The small-scale convection is interpreted to have developed in response to heating of the upper mantle beneath the slow-moving African continental plate. The mean diameter of the polygons is about 1400 km and the width of the linear upwelling zones is 100 to 200 km [Figure 17 in McKenzie and Weiss, 1975]. We are not suggesting that modern Africa is a precise analogue for the Australian (?global) early Proterozoic. However, the concept of small-scale convection does provide a means for producing 1) elongate belts of initial uplift, underplating, heating and eventual extension above the upwelling zones, and 2) broad regions essentially unaffected by these processes (Figure 9). We suggest that extension resulted from gravitational spreading of the hot, uplifted belts that had been thickened significantly by underplating [Houseman and England, 1986]. Rift basins developed during continental extension, which ceased prior to the formation of oceanic crust. Subsequent thermal relaxation and flexural loading produced sag basins overlying the rift-stage. Extension factors (β) of 2 to 4 will give rise to thick basins on highly structured continental crust, which will be identical in many respects to those on modern passive continental margins.

The small scale convection, aided by extensive underplating, volcanism and high level intrusion of mafic magma, together with enhanced heat transport through thinning lithosphere may have been sufficient to cool the upper mantle below the critical thermal state for convection. In addition, the velocity of the continental plate may have increased. It is difficult to determine the duration of the convective phase from the available dating, but assuming that the widespread Sm/Nd model age of about 2000 Ma [Page et al., 1984] represents the onset of mantle melting, and that rifting began at between 1900 and 1950 Ma ago, approximately 100 Ma is a reasonable estimate.

Following the cessation of the small scale rapid convection and continental extension, thermal relaxation has allowed continued subsidence and sediment accumulation for a few tens of Ma. It appears that this thermal subsidence was interrupted by the onset of orogeny, which began with at least a local increase in subsidence rate and widespread flysch deposition. Deformation, meta-

morphism and the widespread I-type magmatism all followed within a very short period (20 Ma).

There are two possible explanations for the essentially synchronous orogeny that affected the intracontinental early Proterozoic Australian basins:

1) The period of local extension which produced the linear basins may have been interrupted by a sudden increase in plate velocity. Stresses propagated through the rapidly moving plate led to shortening within the regions previously thinned during the extension. Reactivation of the extensional fault systems could have allowed shortening at relatively low stresses [Sibson 1985; Etheridge, 1986]. Such a scenario would have given rise to significantly oblique compression in basins whose original extension direction was oblique to the direction of horizontal compression resulting from the rapid plate motion. Oblique compression during the Barramundi Orogeny has been suggested in the Halls Creek domain [Hancock and Rutland, 1984]. However, there is little consistency of tectonic transport directions from domain to domain (Table 1).

2) Extension and thinning of the enriched upper mantle beneath the early Proterozoic basins led to essentially contemporaneous lithospheric delamination during cooling and sediment loading [Bird, 1978, 1979]. This model is attractive because it provides an explanation for both the initial rapid subsidence with flysch sedimentation, and the intense thermal anomaly that accompanied the Barramundi Orogeny. According to the hot mode delamination model of Bird and Baumgardner [1981], the intruded asthenosphere above the delaminating mantle convects, giving rise to very steep thermal gradients in the overlying crust, and leading to widespread crustal melting and low pressure metamorphism. The driving force for delamination decays as the thicker undepleted lithospheric segments on either side of the original basin approach each other.

Our tentative model for this distinctive early Proterozoic orogeny based on 2) above is summarised in Figure 10. In many respects, it is similar to the epicratonic or A-subduction model proposed by Kröner [1983, 1984]. Unlike the conventional actualistic Wilson-cycle model, it explains a) the contemporaneity of process on a continent-wide scale and perhaps even a global scale, b) the onset of orogeny in each terrane immediately after basin formation, and c) the intense crustal thermal anomaly that accompanied the orogeny.

Acknowledgments. We would like particularly to thank our many colleagues in the Bureau of Mineral Resources for extensive discussions and much freely provided information which have contributed sig-

---

Fig. 10. (opposite) Summary of working hypothesis for intracratonic basin development and orogeny in the early Proterozoic of northern Australia. Continental extension is driven by small-scale convection, and the subsequent orogeny by crust-mantle delamination.

nificantly to this paper. Particular thanks are
due to David Blake, Barry Drummond, Rod Page,
Shen-Su Sun, Gladys Warren and Doone Wyborn. We
also acknowledge discussions with and unpublished
information from Mark Fanning, Steve Hancock,
Malcolm McCulloch, and John Parker. The manuscript
was significantly improved by comments from Nick
Arndt, Pat James, Alfred Kroner and two anonymous
reviewers, and by the typing and drafting skills
of Dell Stafford and Tony Convine respectively.
Published with the permission of the Director,
Bureau of Mineral Resources.

References

Anderson, D.L., The Earth as a Planet: paradigms
and paradoxes, Science, 223, 347-355, 1984.

Barton, P., and R. Wood, The tectonic evolution
of the North Sea basin: crustal structure and
subsidence, Geophys. J. R. astr. Soc., 79, 987-
1022, 1984.

Bateman, P.C., and B.W. Chappell, Crystallization,
fractionation, and solidification of the
Tuolumne Intrusive Series, Yosemite National
Park, California, Geol. Soc. Am. Bull., 90,
465-482, 1979.

Bennett, R., R.W. Page, and G.M. Bladon, Catalo-
gue of isotopic age determinations on Australian
rocks, 1966-70, Rep. Bur. Miner. Resour. Geol.
Geophys. Aust., 162, 135pp, 1975.

Bird, P., Initiation of intracontinental subduction in the Himalaya, J. Geophys. Res., 83,
4975-4987, 1978.

Bird, P., Continental delamination and the Colorado Plateau, J. Geophys. Res., 84, 7561-7571,
1979.

Bird, P., and J. Baumgardner, Steady propogation
of delamination events, J. Geophys. Res., 86,
4891-4903, 1981.

Black, L.P., and M.T. McCulloch, Sm-Nd ages of the
Arunta, Tennant Creek, and Georgetown Inliers of
Northern Australia, Aust. J. Earth Sci., 31, 49-
60, 1984.

Black, L.P., T.H. Bell, M.J. Rubenach, and I.W.
Withnall, Geochronology of discrete structural-
metamorphic events in a multiply deformed Precambrian terrain, Tectonophysics, 54, 103-137,
1979.

Blight, D.F., and M.E. Barley, Estimated pressure
and temperature conditions from some Western
Australian Precambrian metamorphic terrains,
Geol. Surv. W. Aust. Ann. Rep., 1980, 67-72, 1983.

Bultitude, R.J., and L.A.I. Wyborn, Distribution
and geochemistry of volcanic rocks in the Duches-
Urandangi region, Queensland, BMR J. Aust.
Geol. Geophys., 7, 99-112, 1982.

Burke, K., J.F. Dewey, and W.S.F. Kidd, Precambrian paleomagnetic results compatible with
contemporary operation of the Wilson cycle,
Tectonophysics, 33, 287-299, 1976.

Chappell, B.W., Source rocks of I- and S-type
granites in the Lachlan Fold Belt, southeastern
Australia. Philos. Trans. R. Soc. London, A310,
693-707, 1984.

Chappell, B.W., and A.J.R. White, Two contrasting
granite types. Pac. Geol., 8, 173-4, 1974.

Christiansen, R.L., and P.W. Lipman, Cenozoic
volcanism and plate-tectonic evolution of the
Western United States, Philos. Trans. R. Soc.
London, Ser. A, 271, 249-284, 1972.

Collins, C.D.N., Crustal structure of the southern
McArthur Basin, northern Australia, from deep
seismic sounding, BMR J. Aust. Geol. Geophys.,
8, 19-34, 1983.

Compston, W., and P.A. Arriens, The Precambrian
geochronology of Australia, Can. J. Earth
Sci., 5, 561-583, 1968.

Dazhong, S., and L. Songnian, A subdivision of the
Precambrian of China, Precambrian Res., 28, 137-
162, 1985.

Dow, D.B., and I. Gemuts, Geology of the Kimberley
Region, Western Australia: The East Kimberley,
Bull. Bur. Miner. Resour. Geol. Geophys. Aust.,
106, 135 pp, 1969.

Drummond, B.J., Seismic constraints on the chemical composition of the crust of the Pilbara
craton, northwest Australia, Rev. Bras.
Geociencias, 12, 113-120, 1981.

Dunn, P.R., K.A. Plumb, and H.G. Roberts, A proposal for time-stratigraphic subdivision of the
Australian Precambrian, J. Geol. Soc. Aust., 13,
593-608, 1966.

Easton, R.M., Stratigraphy of the Akaitcho Group
and the development of an Early Proterozoic
continental margin, Wopmay Orogen, N.W.T., in
Proterozoic Basins of Canada edited by F.H.A.
Campbell, Geol. Surv. Can. Paper, 81-10, 79-95,
1981.

Easton, R.M., Crustal structure of rifted continental margins: Geological constraints from the
Proterozoic rocks of the Canadian Shield, Tectonophysics, 94, 371-390, 1983.

England, P.C., and S.W. Richardson, The influence
of erosion upon the mineral facies of rocks from
different metamorphic environments, J. Geol.
Soc. Lond., 134, 201-213, 1977.

Etheridge, M.A., Reactivation of Extensional Fault
Systems, Phil. Trans. R. Soc. London, in press.

Etheridge, M.A., J.G. Branson, D.A. Falvey, K.L.
Lockwood, P.G. Stuart-Smith, and A.S. Scherl,
Basin forming structures and their relevance to
hydrocarbon exploration in Bass Basin, southeastern Australia, BMR J. Aust. Geol. Geophys., 9,
197-206, 1984.

Etheridge, M.A., J.G. Branson, and P.G. Stuart-
Smith, Extensional basin forming structures in
Bass Strait and their importance for hydrocarbon
exploration, APEA, 344-361, 1985.

Fanning, M., A.H. Blissett, R.B. Flint, K.R.
Ludwig and A.J. Parker, A refined geological
history for the Southern Gawler Craton through
U-Pb zircon dating of acid volcanics, and cor-
relations with Northern Australia, Geol. Soc.
Aust. Abstr., 15 67-68, 1986.

Ferguson, J., Metamorphism in the Pine Creek Geosyncline and its bearing on stratigraphic considerations, in Uranium in the Pine Creek Geosyncline, edited by J. Ferguson and A.B. Goleby,

pp. 91-100, International Atomic Energy Agency, Vienna, 1980.

Ferguson, J., B.W. Chappell, and A.B. Goleby, Granitoids in the Pine Creek Geosyncline, in *Uranium in the Pine Creek Geosyncline*, edited by J. Ferguson and A.B. Goleby, pp. 73-90, International Atomic Energy Agency, Vienna, 1980.

Finlayson, D.M., Seismic crustal structure of the Proterozoic North Australian Craton between Tennant Creek and Mount Isa, *J. Geophys. Res.*, 87, 10569-10578, 1982.

Fumerton, S.L., M.R. Stauffer, and J.F. Lewry, The Wathaman batholith: largest known Precambrian pluton, *Can. J. Earth Sci.*, 21, 1082-1097, 1984.

Gastil, G., The distribution of mineral dates in time and space, *Am. J. Sci.*, 258, 1-35, 1960.

Gellatly, D.C., G.M. Derrick, R. Halligan, and J. Sofoulis, The Geology of the Charnley 1:250 000 Sheet area, Western Australia, *Rep. Bur. Miner. Resour. Geol. Geophys. Aust.*, 154, 88pp., 1974.

Gemuts, I., Metamorphic and igneous rocks of the Lamboo Complex, East Kimberley Region, Western Australia, *Bull. Bur. Miner. Resour. Geol. Geophys. Aust.*, 107, 71pp, 1971.

Geological Society of Australia, Tectonic map of Australia and New Guinea, Scale 1:500 000, *Sydney*, 1971.

Gibb, R.A., Collision tectonics in the Canadian Shield, *Earth Planet. Sci. Lett.*, 27, 378-382, 1975.

Griffin, T.J., A.J.R. White, and B.W. Chappell, The Moruya Batholith and geochemical contrasts between the Moruya and Jindabyne Suites, *J. Geol. Soc. Aust.*, 25, 235-247, 1978.

Hamlyn, P.R., Petrology of the Panton and MacIntosh layered intrusions, Western Australia, with particular reference to the genesis of the Panton chromite deposits, *Ph.D. thesis, University of Melbourne*, Melbourne, 1977.

Hamlyn, P.R., Equilibration history and phase chemistry of the Panton Sill, Western Australia, *Contrib. Mineral. Petrol.*, 33, 21-31, 1980.

Hancock, S.L., and R.W.R. Rutland, Tectonics of an Early Proterozoic geosuture: the Halls Creek Orogenic Sub-province, Northern Australia, *J. Geodynam.*, 1, 387-432, 1984.

Hildebrand, R.S., and S.A. Bowring, Continental intra-arc depressions: A non-extensional model for their origin, with a Proterozoic example from Wopmay orogen, *Geology*, 12, 73-77, 1984.

Hine, R., I.S. Williams, B.W. Chappell, and A.J.R. White, Contrasts between I- and S-type granitoids of the Kosciusko Batholith, *J. Geol. Soc. Aust.*, 25, 219-234, 1978.

Hobbs, B.E., N.J. Archibald, M.A. Etheridge, and V.J. Wall, Tectonic history of the Broken Hill Block, Australia, in *Precambrian Tectonics Illustrated*, edited by A. Kroner and R. Geiling, pp. 353-368, E. Schweizerbart'sche Verlagsbuchhandlung (Nagele u. Obermiller), Germany, Stuttgart, 1984.

Hoffman, P.F., Evolution of an early Proterozoic continental margin: the Coronation geosyncline and associated aulacogens of the southwestern Canadian Shield, *Phil. Trans. R. Soc. Lond., Ser. A*, 273, 547-581, 1973.

Hoffman, P.F., Wopmay orogen: A Wilson cycle of early Proterozoic age in the northwest of the Canadian shield, in *The Continental Crust and its Mineral Deposits*, edited by D.W. Strangway, pp.523-549, *Geol. Ass. Canada, Spec. Paper*, 20, 1980.

Hoffman, P.F., and S.A. Bowring, Shortlived 1.9 Ga continental margin and its destruction, Wopman orogen, northwest Canada, *Geology*, 12, 68-72, 1984.

Hoffman, P.F, and J.C. McGlynn, Great Bear Batholith: a volcano-plutonic depression, in *Volcanic Regimes in Canada* edited by W.R.A. Baragar, L.C. Coleman, and J.M. Hall, *Geol. Ass. Can, Spec. Pap.*, 16, pp 109-192, 1977.

Hoffman, P.F., J.E. King, M.R. St-Onge, and R. Tirrul, Accretionary wedge deformation ca 1.9 Ga: a structural comparison of Wopmay Orogen and Cape Smith Greenstone Belt, northern Canadian Shield, *Geol. Soc. Am. Mem.*, in press.

Holmes, A., *The nomenclature of petrology*, Murby, London, 284p, 1920.

Holmes, A., *Principles of Physical Geology*, Nelson, London, 1288 pp, 1965.

Houseman, G.A., and P.C. England, A dynamical model of lithosphere extension and sedimentary basin formation, *J. Geophys. Res.*, 91, 719-729.

Hynes, A., and R.D. Gee, Geological setting and petrochemistry of the Narracoota Volcanics, Capricorn Orogen, Western Australia, *Precambrian Res.* 31, 107-132, 1986.

Irvine, T.N., and W.R.A. Baragar, A guide to the chemical classification of the common volcanic rocks, *Can. J. Earth Sci.*, 8, 523-548, 1971.

Johnson, J.D., Structural evolution of the Pine Creek Inlier and mineralization therein, Northern Territory, Australia, *Ph.D. Thesis, Monash University*, Melbourne, 1984.

Johnson, R.W., Distribution and major-element chemistry of Late Canozoic volcanoes at the southern margin of the Bismark Sea, Papua New Guinea, *Rep. Bur. Miner. Resour. Geol. Geophys. Aust,* 188, 1974.

Knutson, J., J. Ferguson, W.M.B. Roberts, T.H. Donnelly, and I.B. Lambert, Petrogenesis of the copper-bearing breccia pipes, Redbank, Northern Territory, Australia, *Econ. Geol.*, 74, 814-826, 1979.

Kröner, A., Precambrian mobile belts of southern and eastern Africa - ancient sutures or sites of ensialic mobility? A case for crustal evolution towards plate tectonics, *Tectonophysics*, 40, 101-135, 1977.

Kröner, A., Proterozoic mobile belts compatible with the plate tectonics concept, in *Proterozoic Geology: selected Papers from an International Proterozoic Symposium* edited by G.M. Medaris, C.W. Byers, D.M. Mikelson, and W.G. Shanks, pp. 59-74, *Geol. Soc. Am. Mem.*, 161, 1983.

Kröner, A., Evolution, growth and stabilization of the Precambrian lithosphere, in *Structure and*

Evolution of the Continental Lithosphere, edited by H.N. Pollack and U.R. Murthy, Physics and Chemistry of the Earth, 15 69-106, 1984.

Le Pichon, X, and J.C. Sibuet, Passive margins: a model of formation, J. Geophys. Res., 86, 3708-3720, 1981.

Lipman, P.W., H.J. Prostka, and R.L. Christianson, Cenozoic volcanism and plate tectonic evolution of the Western United States. I Early and Middle Cenozoic, Philos. Trans. R. Soc. London, Ser. A., 271, 217-248, 1972.

McCulloch, M.T., and H.D. Hensel, Sm-Nd isotopic evidence for a major early to Mid-Proterozoic episode of crustal growth in the Australian continent, XXVII Internat. Geol. Conv. Moscow, Vol. II Sections 04,05, 351-352, 1984.

McKenzie, D., Some remarks on the development of sedimentary basins, Earth Planet. Sci. Lett., 40, 25-32, 1978.

McKenzie, D., and N. Weiss, Speculations on the thermal and tectonic history of the Earth, Geophys. J. R. astr. Soc., 42, 131-174, 1975.

Moorbath, S., Age and isotopic evidence for the evolution of continental crust, Philos. Trans. R. Soc. London, Ser.A., 288, 401-413, 1978.

Needham, R.S., and P.G. Stuart-Smith, The relationship between mineralisation and depositional environment in Early Proterozoic metasediments of the Pine Creek Geosyncline, Australas. Inst. Min. Metall., Darwin Conf. 1984, 201-211, 1984.

Needham, R.S., and P.G. Stuart-Smith, Stratigraphy and tectonics of the Early to Middle Proterozoic transition, Katherine-El Sherana area, N.T., Aust. J. Earth Sci., 32, 219-230, 1985.

Needham, R.S., I.H. Crick, and P.G. Stuart-Smith, Regional geology of the Pine Creek Geosyncline, in Uranium in the Pine Creek Geosyncline, edited by J. Ferguson and A.B. Goleby, pp. 1-22, International Atomic Energy Agency, Vienna, 1980.

Oxburgh, E.R., and D.L. Turcotte, Thermal gradients and regional metamorphism in overthrust terrains with special reference to the Eastern Alps, Schweiz. Min. Pet. Mitt., 54, 641-662, 1974.

Page, R.W., Response of U-Pb Zircon and Rb-Sr total rock and mineral systems to low-grade regional metamorphism in Proterozoic igneous rocks, Mount Isa, Australia, J. Geol. Soc. Aust., 25, 141-64, 1978.

Page, R.W., Isotopic record of crustal events in Northern Australia, in Abstracts for the Conference on Tectonics and Geochemistry of early to middle Proterozoic Fold Belts, Darwin, 1985, Bur. Miner. Resour. Geol. Geophys. Aust., Record 1985/28, 1985.

Page, R.W., W. Compston, and R.S. Needham, Geochronology and evolution of the Late-Archean basement and Proterozoic rocks in the Alligator Rivers Uranium Field, Northern Territory, Australia, in Uranium in the Pine Creek Geosyncline, edited by J. Ferguson and A.B. Goleby, pp. 39-68, International Atomic Energy Agency, Vienna, 1980.

Page, R.W., M.T. McCulloch, and L.P. Black, Isotopic record of major Precambrian events in Australia, Precambrian Geology, Proc. 27th Internat. Geol. Cong., 5, 25-72, 1984.

Patchett, P.J., Nd isotopes and the origin of 1.9-1.7 Ga continental crust: South Greenland and Finland, EOS Trans. AGU., 64, 905, 1983.

Patchett, P.J., and D. Bridgwater, Origin of continental crust of 1.9-1.7 Ga defined by Nd isotopes in the Ketilidian terrain of South Greenland, Contrib. Mineral. Petrol., 87, 311-318, 1984.

Peacock, M.A., Classification of igneous rocks, J. Geol., 39, 54-67, 1931.

Phillips, G.N., and V.J. Wall, Evaluation of prograde regional metamorphic conditions: their implications for the heat source and water activity during metamorphism in the Willyama Complex, Broken Hill, Australia, Bull. Mineral., 104, 801-810, 1981.

Pitche, W.S., M.P. Atherton, E.J. Cobbing, and R.D. Beckinsale, Magmatism at a plate edge, the Peruvian Andes, Blackie Halstead Press, London, 1985.

Plumb, K.A., The tectonic evolution of Australia, Earth Sci. Rev., 14, 205-249, 1979.

Plumb, K.A., R. Allen, and S. Hancock, Proterozoic evolution of the Halls Creek Province, Western Australia, Bur. Miner. Resour. Geol. Geophys. Aust., Record 1985/25, 87pp, 1985.

Reymer, A., and G. Schubert, Rapid growth of some major segments of continental crust, Geology, 14, 299-302, 1986.

Roberts, W.M.B., Mineralogy and genesis of White's orebody, Rum Jungle uranium field, Australia, Neues Jb. Miner., Abh. Bd. 94, 868-889, 1960.

Rossiter, A.G., and J. Ferguson, A Proterozoic tectonic model for northern Australia and its economic implications, in Uranium in the Pine Creek Geosyncline, edited by J. Ferguson and A.B. Goleby, pp. 209-232, International Atomic Energy Agency, Vienna, 1980.

Rutland, R.W.R., Tectonic evolution of the continental crust of Australia, in Continental Drift, Sea Floor Spreading and Plate Tectonics: Implications to the Earth Sciences, edited by D.H. Tarling and S.K. Runcorn, pp. 1003-1025, Academic Press, London, 1973.

Rutland, R.W.R., Orogenic evolution of Australia, Earth Sci. Rev., 12, 161-196, 1976.

Rutland, R.W.R., On the growth and evolution of continental crust: a comparative tectonic approach, J. Proc. R. Soc. NSW, 115, 33-60, 1982.

Sheraton, J.W., and L.P. Black, Geochemistry and geochronology of Proterozoic tholeiitic dykes of East Antarctica: evidence for mantle metasomatism, Contrib. Mineral. Petrol., 78, 305-317, 1981.

Sibson, R.H., A note on fault reactivation, J. Struc. Geol. 7,, 751-754, 1985.

Stauffer, M.R., Manikewan: an Early Proterozoic ocean in central Canada, its igneous history and orogenic closure, Precambrian Res., 25, 257-281, 1984.

Streckeisen, A.L., Plutonic rocks. Classification and nomenclature recommended by the IUGS Subcommission on the systematics of igneous rocks, Geotimes, 18, 26-30, 1973.

Sun, S.S., and M.T. McCulloch, Chemical and Nd isotope study of late Archean to middle Proterozoic mafic volcanic rocks in northern Australia, in Abstracts for the conference on the Tectonics and Geochemistry of early to middle Proterozoic Fold Belts, Darwin, 1985, Bur. Miner. Resour. Geol. Geophys. Aust., Record 1985/28, 1985.

Sweet, I.P., Mock, C.M., and Mitchell, J.E., Chemical analyses from the Siegal and Hedleys Creek 1:100 000 Sheet areas, Northern Territory and Queensland, Rep. Bur. Miner. Resour. Geol. Geophys. Aust., 226, 1981.

Thompson, A.B., The pressure-temperature (P,T) plane viewed by geophysicists and petrologists, Terra Cog., 1, 11-20, 1981.

Thorne, A.M., The depositional history of the 2.0 Ga Wyloo Group, Southern Pilbara, Western Australia, Geol. Soc. Aust. Abstr., 15, 189.

Tilley, C.E., Some aspects of magmatic evolution, Quart J. Geol. Soc. Lond., 106, 37-61, 1950.

Warren, R.G., Metamorphic and tectonic evolution of granulites, Arunta Block, Central Australia, Nature, 305, 300-303, 1983.

Weaver, B.L., and J. Tarney, The Scourie Dyke Suite: Petrogenesis and geochemical nature of the Proterozoic Sub-Continental mantle, Contrib. Mineral. Petrol., 78, 175-188, 1981.

Whalen, J.B., R.M. Britten, and I. McDougall, Geochronolgy and geochemistry of the Frieda River Prospect area, Papua New Guinea, Econ. Geol., 77, 592-616, 1982.

White, A.J.R., and B.W. Chappell, Granitoid types and their distribution in the Lachlan Fold Belt, southeastern Australia, in Circum-Pacific Plutonic Terranes, edited by J.A. Roddick, pp. 21-34, Geol. Soc. Am. Memoir, 159, 21-34, 1983.

Wickham, S.M., and E.R. Oxburgh, Continental rifts as a setting for regional metamorphism, Nature, 318, 330-333, 1985.

Williams, S.J., I.R. Williams, and R.M. Hocking, Glenburgh, Western Australia, Sheet SG. 50-6, 1:250 000 Geological Series Explanatory Notes, Geol. Surv. W. Aust., 1983.

Wilson, I.H., Volcanism on a Proterozoic continental margin in northwestern Queensland, Precambrian Res., 7, 205-235, 1978.

Windley, B.F., Precambrian rocks in the light of the plate-tectonic concept, in Precambrian Plate Tectonics, edited by A. Kroner, pp. 1-20, Elsevier, Amsterdam, 1981.

Windrum, D.P., and M.T. McCulloch, Nd and Sr chronology of Strangways Range Granulites: implications for crustal growth and reworking in the Proterozoic of Central Australia, Geol. Soc. Aust. Abstr., 9, 192-193, 1983.

Wyborn, L.A.I., Geochemistry and origin of a major early Proterozoic felsic igneous event of Northern Australia and evidence for substantial vertical accretion of the crust, in Abstracts for the Tectonics and Geochemistry of early to middle Proterozoic Fold Belts, Darwin, 1985, Bur. Miner. Resour. Geol. Geophys. Aust., Record 1985/28, 1985.

Wyborn, L.A.I., and R.W. Page, The Proterozoic Kalkadoon and Ewen Batholiths, Mount Isa Inlier, Queensland: source, chemistry, age, and metamorphism, BMR J. Aust. Geol. Geophys., 8, 53-69, 1983a.

Wyborn, L.A.I., and R.W. Page, Chemical and isotopic constraints on the source regions of the Kalkadoon and Ewen Batholiths, Mount Isa Inlier and their implications for early Proterozoic crustal evolution in Australia, Geol. Soc. Aust. Abstr., 9,, p.191, 1983b.

Wyborn, L.A.I., and R.W. Page, Geochemical Evolution of igneous rocks of the Mount Isa Inlier and significance for mineralisation, Abstracts 14th BMR Symposium, Bur. Miner. Resour. Geol. Geophys. Aust., Record 1985/33, 31-33, 1985.

THE PRECAMBRIAN HISTORY OF THE BALTIC SHIELD

Roland Gorbatschev

Dept. of Geology, Lund University, Sölveg. 13, S-22362 Lund, Sweden

Gabor Gaál

Geological Survey of Finland, SF-02150 Espoo, Finland

Abstract. The Precambrian crust of the Baltic Shield developed during four orogenies between 3.5 and 1.5 Ga ago, and exhibits a general westward younging of major crustal units. Later orogenies contributed little new material to the shield but involved intense crustal reworking in western Scandinavia, particularly during the Sveconorwegian-Grenvillian and Caledonian orogenies. For the Proterozoic, several different continent-collision, continental margin and intracratonic orogenic environments are distinguished. A collisional orogeny is recorded in the north of the shield by a narrow belt of early Proterozoic crust separating two Archaean crustal segments. In the central Baltic Shield, westward growth of continental crust occurred during the Svecokarelian (2.0-1.75 Ga) and Gothian (1.75-1.55 Ga) orogenies. Although these orogenies are transitional into one another, each has its own group of mantle-derived tonalitic-granodioritic plutonites separated by a time interval of about 100-150 Ma. Intracratonic rifting simultaneously disrupted the Archaean parts of the shield. A complex, multistage orogenic development including ocean-continental margin interaction, plate rotation and, eventually, continent collision is proposed for the Sveconorwegian-Grenvillian events 1.25-0.9 Ga ago. The results of recent work suggest that plate tectonic models can also be applied to the late Archaean.

Introduction:
The Baltic Shield in Space and Time

The Baltic Shield in the N of Europe records the evolution of continental crust almost continuously over a period of at least 3000 million years, from the Archaean until less than 400 million years ago, when the latest orogenic nappe thrusting occurred. By and large, the Precambrian crust of the Baltic Shield exhibits a geochronological zonation with younging from east to west (Gorbatschev, 1980; Gaál, 1986). The oldest reliable rock age, established by several methods of dating, is around 3.1 Ga, but Nd-isotopic systematics suggest that crustal material separated from the mantle at least 3.5 Ga ago (Kröner et al., 1981; Jahn et al., 1984).

The Precambrian crust of the Baltic Shield can be subdivided into three domains: The Archaean Domain in the NE, the Svecokarelian Domain in the centre and the Southwest Scandinavian Domain in the W and SW (Figure 1). These domains evolved either during a single orogeny or several superposed orogenies, each of which is characterized by a specific sedimentary, magmatic, and tectono-metamorphic evolution occurring within a distinct period of time.

In the Archaean Domain, most of the continental lithosphere was formed during the Lopian orogeny 2.9-2.6 Ga ago (cf. Keller et al., 1977). The role of possible older Archaean nuclei created by the earliest recognizable, Saamian orogeny ($>$ 3.0 Ga ago) has not yet been adequately assessed. Proterozoic reworking of Archaean rocks was intense in the Belomorian Province (Figure 1) but is otherwise restricted to the vicinity of a narrow Proterozoic mobile belt in Lapland (the "Granulite Belt") and to marginal areas adjacent to the Svecokarelian Domain in the central Baltic Shield.

The Svecokarelian Domain evolved during the Svecokarelian orogeny 2.0-1.75 Ga ago, and subsequent reworking is insignificant. Nevertheless, the Svecokarelian Domain has been intruded by post-orogenic rapakivi granites, and a belt of granites and volcanic rocks is present along its boundary with the Southwest Scandinavian Domain.

The Southwest Scandinavian Domain was formed during the Gothian orogeny 1.75-1.55 Ga ago. Intense, repeated reworking during later orogenies is a hallmark of this realm which therefore features a complex crust resulting from original formation during the Gothian orogeny and deformation, metamorphism and igneous activity during the Hallandian (1.5-1.4 Ga), Sveconorwegian (1.25-0.9 Ga) and Caledonian (0.6-0.4 Ga) orogenies. The Hallandian event was originally

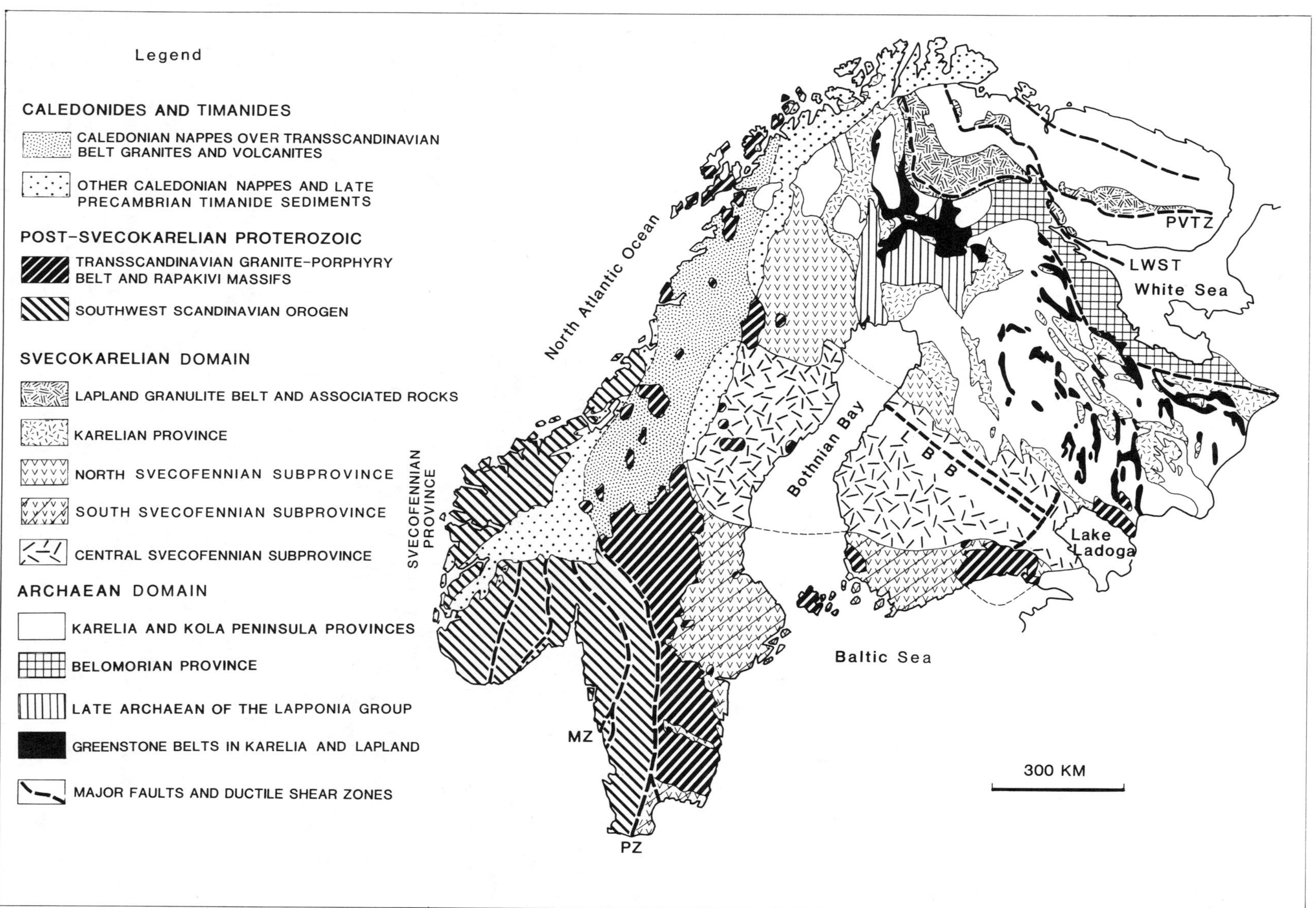

Fig. 1. Schematic geological map of the Baltic Shield. Abbreviations: LBB = Ladoga - Bothnian Bay Tectonic Zone; LWST = Lapland - White Sea Thrust Fault; MZ = "Mylonite" Zone; PBTZ = Pechenga - Varzuga Tectonic Zone; PZ = "Protogine" Zone.

defined in a somewhat more local context by Hubbard (1975). All three events differ from earlier orogenic episodes in that they contributed little new continental crust to the shield. At the present erosional level, the effects of the Hallandian and Sveconorwegian events are restricted to metamorphic and tectonic reworking of older crust and addition of limited amounts of igneous material.

The western margin of the Baltic Shield is covered by a pile of Caledonian nappes overthrust from the west. While Caledonian metamorphism and tectonization were intense in some segments of the Southwest Scandinavian Domain, the volume of Caledonian rocks is small in the autochthonous substratum of the nappe sequence. The central zone of the Caledonian orogeny was, consequently, situated far to the west of present-day Scandinavia.

The Evolution of the Archaean Domain - a Late Archaean Version of Plate Tectonics?

In the ancient nucleus of the Baltic Shield, major interest currently focuses on evidence for the operation of Archaean plate tectonic processes and their compatibility with the evolution of granite-greenstone belt terrains. A salient point in an assessment of the Archaean evolution is the nature and distribution of major Archaean crustal units.

The Archaean Domain comprises three different provinces arranged in a simple pattern (Figure 1). A central, NW-trending segment is occupied by a terrain of high-grade gneisses known as the Belomorian Province (Priyatkina et al., 1975). It is flanked by the Kola Province to the NE and the Karelia Province to the SW. Granite-greenstone belt terrains are well developed in the Karelia Province (Gaál et al., 1978; Rybakov et al., 1981; Krylov et al., 1984), whereas the Kola Province comprises several blocks of high-grade gneisses separated by narrow linear belts of faulting and ductile shearing (Kratz and Platunova, 1978). Direct correlation between the Kola and Karelia provinces is rendered questionable by the recent suggestion that the intervening Proterozoic Lapland Granulite Belt (Figure 1) represents the closure of an early Proterozoic ocean basin during continent collision (Raith et al., 1982; Barbey et al., 1984).

In the Karelia Province, the earliest recognizable continental crust formed during the Saamian orogeny $\geq$ 3.0 Ga ago. This orogeny created a terrain dominated by a tonalitic-trondhjemitic plutonic suite, which forms the rifted substratum of most greenstone belts (Gaál et al., 1978). A later, Lopian suite of granites intruded the greenstone belts and their surroundings 2.8-2.6 Ga ago (Hyppönen, 1983; Krylov et al., 1984). In many parts of Karelia, these younger granites and associated migmatites dominate the present erosional section, but numerous isotopic age data confirm the importance of Saamian plutonism (e.g.

Sobotovich et al., 1963; Lobikov and Lobach-Zhuchenko, 1980; Kröner et al., 1981; Levchenkov et al., 1982, 1985; Krylov et al., 1984; Sergeev et al., 1985a and b). They modify the conclusions on the scarcity of older rocks reached by Bibikova (1984) on the basis of mostly $^{207}Pb/^{206}Pb$ zircon data.

The main period of greenstone-belt formation in Karelia is referred to the Lopian in the Soviet Union (Kratz and Platunova, 1978). It is approximately coeval with or somewhat older than the early Lapponian in northern Finland (Silvennoinen, 1985). However, whereas the Lopian greenstone successions in Karelia occur in relatively limited rifting/sagduction belts (Rybakov et al., 1971; Martin et al., 1984), the Lapponian volcanic-sedimentary sequences have a larger areal extent, forming a specific, komatiite-bearing upper Archaean subprovince (Figure 1). In many cases komatiites were deposited in shallow-water and subaerial environments along lines of faulting (Saverikko, 1983).

Radiometric and stratigraphic evidence suggests that greenstone belt formation did not terminate abruptly at the Archaean-Proterozoic boundary (Lauerma, 1983; Mearns and Krill, 1984). During the early Proterozoic Svecokarelian evolution, greenstone rifts developed on the margins of the Archaean Domain in the northern Baltic Shield (Skiöld and Cliff, 1984). Their relationship with the Svecokarelian orogenic realm is similar to that between the Archaean greenstone belts of Karelia and the Belomorian Province. In both cases, the greenstone belts occur in the rifted forelands of high-grade gneiss belts.

The Belomorian Province is made up of high-grade gneisses derived mainly from Archaean gabbroic-tonalitic-granodioritic plutonic rocks, abundant amphibolites, and aluminous gneisses. Unlike the Archaean province in Karelia, the Belomorian Belt was affected repeatedly by metamorphism under high P/T-conditions (Volodichev, 1975; Priyatkina et al., 1975) best explained by the stacking of crustal slices. Evidence for nappe thrusting towards the west has actually been observed in the field (Glebovitskiy, 1973) and can also be deduced from seismic profiles (Akudinov et al., 1972). Other differences in comparison to Karelia are the absence of early Proterozoic cratonic cover rocks and a relatively greater prominence of Proterozoic metamorphism accompanied by igneous intrusions. This emphasizes the importance of Proterozoic deformation along the continuation of the Lapland - White Sea Tectonic Zone (Figure 1) beneath the waters of the White Sea (Priyatkina et al., 1975). However, a straightforward correlation of the Belomorian with the Proterozoic Granulite Belt of Lapland as suggested by Barbey et al. (1984) ignores the radiometric evidence from this polymetamorphic terrain (Kratz and Platunova, 1978; Tugarinov and Bibikova, 1980). Volodichev (1975) suggests that after an initial phase of low- to moderate-P granulitic metamorphism, presumably about 2.9-2.85 Ga ago, there occurred two periods of high-P metamorphism at 2700 Ma and 2200-1900 Ma.

Traditionally, the Belomorian Province has been regarded as the old nucleus of the Baltic Shield (Sokolov and Heiskanen, 1985), but this interpretation finds no support in U-Pb and $^{207}Pb/^{206}Pb$ zircon datings with ages clustering around 2700+50 Ma (Tugarinov and Bibikova, 1980; Bibikova, 1984). At present, therefore, the Lopian, the Belomorian Belt and some Kola rocks are thought to have evolved in what Gorbunov et al. (1985) call "a single geosynclinal system". On the basis of age and spatial contexts, we suggest that the Belomorian Province is the mobile belt region and the Karelia Province the rifted foreland complex of the Lopian orogeny. The relationship between these two units broadly agrees with a subduction scenario (Gaál, 1986).

The Svecokarelian Domain:
Continental Margin Evolution and Extensive
Formation of New Continental Crust
in the Central Baltic Shield

The largest amount of new continental crust in the Baltic Shield was created during the Svecokarelian orogeny which represents a wide range of different tectonic settings. These comprise a continent collision environment in the Lapland Granulite Belt (Figure 1) of the far north (Barbey et al., 1984), extensive formation of new continental crust in the Svecofennian Province of the central part of the shield (Patchett and Kouvo, 1985; Patchett et al., 1986), and deposition of cover sediments on the Archaean craton in the Karelia Province, associated with mafic volcanism and dolerite intrusions (Meriläinen, 1980; Sokolov and Heiskanen, 1985).

The fundamental problems of Svecokarelian lithogenesis concern the nature of the orogenic process. In turn, these problems relate to questions of crustal configuration during the orogeny and the interaction of mantle differentiation and ensialic reworking in various parts of the Svecokarelian Domain.

Soon after consolidation of the Archaean crust between 2.6 and 2.5 Ga ago, block faulting affected the Archaean Domain which had by then been converted into a landmass known as the Karelian craton (Kharitonov, 1966). Terrestrial, volcanic-sedimentary sequences of the Sumian-Sariolan Group were deposited in NNW- to EW-trending rift zones 2.5-2.3 Ga ago (Meriläinen, 1980). In Lapland, coeval rocks included in the upper Lapponian Group by Silvennoinen (1985) represent a basin facies formed in fault-controlled depressions some of which resemble aulacogens. Where tholeiitic and dacitic volcanic rocks of this Group overlie Archaean greenstone belts, the Archaean-Proterozoic boundary is difficult to identify.

The Sumian-Sariolan was followed by a short depositional hiatus, recorded by weathering profiles, which was in turn succeeded by the craton-cover sedimentary and volcanic rocks of the Jatulian Group (2.3-2.0 Ga; Meriläinen, 1980). The Jatulian cover of the Karelian craton has been invaded by mostly NW-trending swarms of dolerite dykes which relate to rifting at the passive southwestern margin of the Archaean Domain 2.2-2.1 Ga ago (Gaál, 1982 and 1986). In contrast to the Proterozoic realm farther southwest, Proterozoic lithogenesis in the Karelian province of the Svecokarelian Domain was entirely controlled by the presence of an Archaen substratum. Presumably, Archaean continental crust does not extend beyond the Ladoga-Bothnian Bay Tectonic Zone (Figure 1); Kahma, 1978). This inference is confirmed by the results of deep seismic soundings, which indicate a block-like structure and a change of crustal character along the inferred boundary (Korhonen, 1983).

To the north of the Karelian craton, the Lapland Granulite Belt is made up of high-grade metamorphic rocks (Barbey et al., 1982, 1984) rimmed in the south by tholeiitic and calc-alkaline volcanic rocks of island arc affinity (Raith et al., 1982; Pharaoh and Pearce, 1984). Some metasedimentary granulites have been interpreted as flyschoid deposits with mafic volcanic intercalations (Barbey et al., 1982). This association was formed in and along the margins of an oceanic basin, presumed to have been closed during continent collision about 1950 Ma ago (Barbey et al., 1984). The basin contents were thrust towards the west and south along the Lapland - White Sea Thrust Fault (Figure 1), causing high-P metamorphism in the overridden crustal slab (Meriläinen, 1976; Krylova et al., 1982; Bernard-Griffiths et al., 1984). Early Proterozoic volcano-sedimentary sequences and mafic intrusions also occur along the tectonic zones in the Kola Peninsula, and opinion is divided as to whether the Lapland - White Sea Thrust Fault or the Pechenga - Varzuga Tectonic Zone represent the actual suture of collision (Barbey et al., 1984; Berthelsen, 1984), which either separates two previously unrelated Archaean crustal segments or records the closure of a short-lived intracratonic rift basin. Granitic plutonism was relatively subordinate in the Lapland Granulite Belt.

To the southwest of the Karelian Province, the Svecofennian province forms the major part of the Svecokarelian Domain. It comprises three subprovinces featuring a northern volcanic belt along the margin of the Archaean craton, a central Svecofennian subprovince dominated by metagreywackes and metapelites with minor amounts of mafic volcanic rocks, and a southern volcanic belt extending from southern Finland to south-central Sweden (Figure 1).

All parts of the Svecofennian province are abundantly intruded by granitoids. Three main groups can be distinguished geochemically and chronologically (Lundqvist, 1979; Front and Nurmi, 1986; Wilson et al., 1985). The oldest was formed approximately 1.9-1.85 Ga ago. This is a differentiated, I-type suite dominated by tonalites and granodiorites. A second group of granites is approximately 1.82-1.78 Ga old and has an overwhelmingly S-type character. Finally, post-orogenic granites were emplaced into the newly consolidated Svecokarelian crust between 1.80 and 1.55 Ga ago. In regard to age and chemical zonation (Lindh and Gorbatschev, 1984), the post-orogenic Svecokarelian granites are probably related to the Gothian orogeny which created new continental crust farther west in Scandinavia.

The Karelian and Svecofennian rocks were long considered to have been formed during two different orogenies. The concept of a single Svecokarelian orogeny (Rankama and Welin, 1972) arose when it was shown that the granites in the Karelian and Svecofennian provinces were largely coeval. However, subsequent isotopic work demonstrated that the deposition of Karelian supracrustal rocks commenced 2.5 Ga ago, whereas the oldest volcanic rocks in the Svecofennian Province are less than 2.0 Ga old (Kouvo, 1976; Aho, 1979; Welin et al., 1980b; Skiöld and Cliff, 1984). In the light of this evidence a revision of the concept of the Svecokarelian orogeny appears justified. Future terminology should discriminate between the anorogenic Sariolian-Jatulian rocks of the Karelian craton and the rocks of the Svecofennian orogenic complex.

The presence of Archaean crust beneath the Svecofennian Province has repeatedly been suggested on the basis e.g. of old detrital zircons in sediments (Kouvo and Tilton, 1966; Åberg, 1978) and the bimodality and inferred rift character of volcanism in the southern Svecofennian subprovince (Oen et al., 1982; Vivallo and Rickard, 1984). However, the available Nd-isotopic systematics fail to substantiate the existence of an extensive Archaean substratum beneath most of the central and southern Svecofennian subprovinces. In the southern subprovince, most estimates of ϵNd_T are between -0.5 and +1.5 (Patchett and Kouvo, 1985; Patchett et al., 1986). Assuming a depleted mantle source, these data are compatible with an admixture of about 10 per cent of Archaean crustal material to melts newly differentiated from the mantle during the early Proterozoic. The apparently homogeneous distribution of a minor Archaean component over large areas of the Svecofennian Province suggest that it may have been mixed with the upper mantle reservoir by sediment subduction. In the central Svecofennian subprovince, the greywacke-pelite metasediments contain up to 40 per cent Archaean detritus, while the associated, supposedly anatectic, S-type granites comprise a much smaller Archaean component (Patchett et al., 1986; Huhma, 1985). This demonstrates a higher Proterozoic

mantle contribution at depth and militates against an Archaean substratum as well as against continent collision models (e.g. Bowes, 1980) of the Svecokarelian orogeny in the central Baltic Shield which necessarily require the presence of old continental blocks of which there is no immediate evidence. In contrast, the Nd-systematics are in agreement with scenarios suggesting island arc evolution and accretion of the arcs to the Archaean craton (Gaál, 1982; Bowes et al., 1984).

In the northern Svecofennian subprovince and elsewhere along the margin of the Archaean craton, the proportion of Archaean crustal components in Proterozoic granites increases consistently (Wilson et al., 1985; Huhma, 1985), reaching a maximum in anatectic granites intruded within the Archaean Domain (Kouvo et al., 1983).

During the Svecokarelian orogeny the previously passive southwestern margin of the Karelian craton was remobilized and tectonically reactivated. Part of it became the site of the volcanic belt of the North Svecofennian subprovince. Farther SE, imbricate slices of Svecofennian sediments and mafic metavolcanics were translated onto the cratonic margin in a zone where mantled domes of Archaean gneiss also indicate diapiric uprise of basement (Brun et al., 1981). To the NW of Lake Ladoga, thrust nappe displacement of the order of tens of kilometres has been described. Mafic and ultramaric rocks intercalated in the nappes have been interpreted as obducted ophiolite slices associated with a subduction zone along the Ladoga-Bothnian Bay Tectonic Zone (Bowes, 1980; Park, 1983).

The south Svecofennian subprovince has been interpreted as an outlying island arc separating the back-arc central Svecofennian from the open sea and another subduction zone has been supposed to have existed still farther W or SW (Hietanen, 1975; Löfgren, 1979; Rickard, 1979; Loberg, 1980). Such an interpretation is in apparent conflict with the bimodality of volcanism in the southern Svecofennian subprovince, in which rhyolitic to dacitic compositions dominate and andesites are rare (Gorbatschev and Lagerblad, 1985). Geochemistry rather suggests another continental-margin volcanic belt affected by episodes of tensional regime (Oen et al., 1982).

Granulite-facies metamorphism of the high-T/low-P type occurred along the Ladoga - Bothnian Bay Tectonic Zone and at the boundary between the central and southern Svecofennian subprovinces (Gaál, 1982). Characteristically, however, high-pressure regimes are absent in the Svecofennian. This contrasts with the conditions of metamorphism in the Archaean Domain and can be explained either by differences of the tectonic regime - i.e. absence of large-scale crustal stacking - or by different rates of uplift of the two orogenic belts.

In accordance with the trend of the Archaean-Proterozoic boundary in the central Baltic

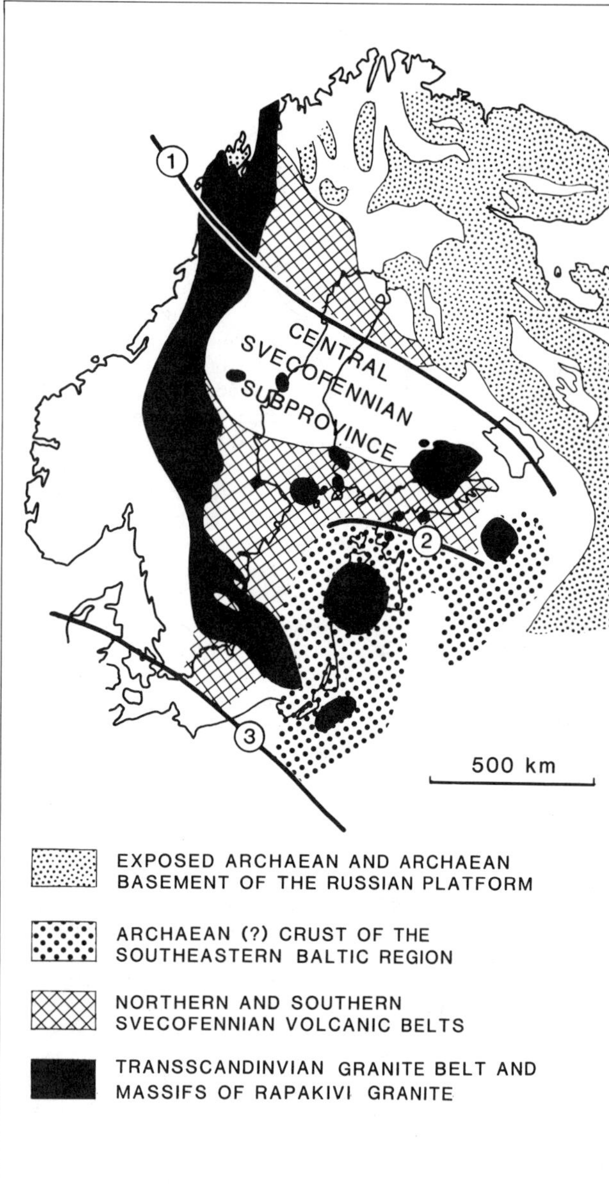

Fig. 2. Spatial relationships between the major tectonic units of the Baltic Shield and adjoining areas: The heavy broken lines indicate: 1 - Southern boundary of Archaean crust; 2 - Inferred boundary between Svecofennian Domain and Archaean or early Proterozoic crust in the SE Baltic region; 3 - Boundary between Baltic Shield and Phanerozoic Domain of Central Europe.

Shield, a general NW-SE trend of the Svecokarelian domain has often been assumed in plate tectonic interpretations (Hietanen, 1975; Rickard, 1979; Bowes, 1980). However, considerations of the wider regional context and palaeogeography (Figure 2) fail to support this assumption.

Extrapolation of the Archaean cratonic margin southwards beneath the Russian Platform defines a N-S trend where the Svecofennian Province forms an embayment in an older continental mass in the east.

Of particular importance is the region SE of the Baltic Sea. The existence of a pre-Svecofennian continental block in this area has been invoked by Edelman and Jaanus-Järkkälä (1983) to explain the tectonics and asymmetry of rock distribution in the South Svecofennian subprovince. In most maps of the basement of the Russian platform, southern Estonia and Latvia are marked as Archaean. However, this classification is based on indirect evidence such as lithological analogies and K-Ar data giving ages between 2100 and 2460 Ma (Koppelmaa et al., 1978). We find that the presence of an older but not necessarily Archaean continental block in the Baltic countries and an early Proterozoic ocean embayment now represented by the Svecofennian Province are in good agreement with available evidence. Such a proposal is also consistent with the configuration of the southern Svecofennian subprovince (Figure 2) and can reconcile its geochemical and isotope geological characteristics simultaneously suggesting the absence of an Archaean basement and a continental-margin character of the volcanism. It also explains the gradual decrease of felsic volcanites toward the east. 2.2 Ga old detrital zircons are, in fact, present in a Svecofennian quartzite in SE Sweden (Åberg, 1978).

Whatever the locations and trends of the earliest Svecofennian subduction zones, a N-S trend of igneous belts became established at least 1.79-1.78 Ga ago during the early stages of magmatic activity in the Transscandinavian Granite-Porphyry Belt (Figure 2). Although field evidence indicates a difference in age between many rocks of the Transscandinavian Belt and the S-type, late orogenic Svecokarelian granites (Gorbatschev, 1980), the earliest plutonic rocks in the Belt have U-Pb zircon ages virtually overlapping with those for the late orogenic Svecofennian granites (Patchett et al., 1986). A common cause for c. 1.8 Ga old magmatic activity in the two regions therefore appears obvious. It may be seen in the beginning or resumption of subduction in a NS-trending zone to the present west of the Transscandinavian Belt with attendant reworking and remelting of Svecofennian crust. In contrast, there is a lapse of time between the formation of the early Svecofennian igneous rocks 1.9-1.8 Ga ago and the intrusion of the late orogenic granites of 1.82-1.78 Ga age. During this time interval, the continental crust of the Svecofennian Province was consolidated and rifted in a regime of E-W directed compression permitting the intrusion of mafic dyke swarms throughout southern Finland and south central Sweden. In consequence, the Svecofennian features two separate maxima of granitic magmatic activity.

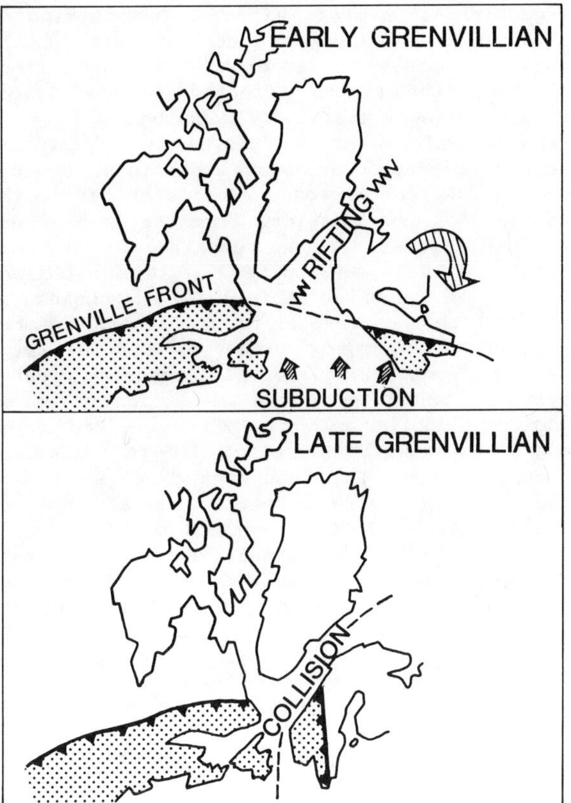

Fig. 3. Reconstruction of Sveconorwegian - Grenvillian continent configurations. The stippled area indicates Grenvillian crust and older crust reworked during the Grenvillian orogeny. Arrow indicates sense of rotation of the Baltic Shield in relation to Laurentia 1190-1050 Ma ago.

The Southwest Scandinavian Domain:
Continental Accretion Moves West

The Southwest Scandinavian and Svecokarelian Domains are separated by the Transscandinavian Granite-Porphyry Belt extending for 1600 km from Arctic Norway to southernmost Sweden. It was formed 1.80-1.65 Ga ago, essentially within previously consolidated Svecofennian crust (Figure 1; Gorbatschev, 1980; Patchett et al., 1986). Farther west, however, relics of Svecofennian rocks disappear, and the formation of most continental crust in southwestern Scandinavia occurred during the Gothian Orogeny c. 1.75-1.55 Ga ago. The generation of new crust involved the emplacement of several suites of predominantly tonalitic-granodioritic plutonic rocks about 1.75-1.65 Ga ago (Welin et al., 1982). These rocks intruded supracrustal formations which show gradually more marked continental-margin characteristics toward the west (Gorbatschev, 1980). Also among the 1.75-1.65 Ga old early Gothian plutonic rocks there is a prominent compositional zonation from W to E. Calcic and calc-alkaline compositions in the west are succeeded by alkali-calcic igneous rocks in the Transscandinavian Belt, while alkaline to alkali-calcic granites characterize the coeval, extraorogenic rapakivi massifs in the interior of the Archaean-Svecokarelian craton (Lindh and Gorbatschev, 1984).

Subsequently there followed several stages of tectonic reworking and granite intrusion. These granites exhibit successively higher initial Sr-isotopic ratios (Lindh, 1982). The formation of Gothian continental crust was completed about 1.5 Ga ago, when rifting permitted the emplacement of mafic dyke swarms (Welin et al., 1980a).

Thereafter, the Southwest Scandinavian Domain underwent an essentially ensialic development. Culminations of igneous activity and metamorphism occurred during the Hallandian event 1.5-1.4 Ga ago (Hubbard, 1975; Welin and Gorbatschev, 1978) and, particularly, during the Sveconorwegian-Grenvillian orogeny 1.25-0.9 Ga ago (Falkum and Petersen, 1980).

The entire Southwest Scandinavian Domain is chararacterized by repeated metamorphism at intermediate to high P/T-conditions. A subhorizontal, penetrative foliation was developed by eastward thrusting and stacking of crustal slices (cf. Lundqvist, 1979). E-W compression against the previously cratonized parts of the Baltic Shield lasted from the Gothian orogeny until the final stages of the Sveconorwegian orogeny, spanning a total period of 800-900 Ma. In this process, large, approximately N-S trending suture-like faults developed in southwestern Scandinavia (the "Protogine" and "Mylonite" Tectonic Zones etc., Figure 1). Lithological breaks along these faults indicate repeated crustal shortening and strike-slip movements accompanied by the intrusion of swarms of mafic dykes. In the boundary zone between the Southwest Scandinavian and Svecokarelian Domains, the latest tectonic activity occurred about 900 Ma ago. At this time the "Protogine" Zone developed into a boundary delimiting a region of Sveconorwegian-Grenvillian K-Ar age resetting and palaeomagnetic overprinting (Bylund, 1981; Stearn and Piper, 1984) in the west. It has often been compared to the Grenville Front of Canada (e.g. Gower and Owen, 1984).

Repeated lateral and vertical displacement along the "Protogine" Zone produced a marked tectonic contrast across this boundary, giving the impression of a sharp, linear demarcation and inviting the application of continent collision models (Eriksson and Henkel, 1983). However, radiometric dating (Welin et al., 1982) does not support the existence of older crustal segments in southwestern Scandinavia. Hypotheses of continent collision along the "Protogine" and "Mylonite" Tectonic Zones are also at variance with the chemical zonation of the early Gothian igneous rocks and the variation of volcanic and sedimentary facies which suggest westward accretion of the continental crust during the Gothian orogeny (Gorbatschev, 1980).

In the Baltic Shield, the Sveconorwegian-Grenvillian orogeny (Falkum and Pedersen, 1980) features two maxima of orogenic activity. During the first stage lasting approximately from 1.25 to 1.15 Ga ago, intense folding and high-grade metamorphism affected southwesternmost Scandinavia. Sveconorwegian plutonic masses were intruded into continental crust formed during the Gothian orogeny. Within the craton farther east, major mafic dyke-sill complexes developed in a zone of tension. However, this early Sveconorwegian development ended before much new continental crust was added to the shield.

The succeeding phase of relative quiescence was marked by faulting, psammitic-pelitic sedimentation in intracratonic basins, and mafic igneous activity.

A second orogenic maximum followed about 1.05 to 0.9 Ga ago, when several phases of deformation resulted in thin-skin nappe thrusting, granite intrusions along predominantly N-S trending faults, and the formation of further mafic dykes.

The complex course of the Sveconorwegian orogeny may be explained by a rotation of the Baltic Shield in relation to Laurentia (Figure 3). This rotation has been dated palaeomagnetically to have occurred between 1190 and 1050 Ma ago (Stearn and Piper, 1984). During the first part of the Sveconorwegian orogeny, the Baltic Shield was affected by plate interaction along a continent-ocean boundary. Rifting between the Baltic Shield and Greenland and rotation of the rifted continental fragments followed. Eventually, the newly formed oceanic basin was closed by continent collision accompanied by compression and thrusting during the second activity peak of the Sveconorwegian orogeny (Figure 3).

Conclusions

Most of the continental crust of the Baltic Shield was generated approximately 3.1 to 1.55 Ga ago during four distinct Precambrian crust-forming events. Each of these events involved initial, massive addition of mantle-derived material by magmatism, particularly by the intrusion of large amounts of tonalitic-granodioritic I-type plutonic rocks with high ϵNd_T values and low $^{87}Sr/^{86}Sr$ initial ratios. In contrast, the Caledonian and the Sveconorwegian-Grenvillian orogenies did not enlarge the Baltic Shield substantially. In general, and irrespective of their age, continent-collision orogenies apparently produced little new continental crust, whereas large masses of new crust were accreted during protracted orogenic processes along continent-ocean boundaries. A point in case is the difference between the collisional Lapland Granulite Belt and the coeval Svecofennian Province of the Svecokarelian Domain. The control exercised by the nature of the orogenic process is also obvious from the absence of high-P/T environments of metamorphism in the Svecofennian Province, in contrast to the occurrence of high P/T-conditions in areas affected by stacking of crustal slices and nappes such as part of the Archaean Domain, the Proterozoic Lapland Granulite Belt, the region affected by the Sveconorwegian orogeny and the Caledonides.

In the Baltic Shield, the early Proterozoic orogenies produced the most significant crustal accretion. Recent isotopic studies emphasize the importance of new additions from the mantle during this process, whereas recycling of Archaean crustal material was limited. Important characteristics of the Baltic Shield are the absence of minor Archaean continental blocks welded together by Proterozoic orogenic processes and the large sizes of contrasting coeval Archaean crustal provinces amongst which the Belomorian resembles a thrusted mobile belt flanked by rifted forelands subsequently affected by calc-alkaline igneous activity. This suggests that plate tectonic processes were operative already in the Archaean. Although the generation of juvenile early Proterozoic continental crust conforms well with plate-tectonic assumptions, increasing knowledge of local tectonic and chemical parameters necessitates a thorough revision of the evolutionary models previously proposed for the Svecofennian Province. The northern and the southern Svecofennian subprovinces probably both represent continental margin igneous belts. Most of the Svecofennian continental crust was formed 1.9-1.85 Ga ago, whereas the late orogenic Svecofennian phase about 1.82-1.78 Ga ago appears to overlap with and is causally closely related to the early stages of the Gothian orogeny.

Thus the Svecokarelian orogeny merged without sharp demarcation into the Gothian orogeny. These two orogenies define, altogether, a sequence of Proterozoic orogenic events lasting about 400 million years. Within this time period, however, the intrusion of juvenile tonalitic-granodioritic plutons only occurred 1.9 to 1.85 and 1.75 to 1.65 Ga ago.

References

Åberg, G., Precambrian geochronology of southeastern Sweden. Geol. Fören. Stockh. Förh., 100, 125-154, 1978.

Aho, L., Petrogenetic and geochronological studies of metavolcanic rocks and associated granitoids in the Pihtipudas area, central Finland. Geol. Surv. Finland Bull., 300, 1-23, 1979.

Akudinov, S.A., Bolgurtsev, N.N., Litvinenko, I.V. and Porotova, G.A., Deep structure of the eastern part of the Karelian region from complex geophysical studies of the Lake Onega – White Sea profile. Geotectonics, 1972:5, 296-297, 1972.

Barbey, P., Capdevilla, R. and Hameurt, J., Major and transition trace element abundance in the khondalite suite of the Granulite Belt of Lapland (Fennoscandia): Evidence for an early Proterozoic flysch-belt. Precambr. Res., 16, 237-279, 1982.

Barbey, P., Convert, J., Moreau, B., Capdevilla, R. and Hameurt, J., Petrogenesis and evolution of an early Proterozoic collisional orogenic belt: The Granulite Belt of Lapland and the Belomorides (Fennoscandia). Bull. Geol. Soc. Finland, 56, 161-188, 1984.

Bernard-Griffiths, J., Peucat, J.J., Postaire, B., Vidal, P., Convert, J. and Moreau, B., Isotopic data (U-Pb, Rb-Sr, Pb-Pb and Sm-Nd) on mafic granulites from Finnish Lapland. Precambr. Res., 23, 325-348, 1984.

Berthelsen, A., The tectonic division of the Baltic Shield. Proc. First Workshop on the European Geotraverse, 13-22. European Science Foundation - European Research Councils, Strasbourg, 1984.

Bibikova, E.V., The most ancient rocks in the USSR territory by U-Pb data on accessory zircons. In: Archaean Geochemistry (A. Kröner, G.N. Hanson and A.M. Goodwin, eds.), 235-250. Springer-Verlag, Berlin-Heidelberg-New York-Tokyo, 1984.

Bowes, D.R., Correlation in the Svecofennides and a crustal model. In: Principles and Criteria of Subdivision of Precambrian in Mobile Zones (F.P. Mitrofanov, ed.). Nauka, Leningrad, 1980.

Bowes, D.R., Halden, N.M., Koistinen, T.J. and Park, A.F., Structural features of basement and cover rocks in the eastern Svecokarelides, Finland. In: Precambrian Tectonics Illustrated, (A. Kröner and R. Greiling, eds.), 147-171. Nägele u. Obermiller, Stuttgart, 1984.

Brun, J.P., Gapais, D. and Le Theoff, B., The mantled gneiss domes of Kuopio (Finland): Interfering diapirs. Tectonophysics, 73, 288-304, 1981.

Bylund, G., Sveconorwegian palaeomagnetism in hyperite dolerites and syenites from Scania, Sweden. Geol. Fören. Stockh. Förh., 103, 173-182, 1981.

Edelman, N. and Jaanus-Järkkälä, M., A plate tectonic interpretation of the Precambrian of the archipelago of southwestern Finland. Geol. Surv. Finland Bull., 325, 4-33, 1983.

Eriksson, L. and Henkel, H., Deep structures in the Precambrian interpreted from magnetic and gravity maps of Scandinavia. Int. Basement Tectonics Assoc. Publ. No. 4, 351-358, 1983.

Falkum, T. and Petersen, J.S. The Sveconorwegian orogenic belt, a case of late-Proterozoic plate collision. Geol. Rundsch., 69, 622-647, 1980.

Front, K. and Nurmi, P.A., Characteristics and geological setting of synkinematic Svecokarelian granitoids in southern Finland, Precambr. Res., 1986 (in press).

Gaál, G., Proterozoic tectonic evolution and late Svecokarelian plate deformation of the central Baltic Shield. Geol. Rundsch., 71, 158-170, 1982.

Gaál, G., 2200 million years of crustal evolution: The Baltic Shield. Bull. Geol. Soc. Finland, 58, 149-168, 1986.

Gaál, G., Mikkola, A. and Söderholm, B., Evolution of the Archean crust in Finland.

Precambr. Res., 6, 199-215, 1978.

Glebovitskiy, V.A., Problems of the Evolution of Metamorphic Processes in Mobile Belts (in Russian). Nauka, Leningrad, 1973.

Gorbatschev, R., The Precambrian development of southern Sweden. Geol. Fören. Stockh. Förh., 102, 129-136, 1980.

Gorbatschev, R. and Lagerblad, B., Hydrothermal alteration as a control of regional geochemistry and ore formation in the central Baltic Shield. Geol. Rundsch., 74, 33-49, 1985.

Gorbunov, G.I., Zagorodny, V.G. and Robonen, W.I., Main features of the geological history of the Baltic Shield and the epochs of ore formation. Geol. Surv. Finland Bull., 333, 17-41, 1985.

Gower, C.F. and Owen, V., Pre-Grenvillian and Grenvillian lithotectonic regions in eastern Labrador - correlation with the Sveconorwegian orogenic belt in Sweden. Can. J. Earth Sci., 21, 678-693, 1984.

Hietanen, A., Generation of potassium-poor magmas in the Northern Sierra Nevada and the Svecofennian of Finland. Jour. Research U.S. Geol. Survey 3 (6), 631-645, 1975.

Hubbard, F.H., The Precambrian crystalline complex of south-western Sweden. The geology and petrogenetic development of the Varberg region. Geol. Fören. Stockh. Förh., 97, 223-236, 1975.

Huhma, H., Nd-isotopic composition and age of Proterozoic basalts from northern Finland. Terra Cognita, 4, 192, 1984.

Huhma, H., Provenance of some Finnish sediments. Geologi, 37, 23-25, 1985.

Hyppönen, V., Ontojoen, Hiisijärven ja Kuhmon kartta-alueiden kallioperä. Summary: Pre-Quaternary rocks of the Ontojoki, Hiisijärvi and Kuhmo map-sheet areas. Geol. Surv. Finland, Explanation to the maps of pre-Quaternary rocks, Sheets 4412 and 4413, 1-60, 1983.

Jahn, B.M., Vidal, P. and Kröner, A., High-chronometric ages and origin of Archaean tonalitic gneisses in Finnish Lapland: A case for long crustal residence time. Contrib. Mineral. Petrol., 86, 398-408, 1984.

Kahma, A., The main sulphide ore belt of Finland between Lake Ladoga and the Bothnian Bay. Bull. Geol. Soc. Finland, 50, 39-44, 1978.

Keller, B.M., Kratz, K.O., Mitrofanov, F.P., All-union conference on the general problems of subdivision of the Precambrian of USSR (in Russian). Sovetskaya Geologiya, 1977:12, 145-149, 1977.

Kharitonov, L.Ya., Structure and Stratigraphy of the Baltic Shield (in Russian). Nedra, Moscow 1966.

Koppelmaa, H.J., Klein, V.M. and Puura, V.A., The metamorphic complex of the crystalline basement of Estonia (in Russian). In: Metamorphic Complexes of the Basement of the Russian Plate, 43-76. Nauka, Moscow, 1978.

Korhonen, H., Seismic studies on geotraverses in Finland. Geol. Fören. Stockh. Förh., 105, 375-376, 1983.

Kouvo, O., Chronology of orogenic events in the Svecokarelides, Finland. 4th Eur. Colloquim on Geochronology, Cosmochemistry and Isotope Geology, Abstracts, 59. Amsterdam, 1976.

Kouvo, O. and Tilton, G.R., Mineral ages from the Finnish Precambrian. J. Geol., 74, 421-442, 1966.

Kouvo, O., Huhma, H. and Sakko, M., Isotopic evidence for old crustal involvement in the genesis of two granites from northern Finland. Terra Cognita, 3, 35, 1983.

Kratz, K.O. and Platunova, A.P., Geological structure (in Russian). In: The Earth's Crust of the Eastern Part of the Baltic Shield (K.O. Kratz, ed.), 54-74. Nauka, Leningrad, 1978.

Kröner, A., Puustinen, K. and Hickman, M., Geochronology of an Archaean tonalitic gneiss dome in northern Finland and its relation with an unusual overlying volcanic conglomerate and komatiitic greenstone. Contrib. Mineral. Petrol., 76, 33-41, 1981.

Krylov, I.N., Levchenkov, O.A., Lobach-Zhuchenko, S.B. and Chekulaev, V.P., Heterogeneity of the structure and development of the Archaean lithosphere of the Karelian granite-greenstone belt region (in Russian). 27th IGC Doklady, 5, 100-106, 1984.

Krylova, M.D., Felix, M. and Fiala, I., Granulite und Orthopyroxene in einem Hochdruck-Granulitkomplex im SW-Teil der Halbinsel Kola. Chemie der Erde, 41, 273-291, 1982.

Lauerma, R., On the age of some granitoid and schist complexes in northern Finland. Bull. Geol. Soc. Finland, 54, 85-100, 1983.

Levchenkov, O.A., Makeev, A.F. and Yakovleva, S.Z., Dating of the crystalline substance of zircon (in Russian). Doklady AN SSSR, 263, 1190-1193, 1982.

Levchenkov, O.A., Lobach-Zhuchenko, S.B., Morozova, I.M. and Chekulaev, V.P., The age of metamorphic complexes in the Archaean of Karelia (in Russian). Problems of Isotope Dating of Metamorphic and Metasomatic Processes, 29-30. Vernadsky Inst. Geochem. Analyt. Chemistry, Moscow, 1985.

Lindh, A., Trends in the Postsvecokarelian development of the Baltic Shield. Geol. Rundsch., 71, 130-140, 1982.

Lindh, A. and Gorbatschev, R., Chemical variation in a Proterozoic suite of granitoids extending across a mobile belt-craton boundary. Geol. Rundsch., 73, 881-893, 1984.

Loberg, B.E.H., A Proterozoic subduction zone in southern Sweden. Earth Planet. Sci. Lett., 46, 287-294, 1980.

Lobikov, B.F. and Lobach-Zhuchenko, S.B., The isotope age of granites of the Palaya Lamba greenstone belt of Karelia (in Russian). Doklady AN SSSR, 250, 729-733, 1980.

Löfgren, C., Do leptites represent Precambrian island arc rocks? Lithos, 12, 159-165, 1979.

Lundqvist, T., The Precambrian of Sweden. Sver. Geol. Unders., C768, 1-87, 1979.

Martin, H., Auvrai, B., Blais, S., Capdevilla, R., Hameurt, I., Jahn, B.M., Piquet, D., Querré, G. and Vidal, Ph., Origin and geodynamic evolution of the Archaean crust of eastern Finland. Bull. Geol. Soc. Finland, 56, 135-160, 1984.

Mearns, E.W. and Krill, A.G., Sm-Nd age for komatiites from the Karasjok Greenstone Belt, northern Norway. Geolognytt Oslo, 20, 38, 1984.

Meriläinen, K., The Granulite Complex and adjacent rocks in Lapland, northern Finland. Geol. Surv. Finland Bull., 281, 1-129, 1976.

Meriläinen, K., Stratigraphy of the Precambrian in Finland. Geol. Fören. Stockh. Förh., 102, 177-180, 1980.

Oen, I.S., Helmers, H., Verschure, R.H. and Wiklander, U., Ore deposition in a Proterozoic incipient rift zone environment: a tentative model for the Filipstad-Grythyttan-Hjulsjö region, Bergslagen, Sweden. Geol. Rundsch., 71, 182-194, 1982.

Park, A.F., Nature, affinities and significance of metavolcanic rocks in the Outokumpu assemblage, eastern Finland. Bull. Geol. Soc. Finland, 56, 25-52, 1983.

Patchett, J. and Kouvo, O., Origin of continental crust of 1.9-1.7 Ga age: Nd isotopes and U-Pb zircon ages in the Svecokarelian terrain of South Finland. Contr. Mineral. Petrol., 1985 (in press).

Patchett, J., Gorbatschev, R. and Todt, W., Origin of continental crust of 1.9-1.7 Ga age: Nd isotopes in the Svecokarelian orogenic terrain of Sweden. Precambr. Res., 1986 (in press).

Pharaoh, T.C. and Pearce, Geochemical evidence for the geotectonic setting of early Proterozoic metavolcanic sequences in Lapland. Precambr. Res., 25, 283-308, 1984.

Priyatkina, L.A., Glebovitskiy, V.A. and Shlayfshteyn, B.A., On the early development stages of the White Sea-Lapland metamorphic belt (in Russian). In: The Eastern Part of the Baltic Shield - Geology and Deep Structure (K.O. Kratz, ed.), 56-69. Nauka, Leningrad, 1975.

Rankama, K. and Welin, E., Joint meeting of the Precambrian Stratigraphy Groups of Denmark, Finland, Norway and Sweden in Turku, Finland, March 1972. IUGS Geolognical Newsletter, 1972:4, 265-267, 1972.

Raith, M., Raase, P. and Hörmann, P.K., The Precambrian of Finnish Lapland: Evolution and regime of metamorphism. Geol. Rundsch., 71, 230-244, 1982.

Richard, D.T., Scandinavian metallogenesis. GeoJournal, 3, 235-252, 1979.

Rybakov, S.I., Kulikov, V.S., Robonen, V.I. and Gorkhovets, V.Ya., Characteristics of the geochemical position of greenstone belts (in Russian). In: Volcanism of the Archaean Greenstone Belts of Karelia (V.A. Sokolov, ed.), 4-11. Nauka, Leningrad, 1981.

Saverikko, M., The Kummitsoiva komatiite complex and its satellites in northern Finland. Bull. Geol. Soc. Finland, 55, 111-139, 1983.

Sergeev, S.A., Arestova, N.A., Presnyakov, S.L., Levchenkov, O.A. and Yakovleva, S.Z., Isotope age of the rocks of the Shilos massif, NE Karelia (in Russian). In: Problems of Isotope Dating of Metamorphic and Metasomatic Processes, 51-52. Vernadsky Inst. Geochem. Analyt. Chemistry, Moscow, 1985a.

Sergeev, S.A., Lobach-Zhuchenko, S.B., Levchenkov, O.A. and Yakovleva, S.Z., Zircon U-Pb dating of the grey gneiss complex of the Vodlozero Block, SE Karelia (in Russian). In: Problems of Isotope Dating of Metamorphic and Metasomatic Processes, 50-51. Vernadsky Inst. Geochem. Analyt. Chemistry, Moscow, 1985b.

Silvennoinen, A., A., On the Proterozoic stratigraphy of northern Finland. Geol. Surv. Finland Bull., 331, 107-116, 1985.

Skiöld, T. and Cliff, R.A., Sm-Nd and U-Pb dating of early Proterozoic mafic-felsic volcanism in northernmost Sweden. Precambr. Res., 26, 1-13, 1984.

Sobotovich, E.V., Grashchenko, S.N., Aleksandryuk, V.M. and Shats, M.N., Determinations of the ages of oldest rocks by lead-isochron and isotope-spectral strontium methods (in Russian). Izvestiya AN SSSR, Ser. Geol., 1963:10, 3-15, 1983.

Sokolov, V.A. and Heiskanen, K.I., Evolution of Precambrian volcanogenic-sedimentary lithogenesis in the south-eastern part of the Baltic Shield. Geol. Surv. Finland Bull., 331, 91-106, 1985.

Stearn, J.E.F. and Piper, J.D.A., Palaeomagnetism of the Sveconorwegian mobile belt of the Fennoscandian Shield. Precambr. Res., 23, 201-246, 1984.

Tugarinov, A.I., and Bibikova, E.V., The Geochronology of the Baltic Shield According to the Results of Zirconometry (in Russian), 132 pp. Nauka, Moscow, 1980.

Vivallo, W. and Richard, D., Early Proterozoic ensialic spreading-subsidence: evidence from the Garpenberg enclave, central Sweden. Precambr. Res., 26, 206-221, 1984.

Volodichev, O.I., The early stage of metamorphism of the rocks of the Belomorian Complex (western White Sea region). In: The Eastern Part of the Baltic Shield - Geology and Deep Structure (K.O. Kratz, ed., in Russian), 43-56. Nauka, Leningrad, 1975.

Welin, E. and Gorbatschev, R., The Rb-Sr age of the Varberg charnockite in Sweden. Geol. Fören. Stockh. Förh., 100, 225-227, 1978.

Welin, E., Lundegårdh, P.H. and Kähr, A.-M., The radiometric age of a Proterozoic hyperite diabase in Värmland, western Sweden. Geol. Fören. Stockh. Förh., 102, 49-52, 1980a.

Welin, E., Wiklander, U. and Kähr, A.-M., Radiometric dating of a quartzporphyritic potassium rhyolite at Hällefors, south central Sweden. Geol. Fören. Stockh. Förh., 102, 269-272, 1980.

Welin, E., Gorbatschev, R. and Kähr, A.-M., Zircon dating of polymetamorphic rocks in southwestern Sweden. Sver. Geol. Unders., C797, 1-34, 1982.

Wilson, M.R., Hamilton, P.J., Fallick, A.E., Aftalion, M. and Michard, A., Granites and early Proterozoic crustal evolution in Sweden: Evidence from Sm-Nd, U-Pb and O isotope systematics. Earth Planet. Sci. Lett., 72, 376-388, 1985.

GEODYNAMIC SIGNIFICANCE OF CONTRASTING GRANITOID TYPES IN NORTHERN SWEDEN

M.R. Wilson †

Swedish Geological Company, Box 801,S-951 28 Luleå, Sweden

B. Öhlander

Department of Economic Geology, University of Luleå,S-951 87 Luleå, Sweden

M. Cuney

Le Centre de Reserches sur la Geologie de L'uranium B.P. 23 54501, Vandoeuvre-les-Nancy Cedex France

P.J. Hamilton

Scottish Universities Research and Reactor Center, East Kilbride, Glasgow, G75 0QU Scotland

Abstract. Early Proterozoic granitoid suites in northern Sweden fall into two main groups, an older phase 1.89-1.84 Ga and a younger phase 1.80-1.75 Ga. Sm-Nd studies on some of the older granitoids indicate that they are derived from sources with short average crustal residence times, while their major and trace element compositions resemble those of intrusive suites from both immature and mature volcanic arcs and from ensialic rifting environments.

In contrast, the younger granitoid suites have been derived from dominantly crustal sources, and show geochemical and isotopic relationships with the rocks of their immediate environments. Their geochemical characteristics resemble those of granites developed in continental collision areas.

The geology, palaeogeography, geochemistry and isotope geology of northern Sweden strongly supports a hypothesis involving an early period of subduction-related magmatism and ensialic rift magmatism in a Cordilleran environment followed by a later period of crustal anatexis related to continental collision.

Introduction

The most common rock types in the Proterozoic of Sweden are granitoid suites of intermediate to acid composition. Information on their timing of intrusion, mode of occurrence, geochemical and isotopic characteristics is of prime importance for understanding the geodynamic evolution of the lithosphere.

------

† deceased

As the result of U-Pb zircon dating at the Naturhistoriska Riksmuseet, Stockholm [Skiöld, 1979a,b, 1981a,b] and the SURRC laboratory in Scotland [Wilson et al., 1985], it is apparent that most plutonic magmatic activity in the early Proterozoic of north Sweden occurred in two periods, 1.89 - 1.84 Ga and 1.80 - 1.75 Ga.

Sm-Nd isotopic studies [Wilson et al., 1985] on granitoid suites from the Arvidsjaur-Boliden-Storuman area (Figure 1) suggest that major components of the magmas of both periods were derived from the mantle shortly before 1.9 Ga and that this region represents a new Proterozoic addition to the continental crust.

In contrast, The Kiruna-Gällivare region contains areas which first developed as continental crust about 2.8 Ga ago, and which are represented today either as Archaean gneiss or as younger granitoids with a major Archaean component.

In this paper we present a summary of the geochemical characteristics of a selection of contrasting granitoid suites from both periods. We attempt to determine the possible sources of the different magmas, and relate the magmatic episodes to the geodynamic evolution of the early Proterozoic in Sweden. Virtually all granitoid suites in northern Sweden yield Rb-Sr isochron ages younger than equivalent U-Pb zircon ages, occasionally by as much as 200 million years [e.g. Skiöld, 1981 a; Wilson et al., 1985]. The Rb-Sr isotopic systems appear to be reset in these intrusions. The geochemical diagrams based upon elements such as Rb, K, Na and Ca (Figures 5, 6, 8 and 9), which are often mobilised under conditions of hydrothermal alteration and metamorphism, therefore must be interpreted with caution.

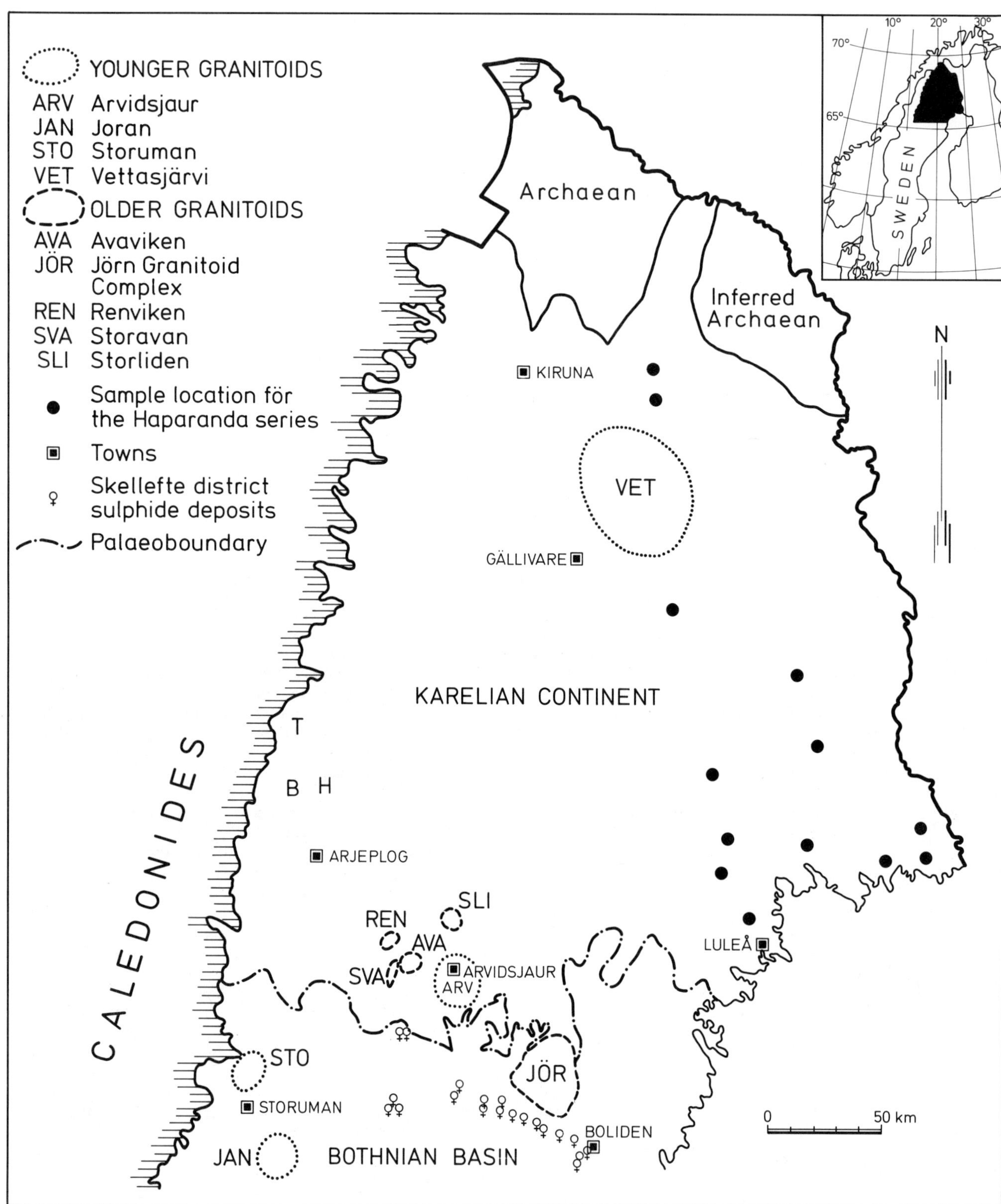

Fig. 1. Precambrian bedrock of northern Sweden showing main crustal provinces and location of granitoid suites discussed in the text. B, T, and H refer to Björntjärn, Tjeggelvas and Hällnäs (see text).

Geological Setting

Although the Svecokarelian of north Sweden is dominated by granitoids, there are sufficient supracrustal suites preserved to allow the subdivision of the region on the basis of palaeoenvironments. The northern part of the area represented in Figure 1 is a dominantly continental environment with some areas of Archaean gneiss in the far north, and is termed the Karelian continent. It is bounded to the south by a narrow volcanic belt, the Skellefte district. To the south of the Skellefte district lies a major province dominated by marine metasediments and basic volcanics, termed the Bothnian basin.

The regional fabric of north Sweden is illustrated by Figure 2, an interpretation of a high-resolution airborne survey. North Sweden is subdivided into a number of large domains with low and homogeneous magnetic susceptibility, representing either Archaean basement (in the far north) or homogeneous granite or migmatite.

These domains are separated by wide zones with banded magnetic anomalies and distinct circular structures interpreted as granitoid diapirs, representing mobile zones between the domains. These mobile zones contain greenstone belts, sediments, volcanics, gabbros and intermediate to acid plutonic suites. They also contain the majority of the region's ore deposits and mineralisations (Fe, Cu, Au, Mo, W, U) [Öhlander and Nisca, 1985].

Metamorphic grade and degree of deformation vary widely throughout the Karelian continent. For example in the Arvidsjaur district, the rocks are of low metamorphic grade and are rarely penetratively deformed. In contrast the supracrustals of the Arjeplog area are strongly deformed and recrystallised in amphibolite facies.

Some of the northernmost domains have given Archaean U-Pb zircon ages [Welin et al., 1971; Skiöld, 1979b], and the more uniform granite domains such as the Vettasjärvi granite, NE of Gällivare, with a U-Pb zircon age about 1.8 Ga [Skiöld, 1979b], are considered by Witschard [1984] to represent remobilised Archaean. Our Sm-Nd data confirms that the Vettasjärvi granite was derived from sources with a major Archaean component.

The mobile belts of the northeastern part of the Karelian continent contain gabbros, granodiorites and granites belonging to the Haparanda suite with zircon ages between 1.90 and 1.85 Ga and with low initial Sr isotope ratios [Skiöld, 1979a, 1981a,b]. The major and trace element geochemistry of the suite is described by Öhlander [1984]. Witschard [1984] proposed that these Haparanda granitoids developed in an intracratonic environment, not related to subduction, and the excess of heat from these mantle-derived magmas caused the extensive remobilisation of basement and formation of widespread younger granites.

The southern part of the Karelian continent comprises a wide zone of granitoids and continental supracrustals, metamorphosed to varying degrees. We have investigated a series of subalkaline to alkaline monzonites and granites in the Arvidsjaur-Storavan district, some 50 km north of the palaeogeographic continental margin. Together with an extensive peralkaline rhyolitic ignimbrite these form a alkali magmatic subprovince, closely related to a major NNE-SSW fault zone and associated with Mo and U mineralisations, [Adamek and Wilson, 1979; Walser and Einarsson, 1982]. The province is accompanied by a negative gravity anomaly.

The rocks of the Skellefte district comprises marine graphitic schists, pelites, greywackes, conglomerates, acid and basic volcanites and sulphide deposits. Granodioritic gneisses in the south of the Skellefte district may represent basement. The environment, petrochemistry and sulphide ore deposits [Rickard and Zweifel, 1975; Claesson, 1982, and in press; Lundberg 1980] are analogous to those formed at Phanerozoic destructive plate margins. On the northern flank of the district the supracrustals are overlain by the continental volcanites of the Karelian continent. Both supracrustal series are cut by the Jörn granitoid complex, which is thought to be a plutonic memeber of the volcanic arc assemblage. Jörn lies immediately west of a major N-S fault zone which displaces the supracrustals. The outer part of the complex consists of granodiorite and is intruded by a more evolved granitic core. The metamorphic grade and degree of deformation of the Skellefte district appears to be relatively low in the north but increases rapidly southwards.

The Bothnian basin comprises high-grade metasedimentary gneisses, migmatites and late, crosscutting granites. Some older granites and gneisses of unknown age also occur. Supracrustals include greywackes, pelites, pillow lavas and an ultramafic breccia. The Bothnian basin may represent slices of flysch and ocean floor in an outer arc accretionary wedge south of the Skellefte district volcanic belt. The contrast in metamorphism and degree of deformation between the Karelian continent, the Skellefte district and the Bothnian basin indicate that the Bothnian basin has been depressed to a considerable depth and later uplifted to its present relative elevation and that much of the strain accomodating this movement is seen in the south part of the Skellefte district.

The similarity between ores of the Skellefte district and those developed in Phanerozoic destructive arcs was first noted by Mitchell and Bell [1973] and developed in detail by Richard and Zweifel [1975] who proposed a model with a northward-dipping subduction of ocean crust. At the same time both Hietanen [1975] and Adamek and Wilson [1977] proposed similar models, based on the chemistry of plutonic associations and on the palaeogeography of the southern margin of the Karelian continent respectively. Further contributions have been made by Adamek and Wilson [1979], Rickard [1979], Lundberg [1980], Wilson [1980], Walser and Einarsson [1982], Claesson

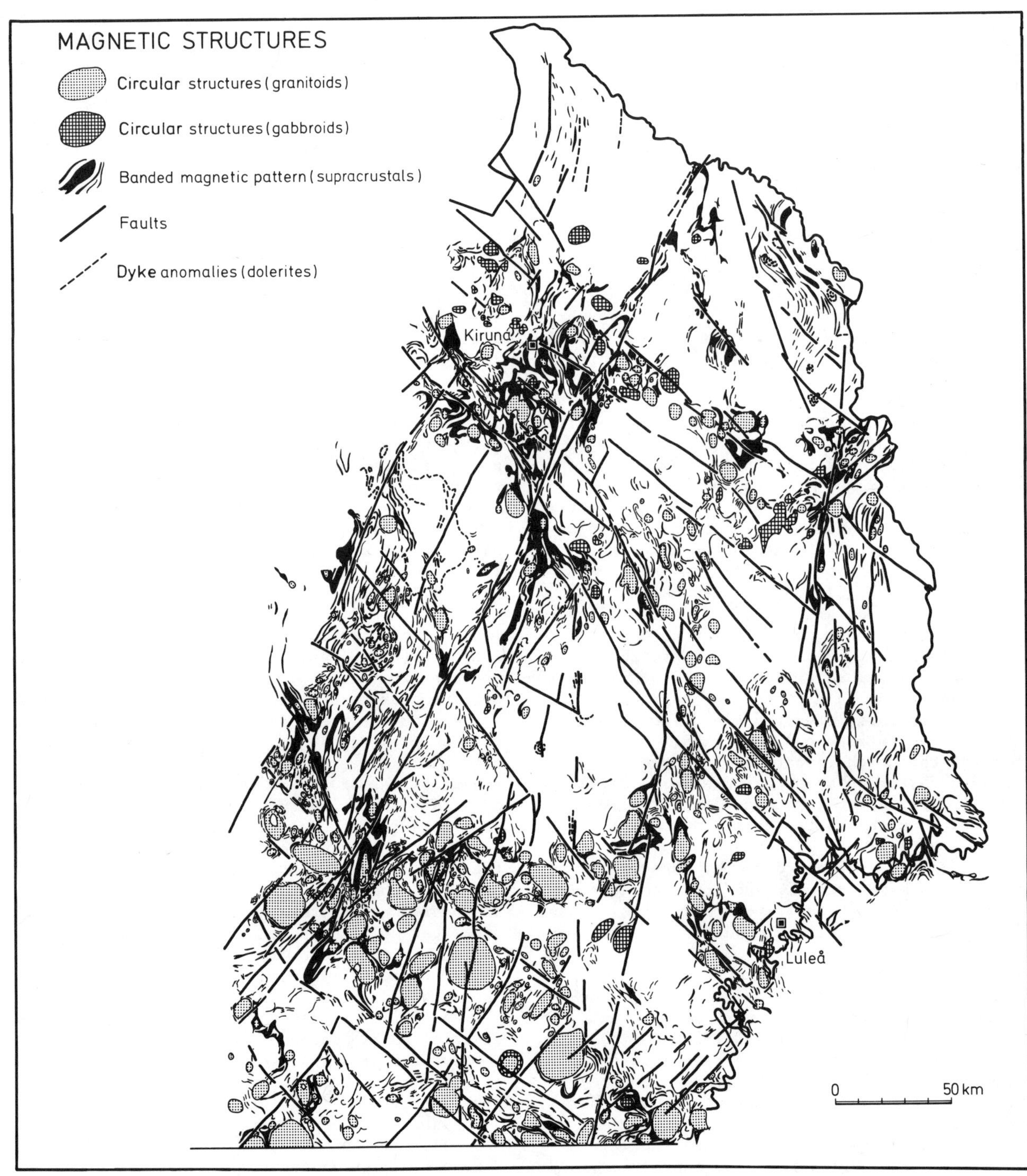

Fig. 2. Interpretation of magnetic structures in the area covered by Figure 1. Compiled by Dan Nisca, SGAB, from airborne survey by SGU.

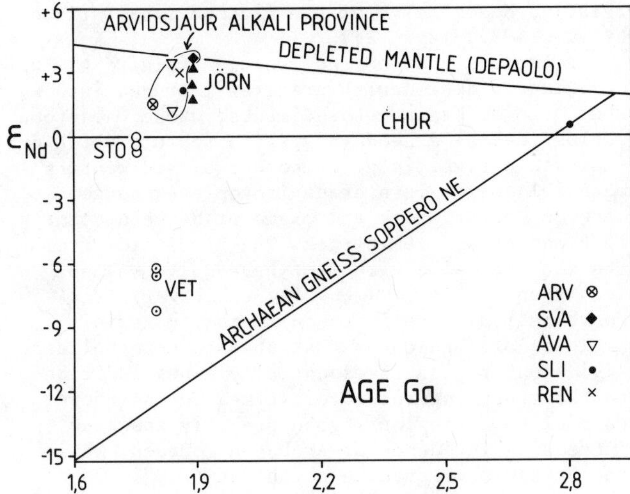

Fig. 3. Epsilon Nd values at T-zircon for selected granitoid suites, northern Sweden. LREE-depleted mantle evolution line after DePaolo [1981]. Archaean evolution line based on our unpublished data from map-sheet Soppero NE, on Archaean gneiss U-Pb zircon dated by Skiöld [1979b].

[1982, and in press], and Pharoah and Pearce [1984].

It is purpose of current work by ourselves and others (L-Å Claesson, W. Vivallo and L. Widenfalk) to study the geochemistry of plutonic suites and stratigraphically well-defined supracrustal successions across this transition as a means of comparing Proterozoic processes with those operative in the Phanerozoic.

Chronology of Plutonic Activity

Although whole-rock Rb-Sr dating [Welin et al., 1970, 1971, 1977; Gulson, 1972] suggested a range of ages down to 1.5 Ga, U-Pb zircon dating [Skiöld, 1979a, 1981a, 1981b; Wilson et al., 1985] indicates a much shorter period of plutonic activity. If the Dobblon volcanics and the Sorsele granitoid suite are excluded (since these may belong to a separate Post-Svecokarelian magmatic phase), Svecokarelian plutonic activity seems to be divided into two periods, 1.89 – 1.84 Ga and 1.80 – 1.75 Ga, although it is not certain how fundamental the apparent gap between 1.84 and 1.80 Ga really is. It is, however, highly significant that there are no indications of plutonic activity prior to 1.9 Ga, although in Finland there is good evidence of basic volcanic activity at several stages in the early Proterozoic.

Selection of Granitoid Suites

In this review we have selected certain representative granitoid suites from a larger number under current investigation. From the older group we have selected the oldest, the Jörn granitoid complex, intruded at 1.89 Ga into the base of the Skellefte district volcanic arc succession and associated with a minor porphyry copper mineralisation and contrast it with four granitoids from the Arvidsjaur district, some 50 km further north. The Avaviken (1.84 Ga) and Storavan granites have irregular forms, while the Storliden granite has a diapiric form. All three are associated with U, Mo or W mineralisations. The Renviken granitoid suite has an irregular form and is associated with a Ni-mineralised gabbro, Storbodsund [Grip, 1978]. We also compare these granites with series of analyses from separate intrusions of the Haparanda suite [Öhlander, 1984].

To illustrate the younger period we contrast (a) the barren Vettasjärvi granite [ca. 1.8 Ga, Skiöld 1981a], from a large low-magnetic area northeast of Gällivare, (b) the barren diapiric Arvidsjaur granite [1.78 Ga, Wilson et al., 1985] from some 50 km north of the Skellefte district, and (c) late tectonic granites [ca. 1.75 Ga, Wilson et al., 1985] from the Storuman area of the metasedimentary Bothnian basin. These Storuman granites vary considerably in character and are relatively rich in U and Th, while W and Sn mineralisations are associated with them. We

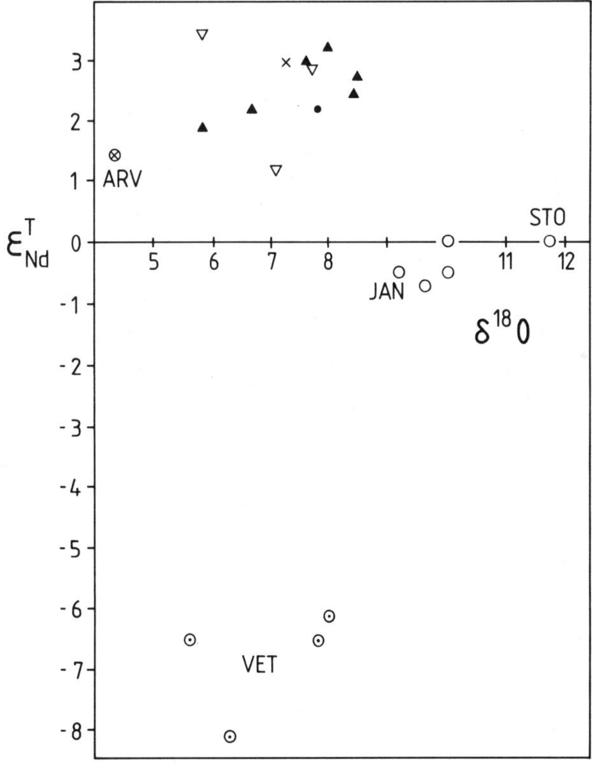

Fig. 4. Epsilon Nd at T-zircon v. oxygen isotopic ratios for granitoid suites from north Sweden. Symbols as in Figure 3.

have sampled a highly evolved diapir at Joran, southeast of Storuman, and a less-evolved two-mica granite in an irregular roof zone adjacent to locally migmatised metasediments north of Storuman.

Isotopic Constraints on Magma Sources

The evolution of epsilon Nd with time for three of the several possible magma sources are illustrated in Figure 3: light rare earth (LREE) depleted mantle [DePaolo, 1981], non-depleted chondritic mantle (CHUR) and Archaean crust. The Archaean evolution line is based on our unpublished data on gneisses U-Pb zircon-dated by Skiöld [1979b].

Nd isotope ratios [Wilson et al., 1985, and work in progress] clearly distinguish between the early and later groups and between the three different late intrusions (Figures 3 and 4).

The older group (not inculding Haparanda, for which no Sm-Nd data is available) shows epsilon Nd values between +3.5 and +1.1, suggesting that their sources were derived rapidly from LREE-depleted mantle with minimal involvement of Archaean material, and only a short average crustal residence time.

Three intrusions north of Arjeplog (Figure 1), Björntjärn and Hällnäs, [Wilson et al., 1985] and the Tjeggelvas gneiss (our unpublished results) give Sm-Nd model ages (T-DM) of 2.3 to 2.2 Ga, greater than those of the Arvidsjaur-Boliden area granitoids (2.1-1.9 Ga), although their intrusion age is uncertain. We suggest that the lower crust in the vicinity of these granitoids probably contains a component of Archaean age, which has been sampled by the magmas during ascent.

Much longer average crustal residence times are indicated by the dramatically different Nd data from the younger granites. The Vettasjärvi granite gives epsilon Nd values of between −6 and −8, indicating a major Archaean component, [Öhlander et al., in press]. If the 2.8 Ga [Skiöld, 1979b] gneisses of northernmost Sweden were reactivated at 1.8 Ga they would give an epsilon Nd value of about −12.

The Arvidsjaur granite gives an epsilon Nd value of +1.5, which is clearly less than expected for depleted mantle at that time, indicating either a mixture of new, mantle-derived material with much older crust or, more likely, reworking of material belonging to the earlier event. The T-DM model age for Arvidsjaur is identical to that of the early granitoids (2.0 Ga).

The Bothnian basin granitoids have epsilon Nd values between 0 and −1 similarly indicating longer crustal residence times. We prefer to interpret this in terms of a dominant source derived from the mantle shortly before 1.9 Ga and mixed with a moderate amount of sediment of Archaean provenance. This is supported by geochemical and isotopic evidence for sedimentary components in these granites, by the presence of Archaean in northern most Sweden and by the

relative homogeneity of the isotope data [Wilson et al., 1985, Table 2].

On the basis of oxygen isotopes (Figure 4), we can clearly distinguish granites intruded into the Bothnian basin metasediments, whose O isotope ratios (delta) exceed +9 $^0/_{00}$, a feature which can indicate a significant component of sedimentary material and/or late stage isotopic exchange between country rock and magma or between country rock and crystallised rock. The granitoids from the continental environment have delta values less than +9 $^0/_{00}$, suggesting relatively little contribution of pelitic sedimentary material.

This data indicates that the Storuman-Boliden-Arvidsjaur region is essentially a new addition to the crust in Proterozoic time. The northern part of the Arjeplog region probably shows an increasing influence of Archaean sources, while the Vettasjärvi granite northeast of Gällivare shows a substantial Archaean component.

Major Element Geochemistry

Each granitoid suite has its own individual geochemical signature, displayed in Figures 5 and 6.

A systematic discrimination can be made in the following way: Firstly we discriminate between suites with a wide range of composition and those with a restricted range (high SiO_2 - content). The wide-ranging suites, though often described as calk-alkaline, demand a more precise description, for example on the basis of the alkali-lime index of Peacock [Peacock, 1931; Brown et al., 1984] or preferably on the basis of their major rock forming minerals, quartz, plagioclase and alkali feldspar [Debon and Le Fort, 1982; Lameyre and Bowden, 1982]. The granitoids can also be discriminated on the basis of their degree of alumina oversaturation [Chappell and White, 1974; Debon and Le Fort, 1982], or their degree of alkalinity (for example with the R1-R2 diagram of De la Roche et al. [1980].

Most, but not all, of the granitoid suites of the early period have a wide range of composition: an exception is Storliden.

Of the wide range suites we can clearly distinguish the outer part of the Jörn granitoid complex from the others through its calcic alkali-lime index (Figure 5). Avaviken and Storavan are calc-alkali or alkali-calcic. Brown [1982] and Brown et al. [1984, Figure 1] illustrate that such calcic granitoids are characteristic of relatively immature volcanic arcs, while granitoids developed within more mature arcs and over thicker continental crust have a lower alkali-lime index.

A similar type of discrimination is made on the triangular plot quartz, alkali feldspar and plagioclase, either using modal data on Streckeisens's QAP diagram [Lameyre and Bowden, 1982], or through a chemical construction [Debon and Le Fort, 1982, Q-P diagram]. On the latter plot (Figure 5c), the Jörn data falls along a

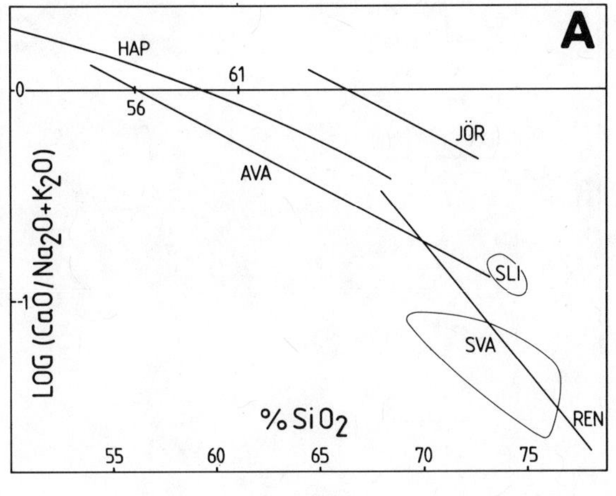

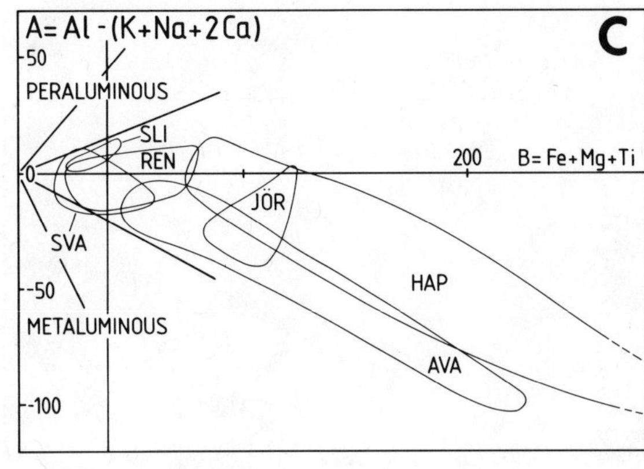

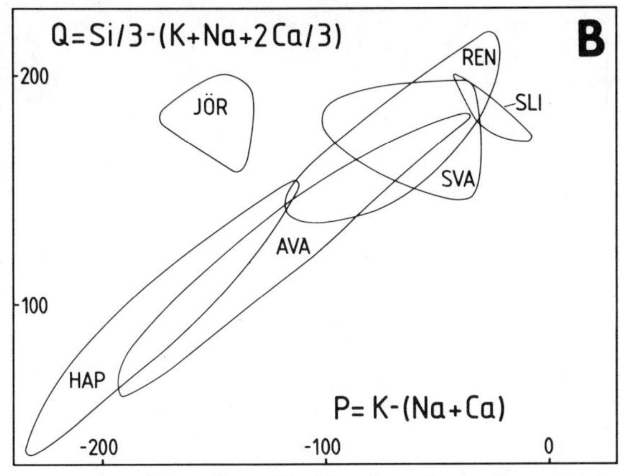

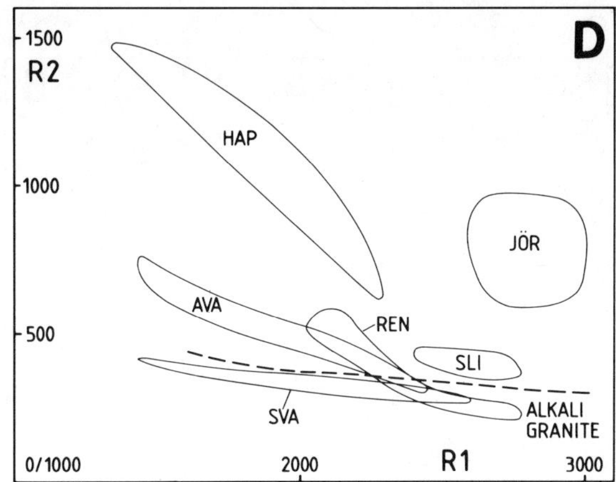

Fig. 5. Plots of major element data from the older granitoid suites. JÖR Jörn granitoid complex, AVA Avaviken suite, SLI Storliden diapir, SVA Storavan granite, REN Renviken suite.
(a) Alkali-lime index [Peacock, 1931].
(b) Q-P diagram [Debon and Le Fort, 1982]. The positions of the main rock-forming minerals are indicated on the inset (see Figure 6b).
(c) A-B diagram [Debon and Le Fort]. See Figure 6c.
(d) R1-R2 diagram [De la Roche, 1980]. R1 = 4Si-11(Na+K)-2(Fe+Ti)
R2 = 6Ca+2Mg+Al. Boundary between alkali granites and granites s.s. marked.

near horizontal line indicating that differentiation involves a change of feldspar type and not a significant change in amount of free quartz. This evolution is typical of granitoid suites passing from tonalite to granite. A somewhat similar evolution trend can be superimposed on a rock through alteration, for example through a metasomatic increase in Si or Ca and it is possible that the outer part of Jörn may have been affected slightly in this way.

A quite different evolution is seen both for the Avaviken and Storavan suites and for various granitoids of the Haparanda suite, which evolve through monzonitic compositions and can be described as subalkaline. The alkaline tendencies of the Avaviken, Storavan and Storliden suites can be seen more clearly on the R1-R2 diagram, Figure 5c.

All these early suites have normal magmatic evolutions trends in terms of alumina saturation. In the "characteristic mineral" plot of Debon and Le Fort [1982] plotting excess alumina against (Fe+Mg+Ti), the samples show normal increase of alumina saturation with increasing evolution, leading to the normal slight degree of "alumina-excess" that is characteristic of highly evolved granites (Figure 6a).

Most of the younger granitoids have restricted ranges of composition, mostly 70-76 % SiO_2. A slightly wider range is seen in the two-mica granite northwest of Storuman. This granite has abundant metasediment xenoliths and it is possible that it is a heterogeneous product of partial assimilation with considerable influence of late stage hydrothermal alteration.

The later granites therefore generally show a

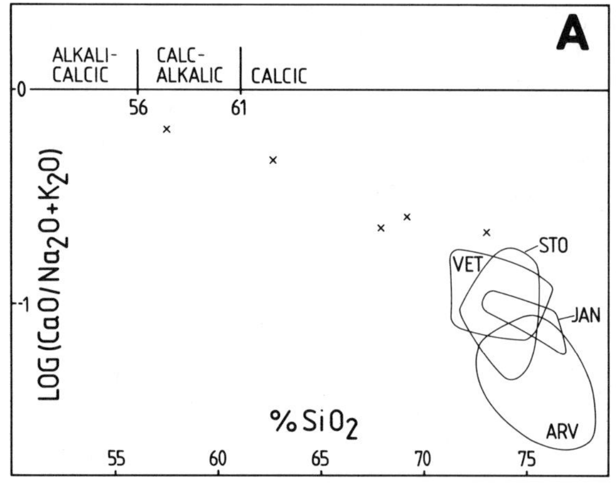

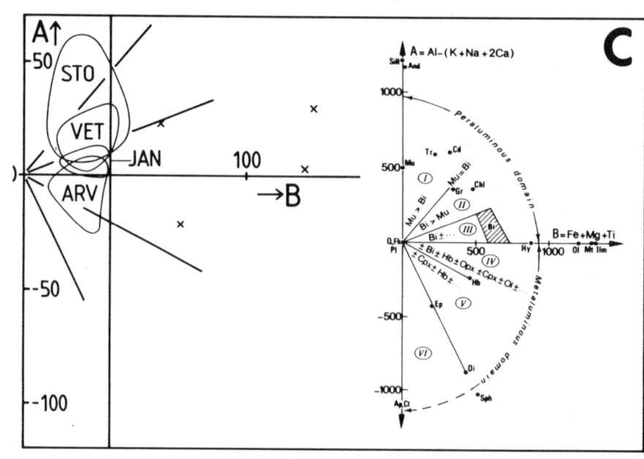

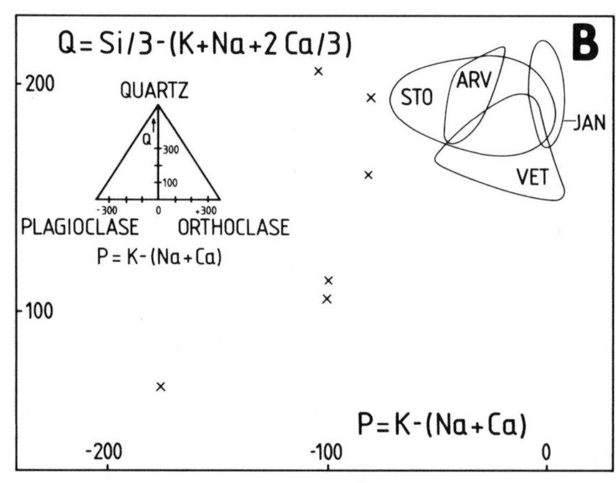

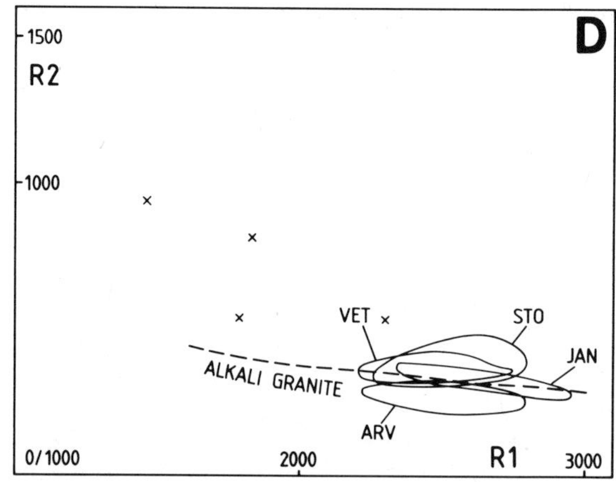

Fig. 6. Plots of major element data from younger granitoid suites.
ARV Arvidsjaur diapir, STO 2-mica granite Storuman, JAN Joran dome Storuman,
VET Vettasjärvi granite.
(a) Alkali-lime index, (b) Q-P, (c) A-B, The positions of the characteristic
minerals are indicated on the inset. (d) R1-R2. See Figure 5d.
Samples from the Storuman 2-mica granite which do not fall within the indicated
area are marked with a cross.

tight cluster of data points on the various major element diagrams (Figures 5 and 6). On the R1-R2 diagram we can distinguish the Arvidsjaur granite on the basis of its alkali character. On the AB diagram the two-mica granite is discriminated as strongly peraluminous, but the Joran granite from the Bothnian basin is not much more peraluminous than Vettasjärvi and Arvidsjaur. This is probably the consequence of evolutionary convergence, since the oxygen isotope composition of Joran is higher than that of the continental granitoids.

Trace Element Geochemistry

Collins et al. [1982] correlate increasing Ga, Nb and Y with increasing alkalinity, suggesting that these elements are incorporated in magmas as a result of high temperature anatexis of residual crustal material that had already been through a partial melting episode. High contents of HFS elements are thus characteristic of so-called "A-type" granites. Brown et al. [1984] argue that in many areas, granite magmatism is initiated in the mantle, initially basic magmas being strongly modified by assimilation and fractional crystal-lisation to give granitoid suites. They therefore suggest that differences in trace-element con-tents, especially HFS elements, may be controlled by the trace-element composition of the initial mantle-derived magmas. The conditions under which subduction zone magmas are initially generated do not favour the incorporation of HFS elements so that the calcic and calc-alkalic granitoids of volcanic arcs have low contents of HFS elements. The processes involved in the generation of basic

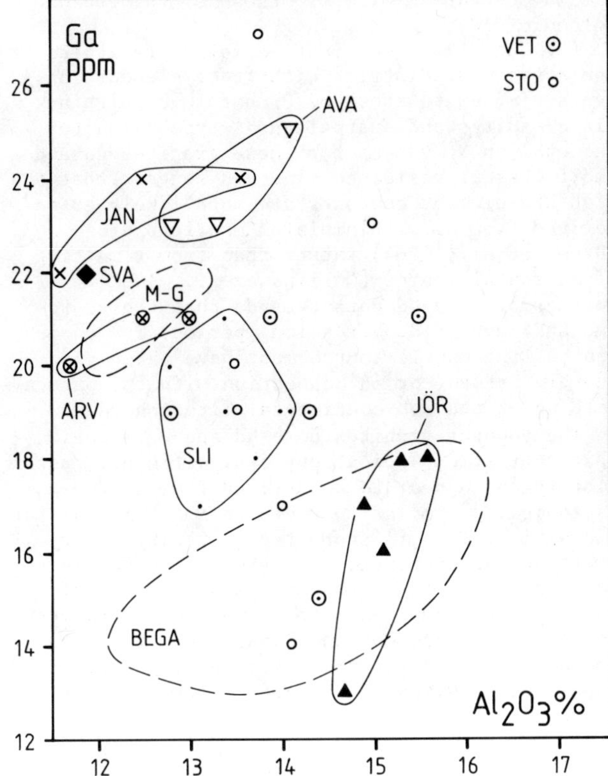

Fig. 7. Ga v. Al₂O₃. Data fields for the I-type
Bega batholith and the A-type Mumbulla-Gabo
batholiths from Australia [Collins et al., 1981]
are indicated for comparison.

magmas in "within-plate" environments allow incorporation of higher concentrations of HFS elements, although it is not certain in many cases
if this enrichment occurs during partial melting
of mantle or is an independant long-term process
in the subcontinental mantle [Pearce, 1982;
Weaver and Tarney, 1983]. These differences in
HFS element concentrations have been utilised by
Pearce et al. [1984] to construct discriminant
diagrams for granitoids from "volcanic-arc" and
"within-plate" environments.

However, "within-plate" environments include
not only the well-known continental rifting
situations, situated far from subduction zones,
but also ensialic spreading environments along
destructive plate boundaries, parallel and inland
from the zones of calc-alkalic magmas [Levi and
Aguirre, 1981; Åberg et al., 1984]. Brown et al.
[1984] demonstrate a gradual transition across
active continental margins from subduction zone
magmas with low contents of HFS elements to
magmas with "within-plate" geochemical characteristics, ie. a subalkaline to alkaline major
element geochemistry and high contents of HFS
elements. The diagrams of Pearce et al. [1984] do
not discriminate between the subalkaline to alkaline granites formed on the continental side of

Cordilleran belts and the very different anorogenic granites formed during continental rifting.

The Jörn complex shows low contents of Ta, Nb,
Yb, Y and Ga (Figure 7). The Haparanda suite is
also low in Y and Nb. Significantly higher Ta, Nb,
Yb, Y and Ga contents are recorded for the
Avaviken, Storavan and Renviken suites conforming
their subalkaline to alkaline chemistry, although
the contents of these elements are not as extreme
as in anorogenic alkali provinces such as
Nigeria [Bowden and Turner, 1974]. The Jörn and
Haparanda suites (Figure 8) lie clearly within
the field of Phanerozoic "volcanic arc" granites,
in marked contrast to the early intrusive complexes in the Arvidsjaur district (Avaviken,
Storliden) which lie in the "within-plate" field.

The younger granites show a variety of trace
element patterns. Vettasjärvi has relatively low
Y and Nb, Arvidsjaur has high Nb, Yb, Ta and Y,
similar to those of the older granitoids in its
vicinity, but the two-mica granite at Storuman
has intermediate Nb and Y contents, while the
Joran granite has distinctly high Y, Nb and Ga.
Of the younger granites, Arvidsjaur and Joran lie
in the "within-plate" field (Figure 8),
Vettasjärvi in the "syn-collision" field and
Storuman overlaps the "volcanic arc", "syn-

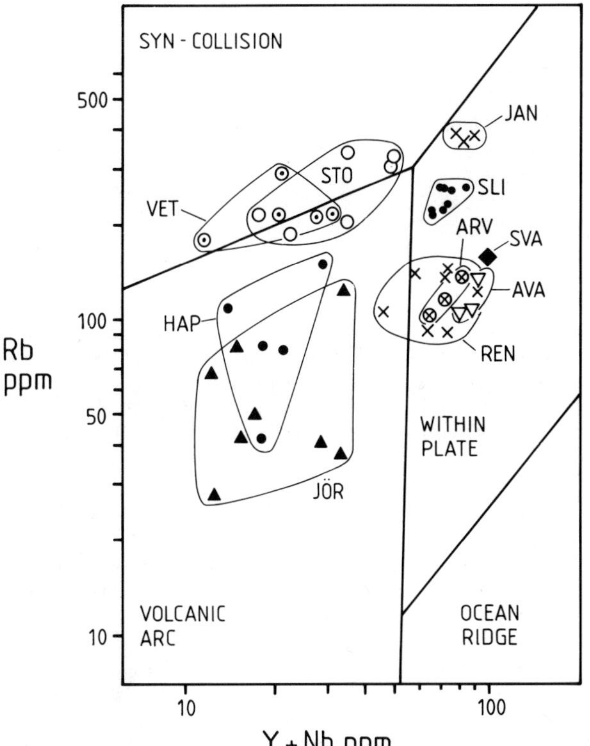

Fig. 8. Rb v. (Y+Nb). Discriminant fields for
granitoids from different Phanerozoic environments after Pearce et al. [1984]. For
Proterozoic granitoids the boundaries probably
should be adjusted to slightly lower abundances.

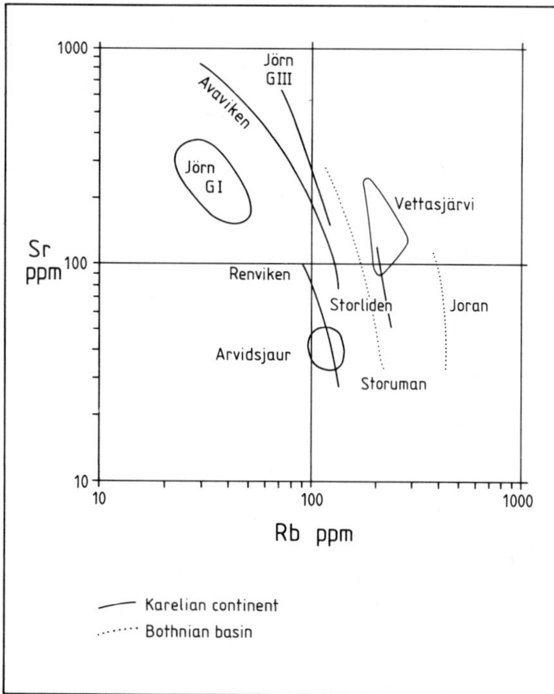

Fig. 9. Rb v. Sr. The outer (GI) and inner (GIII) zones of the Jörn granitoid complex are distinguished.

collision" and "within-plate" fields. These are reasonable positions for granites with major components of crustal origin, derived from a variety of sources.

Rb increases and Sr decreases during the differentiation process. A plot of Rb v. Sr (Figure 9) shows both trends within each suite and systematic differences between suites. The Jörn granitoid complex is distinctive for its high Sr and low Rb, which fits its calcic chemistry. Avaviken has high Sr but intermediate Rb, while Storliden and Arvidsjaur have low Sr and intermediate Rb contents. The Vettasjärvi and Storuman granites have low Sr and high Rb contents.

Integration of Geochemical and Isotopic Data

The older granitoids all appear to have evolved rapidly from sources tapping mantle that had been depleted in LREE over a long period of time. The isotopic evidence for rapid evolution strongly supports the use of trace elements to distinguish tectonic environment. The older granitoids fall into three groups:

(1) The calcic Jörn granitoid complex, within the Skellefte volcanic arc environment, with characteristic low Rb, Y, Nb, Ta and Yb, and high Sr, similar to immature volcanic arc suites.

(2) The calc-alkalic Haparanda suite with Y and Nb contents similar to those of Phanerozoic volcanic arcs.

(3) The alkali-calcic to alkalic granitoids of the Arvidsjaur district with trace element contents similar to those of Phanerozoic "within-plate" suites and Australian "A-type" granites. The epsilon Nd values for these granites demand a short crustal residence time and suggest that the high HFS element contents are more likely to be derived from a "within-plate" mantle source [Brown et al., 1984] rather than from remelting of a crustal source [Collins et al., 1982]. Furthermore the Nd data demands that the source was LREE depleted over a long period. The enrichment of the mantle source must have taken place shortly before, or in connection with the partial melting of the sub-continental lithosphere.

The younger granites have Nd and O isotopic characteristics indicationg a significant crustal history and appear to be derived from sources isotopically similar to rocks in their immediate vicinities. This is supported by their Rb, Sr, Ga, Y and Nb contents. Thus the marine sediments and basalts of the Bothnian basin could supply major source components to the Storuman granites, the Archaean basement and Haparanda-type igneous rocks be sources for the Vettasjärvi granite and the igneous suites of the Arvidsjaur district be sources for the Arvidsjaur granite.

Tectonic Implications

Comparison of the geochemistry of these Proterozoic granites with that of Phanerozoic granites from known tectonic settings, allows the erection of an internally consistent model for the tectonic evolution of northern Sweden, illustrated in Figure 10. However there may well be alternative explanations for the observed geochemical patterns and the model must be regarded as a preliminary hypothesis, especially in the light of possible difficulties of applying Phanerozoic discriminants to Proterozoic rocks.

We suggest that in the Skellefte district, the chemistry of the early granitoid suites together with the palaeogeography could be related to a northerly directed subduction of oceanic crust, the Jörn granitoid complex and the Skellefte district volcanites representing an immature arc developed on a relatively thin continental wedge. The early granitoids of the Arvidsjaur district are geochemically similar to the type of rift magmatism observed along continental margins, inland from the subduction-generated calc-alkalic magmatism. The close relationship of these alkali granitoids to an important zone of NNE-SSW faulting (Figure 2) may be genetically significant: this may be an old transverse break in the belt. The subalkaline Avaviken suite is about 50 Ma younger than the calcic Jörn complex, suggesting that there may be an evolution in time, and not only in space.

The petrogenetic controls on the Haparanda

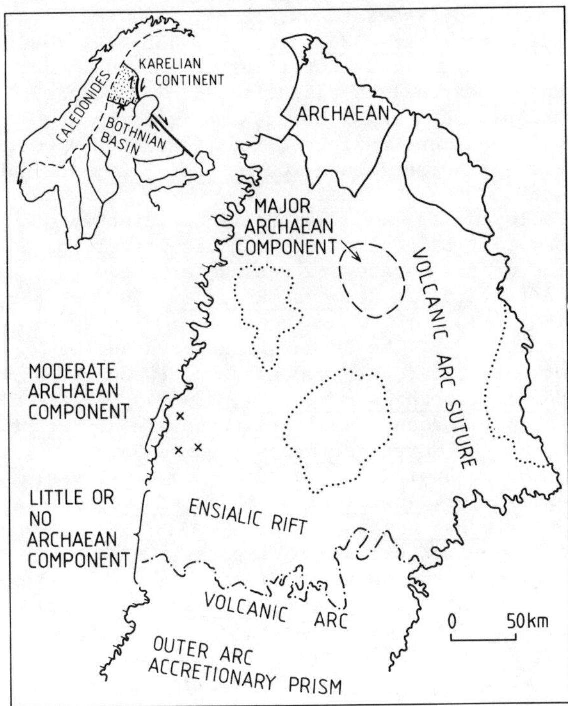

Fig. 10. Sketch to summarise the geodynamic interpretation of the isotopic and geochemical data.

suite are less well constrained. Geochemically the samples are similar to Phanerozoic volcanic-arc granites and suggest a slightly more mature volcanic arc encironment than for Jörn. Further work must critically examine the possibility that the Haparanda suite is subduction-related and not [as suggested by Witschard, 1984] related to rifting of continental crust. The many separate intrusions of Haparanda type must be studied, since the distribution of "volcanic-arc" and "within-plate" types could be a key to the tectonic evolution of northernmost Sweden. Witschard suggests a totally ensialic model for the evolution of the Karelian continent, with relatively little movement between the Archaean blocks. An alternative model might suggest a mosaic of essentially allochthonous terranes of older crust which can have suffered considerable relative movement along transform faults before being assembled into a craton. Such models are under discussion for many parts of the world [see Kerr, 1983] and there is no reason why they should not apply to the Baltic shield. For example the zone of intrusions of Haparanda type seems to form a continuation of the shear-fault dominated Raahe-Lagoda zone in Finland which separates crustal provinces of very different character [Hietanen, 1975, Figure 3, Rickard 1979].

The younger granites have a significant crustal history. At this stage, we do not know whether or not they were accompanied by mantle-derived magmas, or whether they represent a magmatism entirely initiated in the crust. Witschard [1984] suggests that the younger granites relate to the thermal influence of mantle-derived magmas, intruded successively from 1.9 to 1.75 Ga. At present the only evidence for mantle magmas younger than 1.84 Ga is a U-Pb age of 1.8 Ga on zircons from a granite in the contact zone of the Dundret gabbro massif [Skiöld, pers. comm. in Witschard, 1984].

An alternative hypothesis is that the younger granites are the products of a continental collision in the period 1.80-1.75 Ga, which concluded a long period of subduction of ocean floor. This model would not exclude the possibility that mantle-derived magmatism continued down to 1.80 Ga, immediately prior to collision.

Conclusions

The magmatic evolution of northern Sweden between 1.90 and 1.75 Ga can be divided into two dominant stages:

(1) The early stage comprises essentially mantle-derived magmas leading to granitoid suites with generally wide compositional range, which are subdivided into three magma series (a) calcic magmas which could be directly related to subduction, (b) calc-alkalic magmas also similar to those of mature volcanic-arc environments, and (c) subalkaline to alkaline magmas similar to those found in tensional or transverse faulting environments behind a destructive plate margin.

(2) The late stage comprises magmas of more restricted compositional range which are derived dominantly from crustal sources of possibly local provenance and including (a) Archaean sources in the far north, (b) Proterozoic sedimentary sources in the Bothnian Basin.

We therefore suggest that at least part of northern Sweden may have been formed through a typical "Wilson cycle" with subduction of ocean crust under a continental margin giving a volcanic belt with sulphide deposits south of which occurs an accretionary wedge with greywackes and slices of ocean floor. Continental collision at around 1.80-1.75 Ga resulted in imbrication of the sedimentary prism, the depression and metamorphism of deeper parts resulting in extensive anatexis with minimal involvement of new mantle-derived magma.

The Haparanda suite occurs as a series of separate intrusions along a mobile belt within the early Proterozoic continental craton of northernmost Sweden. Although previously regarded as rift-related, it does not show the Nb and Y enrichment normally associated with granitoids formed in a within-plate rifting environment. We speculate that the suite may represent a volcanic arc magmatism, related to the subduction of an ocean floor that has now disappeared due to continental collision. This mobile belt may therefore represent an early Proterozoic suture.

Acknowledgements. This contribution to the International Lithosphere Project is financed by the Swedish Natural Science Research Council (NFR) and the Swedish Geological Company. We also thank CRNS for financing MRW's exchange visit to CREGU. SURRC is supported by the U.K. Natural Environmental Research Council and the Scottish Universities. We thank Dr Phil Potts and Dr John Watson of the UK Open University for trace element analyses.

References

Åberg, G., L. Aguirre, B. Levi, and J.O. Nyström, Spreading-subsidence and generation of ensialic marginal basins: an example from the early Cretaceous of central Chile, in Marginal basin geology: volcanic and associated sedimentary and tectonic processes in modern and ancient marginal basins, edited by B.P. Kokelaar, and M.S. Howells, Special Publication Geological Society of London, 1984.

Adamek, P. M. and M. R. Wilson, Recognition of a new uranium province from the Precambrian of Sweden, in Recognition and Evaluation of Uraniferous Areas. IAEA-TC-25-26, 199-215, IAEA, Vienna, 1977.

Adamek, P. M. and M. R. Wilson, The evoultion of a uranium province in northern Sweden, Phil. Trans. Royal Soc. London, A 291, 335-368, 1979.

Bowden, P., and D. C. Turner, Peralkaline and associated ring-complexes in the Nigerian-Niger province, West Africa, in The Alkaline Rocks, edited by H. Sörensen, pp. 330-351, Wiley, 1974.

Brown, G. C., Calc-alkaline intrusive rocks: their diversity, evolution and relation to volcanic arcs, in Andesites, edited by R. S. Thorpe, pp. 437-461, Wiley, 1982.

Brown, G. C., R. S. Thorpe, and P. C. Webb, The geochemical characteristics of granitoids in contrasting arcs and comments on magma sources, J. Geol. Soc. London, 141, 413-426, 1984.

Chappell, V.W., and A. J. R. White, Two contrasting granite types, Pacific Geology, 8, 173-174, 1974.

Claesson, L.-Å., Stratigraphy and petrochemistry of the ore-bearing Skellefte group volcanites, Geol. Fören. Stockholm Förhand., 104, 378-378, 1982.

Claesson, L.-Å., The geochemistry of early Proterozoic volcanic rocks hosting massive sulphide deposits in the Skellefte district, north Sweden, J. Geol. Soc. London, in press.

Collins, W. J., S. D. Beams, A. J. R. White, and B. W. Chappel, Nature and origin of A-type granites with particular reference to southeastern Australia, Contr. Mineral. Petrol., 80, 189-200, 1982.

Debon, F., and P. Le Fort, A chemical-mineralogical classification of common plutonic rocks and associations, Trans. Roy. Soc. Edinburgh: Earth Sciences, 73, 135-149, 1982.

De la Roche, H., J. Leterrier, P. Grandclaude, and M. Marchal, A classification of volcanic and plutonic rocks using R_1-R_2 diagram and major element analyses; its relationships with current nomenclature, Chem. Geol., 29, 183-210, 1980.

DePaolo, D. J., A Nd and Sr isotopic study of Mesozoic calc-alkaline granitic batholiths of the Sierra Nevada and Peninsular Ranges, California, J. Geophys. Res., 86, 10470-10488, 1981.

Gulson, B. L., The Precambrian geochronology of granitic rocks from northern Sweden, Geol. Fören. Stockholm Förh., 94, 229-244, 1972.

Grip, E., Sweden, in Mineral Deposits of Europe, Volume 1: Northwest Europe, edited by S. H. U. Bowie, A. Kvalheim, and H. W. Haslam, pp. 93-198, Institution of Mining and Metallurgy, London, 1978.

Hietanen, A., Generation of potassium-poor magmas in the northern Sierra Nevada and the Svecofennian of Finland, J. Res. U. S. Geol. Surv., 3, 631-645, 1975.

Kerr, R. A., Suspect terranes and continental growth, Science, 222, 36-38, 1983.

Lameyre, J., and P. Bowden, Plutonic rock types series: discrimination of various granitoid series and related rocks, in Magmatology, edited by R. Brousse and J. Lameyre, J. Volcan. Geothermal Res., 14, 169-186, 1982.

Levi, B., and L. Aguirre, Ensialic spreading-subsidence in the Mesozoic and Palaeogene Andes of central Chile, J. Geol. Soc. London, 138, 75-81, 1981.

Lundberg, B., Aspects of the geology of the Skellefte field, northern Sweden, Geol. Fören. Stockholm Förhand., 102, 156-166, 1980.

Mitchell, A. M. G., and J. D. Bell, Island arc evolution and related mineral deposits, J. Geol., 81, 381-405, 1973.

Öhlander, B., Geochemical analyses of rocks of the Haparanda suite, northern Sweden, Geol. Fören. Stockholm Förhand., 106, 167-169, 1984.

Öhlander, B., and D. H. Nisca, Tectonic control of Precambrian molybdenite mineralization in northern Sweden, Econ. Geol., 80, 505-512, 1985.

Öhlander, B., P. J. Hamilton, and M. R. Wilson, Crustal reactivation in northern Sweden: The Vettasjärvi granite, Precambrian Res., in press.

Patchett, P. J. and D. Bridgewater, Origin of continental crust of 1.9 - 1.7 Ga age defined by Nd isotopes in the Ketilidian terrain of south Greenland, Contrib. Mineral. Petrol., 87, 311-318, 1984.

Peacock, M. A., Classification of igneous rock series, J. Geol., 39, 54-67, 1931.

Pearce, J. A., Trace element characteristics of lavas from destructive plate boundaries, in Andesites, edited by R. S. Thorpe, pp. 525-548, Wiley, 1982.

Pearce, J. A., N. B. W. Harris, and A. G. Tindle,
Trace element discrimination diagrams for the
tectonic interpretation of granitic rocks.
J. Petrol., 25, 956-983, 1984.

Pharoah, T. C., and J. A. Pearce, Geochemical
evidence for the geotectonic setting of early
Proterozoic metavolcanic sequences in Lapland,
Precambrian Res., 25, 283-308, 1984.

Rickard, D. T., and H. Zweifel, Genesis of
Precambrian sulphide ores, Skellefte district.
Sweden, Econ. Geol., 70, 255-274, 1975.

Rickard, D. T., Scandinavian Metallogenesis,
GeoJournal, 3.3, 235-252, 1979.

Skiöld, T., U-Pb zircon and Rb-Sr whole rock and
mineral ages of Proterozoic intrusives on map
sheet Lannavaara, northeastern Sweden, Geol.
Fören. Stockholm Förhand., 101, 131-137, 1979a.

Skiöld, T., Zircon ages from an Archaean gneiss
province in northern Sweden, Geol. Fören.
Stockholm Förhand., 101, 169-171, 1979b.

Skiöld, T., Radiometric ages of plutonic and
hypabyssal rocks from the Vittangi-Karesuando
area, northern Sweden, Geol. Fören. Stockholm
Förhand., 103, 316-328, 1981a.

Skiöld, T., Chronostratigraphic correlations
based on radiometric data from northern Sweden,
Geol. Fören. Stockholm Förhand., 103, 523-524,
1981b.

Walser, G., and Ö. Einarsson, The geological
context of molybdenum occurrences in the
southern Norrbotten region, northern Sweden,
Geol. Rundschau, 71, 213-229, 1982.

Weaver, B. L., and J. Tarney, Chemistry of the
sub-continental mantle: inference from
Archaean and Proterozoic dykes and continental
flood basalts, in Continental Basalts and
Mantle Xenoliths, edited by C. J. Hawkesworth,
and M. J. Norry, pp. 204-229, Shiva, Nantwich,
Cheshire, 1983.

Welin, E., K. Christiansson, and Ö. Nilsson, Rb-
Sr age dating of intrusive rocks of the
Haparanda suite, Geol. Fören. Stockholm
Förhand., 92, 336-346, 1970.

Welin, E., K. Christiansson, and Ö. Nilsson, Rb-
Sr radiometric ages of extrusive and intrusive
rocks in northern Sweden. I, Sveriges geol.
undersökning, C 666, 1-38, 1971.

Welin, E., Ö. Einarsson, B. Gustafsson,
R. Lindberg, K. Christiansson, G. Johansson,
and Ö. Nilsson, Radiometric ages of intrusive
rocks in northern Sweden. II, Sveriges Geol.
undersökning, C 731, 1-21, 1977.

Wilson, M. R., Granite types in Sweden, Geol.
Fören. Stockholm Förhand., 102, 167-176, 1980.

Wilson, M. R., P. J. Hamilton, A. E. Fallick,
M. Aftation, and A. Michard, Granites and
early Proterozoic crustal evolution in Sweden:
evidence from Sm-Nd, U-Pb and O isotope systematics, Earth Planet. Sci. Lett., 72,
376-388, 1985.

Witschard, F., The geological and tectonic
evolution of the Precambrian of northern Sweden
- a case for basement reactivation?,
Precambrian Res., 13, 273-315, 1984.

COMPARATIVE CHARACTERISTICS OF THE LITHOSPHERE OF THE
RUSSIAN PLATFORM, THE WEST SIBERIAN PLATFORM AND THE SIBERIAN
PLATFORM FROM SEISMIC OBSERVATIONS ON LONG-RANGE PROFILES

Ju.A. Burmakov,[1] N.M. Chernyshev,[2] L.P. Vinnik,[1] and A.V. Yegorkin[2]

[1]Institute of Physics of the Earth, Academy of Sciences of the USSR,
10 Bolshaya Gruzinskaya, 123-242 Moscow, USSR

[2]Ministry of Geology of the USSR, Moscow, USSR

Abstract. The paper deals with seismic observations on long-range profiles crossing the Precambrian Russian platform, the West-Siberian platform and the Siberian platform. The observations were carried out on reversed and overlapping profiles facilitating 2-dimensional mapping of the lithosphere. The 2D-velocity cross-sections of the mantle in Siberia were obtained by a linearized inversion of the observed travel-times of the refracted P-waves. The main results of this study are as follows. The velocity-depth distribution in the mantle of the central part of the Siberian platform incorporates some high- and low-velocity layers of very large extent. In the north of the Siberian platform, the subcrustal lithosphere is composed of a series of thin high- and low-velocity layers dipping from the Anabar shield in the east to the Tunguska syneclise in the west. This observation implies that the horizontal layering was formed prior to the subsidence of the Tunguska syneclise (earlier than the Upper Paleozoic) while the development of the syneclise was due to the subsidence of its subcrustal lithosphere. At the southwestern margin of the Siberian platform, a large body with extremely high P-velocity (8.6 km/s at 80 km depth) is found close to the Baikal rift zone. One of the most pronounced discontinuities in the upper mantle of Siberia changes its depths from 270 km beneath the Siberian platform to 200 km beneath the West-Siberian platform. This discontinuity is interpreted as a boundary separating the upper layer of the mantle which is depleted in the low melting-point components and the underlying undepleted mantle. The P-velocity in the uppermost mantle of West-Siberia is anomalously low, being related to the rift zone which was active at the turn of Paleozoic to Mesozoic times. The velocity-depth distribution in the upper mantle of the Russian platform is different from that of the Siberian platform.

Introduction

The observations of refracted and reflected P-waves on long-range profiles provide an opportunity to study the structure of the mantle to a depth of several hundred kilometers. In this paper, we present some results of experiments of this kind that were carried out in the Soviet Union. The profiles (some of them more than 3000 km long) cross the northeastern part of the Precambrian East-European platform (Russian platform), the West-Siberian platform, whose basement is of Paleozoic age, and the Precambrian Siberian platform (Fig. 1).

The bulk of the previously available seismic data on the structure of the subcrustal lithosphere of an old platform was obtained from the Early Rise experiment that was carried out in North America 20 years ago (Roller and Jackson, 1966; Mereu and Hunter, 1969; Iyer et al., 1969; Masse, 1973). In that experiment, underwater explosions in Lake Superior were recorded on several radial long-range profiles. It is now well understood that the lithosphere is usually heterogeneous in 3 dimensions, and in order to resolve its structure more complicated observational systems are required. A drawback of the Early Rise experiment in comparison with the recent Soviet experiments lies in the absence of reversed and overlapping profiles that are necessary for a seismic study of a laterally heterogeneous medium. The station spacings in the Early Rise experiment were somewhat too large while the signal-to-noise ratio at large epicentral distances was often too low.

Although, in our view, the most important data obtained in the USSR from the long-range experiments are those related to the mantle, we start from the short generalized description of the crustal structure in Siberia (section 2). In section 3, we discuss the wave fields that were

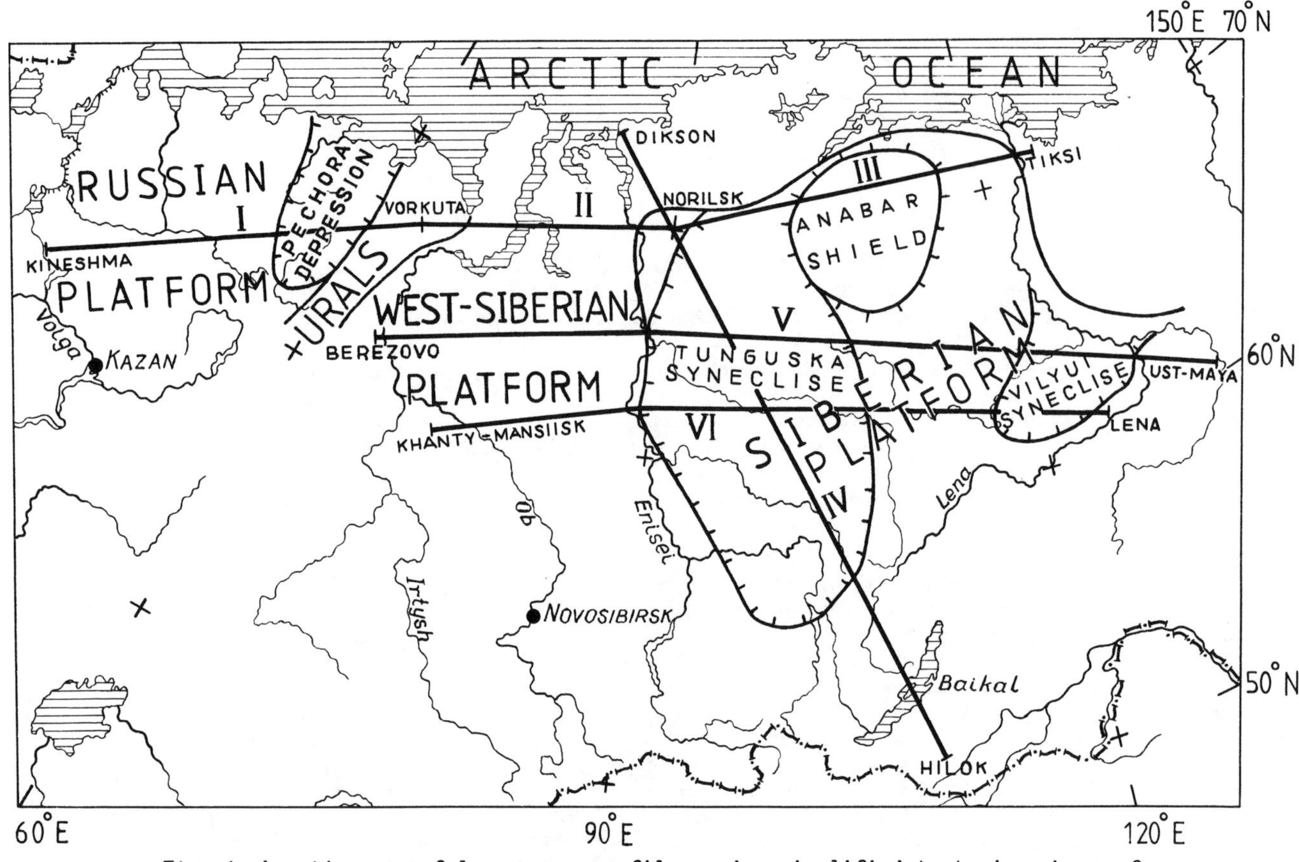

Fig. 1 Location map of long-range profiles and a simplified tectonic scheme of Siberia.

observed on the long-range profiles and, for some profiles, present the corresponding velocity-depth models. In Siberia, the seismic observations indicate a pronounced lateral heterogeneity. In section 4, we outline the method of the determination of the two-dimensional velocity structure from the travel-time data and present the two-dimensional models of the lithosphere in Siberia. Finally, we summarize and discuss our results (section 5).

2. Crustal Structure in Siberia

For detailed crustal studies the powerful explosions whose waves could be recorded very far away from the shot-points were usually complemented in Siberia by a number of more densely spaced, relatively small explosions. The observations on the long-range profiles were carried out with three-component receiver units which also facilitated the detection of the converted phases (e.g. P-S). The data on the crustal structure in Siberia were obtained by combining observations of the refracted, reflected and converted waves.

Fig. 2 shows the relationship between the thickness of the sedimentary layer and the total thickness of the crust. For the West-Siberian and the Siberian platform this relationship is qualitatively the same: a thick sedimentary layer is usually combined with a relatively thin crystalline crust while a thin sedimentary cover corresponds to a thick crystalline layer. At the same time, the average thickness of the crust of the Siberian platform exceeds that of the West-Siberian platform by 5 km. This difference is mainly due to the increased thickness of the crystalline layer of the crust of the Siberian platform. It should be noted that the structure of the crust at the margins of both tectonic units sometimes deviates from the above mentioned regularity. For instance, the crustal structure of the Vilui syneclise of the Siberian platform is close to that of the West-Siberian platform.

The crust of both regions can be divided very roughly into 4 layers with P- velocities ranging from less than 6 km/s in the sediments to 6.1-6.4 km/s in the upper crystalline layer, 6.5-6.8 km/s in the middle layer and 6.8-7.3 km/s in the lower layer. Fig. 3 shows a noteworthy relationship between the thickness of the layer I with the "granitic" velocity and the total thickness of the

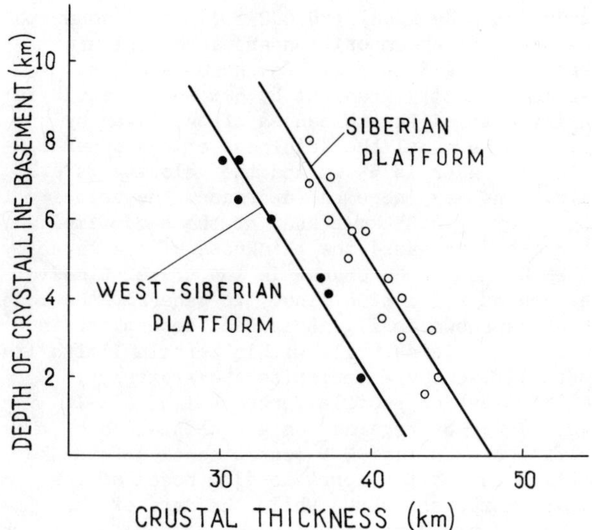

Fig. 2 Relationship between the thickness of
sediments and the crustal thickness in Siberia.
Each point corresponds to a specific tectonic unit
several hundred kilometers in extent.

crystalline crust. There are also variations in
the thickness of the second and third layers but
they are relatively less significant.

The ratio of the average velocities of the P-
and S-waves in the lower crust is somewhat
different for the West-Siberian and the Siberian
platform. Moreover, the dependence of this ratio
on the thickness of the crystalline crust is
markedly different for the two structural units
(Fig. 4). These differences may imply some
compositional differences between the lower crust
of the Siberian and that of the West-Siberian
platform.

After this short review of the most general
features of the crustal structure in Siberia, we

proceed to the upper mantle structures that were
revealed by the long-range experiments.

3. Wave Fields and the Velocity-Depth
Distributions in the Mantle

The structure of the upper mantle of the
Russian platform is illustrated mainly by data
from the profile Kineshma-Vorkuta, 1400 km long
(Profile I, Fig. 1). A relatively small
northeastern segment of the profile crosses the
Pechora depression which does not belong to the
Russian platform although its basement is of
Precambrian age. The records of reasonably high
quality were obtained on the main profile from the
explosion near the southwestern tip of the profile
and on the overlapping profile. The record
section of the main profile is shown in Fig. 5
while that of the overlapping profile can be found
in Vinnik and Ryaboy (1981) and Vinnik and
Yegorkin (1981).

The first arrivals in the record section in
Fig. 5 can be grouped into three refraction lines.
A first line is observed in the epicentral
distance range from about 200 km to 600 km, the
second in the range from 900 to 1100 km, and the
third in the range from 1200 to 1400 km. Each
line is offset from the previous one by a time
delay thus forming an echelon-type structure. A
similar wave field was observed in the overlapping
profile implying that to a first approximation the
lower lithosphere in the region is laterally
homogeneous. The corresponding velocity-depth
model is plotted in Fig. 6. Every refraction line
corresponds to a certain high-velocity layer while
every time offset signifies the presence of a low-
velocity layer.

A model derived from the seismic refraction
data is never unique. In Fig. 6, the thickness of
the upper layer with the 8.4 km/s velocity depends
on the P-velocity gradient. For a number of
reasons, a reliable determination of the gradient
from the seismic data is difficult. The gradient

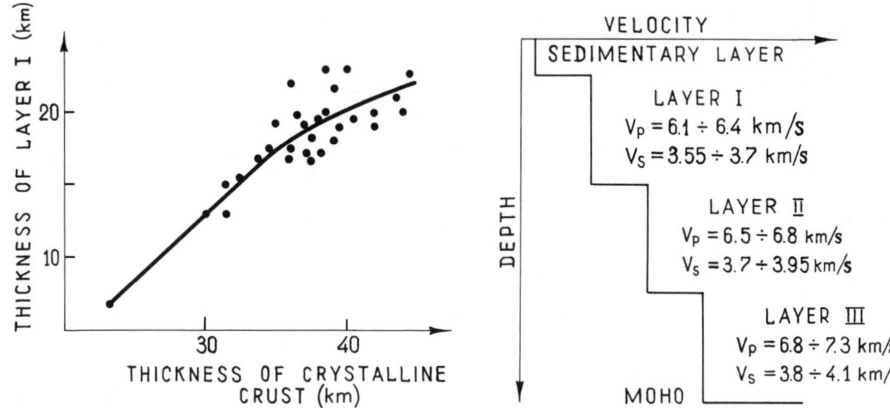

Fig. 3 Schematic representation of the velocity structure of the crust in Siberia and
the relationship between the thickness of the upper crystalline crust.

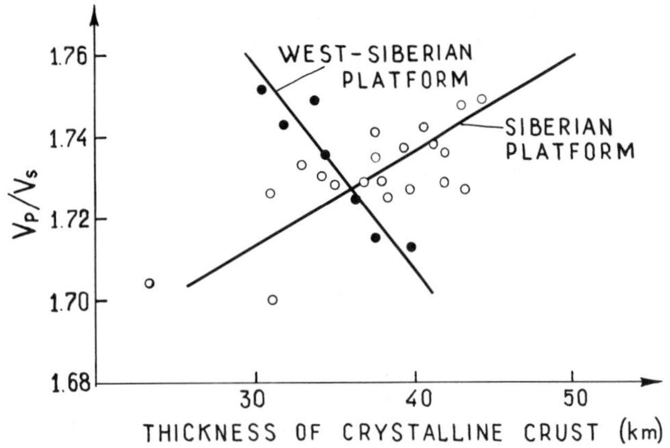

Fig. 4 Dependence of the ratio of the average P- and S-velocities in the crystalline crust on its thickness. Each point corresponds to a specific tectonic unit several hundred kilometers in extent.

assumed in the model ($\sim$0.002 s^{-1}) corresponds to the peridotite composition and a temperature gradient of 2-3 deg/km. An increase of the assumed velocity gradient by 50% would result in an increase of the thickness of the layer by $\sim$5 km. In the model the thickness of the upper low-velocity layer is 53 km and the velocity is 8.27 km/s. One may increase (decrease) the velocity in the layer by 0.05 km/s and, at the same time, decrease (increase) the thickness of the layer by 10 km without much change in the travel times of the second refraction line. In general, the model which is shown in Fig. 6 is well determined in qualitative terms while within certain limits its quantitative characteristics are arbitrary.

The Siberian profile Vorkuta-Tiksi (II-III in Fig. 1) may be regarded as a continuation of the profile I. In Fig. 7 a record section from the profile II and the corresponding model of the upper mantle are shown while the data of the profile I are presented for comparison. The first arrivals in profile II can be well approximated by

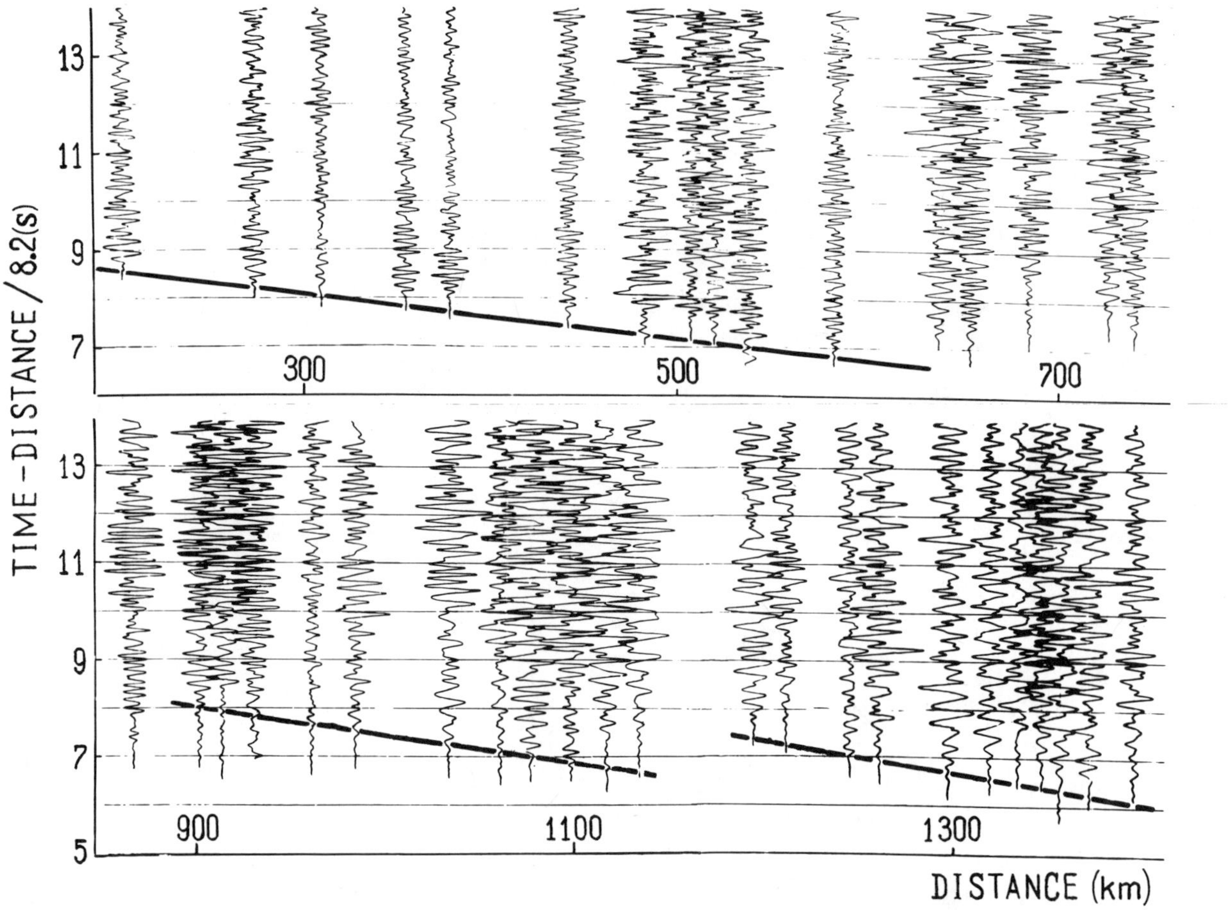

Fig. 5 Record section of the profile Kineshma-Vorkuta (I in Fig. 1). The records in the range from 750 km to 850 km are missing.

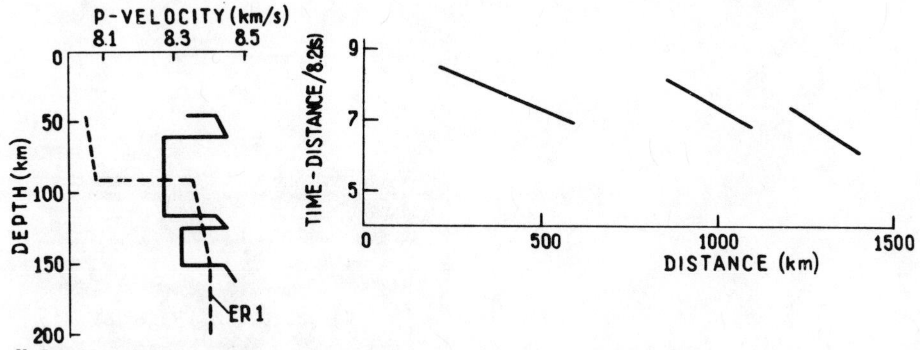

Fig. 6 Velocity-depth model of the upper mantle for the profile I, Kineshma-Vorkuta and the corresponding refraction lines. Model ER1 (Green and Hales, 1969) is shown for domparison (dashed line).

2 lines of correlation, the second line being delayed relative to the first by ~3 s. This delay corresponds to the well-developed low-velocity layer in the depth range between 80 km and 150 km. The differences between the travel times for the Russian plate and the West-Siberian platform are much larger than the possible observational errors. The same is true with respect to the depth-velocity models.

The observational system for profile III, Norilsk-Tiksi crossing the Siberian platform in the North consisted of two reversed profiles 1500 km long. The record sections and the travel-times of the identified refracted phases are shown in Figs. 8 and 9 respectively. The shot-point termed Norilsk was located near the point of intersection of the profiles III and IV (Fig. 1).

In the profile Norilsk-Tiksi, the wave with the apparent velocity of 8.2 km/s is clearly visible as the first arrival in the distance range from 180 km to 440 km. The records in the interval from 440 km to 600 km are missing, but after this gap a phase with a higher apparent velocity appears in the first arrivals. In the range from 700 km to 1200 km, the apparent velocity of the first arrivals returns to the value of 8.2 km/s.

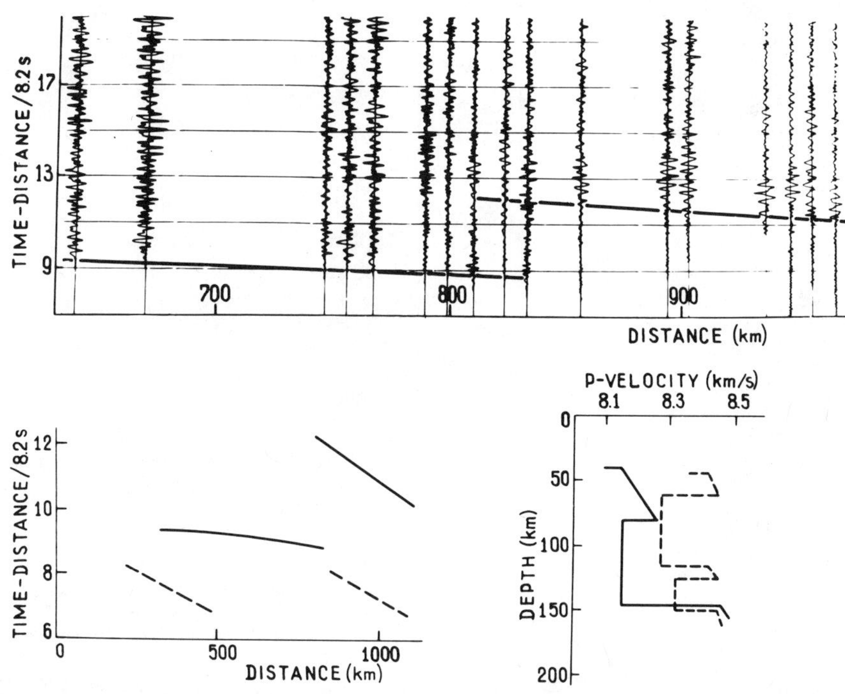

Fig. 7 Record section of profile II, the depth-velocity model for the upper mantle of the West-Siberian plate and the corresponding refraction lines. The data of the Kineshma-Vorkuta profile are shown for comparison (dashed lines).

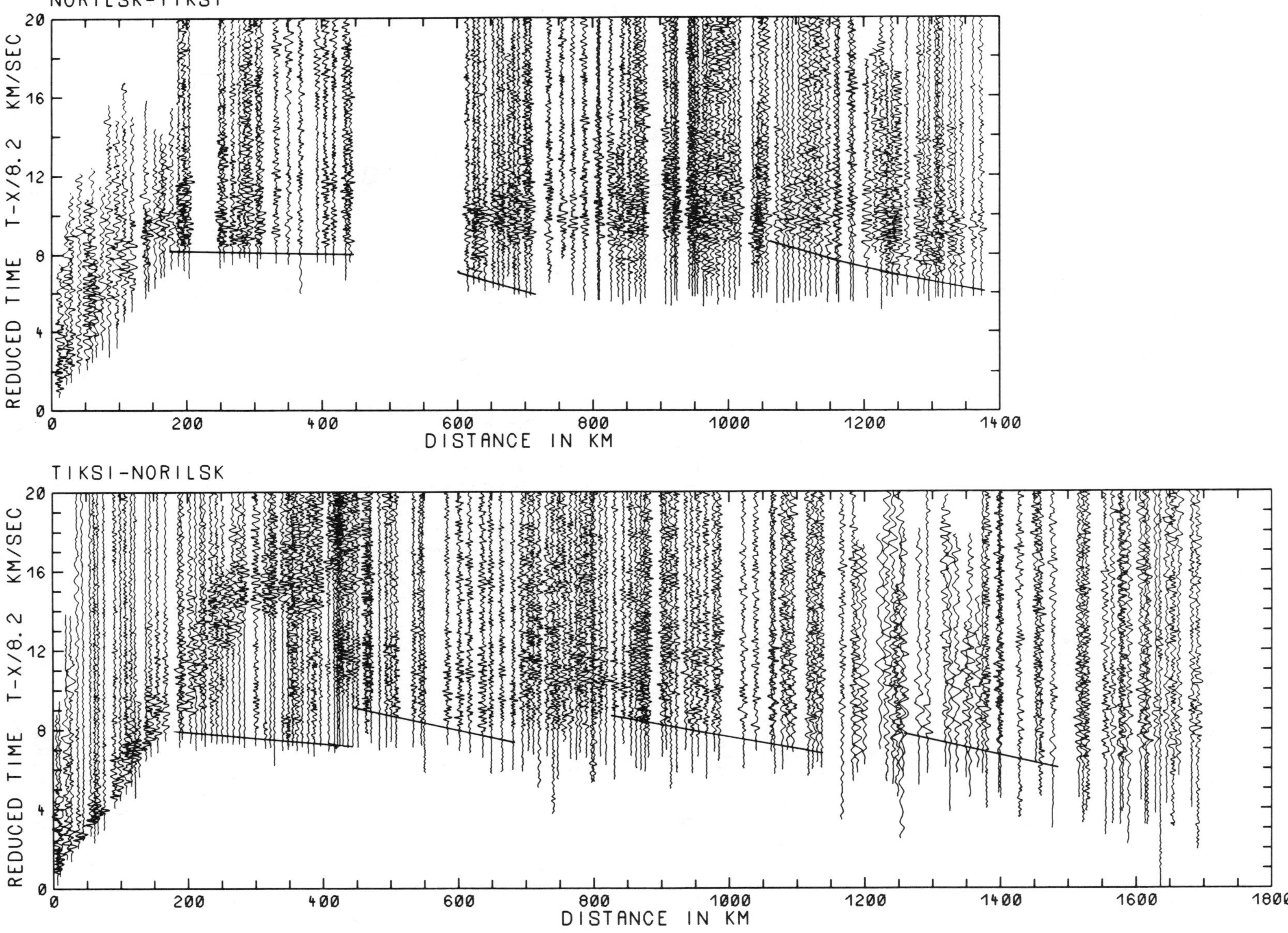

Fig. 8   Record section of profile III.

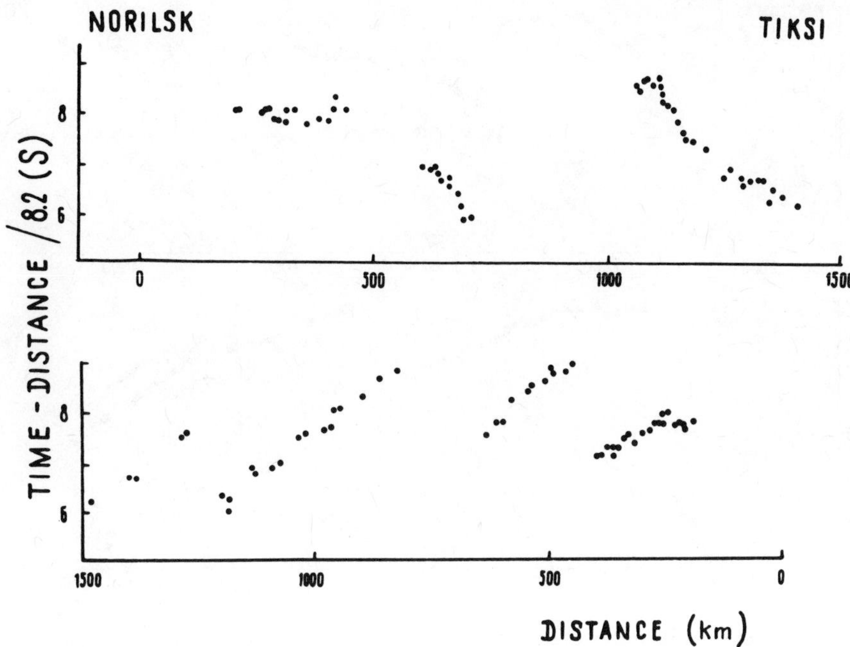

Fig. 9 Travel-times of refracted phases for profile III.

These arrivals are most likely formed by random wave scattering since the 8.2 km/s velocity is too low to be appropriate for any refracted wave. A new phase emerges in late arrivals at a distance of 1050 km and can be correlated with much confidence to the end of the profile. There are some other bursts of energy that are especially strong in late arrivals between 600 km and 1000 km but a continuous correlation of them seems impossible.

In the reversed profile (Tiksi-Norilsk), a similar analysis of the record section reveals the echelon system of the refraction lines. The correlation of the phases at distances less than 1200 km is fairly reliable while at larger distances the wave field becomes more irregular. A comparison of the data of the reversed observations (Fig. 9) shows that they are strongly different. The implication of this is that the velocity structure of the lithosphere is variable along the profile. For this reason, no attempt was made to construct a velocity-depth model.

The next profile termed Dikson-Nilok (IV in Fig. 1) crosses the Tunguska syneclise of the Siberian platform and the Baikal rift zone in a submeridional direction. The seismic waves whose travel times are shown in Fig. 10 were generated by 4 explosions. The shot-point IV was located near the point of intersection of the profiles III and IV. Unfortunately, we are not able to present the record sections, although that of the explosion I without a phase correlation has been published elsewhere (e.g. Yegorkin and Pavlenkova, 1981).

The data shown in Fig. 10, like those of the profile III, provide strong evidence for a lateral heterogeneity along the profile. For instance, in the wave field of the explosion I there is a broad shadow zone in the distance range from 800 km to 1500 km while this peculiarity is absent in the reversed observations (data of explosion IV).

Profile Berezovo-Ust-Maya (V in Fig. 1) is the longest of the existing long range profiles. It crosses the West-Siberian platform in the west and the Siberian platform in the east. The record section of explosion II which was fired near the border between the West-Siberian and the Siberian platform is presented in Fig. 11. To the east of the shot-point, one can recognize in the first arrivals: a strong crustal wave with an apparent velocity of 6.0 km/s (0-200 km), the mantle wave with an apparent velocity of 8.2 km/s (200-700 km) and a phase with an apparent velocity of 8.5 km/s (700-1000 km). The mantle phase with an apparent velocity of 8.65 km/s emerges in late arrivals at the distance of 1000 km and is then recorded as the first arrival to the distance of 1400 km. The interval between approximately 1400 km and 2000 km may be regarded as the shadow zone since the corresponding first arrivals are too weak, slow and irregular. The zone of the irregular first arrivals is followed by well correlatable arrivals with an apparent velocity of 9.0 km/s. The strong phase emerging in late arrivals at 1600 km and coming close to the first arrivals near the end of the record section is associated with the well-known 400-km discontinuity which, due to its large depth, is beyond the scope of this paper.

The record section to the west of the shot-

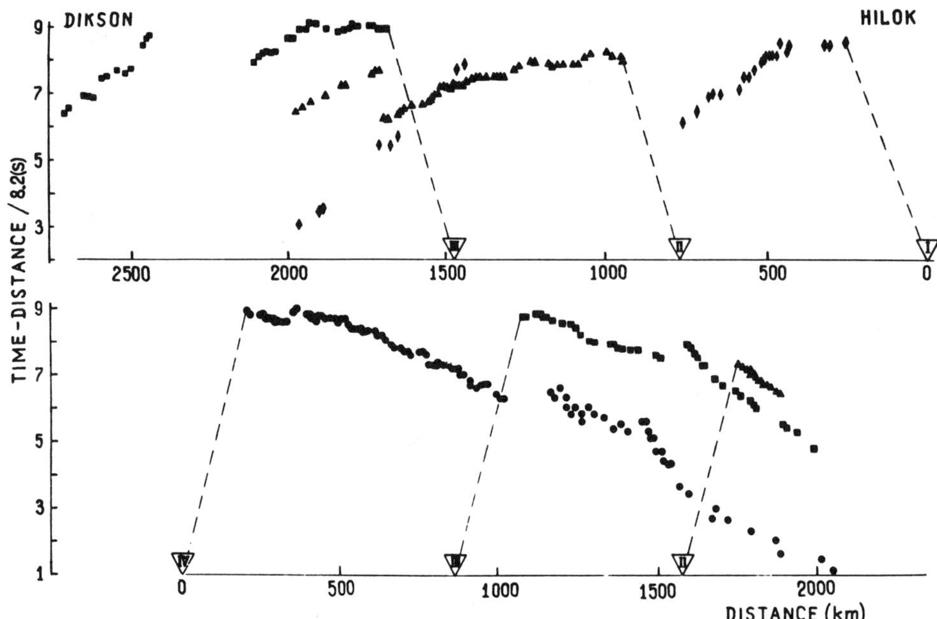

Fig. 10   Travel-times of refracted phases from profile IV, Dikson-Hilok.   The shot-
points are shown as triangles.   The travel-times corresponding to different shot-points
are shown by different symbols.

point is quite different from the eastern one.
Crustal waves, very strong in the eastern part of
the profile, are practically absent from late
arrivals at the distances greater than 200 km.
The implication of this is that the crust-mantle
transition in the West is a smooth zone rather
than a sharp boundary.   The first-arriving mantle
phase in this distance interval is much slower
than that in the eastern part.   The apparent
velocity of the first arrivals increases sharply
in the distance interval from 700 to 900 km but
later returns to the lower values which are
characteristic of shadow zones.

The comparison of the two record sections
testifies to the difference of the structure of
the lithosphere of the West-Siberian plate and
that of the Siberian platform.   On the other hand,
the sequence of phases found in the eastern part
of the record section of the explosion II can be
recognized in every other travel-time data set
(Fig. 12).   This remarkable stability implies
that, in spite of the lateral variations, the
general structure of the depth-velocity
distribution in the upper mantle remains more or
less the same along the whole length of the
profile.

In order to characterize this structure, we
have constructed the model shown in Fig. 13.   The
travel-times of the refracted waves for this model
are close to those obtained from the eastern part
of the record section in Fig. 11.   In other words,
the model is supposed to represent the depth-
velocity distribution in the mantle of the
Siberian platform.   The high-velocity layer at the

100-km depth is responsible for the refraction
line observed in the distance range from 700 to
1000 km while the layer at the 150-km depth
corresponds to the refraction line observed in the
interval from 100 to 1400 km.   The low-velocity
layer which is sandwiched between the two high-
velocity layers accounts for the offset between
the corresponding refraction lines.   The broad
low-velocity layer in the depth interval from 155
to 270 km is responsible for the offset between
the refraction lines in the 100-1400 km and 1900-
2150 km distance ranges.   The boundary at the 270-
km depth corresponds to the latter line.   The
model is based on the abundant and accurate
observational data and there are reasons to
believe that it represents the main features of
the real depth-velocity distribution reasonably
well.   The wave field of the profile VI, Khanty-
Mansilsk-Lena (Fig. 1) is similar to that observed
on the profile V implying that the model in Fig.
13 is representative not only for the profile V
but also for a large area to the south.

4.   2-D Velocity Models of the Lithosphere in Siberia

The travel time data of the profiles III, IV
and V were processed using the technique of
Burmakov et al. (1984).   The lateral velocity
inhomogeneities manifest themselves in travel-time
residuals determined with respect to travel-times
given by laterally homogeneous reference model.
We consider a region in the half-space containing
the family of rays R_i, i = 1, 2, ..., I, of

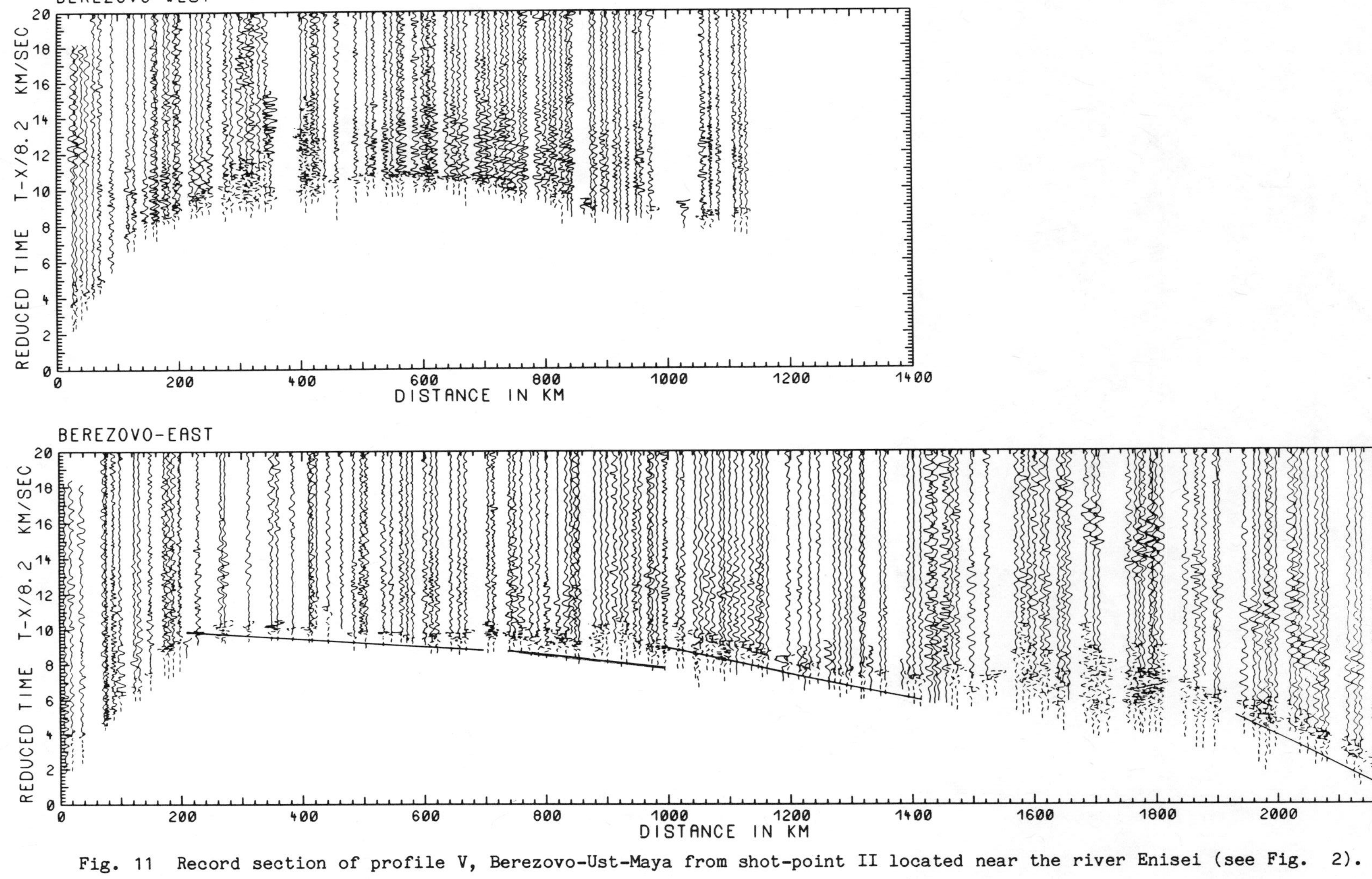

Fig. 11 Record section of profile V, Berezovo-Ust-Maya from shot-point II located near the river Enisei (see Fig. 2).

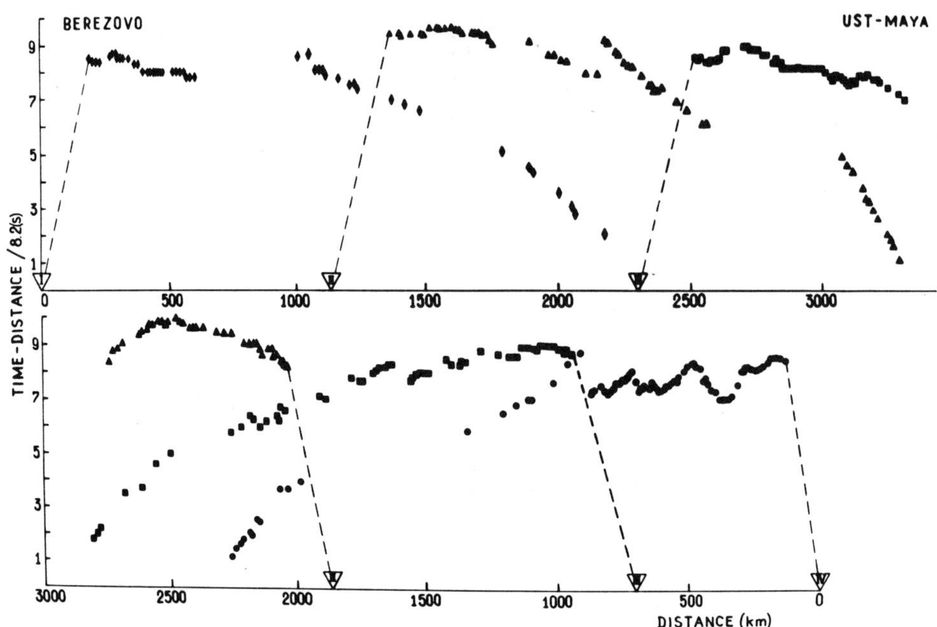

Fig. 12 The same as Fig. 11 but for profile V, Berezovo-Ust-Maya.

refracted body waves. The travel time t_i of a wave propagating along the ray-path R_i is given by

$$t_i = \int_{R_i} S(y,z)dr$$

where $S(y,z)$ is the wave slowness and y and z are the horizontal and the vertical coordinates respectively. Assuming weak lateral heterogeneity, one may write

$$S(y,z) = S_0(z) + S_1(y,z)$$

where S_0 is the unperturbed wave slowness and S_1 is the perturbation which depends on the perturbation of the velocity δV and the unperturbed velocity V_0,

$$S_1 \approx -\,\delta V/V_0^2$$

Then the travel-time residual δt_i can be expressed as

$$\delta t_i \approx \int_{R_{0i}} S_1(y,z)dr \tag{1}$$

where R_{0i} is the raypath corresponding to the reference velocity $V_0(z)$.

$S_1(y,z)$ can be expanded in a two-fold series of orthogonal Chebyshev polynominals $T(u)$:

$$S_1[y(u_1,u_2),z(u_1,u_2)] = \sum_{k=0}^{K-1} \sum_{l=0}^{L-1} P_{kl} T_k(u_1) T_l(u_2) \tag{2}$$

where P_{kl} is the unknown value to be found from the observations. Substituting eq.(2) into eq.(1) we get a set of equations which can be written as

$$A\,X = C \tag{3}$$

where C is the vector of travel-time residuals, X is the vector of unknown elements, and A is rectangular matrix with I rows and M = K L columns.

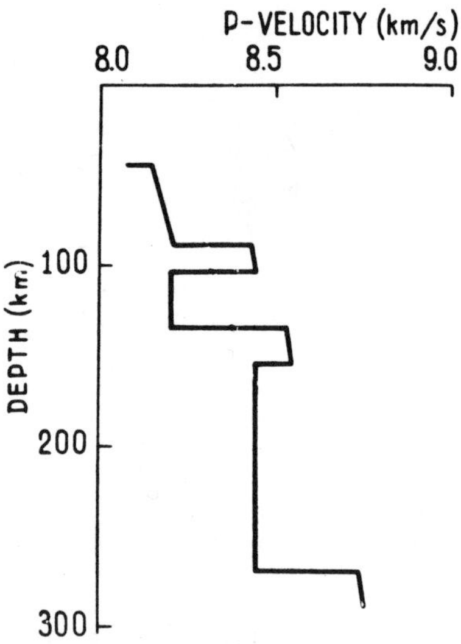

Fig. 13 Velocity-depth model of the upper mantle of the Siberian platform which corresponds to the travel-times for shot-point II of profile V.

184   BURMAKOV ET AL.

The solution of eq.(3) can be found by the method of singular value decomposition (Forsythe, Malcolm and Moler, 1977). Such a solution provides a minimum difference between AX and C in the least-squares sense while at the same time it is relatively insensitive to errors in the input data. A singular value decomposition of the matrix A is given by

$$A = U \cdot D \cdot W^T \qquad (4)$$

where U is an I x I orthogonal matrix, W is an M x M orthogonal matrix and D is an I x M diagonal matrix. The diagonal elements σ_j of the matrix D are called the singular values of A. The matrices U and W are used to transform eq.(3) into an equivalent diagonal set of equations

$$D \, \overline{X} = \overline{C} \qquad (5)$$

The unknown elements $\overline{x}_j$ of the vector $\overline{X}$ can be expressed as

$$\overline{x}_j = \overline{c}_j/\sigma_j \qquad j = 1, \ldots, M. \qquad (6)$$

We assume $\overline{x}_j = 0$ if σ_j. The threshold τ is defined as

$$\tau = \varepsilon \sigma_{max}$$

where ε is the relative error in the input data. Introducing the threshold results in a more reliable determination of X.

To apply this technique to the long-range profiling data, the crust was "stripped off" and the travel-times were recomputed to 40-km depth. The function $V_O = 1/S_O(z)$ was taken to the form

$$V_O(z) = a(1+bz) \qquad (7$$

where z is in km and V_O in km/s. By numerical experiments it was found that the best first order approximation of the depth-velocity distribution could be obtained by assuming a = 8.1 and b = 0.0005 for every region of our study. The values of K and L were assumed to be 7 and thus the number of unknown parameters was 49. The value of ε was taken to be 0.01 on the assumption that the error in the values of the observed travel-time residuals was on the order of 1 per cent. To account for the curvature of the earth surface, the resulting values of V(y,z) were corrected by applying transformations given, for example, by Aki and Richards (1980). The resulting 2-D P-velocity distributions are shown in Figs. 14, 15, and 16. The r.m.s. deviations of the corresponding travel-times from the observed values are reasonably small: 0.34, 0.24, and 0.2 s for the profiles III (Norilsk-Tiski), IV Dikson-Hilok) and V (Berezovo-Ust-Maya), respectively,

The subcrustal lithosphere in the profile Norilsk-Tiksi (Fig. 14) is composed of a series of alternating high- and low-velocity layers dipping to the West. This inclination accounts for a systematic difference in the slopes of the refraction lines in the reversed observational systems (Fig. 9).

The velocity-depth distribution in the Dikson side of the Dikson-Hilok profile may be regarded as normal while the opposite side of the profile is dominated by a large high-velocity body at a shallow depth (Fig. 15). This body manifests itself in the unusually high apparent velocity (8.7 km/s) of the waves that were propagated from shot-point I and II in the northern and the southern directions, respectively (Fig. 10). The underlying low-velocity "tongue" accounts for a broad shadow zone in the wave field of explosion I. Lake Baikal is located over the southern flank of the high-velocity body (Fig. 15). This contradicts other seismic data (e.g. Krilov et al., 1981) indicating relatively low mantle velocities in the Baikal rift zone. The anomalously low velocities in the top of the mantle of the Baikal zone form a strip 200 km wide and this width seems to be somewhat too small to be resolved by the long-range profiling data and/or the available data inversion technique.

The cross-section of the profile Berezovo-Ust-Maya (Fig. 16) shows that the velocities in the mantle of the West-Siberian plate are substantially lower than those of the Siberian platform in the depth range of 50-200 km. One pays for the introduction of the second dimension in the data inversion procedure by a loss of resolution in the first dimension. For this reason, the velocity-depth distribution in the upper mantle of the Siberian platform in Fig. 16 is smoother than in Fig. 13. Nevertheless, the broad low-velocity zone in the depth range of 150-270 km beneath the Siberian platform is equally well pronounced in both models. This zone is absent in the western part of the cross-section in Fig. 16 in agreement with the absence of the pronounced offsets between the corresponding refraction lines in Fig. 12. As a result of this, the 8.6 km/s isoline which corresponds to the 270 km discontinuity in Fig. 16 is at substantially different depths beneath the West-Siberian platform and the Siberian platform.

5. Discussion and Conclusions

The correlations shown in Fig. 2 and 3 may have a direct bearing on the origin and evolution of the continental crust. The systematic difference in the thickness of the crust between the Siberian and West-Siberian platforms (Fig. 2) is probably related to their difference in age. A similar relation can be found, for example, in Europe where the oldest crustal provinces (Svecofennides and Svecokarelides of Fennoscandia) exhibit the greatest crustal thickness (up to 47 km according to Bungum et al., 1980). The negative correlation between the thickness of sediments and that of the

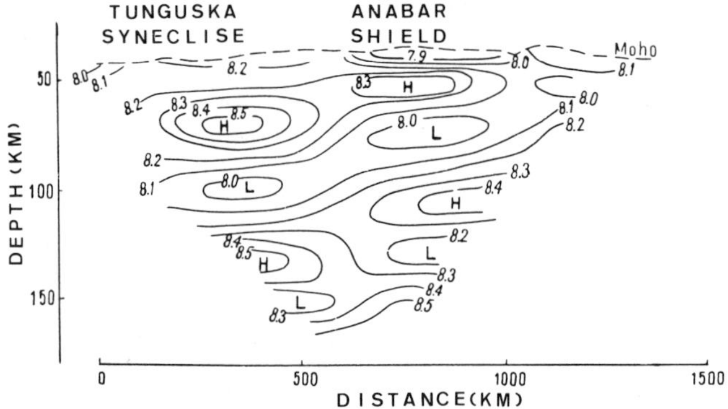

Fig. 14 Two-dimensional velocity model of the upper mantle on profile III. The regions of relatively high- and low-velocity are marked by H and L, respectively.

crystalline layer (Fig. 2) implies, among other possibilities, that the accumulation of sediments in the regions of our study was usually accompanied by a partial destruction of the crystalline crust or its transformation into the upper mantle. The correlation between the thickness of layer 1 with a "granitic" velocity and the total thickness of the crust (Fig. 3) implies that the lateral variations in the thickness of the crystalline crust are mostly due to variations in the thickness of the "granitic" layer. A similar result was obtained for the crust in the southwestern part of the Baltic shield (Vinnik and Kosarev, 1981). These observations modify the view of Penttilä (1969) that the lateral variations in the thickness of the old crystalline crust are controlled by the "basaltic" layer.

The velocity-depth distribution in Fig. 13 may represent the structure of the mantle of a large part of the Siberian platform. In comparison, the models representing the structure of the other old platforms are often either too simple (Green and Hales, 1968; Fig. 6) or too complex (e.g. Ansorge, 1975). There is, however, a set of velocity models for the North American platform for the upper 100 km of the mantle (Ansorge and Mueller, 1971) which are based on the data of the Early Rise experiment and resemble our model in this depth range in almost every detail. This may be regarded as an indication that the deep structure of widely separated old platform areas can be quite similar.

The increase of the P-velocity from 8.1 to 8.4 km/s at a depth of around 90 km (Fig. 13) can be explained by the spinel/garnet transformation (Hales, 1969) and is in broad agreement with petrological data (e.g. Ringwood, 1975). The explanation for the other seismic boundaries in Fig. 13 is more difficult since, contrary to the observations, the garnet pyrolite composition, combined with a suitable geotherm, predicts a monotonous increase of the P-velocity with depth in the depth range of our study (e.g. Leven et al., 1981). One could interpret the high-velocity layers as lenses of eclogite trapped in peridotite, but the extent of the layers seems to be too large for this. Moreover, according to Leven et al. (1981) velocities in the mantle-derived eclogites are slightly too low to match the highest of the observed seismic velocities. One must admit that either the chemical composition of the upper mantle deviates from garnet pyrolite, or some properties of garnet pyrolite remain unknown, or seismic velocity anisotropy is responsible for the complexities in the mantle depth-velocity models. The latter hypothesis seems to be especially appealing (e.g. Fuchs, 1977; Hirn, 1977; Fuchs and Vinnik, 1982; Vinnik and Yegorkin, 1980).

The boundary at a depth of 170 km (Fig. 13) corresponds to the discontinuity which was previously identified in a number of upper mantle seismic studies. In the preliminary reference Earth model (PREM) by Dziewonski and Anderson

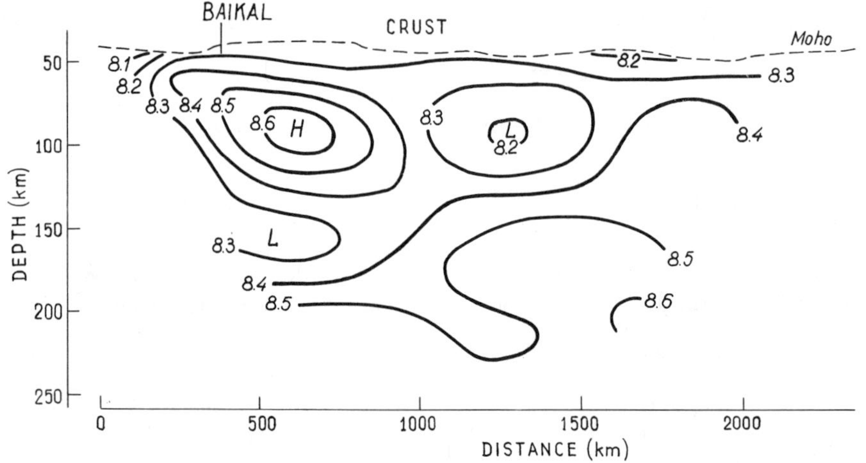

Fig. 15. The same as Fig. 14 but for profile IV, Dikson-Hilok.

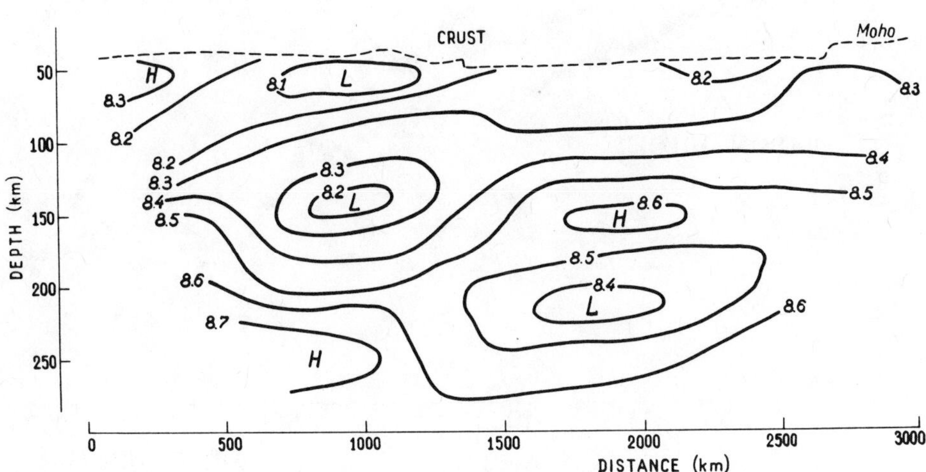

Fig. 16. The same as Fig. 14 but for profile V, Berezovo-Ust-Maya.

(1981) this discontinuity is placed at a depth of 220 km. According to Leven et al. (1981) neither a phase transition nor a compositional change can explain this feature of the continental upper mantle but it can be interpreted in terms of deformation-induced azimuthal velocity anisotropy in a zone of possible decoupling of the continental lithosphere from the underlying mantle. In support of this hypothesis, they discuss the seismic observations in Australia and emphasize, among other facts, that the direction of their profile (in the direction of highest velocities) is close to the direction of plate movement over the underlying mantle. Unfortunately for the hypothesis of anisotropy, the model in Fig. 13 corresponds to the profile V in Siberia (Fig. 1) which is roughly perpendicular to the direction of plate movement. It is also not clear whether the hypothesis of anisotropy can explain the change of depth of this discontinuity from approximately 200 km beneath the young West-Siberian plate to 270 km beneath the old Siberian platform (section 4). On the other hand, this observation sems to be consistent with the hypothesis that the 220-km discontinuity separates the upper layer of the mantle which is depleted in the low melting-point components from the underlying relatively undepleted mantle. The layer of depleted upper mantle (tectosphere) should be especially thick in the old and stable continental regions (Jordan, 1979), in agreement with our observation. The possibility of anisotropy in the zone of transition from the tectosphere to the undepleted mantle is not ruled out but there is no clear evidence of it in our data.

Although the velocity-depth distribution in the uppermost mantle of the central part of the Siberian platform (profile V) is fairly similar to those found in some studies of the North American platform (Ansorge and Mueller, 1971) the deviations from it can be found even in a close neighborhood of the profile V. In the upper mantle of the Russian platform, for instance, the layer with velocities around 8.4 km/s is found immediately below the crust (Fig. 6). According to Ringwood (1975), the pressure of transition from spinel peridotite to garnet peridotite is very sensitive to chemical composition, and this could explain the pronounced difference in the depths of this presumably garnet-peridotite layer beneath the Russian and the Siberian platforms.

The subcrustal lithosphere in the northern margin of the Siberian platform is composed of a number of relatively thin high- and low-velocity layers dipping in the direction from the Anabar shield in the East to the Tunguska syneclise in the West (Fig. 14). The differential vertical movements of the Anabar shield and the Tunguska syneclise started in Carboniferous time (Staroseltsev et al., 1970). The subsidence of the basement of the Tunguska syneclise was accompanied by the accumulation of Upper-Paleozoic sediments and Triassic volcanic deposits (mostly, flood basalts). A combined analysis of the tectonic evidence and the seismic data shown in Fig. 14 suggests the thin-layered structure of the subcrustal lithosphere in the North of the Siberian platform was formed prior to the subsidence of the Tunguska syneclise. At that time, the layering of the subcrustal lithosphere would have been nearly horizontal. The subsequent development of the Tunguska syneclise was due to the subsidence not only of the surface of the crust but of the lithosphere as a whole. This scenario can readily explain the peculiar velocity structure of the lithosphere exposed in Fig. 14.

A remarkable feature in the lithosphere of the

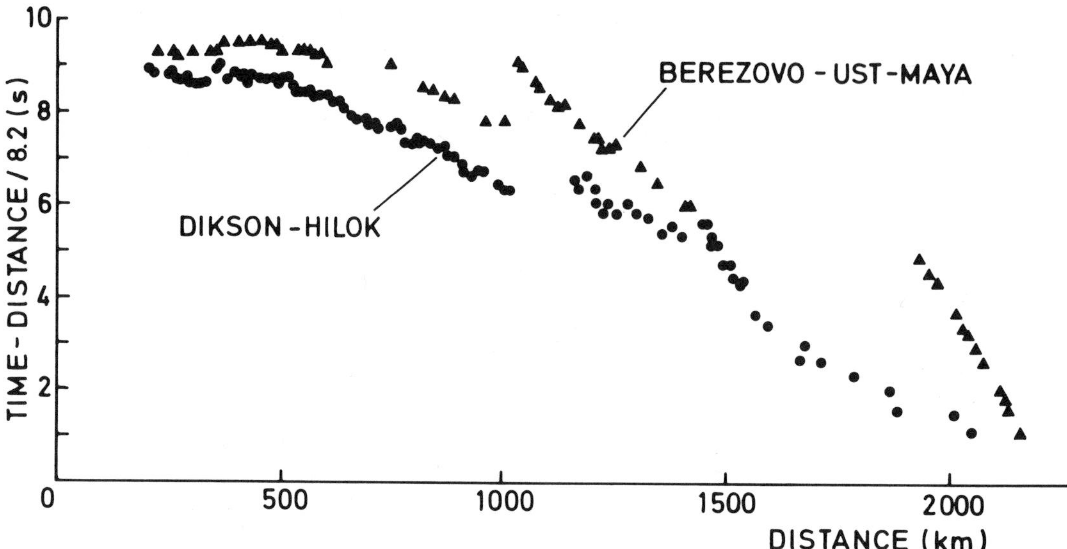

Fig. 17 Comparison of the travel-times of the refracted waves that were propagated in
the lithosphere of the Siberian platform. The data from profile IV, Dikson-Hilok
(shot-point IV) and profile V, Berezovo-Ust-Maya (shot-point II), are shown by filled
circles and triangles respectively.

Dikson-Hilok profile is the shallow high-velocity
body near the Baikal rift zone (Fig. 15). The
origin of this body is enigmatic. There is no
evidence of its existence in the data of a profile
running roughly parallel to the Dikson-Hilok
profile to the North of Lake Baikal (Krilov et
al., 1981). This observation may imply that the
formation of the Baikal rift could be related to
the presence of the high-velocity body.

The West-Siberian platform is characterized by
upper mantle velocities that are very low in
comparison to those of the Russian plate (Fig. 7)
and the Siberian platform (Fig. 16). The region
of the low mantle velocities corresponds to the
system of rifts that was active at the Paleozoic-
Mesozoic boundary (Sukov and Gero, 1981). The
existence of the low-velocity anomaly suggests
that the temperature in the depth range of 50-150
km may still be anomalously high and/or the
composition of the upper-mantle rocks is different
from that in the neighboring regions. It is
interesting to see that the anomaly has no
continuation in the Siberian platform although the
upper mantle of the latter was strongly active at
approximately the same time, as shown by the
presence of thick cover of Triassic basalts in the
upper crust of the Tunguska syneclise.

As we have already mentioned, it is possible
that the seismic velocities in the upper mantle
are dependent on the direction of wave
propagation. This phenomenon is well documented
for the oceanic upper mantle but is relatively
unknown in studies of the continental mantle. For
a comprehensive test of this hypothesis it is
desirable to have a number of profiles crossing
the same area in many directions. This test is
impossible now in Siberia but we can compare the
travel-time data from the profiles crossing the
relatively homogeneous part of the Siberian
platform in two approximately perpendicular
directions (Fig. 17). The comparison shows that
there are differences indeed, and they are fairly
large at distances exceeding 1000 km.
Unfortunately, it is impossible to conclude from
these data whether the difference is due to
azimuthal anisotropy or lateral heterogeneity.
The development of a simpler method of measuring
anisotropy in the lithosphere (Vinnik et al.,
1984) may clarify this problem in the near future.

In conclusion the seismic observations on the
long-range profiles have shed some light on the
deep structure of the continental lithosphere, but
much more should be done in order to understand
its fine structure.

Acknowledge   . This paper was completed
during an ext   ed visit by one of us (L.P.V.) to
the Federal Republic of Germany, and we appreciate
exciting discussions with K. Fuchs, A. Kröner, R.
Kind and many other German scientists.

References

Aki, K., and P.G. Richards, Quantitative
    seismology, W.H. Freeman and Company, San
    Francisco, 1980.
Ansorge, D., Die Feinstruktur des obersten
    Erdmantels unter Europa und dem mittleren Nord-
    amerika, PhD-Thesis, Universität Karlsruhe,
    111p., 1975.
Ansorge, D., and St. Mueller, The fine structure
    of the upper mantle in Europe and in North

America, XII Assemblee Generale de la Commission Seismologique Europeenne, Observatoire Royale des Belgique, Communications, Serie A-No. B, 1971.

Bungum, H., S.E. Pirhonen, M., and E.S. Husebye, Crustal thickness in Fennoscandia, _Geophys. J.R. astr. Soc._, 63, 759-774, 1980.

Burmakov, Ju.A., A.V. Treussow, and L.P. Vinnik, Determination of three-dimensional velocity structure from observations of refracted body waves, _Geophys.J.R. astr. Soc._, 79, 285-292, 1984.

Dziewonski, A.M. and D.L. Anderson, Preliminary reference Earth model, _Phys. Earth Planet. Inter._, 25, 297-365, 1981.

Forsythe, G.E., M.A. Malcolm, and C.B. Moler, _Computer Methods for Mathematical Computations_, Prentice-Hall, Engelwood Cliffs, New Jersey, 1977.

Fuchs, K., Seismic anisotropy of the subcrustal lithosphere as evidence for dynamical processes in the upper mantle, _Geophys. J.R. astr. Soc._, 49, 16-179, 1977.

Fuchs, K., and L.P. Vinnik, Investigation of the subcrustal lithosphere and asthenosphere by controlled source seismic experiments on long range profiles, in _Continental and oceanic rifts, Geodyn. Ser._, 8, 81-89, 1982.

Green, H.D., and A.L. Hales, The travel times of P-waves to 30 in the central United States and upper mantle structure, _Bull. seism. Soc. Amer._, 58, 267-289, 1968.

Hales, A.L., A seismic discontinuity in the lithosphere, _Earth Planet. Sci. Lett._, 7, 44-46, 1969.

Hirn, A., Anisotropy in the continental upper mantle: possible evidence from explosion seismology, _Geophys. J.R. astr. Soc._, 49, 49-58, 1977.

Iyer, H.M., L.C. Pakiser, D.J. Stuart, and D.H. Warren, Project Early Rise: seismic probing of the upper mantle, _Journ. Geophys. Res._, 74, 4409-4441, 169.

Jordan, T.H., Structural geology of the Earth's interior, _Proc. of the Nat. Acad. Sci. USA_, 76, 4192-4200, 1979.

Krilov, S.V., M.M. Mandelbaum, B.P. Mishenkin, Z.P. Mishenkina, G.V. Petrik, and V.S. Seleznev, _Deep structure of Baikal (from seismic data)_, Nauka, Siberian branch, Novosibirsk (in Russian), 1981.

Leven, J.H., I. Jackson, and A.E. Ringwood, Upper mantle seismic anisotropy and lithospheric decoupling, _Nature_, 289, 234-239, 1981.

Masse, R.P., Compressional velocity distribution beneath central and eastern North America, _Bull. seism. Soc. Amer._, 63, 911-935, 1973.

Mereu, R.F., and J.A. Hunter, Crustal and upper mantle structure under the Canadian Shield from Project Early Rise data, _Bull. seism. Soc. Amer._, 59, 147-165, 1969.

Perttilä E., A report summarizing on the velocity of earthquake waves and the structure of the earth's crust in Baltic Shield, _Geophysica_, 10, 1-13, Helsinki, 1969.

Ringwood, A.E., _Composition and petrology of the earth's mantle_, McGraw-Hill, New York, 604 pp., 1975.

Roller, J.C., and W.H. Jackson, Seismic wave propagation in the upper mantle: Lake Superior, Wisconsin, to central Arizona, _Journ. Geophys. Res._, 71, 5933-5941, 1969.

Staroseltsev, V.S., V.M. Lebedev, and A.V. Khomenko, Structure and the history of formation of the Tunguska syneclise, in _Tectonics of Siberia_, Nauka, Moscow (in Russian), 1970.

Surkov, V.S., and O.G. Gero, Basement and the development of the platform cover of the West-Siberian plate, Nauka, Moscow, 143 pp. (in Russian), 1981.

Vinnik, L.P., and G.L. Kosarev, Determination of the parameters of the earth's crust from observations of teleseismic body waves, _Proc. (Dokl.) Acad. Sci. USSR_, 261, 1091-1095 (in Russian), 1981.

Vinnik, L.P., and V.Z. Ryaboy, Deep structure of the East-European platform according to seismic data, _Phys. Earth Planet. Interiors_, 25, 27-37, 1981.

Vinnik, L.P., and A.V. Yegorkin, Wave fields and models of the lithosphere-asthenosphere according to seismic observations in Siberia, _Proc. (Dokl) Acad. Sci. USSR_, 250, 319-323 (in Russian), 1980.

Vinnik, L.P., and A.V. Yegorkin, Low-velocity layer in the mantle of the old platforms according to the seismic observations on long-range profiles, _Bull. (Izv.) Acad. Sci. USSR, Earth Physics_, 12, 12-18 (in Russian), 1981.

Vinnik, L.P., G.L. Kosarev, and L.I. Makeyeva, Anisotropy of the lithosphere from observations of waves SKS and SKKS, _Proc. (Dokl.) Acad. Sci. USSR_, 278, 1335-1339, 1984.

Yegorkin, A.V. and N.I. Pavlenkova, Studies of mantle structure of teh USSR territory on long-range seismic profiles, _Phys. Earth Planet. Inter_, 25, 12-26, 1981.

GEOELECTRICAL AND PALAEOMAGNETIC STUDIES ON THE BUSHVELD COMPLEX

J. H. de Beer,[1] R. Meyer,[1] and P. J. Hattingh[2]

[1]Geophysics Division, National Physical Research Laboratory of the Council for Scientific and Industrial Research, P O Box 395, Pretoria, 0001, South Africa

[2]Geology Department, University of Pretoria, Pretoria, 0002, South Africa

Abstract. This report deals with a palaeomagnetic study of the layered mafic sequence of the Bushveld Complex and a Schlumberger sounding survey of the acid phase of the igneous complex. The palaeomagnetic study comprises data from 88 sites of all zones of the mafic layered sequence of both the eastern and western Bushveld Complex. The layered sequence displays a striking palaeomagnetic polarity pattern and it can be subdivided into three magnetic polarity zones. The palaeomagnetic data clearly show that the subsidence in the lower zones of the complex occurred before the emplacement and cooling of the Upper Zone. Up to November 1984 a total of 124 deep Schlumberger soundings have been carried out on the complex and the strata underlying it. The sounding data show a clear geoelectrical stratification and two conductive marker horizons can be identified. One is a graphitic shale horizon in the floor rocks to the complex and the other is the magnetite-bearing Upper Zone of the layered mafic sequence. The interpretation of the geoelectrical results for the two main bodies of acid rocks in the complex shows that the granitic rocks have a maximum thickness of less than 2.7 km in the eastern lobe and less than 4 km in the western compartment. In general there is a good agreement between the thicknesses determined from geoelectrical soundings and those obtained from the modelling of gravity data.

Introduction

The geoelectrical work discussed in this report forms part of the South African National Geoscience Programme and the research on this aspect was carried out by the first two authors. This project is still in progress. The palaeomagnetic investigation formed part of the research for a doctoral dissertation submitted by the third author (Hattingh, 1983). Both research projects were aimed at a better understanding of the gross structure and emplacement history of the Bushveld Complex.

The Bushveld Complex consists of a large array of plutonic to volcanic igneous rocks which under-

lie an area of about 65 000 km^2 in the central Transvaal, South Africa (Fig. 1). This makes it the largest known intrusion of its kind.

The complex is not only unique because of its size, but also because of the significance of its diverse suite of economically important ores and the regularity and persistence of the layering in the basic and ultrabasic rocks. Gravity data (Cousins, 1959; Smit et al., 1962) demonstrated that the Bushveld Complex is not a simple lopolith and the main complex is now regarded as consisting of four lobes, a western, a northern, an eastern and a southeastern compartment (Fig. 1). All these lobes are situated within the depositional basin of the Transvaal Sequence and the rocks of this sequence form the framework into which the Bushveld Complex was intruded.

The previous geophysical studies of the complex include one of the first palaeomagnetic investigations in South Africa by Gough and Van Niekerk (1959) on the layered mafic sequence. Their limited data provided evidence that the magnetization of the Main Zone was acquired with the igneous layering in a horizontal position.

The interpretation of gravity and magnetic data (Cousins, 1959; Smit et al., 1962; Biesheuvel, 1970; Van der Merwe, 1976; Walraven and Darracott, 1976; Molyneux and Klinkert, 1978; Barrett et al., 1978; Hattingh, 1980) showed that the complex consists of separate, overlapping intrusions (Hunter, 1975, 1976; Tankard et al., 1982). The published gravity interpretations for the eastern lobe (Molyneux and Klinkert, 1978; Hattingh, 1980) indicate a maximum thickness of about 5 km for the layered mafic sequence and 2.5 to 6 km for the granites. These estimates are affected by the inherent nonuniqueness of gravity interpretations. The situation is further complicated by the very complex setting of the Bushveld intrusions in the Transvaal basin which in suboutcrop has an unknown thickness and distribution of units of different densities. The regional component of the gravity field is therefore very difficult to determine. It is nevertheless clear that the geophysically determined maximum

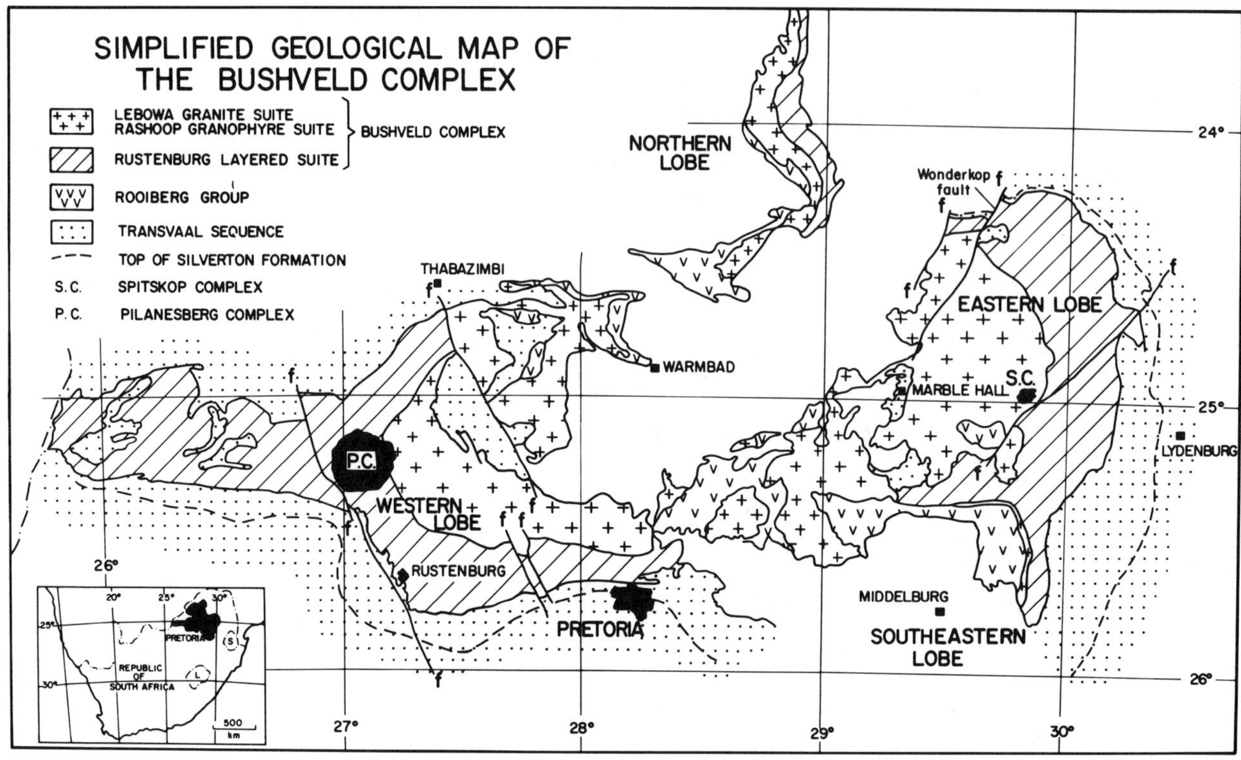

Fig. 1. Simplified geological map of the Bushveld Complex as compiled from the 1970 edition of the 1:1 000 000 Geological Map of the Republic of South Africa (Geological Survey, Department of Mines of the Republic of South Africa).

thicknesses for the eastern lobe are markedly thinner than the geologically estimated average thickness for the eastern, western and northern mafic sequences of 8 000 ± 750 m (Vermaak and Lee, 1981).

The palaeomagnetic study reported here involves the mafic layered sequence and the geoelectrical study will deal with results obtained on the acid phase of the Bushveld Complex.

General Geology

The early Proterozoic sedimentary and volcanic strata of the Transvaal Sequence occupy a basin in the Archaean granite gneiss of the Kaapvaal Craton. Over most of the area these strata form the floor to the Bushveld Complex. For the present discussion only the Pretoria Group is of importance. This group consists predominantly of quartzite and shale and also includes a prominent andesitic volcanic unit as well as a thick basalt formation in the eastern Transvaal. Numerous diabase sills are present in the succession. A prominent shale formation (the Silverton Shale Formation) containing an electrically extremely conductive graphitic shale member occurs stratigraphically about 4.5 km from the top in the eastern Transvaal and 1.5 km from the top in the central Transvaal (South African Committee for Stratigra-

phy (SACS), 1980). This graphitic shale represents a very significant geoelectrical marker horizon that can be recognized throughout the Transvaal basin in the western, central and eastern Transvaal. This marker occurs in the floor to the Bushveld Complex. The succession above this marker horizon consists mostly of electrically resistive quartzite and basalt.

The mafic and ultramafic rocks of the layered sequence or Rustenburg Layered Suite (SACS, 1980) comprise a variety of rock types including dunites, pyroxenites, harzburgites, anorthosites, gabbros, norites, magnetite gabbros and diorites. Although SACS (1980) introduced a lithostratigraphic subdivision for the layered sequence, in this paper the well-entrenched zonal subdivision first proposed by Hall (1932) and largely based on petrographic considerations, will be used. A simplified stratigraphic column as pertaining to the eastern and western lobes and described in this paper is shown in Table I. A more detailed description can be found in, for example, Tankard et al. (1982).

The informal zonal subdivision consists from the lowermost zone of:
The Marginal Zone: Fine-grained noritic rocks characterized by metasedimentary inclusions and occurring between various rock units of the layered suite and the country rocks.
The Lower Zone: The important lithologic units in

TABLE I. Simplified Stratigraphic Column for the Bushveld Complex
With Associated Electrical Resistivities

Geology		Resistivity (ohm.m)
Lebowa Granite Suite	Nebo Granite (East)	6 000–40 000
	(West)	3 500–15 000
Rashoop Granophyre Suite		5 000–15 000
Rustenburg Layered Suite	Upper Zone	100–800
	Main Zone	>8 000
	Critical Zone	
	Lower Zone	1 000–2 000
	Marginal Zone	
Rooiberg Group		1 500–2 000
Pretoria Group	Dullstroom Basalt Formation (only Eastern Transvaal)	5 000–15 000
	Several formations comprising quartzites, shales and hornfels (in Central and Eastern Transvaal only)	
	Magaliesberg Quartzite Formation	$\sim$10 000
	Silverton Shale Formation	200–3 000
	Graphitic Shale Member	<10

this zone are norite, feldspathic pyroxenite and harzburgite.

The Critical Zone: Pyroxenite interlayered with dunite and chromitite overlain by alternating layers of chromitite, pyroxenite, norite and anorthosite. The well-known Merensky Reef forms part of this zone.

The Main Zone: A succession of relatively homogeneous rocks consisting mostly of gabbro and norite.

The Upper Zone: The appearance of cumulus magnetite marks the base of this zone that consists of diorite, olivine diorite, magnetite gabbro with layers of magnetite, anorthosite and olivine gabbro.

The final rock suite of importance to this study is the acid phase of the Bushveld Complex which comprises the Lebowa Granite Suite (formerly collectively known as the Bushveld granites) and the Rashoop Granophyre Suite (Fig. 1). Of the first-mentioned suite of rocks the most important is the Nebo Granite which forms the bulk of this unit (SACS, 1980). This granite formed the target of the geoelectrical study discussed in this paper. It displays a crude stratiform layering consisting of coarse-grained grey granite at the base grading upward through medium-grained grey and red granite, then through red granophyric granite to granophyre. A full sequence is not always present. The Nebo Granite is largely confined to an area within the four main occurrences of the Rustenburg Layered Suite. The Rashoop Granophyre Suite occurs as a fine-grained facies in the upper part of the Bushveld acid phase, and also at the contact between the mafic sequence and the felsites. The stratigraphic position of the felsites of the Rooiberg Group is still under debate. This rock unit is not of importance to the present discussion.

The Palaeomagnetism of the Mafic Portion of the Bushveld Complex

This investigation was initiated to extend the study of Gough and Van Niekerk (1959) and to relate the new palaeomagnetic results from the mafic layered sequence to more modern concepts of the genesis and structure of the complex (Hunter, 1975; Hunter and Hamilton, 1978; Von Gruenewaldt, 1979; Sharpe, 1980; Sharpe and Bahat, 1981).

For the purpose of this investigation 88 sites of all zones of the layered sequence of both the eastern and western Bushveld Complex were sampled (Fig. 2). Several samples were found to be extensively altered and yielded inconsistent palaeomagnetic data at site and sample levels. Magnetization directions were however recovered from 69 sites.

Palaeomagnetism of the Critical Zone

Twenty sampling sites are situated in the Critical Zone. The rock types sampled consist of pyroxenite, anorthositic norite, norites and pegmatitic feldspathic pyroxenite of the Merensky Reef. The NRM's of samples of pyroxenite are characterized by a low intensity of magnetization (generally below 10×10^{-3} Am^{-1}) with inconsistent and unstable magnetization directions at sample and site levels. As a result, data from 12 sampling sites had to be rejected.

The mean intensity of the natural remanent mag-

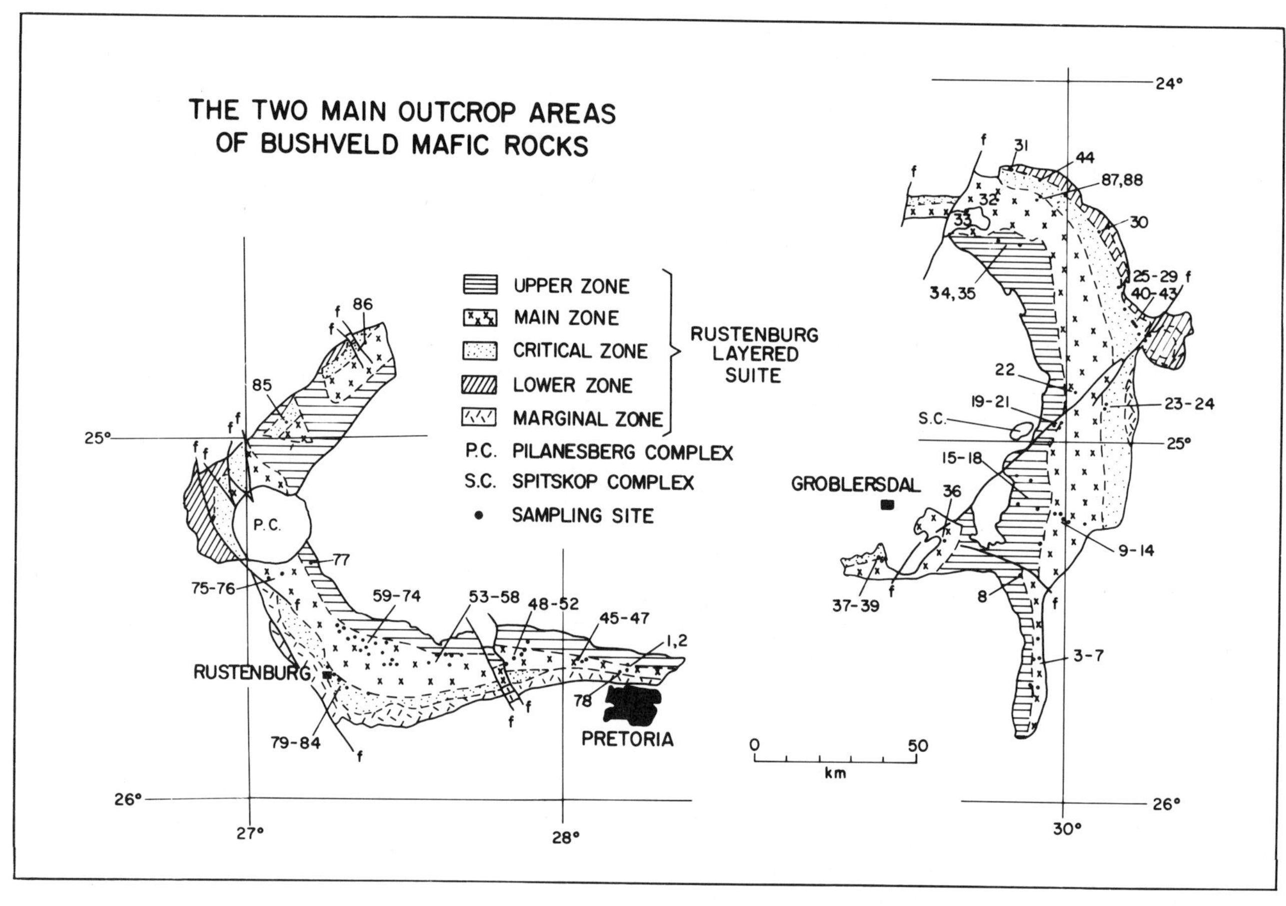

Fig. 2. Simplified geological map of the two main outcrops of the layered mafic sequence of the Bushveld Complex showing the sampling sites for the palaeomagnetic study.

netization of all the specimens from the Critical Zone is 27×10^{-3} Am^{-1}. After stepwise demagnetization of pilot specimens, bulk alternating field (AF) demagnetization of specimens was done. This yielded consistent magnetization directions from eight sites (Fig. 3a, Table II) of which six (sites 79, 80, 81, 83, 84, 87) group together to form a group with dual polarity with the following statistics: group = BCZ1; N = 6; D = 194.4°; I = −43.6°; α_{95} = 16.0°; k = 18 with the corresponding palaeomagnetic north pole position at latitude 37°S; longitude 135°W and polar error (dp, dm); 12.5° and 19.9° respectively.

With the igneous layering of the Critical Zone in a horizontal position group BCZ1 changes to group BCZ2R with the following statistics: N = 6; D = 195.6°; I = −40°; α_{95} = 10.9°; k = 38 and the palaeomagnetic north pole position at latitude 39.5°S; longitude 133°W with polar error (dp, dm) 7.8° and 13.1° respectively.

Although group BCZ2R is an improvement on group BCZ1, the improvement is not significant at the 95 per cent confidence limit. However, at five of the sites, the dip and dip directions of the igneous layering are similar and one cannot con-

clude that the result of the fold test is insignificant. The result is inconclusive due to the lack of evenly distributed sampling sites in the Critical Zone.

On the basis of geological and mineralogical evidence (Hattingh, 1983), as well as the position of the palaeomagnetic pole, calculated for group BCZ2R, on the APW path for southern Africa, the mixed polarity of these two groups is provisionally accepted to be the result of a self-reversal of magnetization, with the mean direction quoted for each group corresponding to the polarity exhibited by the majority of sites in the respective group. The mean magnetization directions of these groups both differ in polarity with respect to the magnetization direction given by Gough and Van Niekerk (1959), for the Main Zone of the layered sequence.

Palaeomagnetism of the Main Zone, Eastern Bushveld Complex

The sampling pattern yielded 20 sampling sites, covering subzones A, B and C of the Main Zone in the eastern Bushveld Complex. Gabbro, norite and

194   DE BEER ET AL.

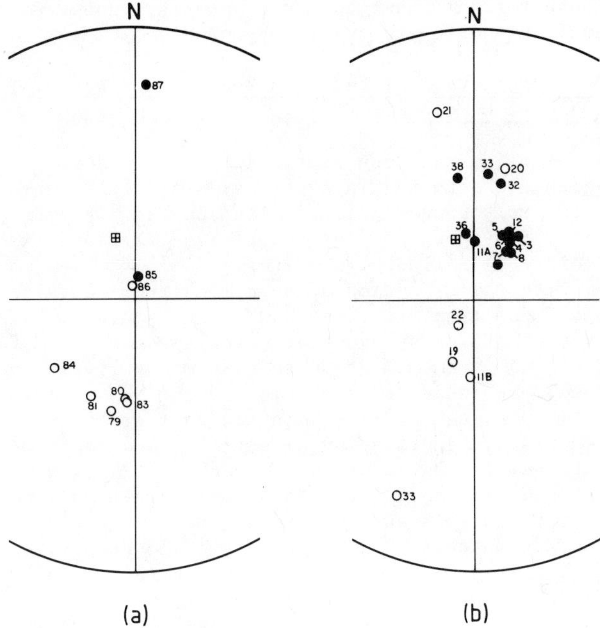

Fig. 3. Stereographic (equal angle) projection
of magnetization directions of sites in (a) the
Critical Zone and (b) the Main Zone eastern Bushveld Complex, after AF demagnetization. Solid
circles on the lower hemisphere, open circles on
the upper hemisphere. Present day magnetic field
direction upwards, indicated by ⊞ .

anorthosite, exhibiting very little, if any alter-
nation, were sampled. With the aid of AF and
thermal demagnetization techniques, stable and consistent magnetization directions from 18 sites
were recovered.

TABLE II. Magnetization Directions of Sites in
the Critical Zone After Bulk Alternating Field
Demagnetization

Site	Alternating field (mT)	D	I	α_{95}	k
79	30	190.7°	−44.5°	12.1°	56
80	30	184.5°	−50.3°	12.6°	54
81	30	203.2°	−47.6°	24.4°	10
83	30	182.9°	−49.0°	20.6°	14
84	30	229.5°	−48.4°	13.5°	33
85	50	9.5°	80.3°	10.2°	57
86	20	355.5°	−84.0°	11.5°	43
87	30	2.7°	14.0°	7.5°	107

D is declination of the magnetization direction,
  measured from north in a clockwise direction.
I is the inclination of the magnetization direction; downward is positive, upwards is negative.
α_{95} is the radius of the circle of 95 per cent
  confidence.
k is the estimate of the precision parameter k.
  The above convention is also followed for Tables
  III to V.

TABLE III. Magnetization Directions of Sites
in the Main Zone, Eastern Bushveld Complex
After Bulk Alternating Field Demagnetization

Site	Alternating field (mT)	D	I	α_{95}	k
3	10	36.2°	57.6°	2.2°	1775
4	20	29.6°	59.3°	4.4°	444
5*	20	27.5°	58.2°	−	−
6	10	26.4°	60.6°	6.2°	219
7*	20	32.6°	60.8°	−	−
9	20	35.7°	64.4°	8.7°	111
11A*	40	1.4°	65.0°	−	−
11B*	40	182.4°	−58.4°	−	−
12	60	38.4°	64.2°	4.8°	695
14	60	35.8°	71.5°	19.4°	23
19	20	198.6°	−62.9°	17.0°	53
20	40	13.8°	−37.2°	7.6°	258
21	20	348.7°	−19.3°	11.0°	70
22	20	211.5°	−77.5°	9.9°	86
32	10	13.8°	41.9°	8.4°	120
33	20	202.7°	−14.7°	26.0°	23
36	0	354.0°	61.9°	4.2°	496
37	10	6.7°	39.5°	7.8°	140
38	20	352.5°	40.8°	10.5°	76

*N = 2

The mean intensity of the NRM of all specimens
from the Main Zone in the eastern Bushveld Complex
is 3 395 x 10^{-3} Am^{-1}. After bulk AF demagnetization 13 sites show mean magnetization directions
with positive inclinations and six exhibit magnetization directions with negative inclinations
(Fig. 3b, Table III).
All the sites with reversely magnetized directions (positive inclinations) are situated in subzone B of the Main Zone and group together to form
group BMZ1 with the following statistics: N = 13;
D = 18.8°; I = 58.5°; α_{95} = 7.0°; k = 36. With the
igneous layering in a horizontal position this
changes to: group = BMZ2R; N = 13; D = 10°; I = 64.7°;
α_{95} = 4.2°; k = 100. This improvement after structural folding is significant at the 99 per cent
confidence limit, indicating that subzone B acquired its magnetization with the igneous layering in a horizontal position. The palaeomagnetic
pole position corresponding to group BMZ2R is
situated at latitude 17.3°N, longitude 35.7°E
with polar error (dp, dm) 5.4° and 6.4° respectively.
With the exception of site 11B, all sites with
negative inclinations of magnetization originate
from subzone C. Three sites from subzone C (sites
19, 22, 23) can be grouped together after bulk AF
demagnetization to form group BMZ3 with the following statistics: N = 3; D = 202.7°; I = −52.3°; α_{95} =
54°; k = 6, and with the igneous layering in a horizontal position: group = BMZ4R; N = 3; D = 168.9°;
I = −55°; α_{95} = 20°; k = 37. This improvement after
structural folding is significant at the 90 per

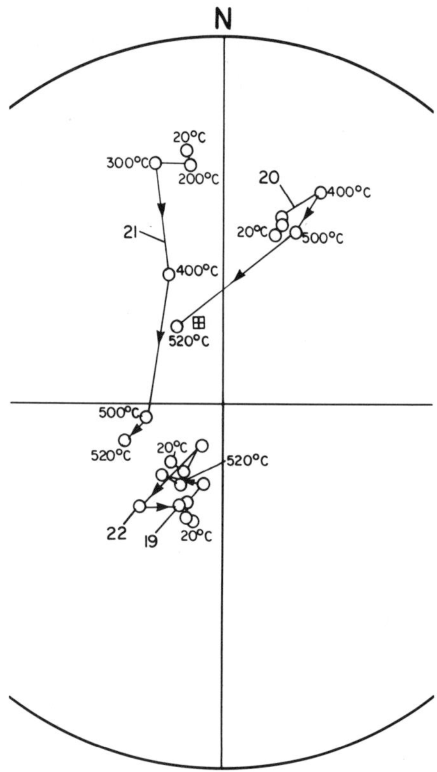

Fig. 4. Continuous thermal demagnetization response of specimens from sites 19, 20, 21 and 22. Plotting convention as in Fig. 3.

cent confidence limit. Groups BMZ3 and BMZ4R can however only be accepted on evidence yielded by thermal demagnetization of specimens from two sites in subzone C (sites 20, 21), which failed to group with groups BMZ3 and BMZ4R. During continuous thermal demagnetization of specimens from these sites the magnetization directions change towards the mean magnetization direction of group BMZ3 (Fig. 4).

The palaeomagnetic pole position corresponding to group BMZ4R and thus representing subzone C of the Main Zone in the eastern Bushveld Complex is situated at 28°S, 161.7°W with the polar error (dp, dm) 20.1° and 28.6° respectively.

In the absence of any evidence for self-reversal the magnetic polarity difference between subzones B and C suggests reversal of the earth's magnetic field during the cooling of the Main Zone after emplacement.

Subzone A at the base of the Main Zone is represented by only one site, site 11, which contains mixed polarities of magnetization. One direction is approximately parallel to the mean direction yielded by subzone B, and the other direction parallel to the mean magnetization of subzone C. Experimental evidence exists (Hattingh, 1983) which supports both a geomagnetic field reversal and a self-reversal of magnetization as

cause for the observed mixed polarity, however, available data are still inconclusive.

Palaeomagnetism of the Main Zone in the Western Bushveld Complex

Sampling was done at 34 sites, the majority of which are situated in quarries. The resultant relatively unweathered samples enabled the extraction of stable and consistent magnetization directions from all sampling sites.

With the exception of specimens from two sites, specimens from all sites had a fairly high intensity of NRM. The average intensity of the NRM is $2\ 078 \times 10^{-3}$ Am^{-1}. Stepwise AF demagnetization of specimens indicated a multicomponent NRM. The primary magnetization dominates over small hard secondary magnetization directions, and was isolated with the aid of bulk AF demagnetization at optimum AF values for each site, as well as with vector analyses and the aid of remagnetization circles (Hattingh, 1983, Hattingh, in preparation).

The mean primary magnetization direction of the sites with reversed magnetization directions (positive inclinations), (Fig. 5a, Table IV) is as follows: group BMZ5; N = 32; D = 336.1°; I = 82.0°; $\alpha_{95} = 3.4°$; k = 58. With the igneous layering in a horizontal position, this changes to: group BMZ6R; N = 32; D = 357.8°; I = 70.9°; $\alpha_{95} = 2.8°$; k = 80, with corresponding palaeomagnetic pole position: 9.2°N; 27.3°E with polar error (dp, dm) 4.2° and 4.9° respectively.

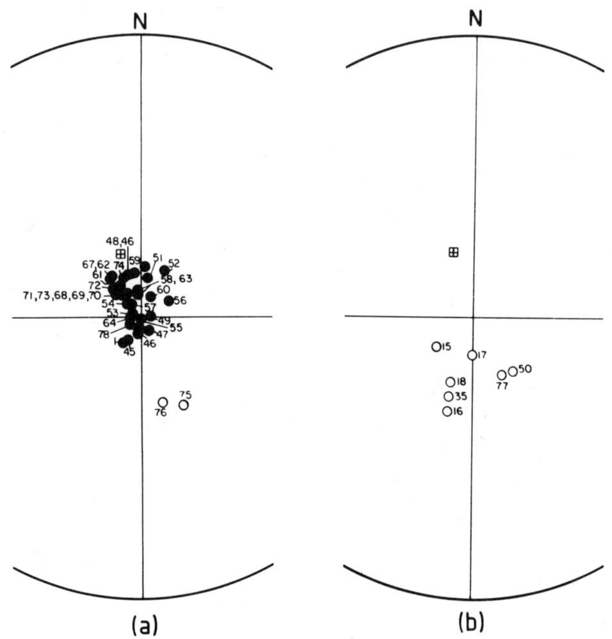

Fig. 5. Magnetization directions from sites in a) the Main Zone, western Bushveld Complex and b) in the Upper Zone, after bulk AF demagnetization. Plotting convention as in Fig. 3.

Site	Demagnetiza-tion field (mT)	D	I	α_{95}	k
1	20	214.8°	77.8°	8.0°	236
2	40	193.1°	86.4°	4.7°	386
45	60	209.9°	79.6°	4.2°	483
46	40	182.7°	83.9°	4.7°	256
47	50	141.3°	84.1°	3.9°	555
48	60	352.5°	71.8°	13.1°	50
49	30	87.2°	85.8°	1.6°	2970
51	30	11.1°	73.5°	20.6°	20
52	30	27.0°	68.8°	5.6°	268
53	10	303.0°	86.6°	4.9°	358
54	20	320.0°	82.7°	3.9°	532
55	20	185.5°	89.6°	3.7°	605
56	30	61.0°	76.6°	3.6°	640
57	20	334.0°	84.0°	2.9°	973
58	40	353.6°	78.7°	4.7°	388
59	20	5.0°	69.5°	3.4°	731
60	60	25.2°	80.4°	2.6°	1226
61	70	330.2°	74.6°	6.3°	213
62	30	336.0°	72.7°	1.8°	2650
63	50	354.6°	80.6°	6.6°	197
64	20	270.0°	85.9°	2.2°	1778
65	30	328.6°	75.0°	6.0°	231
66	30	342.2°	72.1°	1.3°	5304
67	30	323.2°	70.6°	2.4°	1425
68	30	319.5°	77.8°	2.9°	950
69	30	332.4°	79.4°	3.0°	892
70	30	323.4°	79.2°	2.2°	1676
71	30	312.3°	76.4°	2.4°	1503
72	30	317.2°	74.2°	3.3°	775
73	30	317.8°	76.7°	4.9°	346
74	30	325.9°	70.1°	5.1°	314

The magnetization direction obtained from group
BMZ6 is, however only representative of the top
1 800 m of the Main Zone in the western Bushveld
Complex, because all 32 sampling sites in group
BMZ6 are situated in this top part of the Main
Zone. Two sampling sites (sites 75, 76) situated
below that level are normally magnetized and group
together, after bulk AF demagnetization (Fig. 5a).
This indicates a polarity transition approximately
1 000 m above the base of the Main Zone in the
western Bushveld Complex. Evidence from theoretical calculations (Hattingh, 1983) indicates that
there is a possibility that this polarity transition could be due to a self-reversal of magnetization, but the results are inconclusive.

The Palaeomagnetism of the Upper Zone

Nine sampling sites are situated in the Upper
Zone of the layered sequence. Seven of these
sites yielded consistent magnetization directions.
The average intensity of the NRM of specimens
from the Upper Zone is $6\,070 \times 10^{-3}$ Am^{-1}. Although
this is lower than that of specimens from the Main
Zone in the eastern Bushveld Complex, the range of
intensities is remarkably broad and specimens with
intensities as high as $41\,000 \times 10^{-3}$ Am^{-1} and as
low as 500×10^{-3} Am^{-1} were measured.

Although the NRM directions of specimens were
consistent at sample level, there is no consistency of directions between sites. After bulk AF
demagnetization, consistency between seven sites
was obtained (Fig. 5b, Table V). These seven
sites group together to form group BUZ1 with the
following statistics: $N = 7$; $D = 184.3°$; $I = -66.0°$;
$\alpha_{95} = 11.0°$; $k = 31$ and with the corresponding
palaeomagnetic pole position situated at 16.1°S
and 148.5°W and polar error (dp, dm) 14.7° and
18.0° respectively.

With the igneous layering of the Upper Zone in
a horizontal position the statistics of group
BUZ1 change to: $N = 7$; $D = 175.1$; $I = -60.9$; $\alpha_{95} =$
10.6°; $k = 33$. The small improvement in k from 31
to 33 is however not significant at the 95 per
cent confidence limit. This indicates that the
Upper Zone had acquired its NRM with the igneous
layering in its present position.

Because of the abundance of large size grains
of magnetite in all specimens, Lowrie–Fuller tests
(Lowrie and Fuller, 1971) as modified by Dunlop
et al. (1973) were applied to determine the magnetic domain characteristics of the NRM. These
tests indicated the presence of both multidomain
(MD) and single domain (SD) magnetite grains in
specimens. It also indicated that the stable and
consistent magnetization directions obtained by
bulk AF demagnetization is carried by SD grains.

Palaeomagnetic Subdivision of the
Layered Sequence

The palaeomagnetic polarity pattern is a striking feature of the palaeomagnetism of the mafic
layered sequence of the Bushveld Complex. Although
a resemblance exists, there is no close relation
between either the zonal or lithostratigraphic
subdivision of the layered sequence and the observed polarity pattern, which is basically a
chronostratigraphic subdivision. This apparent un-

TABLE V. Magnetization Directions of Sites in
the Upper Zone, After Bulk Alternating Field
Demagnetization

Site	Alternating field (mT)	D	I	α_{95}	k
15	40	231.1°	-71.0°	9.1°	101
16	40	194.8°	-52.4°	17.9°	27
17	20	180.5°	-75.1°	23.9°	16
18	40	198.3°	-62.9°	6.1°	224
35	20	196.5°	-57.8°	6.7°	184
50	10	141.5°	-62.8°	7.7°	98
77	50	150.4°	-63.8°	35.0°	13

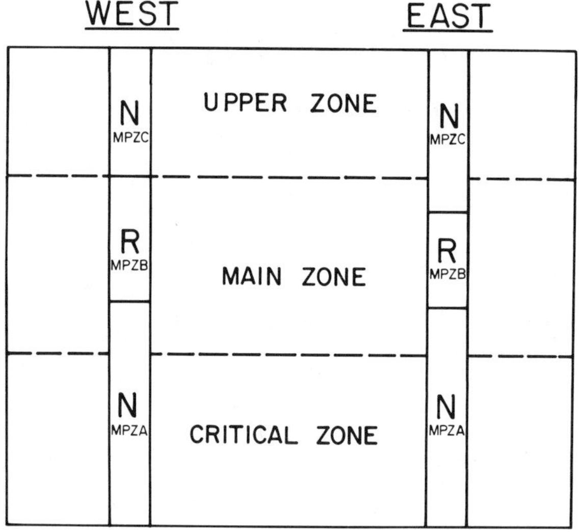

Fig. 6. The zonal subdivision of the mafic layered sequence showing the magnetic polarity pattern observed in the western and eastern Bushveld Complex (N = normal magnetization defined here as a magnetization with the same sense of inclination as the present-day magnetic field in southern Africa, R = reversed magnetization).

conformal relationship may be pertinent to the evolution of the layered sequence, but it also lends itself to a clear palaeomagnetic subdivision of the sequence.

The magnetic polarity pattern of the layered sequence in the western and eastern Bushveld in terms of either a normal magnetization direction (defined here as a magnetization direction in the same sense as the present-day magnetic field in southern Africa), or a reversed magnetization direction is shown in Fig. 6.

The relationship of the palaeomagnetic pole positions of the Bushveld Complex with respect to the apparent polar wander path for southern Africa will not be discussed in this report but will be dealt with elsewhere (Hattingh, in preparation).

Schlumberger Sounding Survey

The programme of geoelectrical soundings was undertaken after it was realized that the graphitic zone in the Silverton Shale Formation that underlies the Bushveld Complex forms an ideal geoelectrical marker horizon. This fact combined with the layered nature of the complex as well as the generally low dip angles of the layers (10–25°) makes it an ideal target for deep Schlumberger soundings.

Schlumberger Sounding Method

The Schlumberger sounding technique as used in this investigation corresponds in most aspects to the standard method as described for example by Kunetz (1966) or Koefoed (1979) except that current electrode spacings of up to 60 km were used. To enable readers not familiar with the method to assess the results, certain characteristics of the method merit some comment.

As the distance between the current electrodes (AB) increases during the sounding, so does the volume of earth that affects the electric field as observed over the distance between the measuring electrodes (MN). The apparent resistivities observed therefore involve large volumes of earth and the resistivities obtained by interpretation of the data are average values of large volumes. As the AB distance and depth of investigation increases this average is taken over larger and larger volumes.

The thickness and resistivity of a thick geoelectrical bed that is near the surface can be determined uniquely from a sounding curve. As the depth to a bed increases and its relative thickness (ratio of thickness to depth) diminishes, it becomes increasingly difficult to separate its effect from those of the neighbouring beds unless its resistivity differs drastically from that of the neighbouring beds. Even under optimum conditions its thickness and resistivity cannot be determined separately. Where the bed is more resistive than the neighbouring beds, it is typified by its transverse resistance T which is the product of its thickness and its resistivity; where the bed is more conductive than the neighbouring beds, it is characterized by its longitudinal conductance, S, which is the ratio of its thickness and its resistivity. In practice this means that two equally deep, resistive beds of different thicknesses and resistivities, but with the same T-value can give rise to virtually identical sounding curves. This is referred to as T-equivalence. Similarly, two equally deeply buried conductive layers with different thicknesses and resistivities but with the same S-value give rise to S-equivalent sounding curves. This principle of equivalence can apply even to relatively thick beds if there is a large resistivity contrast between the bed and its neighbours. In the case of T-equivalence the minimum resistivity of the resistive bed and its maximum thickness can be determined while for S-equivalence the maximum thickness and resistivity can be established.

When the target horizon has a resistivity that is intermediate to that of the layers above and below it, its effect on the data is suppressed and only becomes clear when it becomes very thick. This is referred to as the principle of suppression.

The averaging effect causes electrical sounding data to reflect the presence of ensembles of beds of distinct geoelectrical character at depth rather than individual horizons. This means that as a rule the problem of deriving the resistivity distribution with depth from a single sounding curve is indeterminate and that there exists a large family of equivalent solutions that will

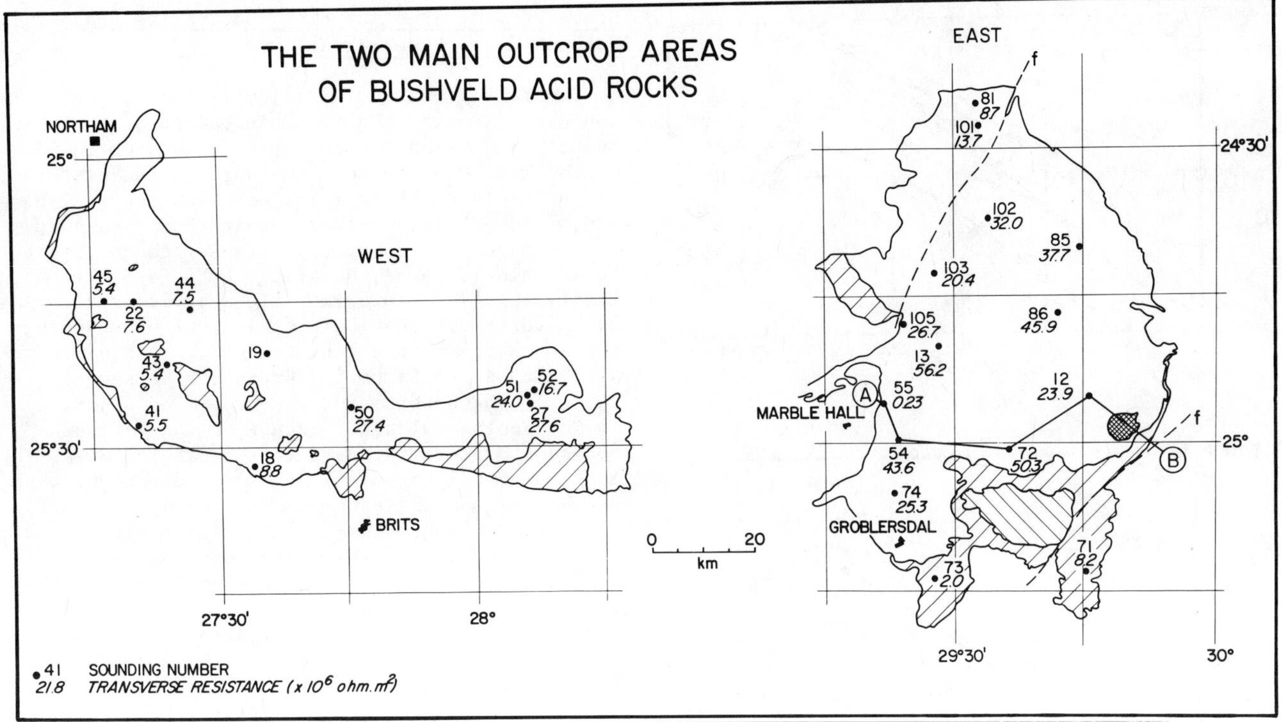

Fig. 7. The positions of the Schlumberger sounding centres on the two main bodies of Bushveld acid rocks showing the sounding numbers and the transverse resistance of the acid rocks at each sounding site. The areas underlain by felsites (hatching slanting from left to right) and granophyres (hatching slanting from right to left) are also indicated. The position of the Spitskop Complex (Fig. 1) is indicated by crossed hatchings in this figure as well as Figs 9 and 10. Section line AB (Fig. 11) is shown in this figure as well as Figs 9 and 10.

fit the field data. To decrease the number of acceptable models constraints are used such as the simultaneous interpretation of data from several neighbouring sounding centres, geological considerations and where possible, borehole information at a sounding centre. The last two constraints are rather weak when dealing with deep sounding data and then the most important constraint is offered by a comparative study of the common characteristics and progressive deformation of data from a series of neighbouring electrical soundings. The final model for the earth underneath each sounding site incorporates all the constraints offered by all the available sounding data and geological information in the vicinity. In addition, this approach constrains the number of beds that has to be incorporated into the model as well as their thicknesses and resistivities over and above the limits imposed by equivalence and suppression.

Van Zijl (1977) discusses the application of the deep Schlumberger sounding technique to studies of the deeper crust in southern Africa.

Geoelectrical Study of the Bushveld Complex

At the time of writing the geoelectrical study of the Bushveld Complex is still in progress. Up to November 1984 a total of 124 deep electrical soundings (average current electrode spacing = 20 km) have been carried out on the various components of the intrusive complex and the sedimentary and volcanic strata underlying the complex. All the soundings except one were carried out by expanding the current emission lines along roads. For one sounding an uncommissioned power line was used as current emission line.

Table I gives the resistivity range for the various lithological units as deduced from the sounding curves. It is clear that the graphitic shale member of the Silverton Formation is the most conductive unit in the succession. A detailed study on outcrops of this member some 15 km east of Pretoria (Fig. 1) showed that at this locality the unit has a total longitudinal conductance of more than 1 000 siemens and a resistivity of less than 0.2 ohm.m. The overlying shales have a resistivity of 200 to 3 000 ohm.m and higher up in the succession the more quartzitic strata with interlayered lava, diabase and shale have a resistivity that ranges from 5 000 – 15 000 ohm.m. Most of the Rustenburg Layered Suite has electrical resistivities in excess of 1 000 ohm.m except for the Upper Zone where the magnetite content causes a lower resistivity. In this zone the resistivity ranges from 100 to

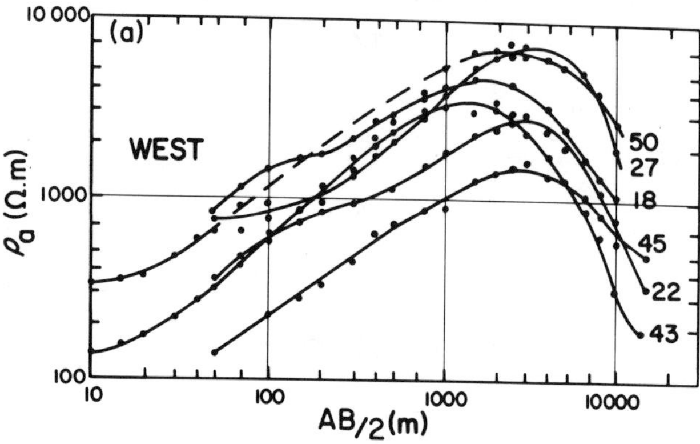

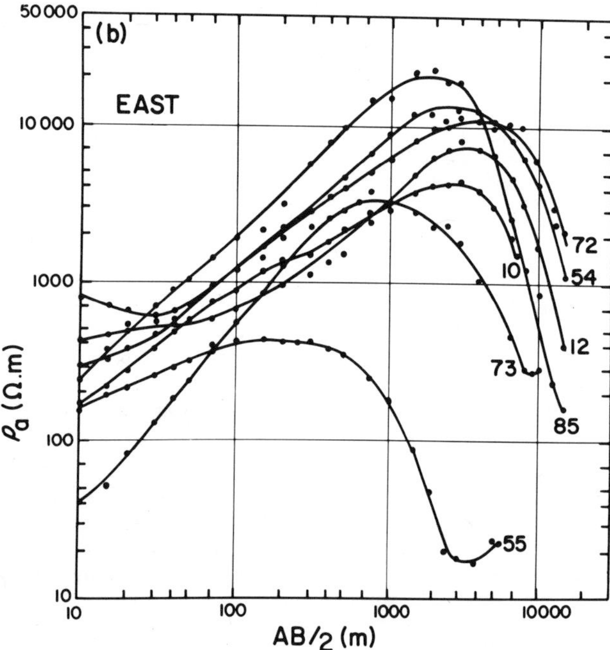

Fig. 8. Representative sounding curves on the
Nebo Granite in a) the western and b) the eastern
compartments. The dots represent the field data
and the curves one-dimensional models that fit the
data. The sounding positions corresponding to the
numbers are shown in Fig. 7. Note that the final
branch of all the sounding curves clearly shows a
conductive zone below the resistive granite.

800 ohm.m. The Rashoop Granophyre has a resistivity of between 5 000 and 15 000 ohm.m, while
the resistivity of the Nebo Granite ranges between
3 500 and 40 000 ohm.m.

In this report only results obtained on the
outcrops of the acid phase of the Bushveld Complex will be discussed in detail, because the interpretation of the other data is still in progress.

Geoelectrical Study of the Nebo Granite and Rashoop Granophyre Suites

Figure 7 gives the distribution of the 26
sounding sites on the two main bodies of Nebo
Granite and Rashoop Granophyre. Sounding curves
on the Nebo Granite from the western and eastern
compartments of the complex are shown in Fig. 8a
and b respectively. The sounding curves are in
all instances T-equivalent curves with an upper
less resistive zone, underlain by a high resistivity zone which overlies a conductive zone. As
was pointed out above, for this type of sounding
curve the transverse resistance T, the minimum resistivity and maximum thickness of the second
layer can be determined.

The geological significance of the various
zones of different resistivity is as follows. The
near-surface low resistivity zone represents the
weathered zone while the underlying high resistivity zone reflects the unweathered granitic rocks.
The zone of lower resistivity that underlies the
granites is somewhat problematic. The two lithological units with the lowest resistivities in
Table I are the Silverton Formation which includes
the graphitic shale horizon and the Upper Zone of
the Rustenburg Layered Suite. Unfortunately it
is impossible to discern on stratigraphic grounds
which one forms the conductive third layer underlying the acid rocks for the different sounding
graphs. For example, in the case of electrical
sounding ES 55 that was measured on the granites
close to and parallel to the contact between the
granite and Silverton Formation, the conductor is
almost certainly the graphitic marker horizon.
At ES 71 situated on the Rashoop Granophyre on a
tongue of granite jutting into the mafic sequence,
surrounded and directly underlain by the Upper
Zone, the conductor is certainly the magnetite-
bearing rocks of the Upper Zone. When it comes
to sounding sites removed from the contacts, it
is impossible to decide on the basis of the stratigraphical and geoelectrical data which of the
conductive zones underlie the granite directly. A
study using all other available geophysical data
is underway to try and solve this problem.

For T-equivalent curves as measured on the
rocks of the acid phase of the Bushveld Complex,
it is possible to obtain a qualitative interpretation in terms of the transverse resistance of
the assemblage of high resistivity beds situated
between the low resistivity surface layer and the
final bed. Figure 7 shows the value of the transverse resistance as determined at each sounding
site. In the eastern lobe the largest value
$(56.2 \times 10^6 \text{ ohm.m}^2)$ is observed at ES 13 and the
lowest $(0.2 \times 10^6 \text{ ohm.m}^2)$ at ES 55 right on the
edge of the granitic body.

On the granite outcrops of the western lobe the
maximum transverse resistance for this resistive
zone is less than $30 \times 10^6 \text{ ohm.m}^2$ and the higher
values all occur in the eastern part of this gra-

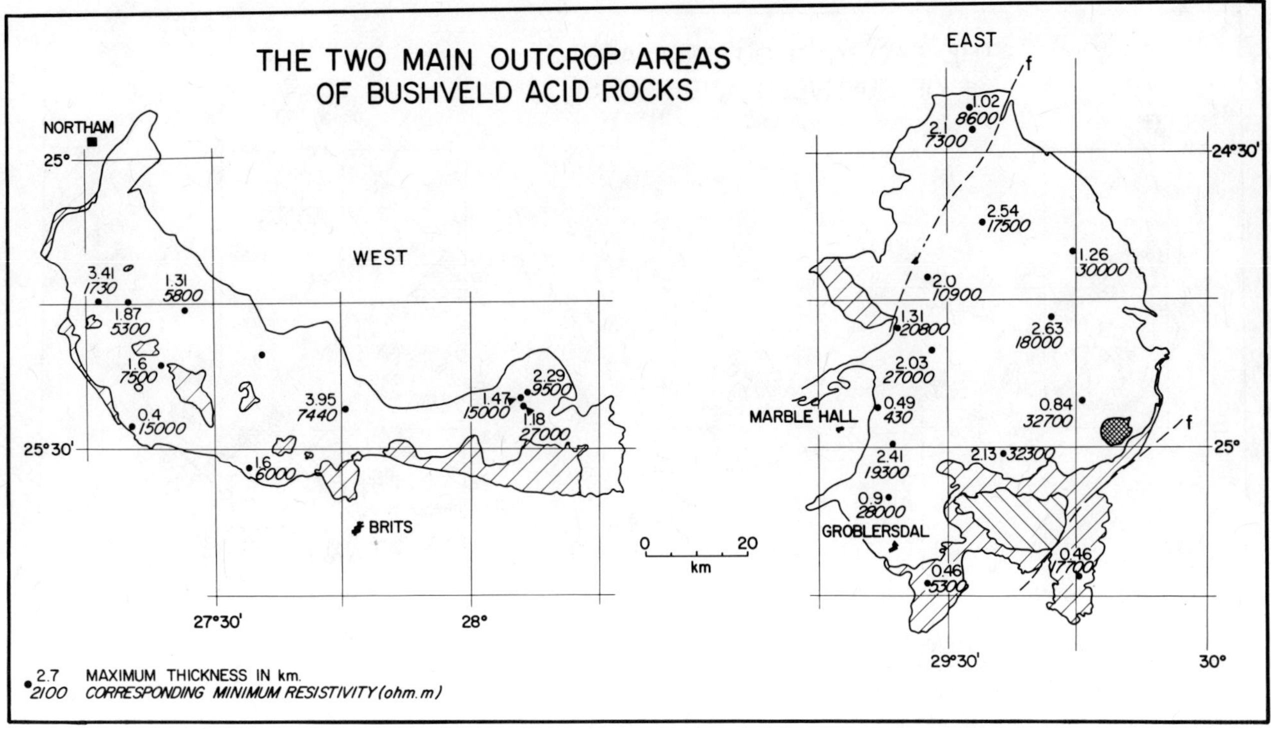

Fig. 9. Map showing values for the maximum thickness and corresponding minimum resistivity for the T-equivalent resistive zone of the acid phase of the Bushveld Complex.

nitic body. The minimum values for T are around 5.5×10^6 ohm.m^2. It is clear that the maximum and average transverse resistance for the granites in the eastern lobe are about double the values observed in the western lobe.

If the resistivity of the granite were to be the same at all sounding sites, the variation in the T-value would have reflected a variation in thickness, but since the granite is not entirely homogeneous, the resistivity of the resistive layer varies from sounding site to sounding site.

Despite the uncertainty regarding the identification of the lithology of the conductor, the geoelectrical data are still ideal to study the distribution of the maximum thickness and minimum resistivity of the granite. To obtain a maximum estimate for the thickness of the resistive zone, it is assumed that the granite consists of a single layer, that the minimum resistivity applies and also that the granite is directly in contact with the conductor. This last assumption may not always be the case, but when a non-granitic, suppressed layer occurs between the granite and the conductor, the true maximum thickness for the granite will be less than the estimate using the above assumption. If the granite consists of two layers, and a resistive zone overlies a less resistive zone, the suppressed layer will of course also be granitic and in such an instance the es-

timate for the maximum thickness may be too low. Although the granites of the Bushveld show a crude stratiform layering, the geoelectrical data obtained so far do not give any clear indication of a concomitant geoelectrical layering within the granitic zone.

To obtain the maximum thickness, a routine based on singular value decomposition was used, as outlined by Johansen (1977). The maximum thickness and minimum resistivity for the T-equivalent resistive zone as given in Fig. 9 therefore represent the maximum and minimum values for the respective parameters for a model that still give a fit to the field data within a maximum specified r.m.s. error of 5 per cent.

Figure 9 shows that in the eastern compartment the maximum thickness for the Nebo Granite is about 2.6 km, while it is almost 4 km in the western lobe. This is in contrast with the higher T-values in the eastern lobe as compared to the western lobe. The reason for this is the difference in minimum resistivities for the resistive zone in the two areas. The representative sounding curves in Fig. 8 clearly show that the minimum resistivity of the resistive zone for the western lobe is as a rule markedly less than the minimum for the curves representative of the eastern lobe. The minimum resistivities for the resistive granites as given in Fig. 9 are in general

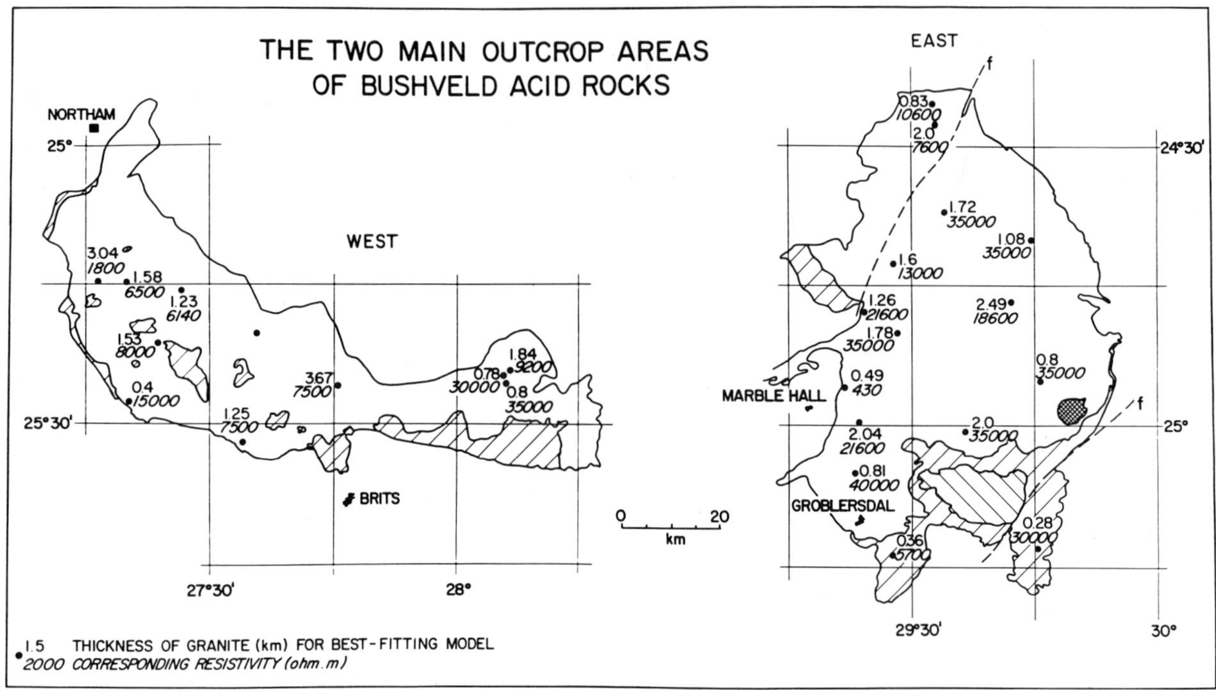

Fig. 10. Map showing values for the thickness and resistivity parameters of the best-fitting
models for the acid phase of the Bushveld Complex.

below 10 000 ohm.m in the western lobe and above
10 000 ohm.m in the eastern lobe. The cause of
this marked difference in the minimum resistivity
between the Nebo Granite in the western and eastern parts of the complex is not entirely clear,
but the western lobe is characterized by the occurrence of numerous diabase dykes intruded into the
granite.

Figure 10 gives the thickness and resistivity
distribution of the granites for the best-fitting
models. In some instances the resistivity of the
resistive zone had to be constrained to obtain
realistic values in the modelling process. From
a study of the sounding curves a resistivity of
35 000 – 40 000 ohm.m is regarded to be a realistic
upper limit for the resistive zone of the eastern
lobe. In the western lobe an upper limit of
15 000 ohm.m was chosen at ES 41, 30 000 ohm.m at
ES 51 and 35 000 ohm.m at ES 27. The thickness of
the best-fitting models for the Nebo Granite in
the eastern lobe of the complex varies from 280 m
at ES 71 on the granophyre to 2 490 m at ES 86.
Five soundings (ES 13, 54, 72, 86 and 101) in the
central zone of this granitic body yielded maximum thicknesses of 2 km or more to the bottom of
the resistive zone. Towards the edges the granite
gets thinner. Except where the acid rocks become
thin (e.g. ES 55 and 71) the resistivity of this
layer is in excess of 10 000 ohm.m. Figure 10
also shows that the resistivity of the granite
layer for the best-fitting models in the western
lobe is markedly lower. Except for two curves
mentioned before the resistivity of this zone is

below 10 000 ohm.m in this area. The thickness of
granite for the best-fitting models is relatively
thin in the western part and increases to around
3.6 km towards the centre of the granite body in
the western lobe of the complex.

Figure 11 presents an interpretation for the
geoelectrical data along a Section AB (Fig. 7)
through the granitic rocks of the eastern lobe
of the Bushveld Complex. On this section the
thickness for the acid rocks for the best-fitting
models as well as the maximum thickness are indicated. According to the interpretation of the
Schlumberger sounding data the thickness of the
granitic intrusion increases from east to west
along this section with the thickest part of the
body near the western edge. At ES 55 the conductor below the resistive zone is the graphitic
shale marker in the Silverton Formation. At ES 12
it is most probably the Upper Zone of the Rustenburg Layered Suite, because xenoliths derived
from this zone occur in the alkaline rocks of the
nearby Spitskop Complex (Fig. 1) that intruded
through the Nebo Granite (Verwoerd, 1967). At
ES 54 and ES 72 the nature of the conductor is
still uncertain.

Discussion of Geoelectrical Results

The geoelectrical study discussed here represents the first artificial source geophysical investigation aimed at determining a regional structural model for the Bushveld Complex. The size of
the complex combined with the near-horizontal atti-

202   DE BEER ET AL.

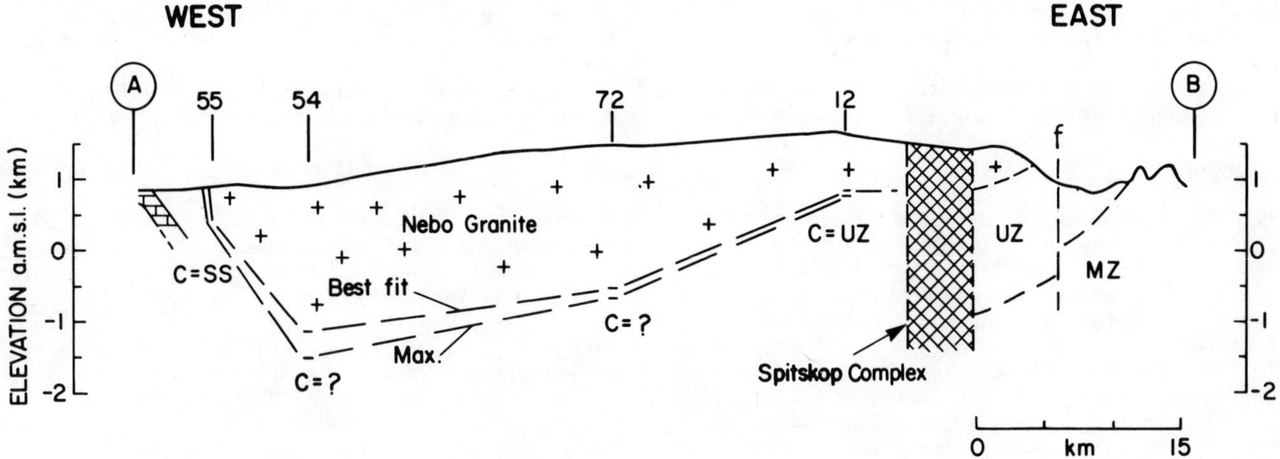

Fig. 11. Geological section AB (Fig. 7) through the acid rocks of the Eastern Bushveld as based on an interpretation of the Schlumberger sounding data. The maximum depth to the conductor underlying the resistive acid rocks and the same parameter for the best-fitting models are indicated (C = conductor, SS = Silverton Shale Formation, UZ = Upper Zone and MZ = Main Zone).

tude of the igneous layering and the presence of electrically conductive marker horizons in the generally very resistive igneous environment makes this intrusive structure an ideal target for this type of study.

The more than 120 Schlumberger soundings carried out up to the time of writing provide a clear geoelectrical stratigraphy for the Bushveld Complex. In this stratigraphy the graphitic shale that occurs in the Silverton Formation of the Pretoria Group proved to be the most conductive unit and a most useful marker horizon. The second conductive marker is the Upper Zone of the Rustenburg Layered Suite. This zone is not as conductive as the graphitic shale and has a lower longitudinal conductance, S, but is still much more conductive than the younger granites and older Main Zone gabbros and norites.

The interpretation of the geoelectrical results for the acid phase of the complex shows that the granitic rocks have a maximum thickness of less than 2.7 km in the eastern lobe and less than 4 km in the western compartment. The best-fitting models tend to be in general agreement with the interpretation of the gravitational data in the area (Table VI). Although the conductive horizon underlying the granite in many instances cannot be identified uniquely, it nevertheless proved an

TABLE VI. Comparison Between Geoelectrical and Gravitational Results

| ES Number | Geoelectrical Results | | Gravitational Results |
| | Thickness (km) | | Thickness (km) |
	Maximum	Best Fit	
13	2.0	1.8	2.2[2]
43	1.6	1.5	1.0[1]
44	1.3	1.2	1.6[1]
54	2.4	2.0	2.0[2]
74	0.9	0.8	1.5[3]
102	2.5	1.7	1.2[2]

[1]Walraven and Darracott (1976)
[2]Molyneux and Klinkert (1978)
[3]Hattingh (1980)

ideal marker horizon for the determination of the maximum thickness of the Bushveld acid phase.

Constraints on the Gross Structure and Emplacement History of the Bushveld Complex

Palaeomagnetic Constraints

1. The igneous layering of the Main Zone and possibly also of the Critical Zone in the areas sampled in sufficient detail was originally horizontal. The present dip was acquired after these zones had cooled to below their respective Curie temperatures.

2. The results indicate that the Upper Zone acquired its remanent magnetization with its igneous layering in its present orientation. This, together with the first constraint, mean that the subsidence of the lower zones in the Complex occurred before the emplacement and cooling of the Upper Zone.

3. The stratigraphic levels at which polarity changes occur do not correspond with the recognized lithological subdivisions and the polarity reversals occur at different levels in the eastern and western Bushveld Complex. This might be taken to imply that the magma chamber(s) of the eastern and western Complex were supplied with similar batches of magma - but at different times. These data are consistent with recent models of the magmatic evolution of the Bushveld in which large inflows of gabbroic magma occurred at the base of the chambers and superelevated the pre-existing magmas which later crystallized to give the sequence above the pyroxenite marker (Sharpe, in press).

Geoelectrical Constraints

1. The maximum thickness estimates for the acid phase of the complex is an important structural constraint.

2. The presence of the conductive zone underlying the Nebo Granite and Rashoop Granophyre in the areas discussed form an important stratigraphical constraint. If additional data can help to resolve the present ambiguity about the identity of the conductor it can form a major constraint in the reconstruction of the tectonic history of the igneous complex.

3. The marked difference in the minimum resistivity between the Nebo Granite in the western part and the eastern part of the complex would seem to be due to more fracturing and dyke intrusions that lowered the resistivity of the granite in the western lobe. This factor places a constraint on the post-Bushveld tectonic history of the complex.

Acknowledgements. The first two authors would like to thank all members of the Geophysics Division of the National Physical Research Laboratory that contributed to the geoelectrical study. Financial assistance for this research is provided through the South African National Geoscience Programme. The third author would like to thank the University of Pretoria for a research grant that made the palaeomagnetic study possible. Mrs M van Wyk is thanked for typing the manuscript.

References

Barrett, D.M., Jacobsen, J.B.E., McCarthy, T.S., and Cawthorn, R.G., The structure of the Bushveld Complex south of Potgietersrus, as revealed by a gravity survey. Trans. geol. Soc. S. Afr., 81, 271-276, 1978.

Biesheuvel, K., An interpretation of a gravimetric survey in the area west of the Pilanesberg in the Western Transvaal, Spec. Publ. Geol. Soc. S. Afr., 1, 266-282, 1970.

Cousins, C.A., The structure of the mafic portion of the Bushveld Igneous Complex, Trans. geol. Soc. S. Afr., 62, 179-201, 1959.

Dunlop, D.W., Hanes, J.A. and Buchan, K.L., Indices of multidomain magnetic behaviour in basic igneous rocks: alternating field demagnetization, hysteresis and oxide petrology, J. Geophys. Res., 78, 1387-1393, 1973.

Gough, D.I., and Van Niekerk, C.B., A study of the palaeomagnetism of the Bushveld gabbro, Phil. Mag., 4, 126-136, 1959.

Hall, A.L., The Bushveld Igneous Complex of the central Transvaal, Mem. Geol. Surv. S. Afr., 28, 54 pp., 1932.

Hattingh, P.J., The structure of the Bushveld Complex in the Groblersdal-Lydenburg-Belfast area of the Eastern Transvaal as interpreted from a regional gravity survey, Trans. geol. Soc. S. Afr., 83, 125-134, 1980.

Hattingh, P.J., A palaeomagnetic investigation of the layered mafic sequence of the Bushveld Complex, D.Sc thesis (unpublished), Univ. of Pretoria, 177 pp, 1983.

Hattingh, P.J., The palaeomagnetism of the main zone of the Bushveld Complex (in prep.).

Hunter, D.R., The regional setting of the Bushveld Complex (An adjuct to the Provisional Tectonic Map of the Bushveld Complex). Econ. geol. res. Unit, Univ. of the Witwatersrand, 18 pp, 1975.

Hunter, D.R., Some enigmas of the Bushveld Complex, Econ. Geol., 71, 229-248, 1976.

Hunter, D.R., and Hamilton, P.J., The Bushveld Complex, 107-173. In: Tarling, D.H. (ed). Evolution of the Earth's Crust, Academic Press, London, 1978.

Johansen, H.K., A man/computer interpretation system for resistivity soundings over a horizontally stratified earth, Geophys. Prosp., 25, 667-691, 1977.

Koefoed, O., Geosounding Principles, 1. Resistivity sounding measurements, Elsevier Scientific Publishing Co., Amsterdam, 276 pp, 1979.

Kunetz, G., Principles of direct current resistivity prospecting, Gebrüder Borntraeger, Berlin, 103 pp, 1966.

Lowrie, W., and Fuller, M., On the alternating field demagnetization characteristics of multi-

domain thermoremanent magnetization of magnetite, J. Geophys. Res., 76, 6339-6349, 1971.

Molyneux, T.G., and Klinkert, P.S., A structural
interpretation of part of the eastern Mafic Lobe
of the Bushveld Complex and its surrounds,
Trans. geol. Soc. S. Afr., 81, 359-368, 1978.

Sharpe, M.R., Evolution of the Bushveld magma
chambers. Inst. geol. Res. Bushveld Complex,
Univ. of Pretoria, Res. Rep., 30, 57 pp, 1980.

Sharpe, M.R., Strontium isotopic evidence for preserved density stratification from the main zone
of the Bushveld Complex, South Africa, Nature,
(in press).

Sharpe, M.R., and Bahat, D., The Great Dyke - a
possible expression of combined hydrofracturing
and Hertzian fracture. Inst. geol. Res. Bushveld Complex, Univ. of Pretoria, Res. Rep., 26,
13 pp, 1981.

Smit, P.J., Hales, A., and Gough, D.I., The gravity survey of the Republic of South Africa,
Handb. Geol. Surv. S. Afr., 3, 484 pp, 1962.

South African Committee for Stratigraphy (SACS),
Stratigraphy of South Africa, Part 1 (Comp.
L.E. Kent). Lithostratigraphy of the Republic
of South Africa, South West Africa/Namibia, and
the Republics of Bophuthatswana, Transkei and
Venda, Handb. Geol. Surv. S. Afr., 8, 690 pp,
1980.

Tankard, A.J., Jackson, M.P.A., Eriksson, K.A.,

Hobday, D.K., Hunter, D.R., and Minter, W.E.L.,
Crustal evolution of South Africa - 3.8 billion
years of Earth History, Springer Verlag, Berlin,
523 pp, 1982.

Van der Merwe, M.J., The layered sequence of the
Potgietersrus limb of the Bushveld Complex,
Econ. Geol., 71, 1337-1351, 1976.

Van Zijl, J.S.V., Electrical studies of the deep
crust in various tectonic provinces of southern
Africa. In: Heacock, J.G. (ed). The Earth's
crust, Am. Geophys. Un. Monogr., 20, 470-500.
1977.

Vermaak, C.F. and Lee, C.A., Bushveld and kindred
complexes. In: D.R. Hunter (ed). The Precambrian of the Southern Hemisphere, Developments
in Precambrian Geology, 2, Elsevier, Amsterdam
882 pp, 1981.

Verwoerd, W.J., The carbonatites of South Africa
and South West Africa, Handb. Geol. Surv. S.
Afr., 6, 452 pp, 1967.

Von Gruenewaldt, G., A review of some recent concepts of the Bushveld Complex, with particular
reference to sulfide mineralization, Can,
Mineral., 17, 133-256, 1979.

Walraven, F., and Darracott, B.W., Quantitative
interpretation of a gravity profile across the
western Bushveld Complex, Trans. geol. Soc. S.
Afr., 79, 22-26, 1976.

THE STRUCTURAL-STRATIGRAPHIC DEVELOPMENT OF PART OF THE NAMAQUA METAMORPHIC
COMPLEX, SOUTH AFRICA - AN EXAMPLE OF PROTEROZOIC MAJOR THRUST TECTONICS

G. van Aswegen, D. Strydom, W.P. Colliston, H.E. Praekelt, A.E. Schoch, H.J.
Blignault, B.J.V. Botha, and S.W. van der Merwe

Department of Geology, University of the Orange Free State,
P.O. Box 339, Bloemfontein 9301, South Africa

Abstract. The gneisses, metavolcanics and
metasediments of the western Namaqua mobile belt,
South Africa, underwent horizontal tectonism
resulting in extreme deformation and metamorphic
transformation. Detailed and regional mapping has
resulted in the recognition of several stratigraphic groups and suites: the Orange River Group,
Okiep Group and Bushmanland Group, intruded by the
Vioolsdrif Suite, Gladkop Suite, Little Namaqualand Suite and Spektakel Suite in the period 1100
Ma to 1950 Ma ago. Large tectonic domains are
separated by major thrust zones and extensive
subvertical shear zones: the Groothoek, Skelmfontein and Geselskapbank thrust zones, and the
Buffels River, Ratelpoort North, Steenbok and
Tantalite Valley shear zones. Sections across
specific structures at Geselskapbank, Dabenoris,
Haramoep, Aggeneys and Pella reveal that the
thrusting operated from the northeast, and that
the distances of movement may exceed 100 km in
some cases.

Introduction

The Proterozoic rocks of Namaqualand and
Bushmanland (Figures 1 and 2) consist of gneisses,
metavolcanics and metasediments. The metasediments
have received scant attention in the past because
of structural complexities [Gevers et al., 1937;
Joubert, 1971; Clifford et al., 1975; Kröner and
Blignault 1976], but recently, investigations have
been prompted by the discovery of economic strata-
bound sulphide deposits in the early 1970's [Jou-
bert, 1972, 1974; Rozendaal, 1977; Moore, 1977;
Blignault et al., 1983; Colliston, 1983]. Geological maps of parts of the region include the
1:100 000 sheets by P.J. Joubert [1975] ("Geology of the Pofadder and Aggeneys regions", 2
sheets, Precambrian Research Unit, University of
Cape Town), and D. Strydom [1982] ("Geselskapbank-Areb", 1 sheet, Dept. of Geology, University
of the Orange Free State), as well as the
1:250 000 map of the "Namaqualand Geotraverse" in
Blignault et al., [1983]. The results of detailed
investigations of areas such as Gamsberg and the
Aggeneys Mountains are given in several M.Sc.

theses (A. Rozendaal 1975, R.D. Lipson 1978, F.
Martens 1979, D.P. Stedman 1980 and H.E. Praekelt
1983, respectively at the Universities of Stellen-
bosch, Orange Free State, Witwatersrand, Witwa-
tersrand and Orange Free State), and Ph.D. theses
(A. Rozendaal 1982, D. Strydom 1985, at the
Universities of Stellenbosch and the Orange Free
State).

The supracrustal rocks (glassy metaquartzite,
biotite sillimanite schist, amphibolite, mafic and
felsic gneiss) have been subdivided into several
groups, namely the Orange River Group, the Okiep
Group and the Bushmanland Group [Blignault 1980]
as well as numerous subgroups and formations. The
plutonites and metaplutonites (granites and orthogneisses) have likewise been subdivided into
suites, such as the Vioolsdrif Suite [Blignault,
1980] of about 1950 Ma age [Reid, 1977], and the
Spektakel Suite [Marais and Joubert, 1980b] of
approximately 1100 Ma age [Clifford et al., 1975].

The dominantly sub-horizontal attitude of the
rocks has led, in the past, to an oversimplified
but popular interpretation of the stratigraphy
which involved a regionally distributed sequence
of augen gneiss ("basement"), "pink gneiss",
schist, quartzite, amphibolite and mafic gneiss
[Ryan et al., 1982]. It is the purpose of this
paper to demonstrate that this simplistic sequence
is incorrect [Praekelt et al., 1983; Praekelt et
al., 1984], and that the horizontal structure is
mainly the result of thrust tectonics [Colliston
et al., 1984; Strydom, 1984]. The stratigraphy of
the area proves to be very involved and is complicated by several episodes of major thrusting, followed by subvertical folding and shearing.

General Lithostratigraphic Features

The distribution of the major stratigraphic
units recognized at present is illustrated in
Figure 2. The stratigraphic classification is
complicated by structural disordering which will
be discussed below.

The lithologic compositions of the Okiep and
Bushmanland Groups (grouped together as "metasedimentary sequence" in Figure 2) are similar when

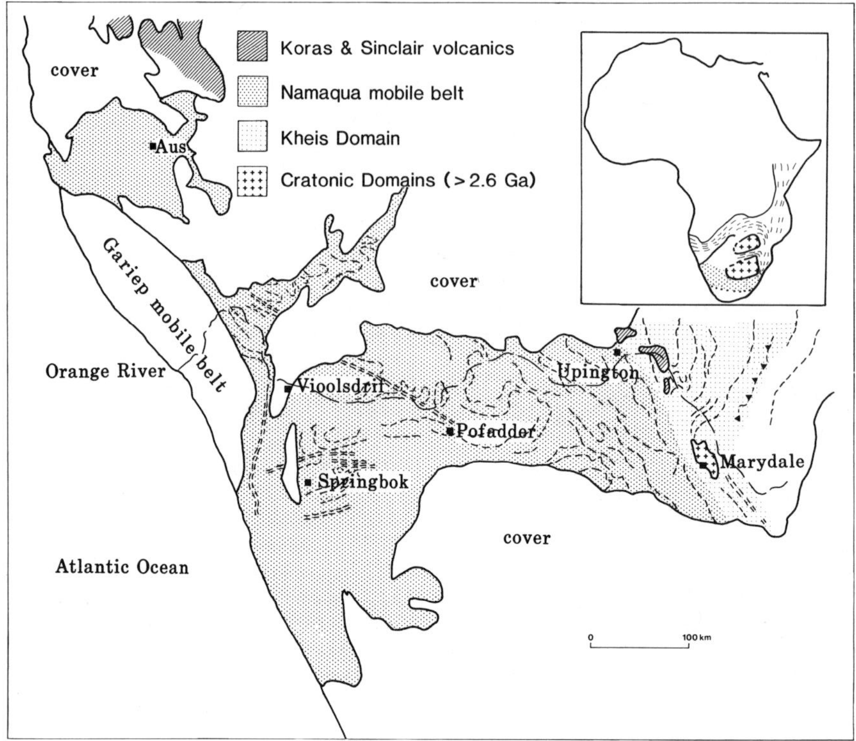

Fig. 1. Locality map.

seen in terms of regional resolution, but many differences emerge when detailed investigations are undertaken. The Okiep Group is divided into the Een Riet and Khurisberg Subgroups by some authors [Joubert et al., 1980], but more recent work has indicated that part of the latter belongs to the Bushmanland Group ("Geselskapbank-Areb", 1:100 000 geological map by D. Strydom, 1982, Dept. of Geology, University of the Orange Free State).

The stratigraphic succession defining the Bushmanland Group has recently been modified. The original subdivision [Colliston, 1979; Blignault, 1980], has the Aggeneys Subgroup at the base (biotite sillimanite schist, quartzite, iron formation, quartz muscovite schist and conglomerate), followed by metavolcanics of the Hom and Guadom Subgroups, with the Pella Subgroup (similar lithology to the Aggeneys Subgroup) at the top. The most recent compilation [Blignault et al., 1981; quoted in Blignault et al., 1983], combines the Pella Subgroup and the Aggeneys Subgroup under the latter name. Blignault et al. [1983] also amalgamate the volcanics of the Hom and Guadom Subgroups, which structurally underlie the Aggeneys Subgroup, under the term Haib Subgroup. Other adaptations can be anticipated, stimulated by the results of increasingly more detailed field observations which are in progress. We have decided to employ a modified version of the original proposal [Blignault, 1980] mentioned above.

Regional Tectonic Style

The disposition of the major supracrustal units and metaplutonites is controlled by easterly trending thrust zones and subvertical shear zones (Figures 2 and 3). The sigmoidal traces of the Groothoek and Skelmfontein thrust zones [Blignault, et al., 1983; Strydom, 1984; Colliston et al., 1984] indicate large scale folding and shearing superimposed on the thrusting. The extremities of the area under consideration are demarcated by other regional structures such as the Buffels River, Steenbok [Blignault et al., 1983] and Tantalite Valley [Beukes and Botha, 1975-1976] shear zones which will not be discussed in this paper.

The simplified map (Figure 2) accommodates the possible equivalence of the Okiep and Bushmanland Groups and of the Gladkop and Vioolsdrif Suites [Van Aswegen, 1983, 1984]. These problems of correlation can only be solved by much detailed work in areas of good exposure. The present state of knowledge concerning correlation and structural evolution will be illustrated below with data from small areas.

Example Structures

A simplified geological map of the region around Geselskapbank is shown in Figure 5 (identified by a box in Figure 2). The sections CD and EF (Figures 2 and 5) show the juxtaposition of

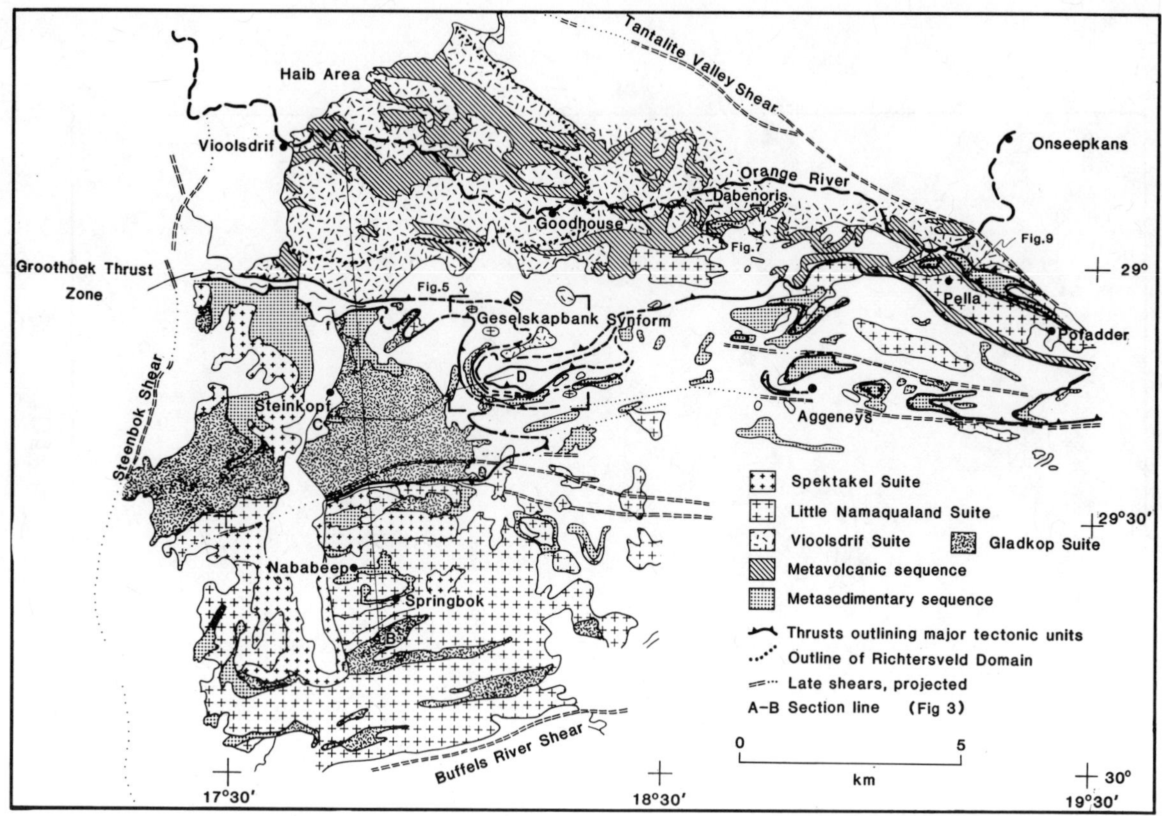

Fig. 2. Simplified geological map of part of the Namaqua mobile belt (the areas indicated are shown in greater detail in Figures 5, 7 and 9).

different stratigraphic units belonging to the Bushmanland Group and the Little Namaqualand and Gladkop Suites. The upper position in the structural column is occupied by units (the Geselskapbank Formation and the Naab Suite) with lithologies that cannot be matched elsewhere in the entire Bushmanland region. In the Naab Suite there are, for example, mafic dykes (dolerites and lamprophyres) which are rare in the rest of the region. The Geselskapbank Formation exhibits a partly retrogressed granulite facies mineralogy in contrast to the underlying rocks that belong to the amphibolite facies.

The most recent syntheses reveal that the Geselskapbank region exhibits a number of thrust sheets [Strydom, 1984]. Geometric analyses, including controlled sections as exemplified by Figure 6 and the application of the Ramsay and Graham technique to the Groothoek and Skelmfontein thrust zones [Blignault et al., 1983] has shown that tectonic movement must have been from the northeast to the southwest. This conclusion is

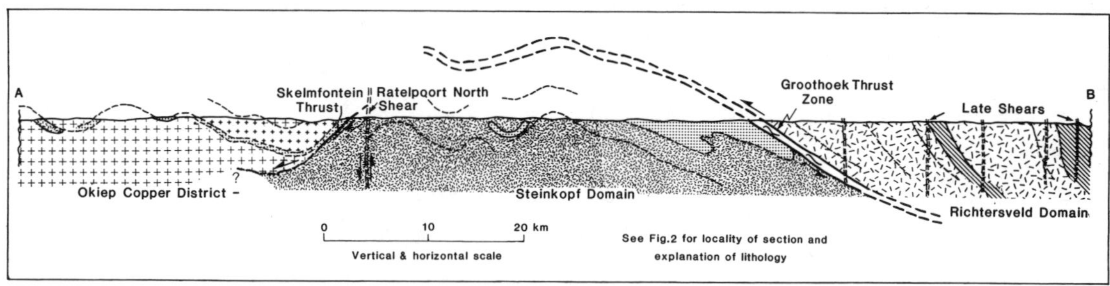

Fig. 3. Controlled geological section from south (A) to north (B), depicting the geometric relationships between the three major tectonic domains. The position of the section is shown in Figure 2. Adapted from Blignault et al. [1983].

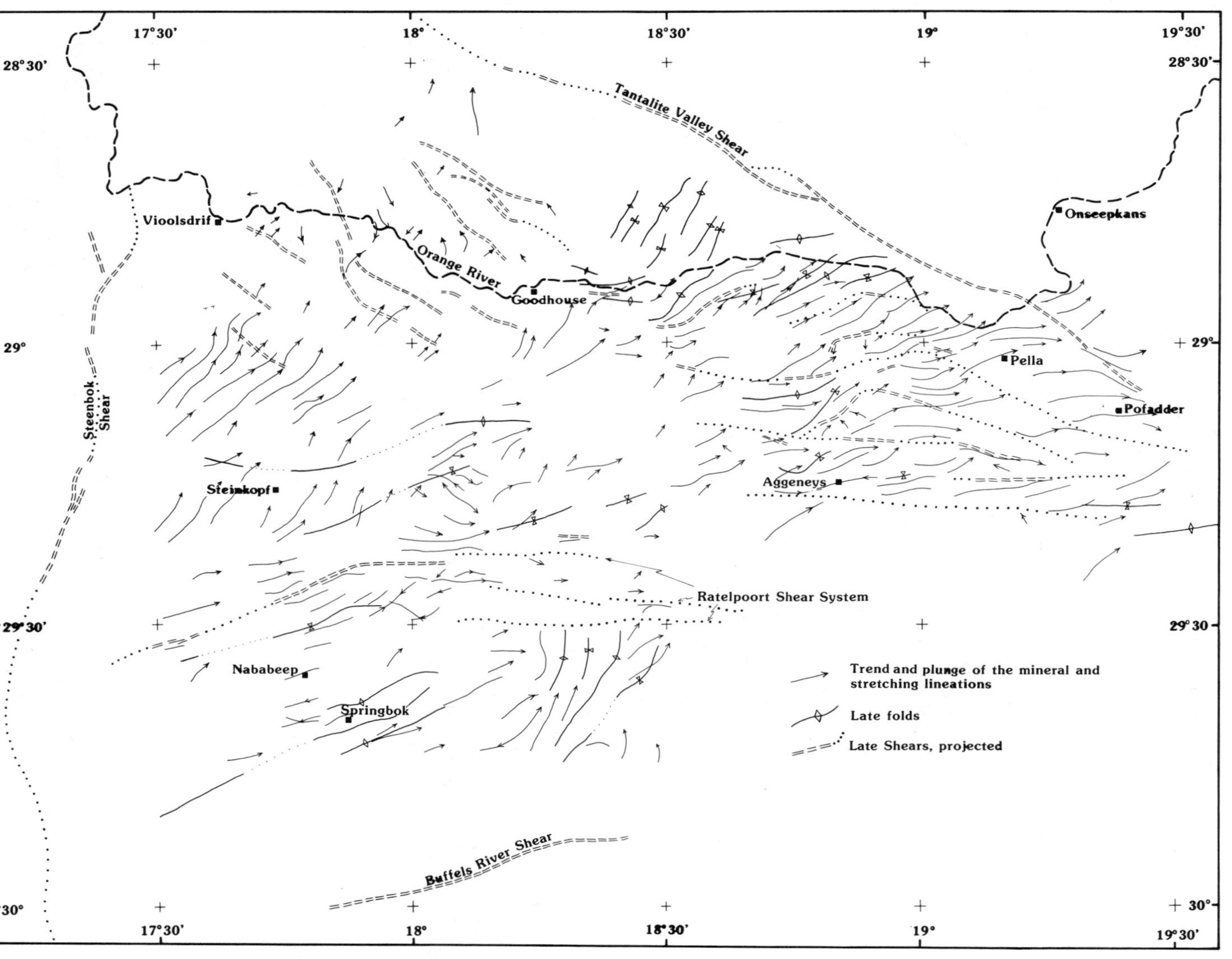

Fig. 4. Regional lineation pattern (after Blignault et al. [1983]).

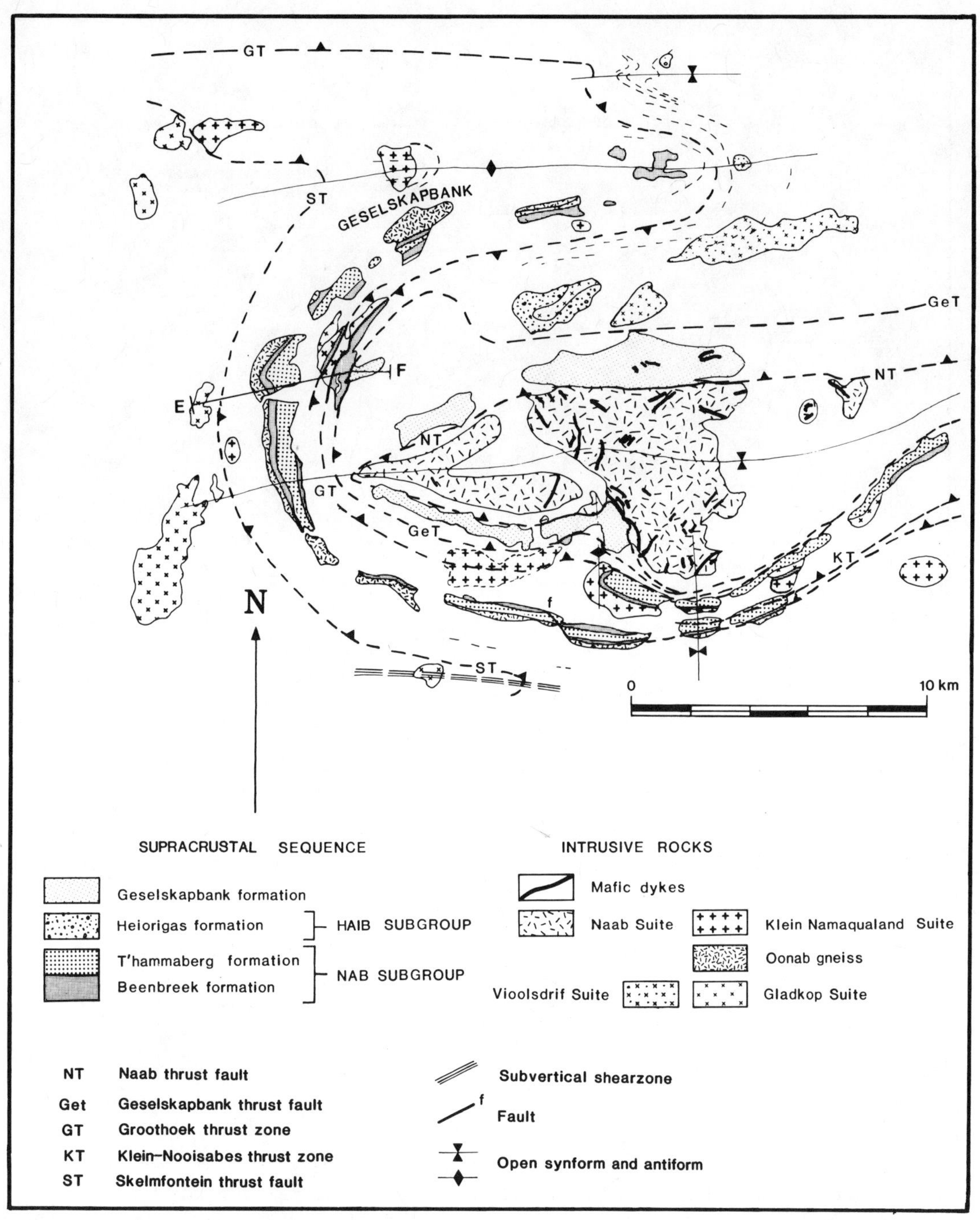

Fig. 5. Geological map of part of Geselskapbank synform.

corroborated by the regional lineation pattern (Figure 4). Distances of movement in the order of 100 km have been calculated (for example Blignault et al., [1983], p. 22).

The second example is taken from the north of the area under discussion (indicated by a box in Figure 2). A simplified structural-stratigraphic map of the region around Dabenoris is shown in Figure 7 (after Blignault et al., [1983]). Section GH (Figure 8) shows a regional sole thrust plane and associated imbricate structure, as well as isoclinal folding of the supracrustal rocks. The strain associated with the thrusting was inhomogeneously distributed throughout the thrust sheets (dermal tectonics) rather than as in the familiar Rocky Mountains type of deformation (epidermal

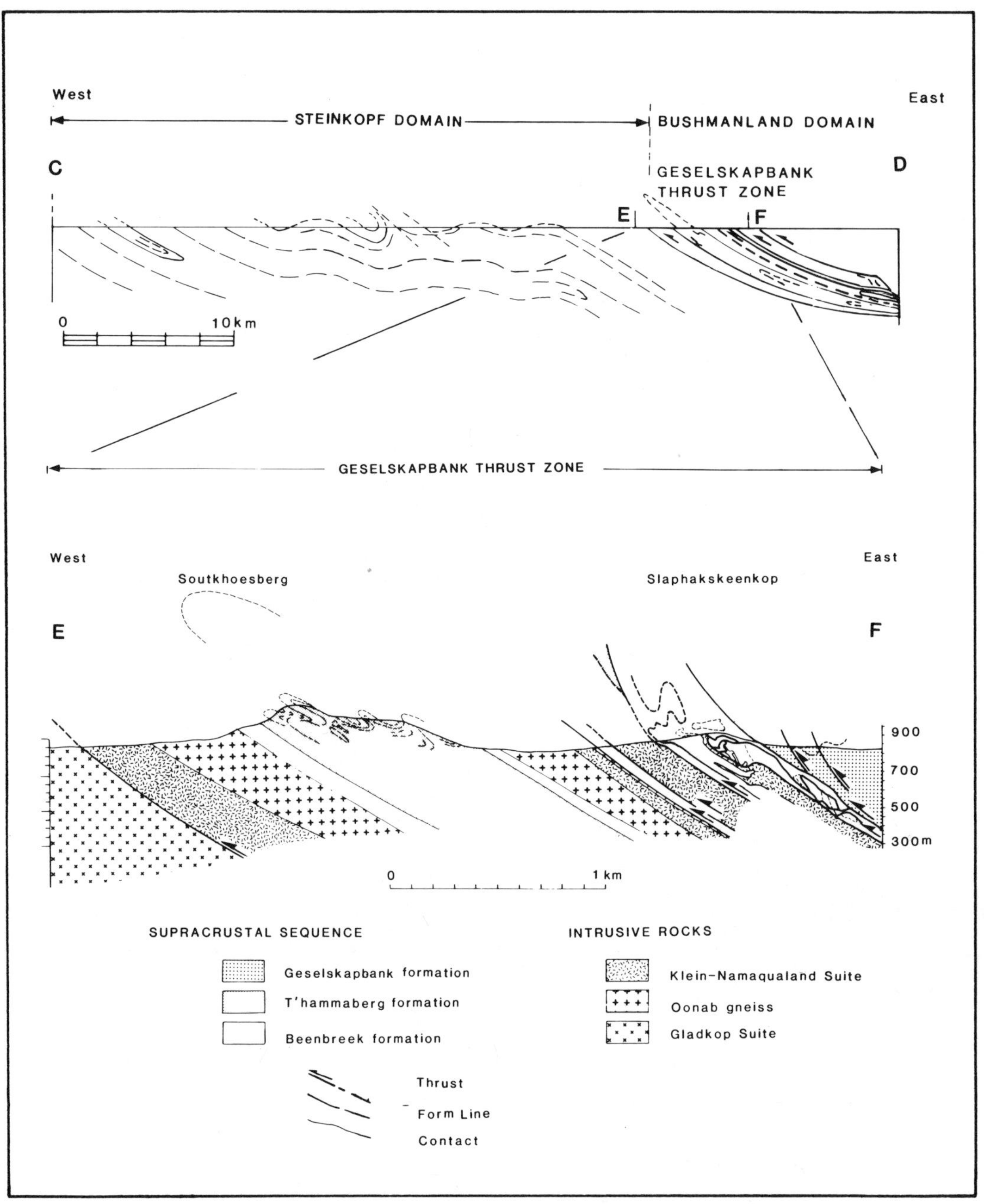

Fig. 6. Controlled cross sections (CD and EF) illustrating relationships between the Steinkopf Domain and the Bushmanland Domain. The locations of the sections are shown in Figures 2 and 5.

tectonics). A notable feature resulting from the high strain is a macroscopic sheath fold shown on the map (Figure 7). The axis of the sheath fold is parallel to the thrust movement direction [Blignault et al., 1983], i.e. parallel to section GH.

In the Aggeneys Mountains (Figure 2), several thrust sheets have now been identified [Colliston et al., 1984]. This new structural model replaces the older view of a static stratigraphic sequence involving basal "pink gneiss", biotite sillimanite schist, metaquartzite, iron formation (with mineralization), quartz-muscovite schist and amphibolite at the top. The contact between the "pink gneiss" and the metasediments proves to be a sole thrust into which small thrust planes merge. The structural architecture of the Aggeneys

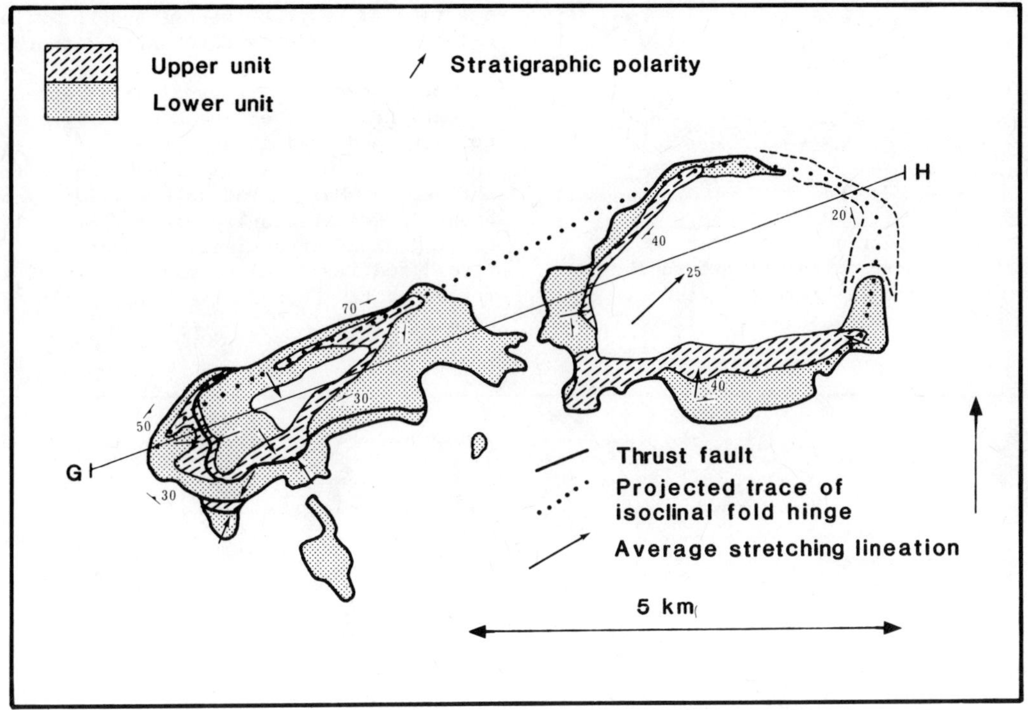

Fig. 7. Simplified geological map of the Dabenoris Mountains (after Blignault et al. [1983]). Note the curved nature of the projected trace of the isoclinal fold hinge, outlining a mega sheath fold.

Mountains and environs is complicated by modified duplexes and major fold nappes. The nearby metasediment occurrence at Gamsberg [Rozendaal, 1977; Colliston, 1983; Odling, 1983] is closely related to the Aggeneys Mountains in terms of structure and stratigraphy, but the details are still obscure at present.

At Haramoep, to the north of the Aggeneys Mountains (Figure 2), a regional thrust plane separates various formations of the Bushmanland Group. Mapping has disclosed structures such as isoclinal folding in the footwall and a series of imbricate slices in the hanging wall [Colliston et al., 1984]. The major thrust (Wortel thrust) can be followed for a distance of at least 100 km along strike. The stratigraphic succession in the hanging wall (Wortel Formation) is interpreted to have moved for a minimum of 40 km in a southwesterly direction.

A section accross strike east of Pella (Figures 2 and 9) shows metasediments (the Pella sequence) forming a structural wedge within the metavolcanics of the Haib Subgroup. The contact between the metasediments and the metavolcanics represents a regional thrust which may be traced out for approximately 90 km on strike. Imbricate structures within the wedge form a duplex.

It has already been shown [Blignault et al., 1983] that large scale thrusting involving predominantly granitoid materials was the principal tectonic style in the Namaqualand region to the west. These thrust sheets were hot and ductile at

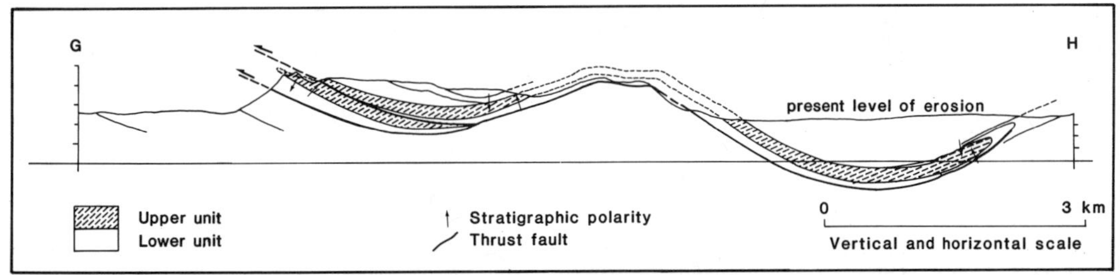

Fig. 8. Controlled cross section GH through the Dabenoris structure, parallel to the movement direction. The location of the section is indicated in Figure 7.

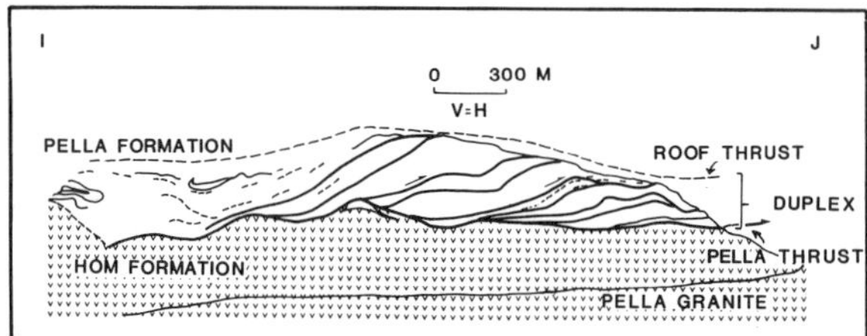

Fig. 9. Controlled section across part of the Pella mountains from northwest (I) to southeast (J), illustrating detailed thrust features. The location of the section is shown in Figure 2.

the time of nappe transport and the highest metamorphic grade is associated with the augen gneiss sheets. It is suggested that the large volume of syntectonic granitoids represents a primary heat source. The dominantly horizontal contacts and foliation of the various granitoid units is the most striking structural feature of the region [Marais and Joubert, 1980a, 1980b]. Specific granitic units are correlatable with finer grained granoblastic tectonites ("granulites" and leptites) formerly regarded as separate metamorphites [Benedict et al., 1964; Holland and Marais, 1983].

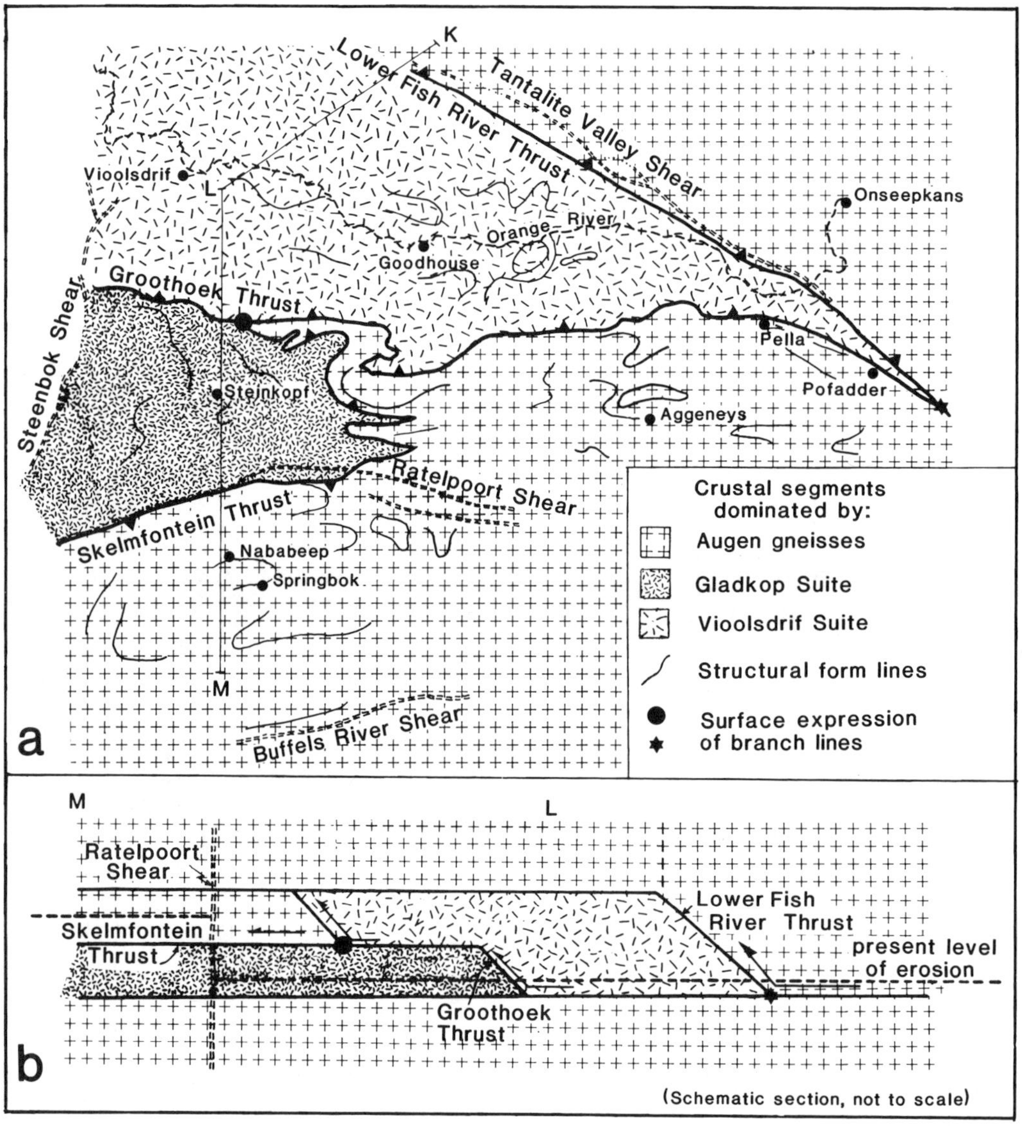

Fig. 10. Sketch map (a) of the juxtaposition of diffferent tectonic domains as a result of regional thrusting. Only the largest thrust zones are shown. The schematic cross section (B) illustrates the structural geometry prior to late folding. The present southerly dip of the Skelmfontein thrust zone is the result of late folding.

Conclusions

It is concluded that the Namaqua orogeny was characterized by large scale deep seated thrusting. The more conspicuous structural features, including open folds with wavelengths of 5 to 30 km, and major subvertical shear zones are regarded as less important than thrusts for an understanding of the present distribution of stratigraphic units. The major regional thrust zones demarcate well-defined tectono-stratigraphic domains (Figure 10). The authors believe that the apparent dominance of the open folds and late shears caused previous geological syntheses to miss the importance of horizontal tectonics.

The recognition of major horizontal thrusting episodes has resulted in a completely new conceptual model for the Bushmanland region. The thrusting was accompanied by severe deformation in the sense that many thrust planes were isoclinally folded.

Acknowledgements. The authors are indebted to the University of the Orange Free State Research Fund, to the Council for Scientific and Industrial Research, to the Geological Survey of South Africa, and to several mining companies for financial aid and access to properties (Goldfields of South Africa; O'okiep Copper Company; Anglo American Corporation; Black Mountain Mineral and Dev. Co.). Mrs. P. Swart typed the manuscript, and the figures were prepared by Mr. A. Felix and Mr. J.R. Stallenberg.

References

Benedict, P.C., Wiid, D., Cornelissen, A.K., and Staff, Progress report on the geology of the O'okiep Copper District, in The Geology of Some Ore Deposits of Southern Africa, 2, edited by S.H. Haughton, pp. 239-302, Geol. Soc. S. Afr., 1964.

Beukes, G.J. and Botha, B.J.V., The Tantalite Valley mega shear zone in the region to the south of Warmbad, Southwest Africa (in Afrikaans). Ann. Geol. Surv. S. Afr. 11, 247-252, 1975 - 1976.

Blignault, H.J., Bushmanland Group. Chapter 5.1.4 in Handbook geol. Surv. S. Afr., 8, edited by L.E. Kent, pp. 268-274, South African Committee for Stratigraphy (SACS), 1980.

Blignault, H.J., Van Aswegen, G., Van der Merwe, S.W. and Colliston, W.P., The Namaqua Geotraverse and Environs: Part of the Proterozoic Namaqua Mobile Belt, in Namaqualand Metamorphic Complex, edited by B.J.V. Botha, pp. 1-29, Geol. Soc. S. Afr. Spec. Publ., 10, 1983.

Clifford, T.N., Gronow, J., Rex, D.C. and Burger, A.J., Geochronological and petrogenetic studies of high grade metamorphic rocks and intrusives in Namaqualand, South Africa, J. Petrol., 16, 154-188, 1975.

Colliston, W.P., The stratigraphy of the Namaqualand Metamorphic Complex in Bushmanland (abstract), Geol. Soc. S. Afr. 18th Congr., 1, 85 - 105, 1979.

Colliston, W.P., Stratigraphic and depositional aspects of Proterozoic metasediments of the Aggeneys Subgroup at Pella and Dabenoris, in Namaqualand Metamorphic Complex, edited by B.J.V. Botha, pp. 101-110, Geol. Soc. S. Afr. Spec. Publ., 10, 1983.

Colliston, W.P., Praekelt, H.E., Strydom, D., Van Aswegen, G., Blignault, H.J. and Schoch, A.E., The recognition of low angle thrusts in Central Bushmanland, Namaqua Mobile Belt (abstract), Geol. Soc. S. Afr. 20th Congr. (Potchefstroom), 1, 29 - 32, 1984.

Gevers, T.W., Partridge, F.C. and Joubert, G.K., The pegmatite area south of the Orange River in Namaqualand. Mem. Geol. Surv. 31, 172 pp., Geol. Surv. S. Afr., Pretoria, South Africa, 1977.

Holland, J.G. and Marais, J.A.H., The significance of the geochemical signature of the Proterozoic gneisses of the Namaqualand metamorphic complex with special reference to the Okiep Copper District, in Namaqualand Metamorphic Complex, edited by B.J.V. Botha, pp. 83-89, Geol. Soc. S. Afr. Spec. Publ.,10, 1983.

Joubert, P., The regional tectonism of the gneisses of part of Namaqualand, Bull. Precambrian Res. Unit, 10, 220 pp., Univ. Cape Town, South Africa, 1971.

Joubert, P., Geological Survey of part of Namaqualand and Bushmanland, Ann. Rep. of the Precambrian Res. Unit, 4-11, Univ. Cape Town, South Africa, 1972.

Joubert, P., The gneisses of Namaqualand and their deformation. Trans. geol. Soc. S. Afr. 77, 339 - 345, 1974.

Joubert, P., Marias, J.A.H., Van Aswegen, G., and Van der Merwe, S.W., O'okiep Group. Chapter 5.15 in Handbook Geol. Surv. S. Afr., 8, edited by L.E. Kent, pp. 275-281, South African Committee for Stratigraphy (SACS), 1980.

Kröner, A. and Blignault, H.J., Towards a definition of some tectonic and igneous provinces in western South Africa and southern South West Africa, Trans. Geol. Soc. S. Afr., 79, 232-238, 1976.

Marais, J.A.H. and Joubert, P., Little Namaqualand Suite, Chapter 5.1.9 in Handbook Geol. Surv. S. Afr. 8, edited by L.E. Kent, pp. 294-304, South African Committee for Stratigraphy (SACS), 1980a.

Marais, J.A.H., and Joubert, P., Spektakel Suite, Chapter 5.1.12 in Handbook Geol. Surv. S. Afr. 8, edited by L.E. Kent, pp. 314-316, South African Committee for Stratigraphy (SACS), 1980b.

Moore, J.M., The geology of Namiesberg, Northern Cape, Bull. Precambrian Res. Unit, 20, 69 pp., Univ. Cape Town, South Africa, 1977.

Odling, N.E., The structure of Gamsberg, Namaqualand, N.W. Cape - an intermediate report, in Eighteenth - twentieth annual reports, edited by P.J. Joubert, pp. 76-104, Precambrian Research Unit, Univ. Cape Town, South Africa, 1983.

Praekelt, H.E., Colliston, W.P. and Schoch, A.E., The stratigraphic interpretation of a highly deformed Proterozoic region in Central Bushmanland, South Africa: First correlation of structurally separated metasediments of the Aggeneys Subgroup, Precambrian Res., 23, 177 - 185, 1983.

Praekelt, H.E., Colliston, W.P. and Strydom, D., Stratigraphic correlation of metasediments of the Proterozoic Aggeneys Subgroup, Bushmanland, Namaqua Mobile Belt (abstract), Conference on Middle to Late Proterozoic Lithosphere Evolution, Precambrian Research Unit, Univ. Cape Town, 41 - 42, 1984.

Reid, D.L., Geochemistry of Precambrian igneous rocks in the lower Orange river region, Bull. Precambrian. Res. Unit, 22, 397 pp., Univ. Cape Town, South Africa, 1977.

Rozendaal, A., Geological structure of the Gamsberg zinc deposit, Namaqualand, South Africa, Ann. Univ. Stellenbosch, A1 (2), 1-105, 1977.

Ryan, P.J., Lawrence, A.L., Lipson, R.D., Moore, J.M., Patterson, A., Stedman, D.P. and Van Zyl, D., The Aggeneys Base Metal Sulphide Deposits, Namaqualand, South Africa, Inf. Circ. Econ. Geol. Res. Unit, 160, 33 pp., Univ. Witwatersrand, South Africa, 1982.

Strydom, D., Techniques employed during the structural-stratigraphic mapping of the Proterozoic Rocks of Western Bushmanland (abstract), Geol. Soc. S. Afr. 20th Congress (Potchefstroom), 1, 142-144, 1984.

Van Aswegen, G., The Gladkop Suite - the grey and pink gneisses of Steinkopf, in Namaqualand Metamorphic Complex, edited by B.J.V. Botha, pp. 31-44, Geol. Soc. S. Afr. Spec. Publ., 10, 1983.

Van Aswegen, G., The significance of the Gladkop Suite for the "Basement Problem" in the Namaqua Mobile Belt (abstract), Geol. Soc. S. Afr. 20th Congr. (Potchefstroom), 1, 159 - 162, 1984.

CRUSTAL EVOLUTION OF THE NORTHERN KIBARAN BELT, EASTERN AND CENTRAL AFRICA

Jean Klerkx, Jean-Paul Liégeois, Johan Lavreau, and Werner Claessens

Department of Geology and Mineralogy, Royal Museum for Central Africa, 1980 Tervuren (Belgium)

Abstract. The Kibaran belt in Burundi is composed of pelitic and quartzitic rocks intruded by large amounts of granites. During an early phase of its evolution the belt has undergone a major tectonic event, resulting in a regional horizontal foliation and a decollement of the sedimentary cover over its basement. Large amounts of granitic magmas associated with mafic intrusives have been intruded contemporaneously with the tectonic deformation. These early structures are considered to result from extension of the lithosphere. Granitic magmas were formed by melting of the lower crust as a result of heat transfer from mafic magmas generated during the extension. The early granitic magmatism started around 1350 Ma ago and reached its paroxysm about 1260 Ma ago. A compressive phase, associated with granitic magmatism and causing upright NE-SW oriented folding, occurred around 1180 Ma ago. Mafic and ultramafic intrusions as well as alkaline granites are associated with a phase of lateral shear affecting the area around 1100 Ma. The Kibaran belt is interpreted to have evolved entirely in an intracontinental environment. It started as an extensional basin, evolving into an extension belt with intensive granitic and mafic magmatism. The late compressive phase, and particularly the late shear, are considered to result from continental collision to the SE which occurred in the southern Malawi-Moçambique belt during the same period. Collision in the south resulted, in the northern Kibaran belt, in delamination of the lithosphere, intrusion of asthenosphere into the continental lithosphere and strike-slip movements; this resulted, in turn, in the formation of shear zones in the upper crust and intrusion of ultramafic and alkaline magmas.

Introduction

The Kibaran orogeny affected large areas of central, eastern and southern Africa. It occurred during the Middle Proterozoic, starting probably around 1400 Ma with the first phases of basin formation and associated magmatism, whereas the maximum intensity of deformation occurred around 1100 Ma. Although the Kibaran orogeny was initially defined in the Kibara Mountains of Shaba (Zaire), it has now been recognized that several linear and mostly parallel belts of Kibatan age exist in the eastern part of Africa (Fig. 1). The northern of the belts of this age, which is considered here, extends in a NE direction from Shaba (Zaire) through Burundi and Tanzania to Rwanda, where it swings to the NW, ending in Uganda and northern Zaire.

Over its whole length the belt is composed predominantly of low-grade pelites and quartzites with only minor calcareous and volcanic sequences. This assemblage constitutes the Burundi or Karagwe-Ankole Supergroup. Large intrusions of granitoid rocks are common as well as mafic and ultramafic bodies. The area investigated in Burundi is representative, from the point of view of lithology, structure and magmatism, of that section of the belt which extends through Tanzania, Rwanda and Uganda.

The aim of the paper is to give an overview of the main structural and magmatic characteristics of the belt which is considered here as an intracontinental fold belt.

Kibaran structural and magmatic events in this region are long-lived, extending from 1350 to 1100 Ma (Klerkx et al., 1984). In our interpretation the belt acquired its main structural and magmatic characteristics during an early phase of décollement tectonics, occurring between 1350 and 1250 Ma ago; during this major phase of deformation, granitic intrusions were emplaced in different phases.

Décollement tectonics and associated magmatism are discussed as resulting or from compressional or extensional processes. Arguments are presented favouring the extension hypothesis.

During a later period (around 1180 Ma ago), limited shortening occurred with the development of upright folds. A later shear event occurred around 1100 Ma and was responsible for the development of intense but narrow shear zones.

Brief Geological Outline of the Region

The Kibaran belt in this region (Fig. 2) is separated from its foreland - the Tanzanian craton - to the east by flat-lying sediments (Malagarasian in Burundi, Bukoban in Tanzania) of Upper Proterozoic age. In east and south west Burundi the Burundi Supergroup overlies the Archaean basement

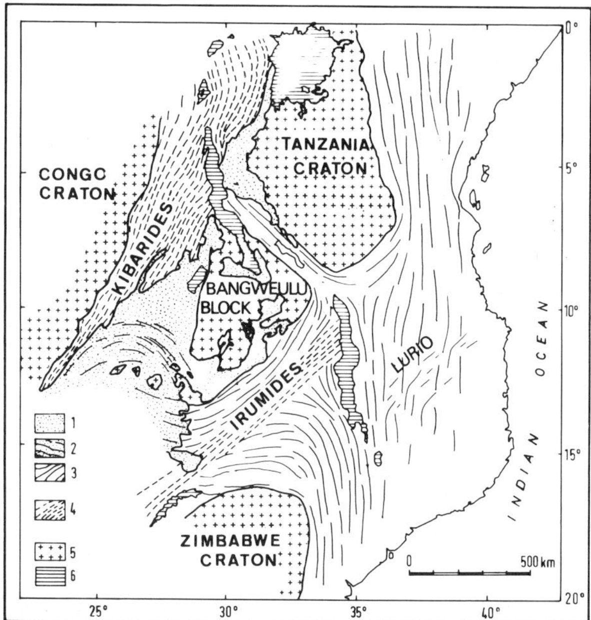

Fig. 1. Location of the Kibaran belt in central
and eastern Africa (modified after Cahen and
Snelling, 1984). 1. Tabular Upper Precambrian;
2. Folded Upper Precambrian; 3. Formations older
than the Upper Precambrian affected by Upper Precambrian events; 4. Kibarides and Irumides; 5.
Zones cratonized since at least 1800 Ma ago;
6. Lakes in the Eastern Rift.

(Demaiffe and Theunissen, 1978) which is composed
of mainly migmatitic gneisses locally containing
granulite facies remnants. These gneisses suffered
a complex structural evolution with extensive reworking during the Kibaran (Nzojibwami, 1984). The
Burundi Supergroup consists mainly of pelitic rocks
with quartzitic intercalations of various thicknesses. The lower and middle divisions of the Burundian
pile reach a thickness of 8 to 10 km; they start
with a sequence of quartzites more than 1 km in
thickness, followed by mainly pelitic sediments.
Thin layers of calcareous sediments are restricted
to the NW part of the region. Volcanic sequences
are rare : a thin intercalation (a few ten of
meters) of dacitic to rhyodacitic volcanoclastic
rocks is known in the lower part of the sedimentary sequence in the eastern part of the belt.
More widespread are basic volcanics situated in
the upper part of the middle Burundian which are
continuously present in western Burundi but are
absent in the east. Locally associated with these
mafic volcanics are volcanics of acid to intermediate composition. They have recently been studied
in detail in some sections by Ntungicimpaye (1984a;
1984b) who describes them as flows and pyroclastic
rocks with tholeiitic composition.

The lower and middle Burundian sediments are
mature, well sorted and mainly fine grained sediments. The upper division of the Burundian, 3 to
4 km thick, also comprises quartzites and pelites,
but the arenites are often immature, generally
badly sorted and commonly contain conglomeratic
layers.

Granitic rocks are concentrated in the western
part of Burundi where they form extensive and complex intrusions that are always associated with
smaller mafic intrusions. They are intrusive mainly in the lower part of the Burundian sequence. In
the eastern part of the country granitic intrusions
are isolated and form more homogeneous bodies.

Large mafic and ultramafic bodies are restricted to the eastern part of the region. A set of
large intrusions are elongated along a NE direction, parallel with the major Kibaran structural
trend. Ultramafic rocks (peridotites) are always
associated with layered intrusions of gabbros,
norites, leuconorites and anorthosites. This linear
alignment of mafic intrusions extends northwards
into Tanzania as far as Lake Victoria.

The metamorphism, which is incipient in the
east, increases westwards and reaches its maximum
of intensity in the vicinity of granitoid complexes. Plurifacial metamorphic mineral assemblages comprising andalusite, staurolite, chloritoid,
biotite and garnet exist in regions where granitoids are abundant.

Several superimposed mineral parageneses have
been observed (Willems, 1985) : the most common
association is andalusite-muscovite, belonging to
the cordierite amphibolite facies, with superimposed crystallisation of sillimanite in the high-
T amphibolite facies. The association staurolite-
chloritoid appears as a retrograde paragenesis.
Locally (SW Burundi) the association staurolite-
almandine-kyanite-biotite has been observed as a
remnant of a first phase of metamorphism.

What concerns the distribution of the metamorphism, it is worth noting that there is no
gradual increase in metamorphism, although the
metamorphic grade is higher in the W. Increased
metamorphism is restricted to particular areas in
association with the presence of abundant granitic
intrusions.

The structural evolution of the Kibaran belt is
characterized by three successive phases of deformation (Theunissen, 1984) : the first phase of horizontal deformation (D1) is particularly well expressed in the more strongly metamorphosed parts
of the western region. It is documented mainly by
the development of bedding-parallel foliation and
by local small intrafolial folds. In the regions
of intense granitic plutonism, structures related
to thin-skinned thrusting are observed. In the
less metamorphosed eastern region the regional
schistosity is much less developed and the D1 formation is mostly expressed by a décollement of the
cover over the Archaean basement, resulting in the
mylonitisation of the basement at the contact. Details of this phase of deformation will be given
in a later section. In the western region numerous
granite intrusions were emplaced or foliated parallel to the D1 foliation.

The second phase (D2) produced open, upright
folds mainly oriented NE-SW, locally swinging to

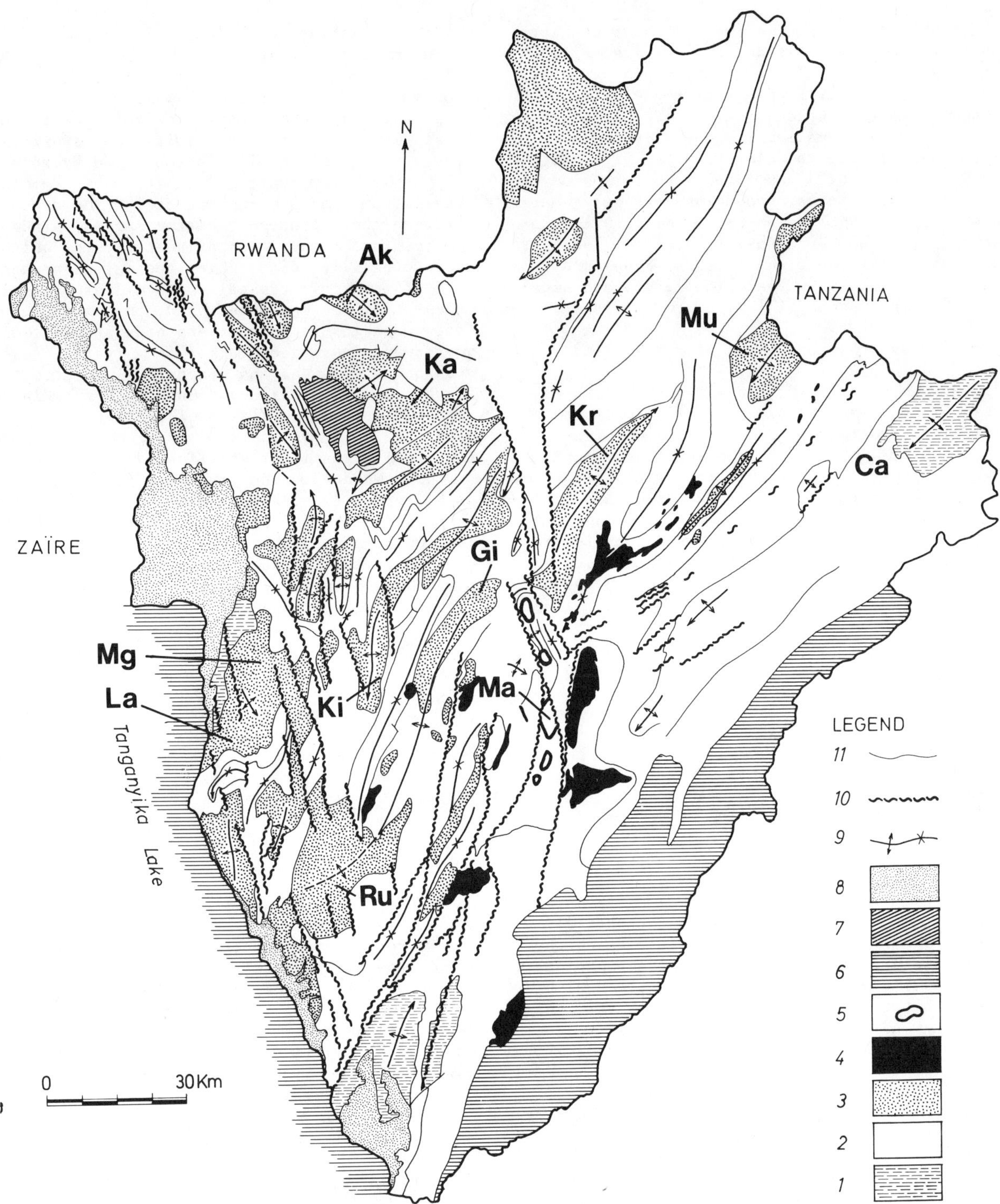

Fig. 2. Major structural and magmatic characteristics of the Kibaran belt in Burundi.
1. Archaean; 2. Burundian metasediments; 3. Kibaran granitoids; 4. Mafic and ultramafic intrusions; 5. Late-Kibaran alkaline intrusions; 6. Malagarasian (post-Kibaran)
sediments; 7. Post-Kibaran alkaline complex; 8. Cainozoic; 9. Principal axes of upright
folding (D2); 10. Principal late-Kibaran shear zones (D2'); 11. Stratigraphic limits and
structural trends. Abbreviations on the map refer to the location of analysed samples
(see also Fig. 3 to 7) : Ca : Cankuzo, Ka : Kayanza, Ru : Rumeza, Mu : Muramba, Ak :
Akanyaru, La : Lake, Mg : Mugere, Ki : Kiganda, Kr : Karuzi, Ma : Makebuko, Bu : Bururi.

NW-SE. This deformation is the most obvious expression of the Kibaran belt and defines its major physiographic features. The regular trends of these folds are deflected around granite-gneiss domes which were mostly emplaced during the D1 deformation. The D2 deformation is generally not sufficiently penetrative to obliterate the D1 structures.

A late shear (D2') was locally superimposed on the earlier structures, producing vertical shear zones oriented NE-SW or NW-SE. Alkali granite intrusions (Tack and De Paepe, 1983; Tack, 1984) are spatially associated with the shear zones and in places are affected by them. The nature and significance of this shear will be discussed later.

Granitoids Associated With D1 Deformation

Two types of early Kibaran granitoids can be distinguished. The first type (Gr 1) consists of biotite granite forming relatively homogeneous batholiths. Their relations with the surrounding rocks are generally difficult to define because of the intense deformation which affected both the granites and the surrounding sediments. The granitic rocks are commonly mylonitic and locally are phyllonites. Their strong deformation attests to emplacement early in the Kibaran tectonic evolution, even possibly before the onset of D1 deformation.

The second type (Gr 2) is much more abundant and, particularly in West-Burundi, constitutes large batholitic complexes which are heterogeneous in texture but whose composition is nearly constant (usually two mica granites). Another feature in common is that their magmatic or metamorphic foliation is always parallel with the bedding and the schistosity of the adjoining sediments. Their heterogeneous aspect results from variations in texture, from mylonitic to a primary parallel alignment of idiomorphic feldspar phenocrysts. Another common character of the Gr 2 granitoids is the presence of metasedimentary inclusions, mostly quartzite. Some granites even contain parallel, decimetre-scale lenses of quartzite, indicative of the partial assimilation of the metasediments into which the granites were intruded. Mafic rocks, amphibolites or amphibole gabbros are always associated with these granitoids.

Evidence for Synkinematic Emplacement of the Gr 2 Granitoids

Whereas the Gr 1 granites occur in homogeneous bodies which, according to their mylonitic texture, are probably pre-kinematic with respect to the Kibaran D1 deformation, the Gr 2 granitoids present many characteristics in favour of a synkinematic emplacement with respect to this deformation. As stated above, they are often made up of different units of texturally different granitoids, but whose parallel contacts are concordant with the overlying sediments. Their internal texture is also always parallel with the subhorizontal D1-related foliation in the metasediments.

If the internal structure of some Gr 2 grani-

toids is doubtless of metamorphic origin (gneissic or mylonitic texture), it is however of magmatic origin in most granitoids. This is best expressed by the granites with well oriented idiomorphic feldspar phenocrysts swimming in an entirely magmatic textured matrix. The synkinematic character of this family is confirmed by some of the porphyritic granites which present a protoclasis (cataclasis occurring during the crystallisation and emplacement of the magmas). Indeed, in these rocks the feldspar phenocrysts, which are macroscopically idiomorphic, consist of an aggregate of isometric crystals with different optic orientation. On the other hand, the matrix around these cataclastic phenocrysts consists of elongated quartz and undamaged plagioclase crystals associated with oriented micas and presents a magmatic texture. The two micas, biotite and muscovite form large and intimely intermixed flakes. They also are oriented along the D1 direction and are considered as resulting from magmatic crystallisation.

The presence in the granitoid complexes of rocks of different texture but with a constant foliation parallel to that of the surrounding metasediments and the presence within these complexes of rocks whose orientation has a magmatic origin suggest that these granitoids were emplaced and partially deformed during the phase of deformation resulting in the regional foliation of the metasediments (D1). This horizontal movement must have persisted over the whole period of time of granite emplacement. This means that the granites emplaced early in this process were already consolidated and have been fracturated and eventually recrystallised under the influence of the persistent movement; the latest phases of granite have not undergone this fracturation and have still preserved their magmatic texture.

The Nature of the Granitic Intrusions

Most of the Gr 1 granitoids have granite compositions and few are granodiorites (Fig. 3a). The rocks are generally coarse grained, contain large microcline and zoned plagioclase crystals and are rich in biotite. Accessory minerals include sphene, zircon and apatite. Granodioritic rocks are found as inclusions in the granites; they are often finer grained, porphyritic and contain large amounts of biotite. These rocks are often mylonitized and then contain secondary muscovite.

The Gr 2 granitoids, although very variable in texture, present a homogeneous mineralogy; these leucocratic granitoids (Fig. 3b) contain large and abundant feldspar phenocrysts, are rich in muscovite and contain variable amounts of biotite. They are also characterized by numerous inclusions of metasedimentary rocks, particularly quartzites, and locally contain decimetre-scale parallel lenses of quartz which are considered remnants of incompletely resorbed metasediments. Tourmaline is ubiquitous. This attests to an interaction of the granitic liquid with the surrounding sediments.

Chemically, although all the investigated gra-

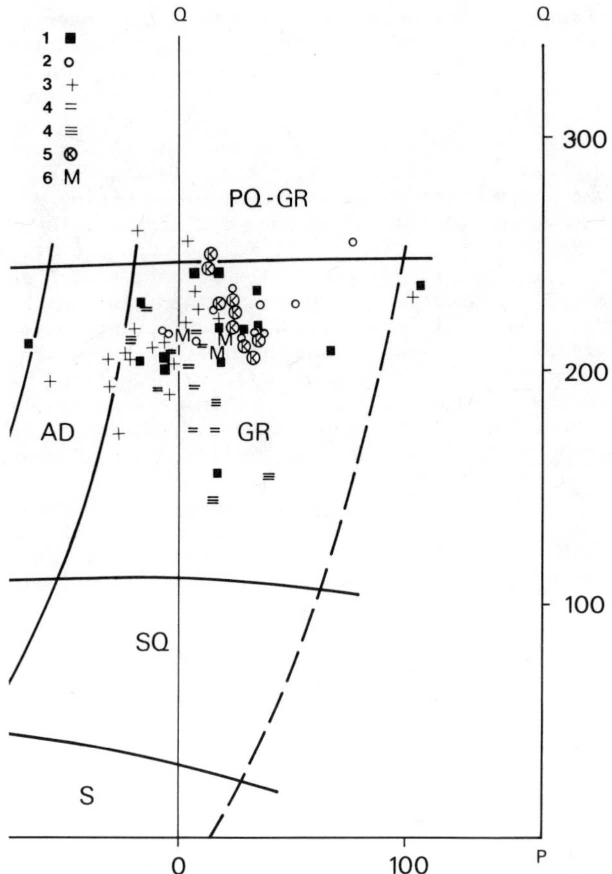

Fig. 3a. The Kibaran granitoids in the chemical-mineralogical classification for plutonic rocks of Debon and Le Fort (1983). Most granitoids are real granites s.s., mainly quartz-rich, some of them per-quartzose. Adamellitic or other types are scarce. Symbols : 1 : Ka and Ru (Gr 1), 2 : Mu and Kr (Gr 2), 3 : La (Gr 2), 4 : Mg (Gr 2), 4' : also Mg, samples from single outcrop, 5 : Ki and Gi (syn-D2), 6 : Ma (syn-D2'). PQ-GR = perquartzose granites, GR = granites, AD = adamellites, SQ = quartz syenites, Q = Si/3 -K-Na-2/3Ca. P = K-Na-Ca.

nites have a peraluminous composition, it is stronger in the Gr 2 granites. Another aspect of the Gr 2 granites is their very variable composition and generally high normative quartz content. This is expressed in the Q-Ab-Or diagram (Fig. 5) by the large area covered by the granite composition and the shifting of the points towards the Q pole relatively to eutectical.compositions.

To evaluate the dispersion of the chemical composition, the rock compositions have been plotted on a c versus q/or+ab+or diagram (normative compositions) (Fig. 4). In this diagram c expresses the peraluminous character while q/or+ab+an represents the deviation of the compositions away from the ternary granitic minimum composition as result of high quartz content. In this diagram the Gr 2 granites are dispersed over a large area, varying from

compositions close to common S-type granites such as the Manaslu leucogranite (Le Fort, 1981) towards granites strongly enriched in corundum and quartz. The extreme values are close to those of metasediments. In this diagram individual massifs appear to form separated lineages. This is best expressed in the Mugere granite; some compositions fall in the field of low peraluminosity whereas others move to compositions enriched in Al and Si. The high peraluminosity is interpreted as the result of a contamination of the granitic magma by metasediments, locally present as residual lenses. The field occurrence of the Gr 2 granites (sheets intercalated in the metasediments) and the spatial variation in composition of one single massif suggest that the contamination with the metasediments occurred essentially during the emplacement of the granites. This implies that it is difficult to deduce the magmatic source composition from the rock composi-

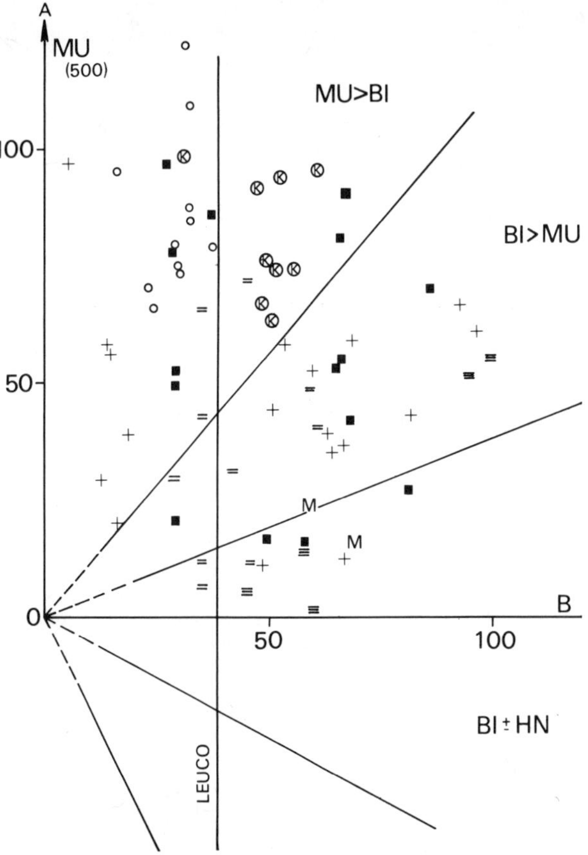

Fig. 3b. The Kibaran granitoids in the chemical-mineralogical classification for magmatic associations of Debon and Le Fort (1983). All granites are peraluminous, with variable muscovite to biotite ratios. No definite trend can be detected, neither between granite units, nor within a particular one. This feature is attributed to the effects of contamination by sediment assimilation (see text for explanation). MU = muscovite, BI = biotite, HN = hornblende. A = Al-K-Na-2Ca, B = Fe+Mg+Ti.

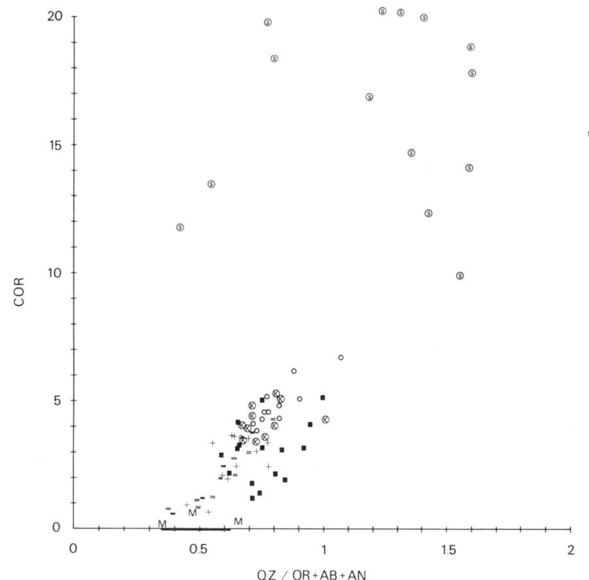

Fig. 4. The Kibaran granitoids compared to a variety of sediments of the Burundi Supergroup, likely to represent contaminants. The diagram expresses (a) the peraluminosity due to the assimilation
of Al-rich sediments giving rise to corundum-nor-
mative rocks (b) the perquartzosity due to the
assimilation of quartz-rich rocks, giving rise to
a quartz to feldspar ratio above normal (heavy line
in diagram) eutectic values.
COR, QZ, AB, AN and OR are CIPW-normative values.
Symbols as in Fig. 3a.

tion. Nevertheless, the contamination trend leads
to estimate a non peraluminous original magma.

Age of the Gr 1 and Gr 2 Granites

In earlier papers age data on magmatic events in
the Kibaran belt have been presented for Shaba
(Zaire), Rwanda and Uganda (for an overview see
Cahen and Snelling, 1984). The ages of the Kibaran
tectonic events have been deduced from these data
in conjunction with preliminary structural studies.

According to the previous studies the ages of
magmatism in the Kibaran belt range between early
and syntectonic granites; i.e. between 1366 $\pm$ 32
Ma and 1289 $\pm$ 31 Ma (data obtained on granites from
Rwanda and Uganda) (Cahen et al. 1967, 1972; Vernon-
Chamberlain and Snelling, 1972) and 970-990 Ma for
tin-bearing leucocratic granites of Shaba and
Rwanda (Gérards and Ledent, 1970; Cahen and Ledent,
1979; Lavreau and Liégeois, 1982). However, as these
late granites intrude Upper Proterozoic sediments in
Zaire, they rather belong to the Katangan orogenic
event than the Kibaran orogeny (Cahen and Ledent,
1979; Lavreau and Liégeois, 1982).

New age data on early granites in Burundi have
recently been given by Klerkx et al. (1984), based
on detailed field and isotopic investigations of
these granites. A preliminary minimum zircon age

(Ledent, 1979) obtained on the Gr 1 type-massif
(Rumeza) suggests that this intrusion was emplaced
around 1325 Ma ago (two zircon fractions with t
$^{207}Pb/^{206}Pb$ = 1335 Ma and 1319 Ma). A Rb-Sr isochron (Klerkx et al. 1984) (Fig. 6b) gives an age
of 1268 $\pm$ 44 Ma for the same body. As this granite
is highly cataclastic and often mylonitic, this
age is interpreted as a product of resetting during
the D1 deformation. A second Gr 1 granite, the
Kayanza pluton, gives an Rb-Sr isochron age of
1330 $\pm$ 30 Ma (Fig. 6c). As the Kayanza pluton is
much less affected by mylonitisation than the Rumeza massif, this age probably corresponds to the
age of emplacement of the granite. The age data
indicate that the Gr 1 granites were emplaced before D1 deformation; some of them were intensely
mylonitized by this event (Rumeza), which resulted
in a resetting of the ages, while others (Kayanza)
were less affected.

The beginning of sedimentation in the Burundian
basin is estimated to have occurred not long after
1400 Ma ago. This evaluation is based on a Rb-Sr
age obtained on rhyodacitic volcanic rocks (location Cankuzo, see Fig. 2) interbedded with the
lower Burundian sediments in eastern Burundi where
the D1 deformation and metamorphism are absent.
These volcanic rocks give an isochron age of 1353
$\pm$ 46 Ma (Fig. 6a), which probably corresponds to

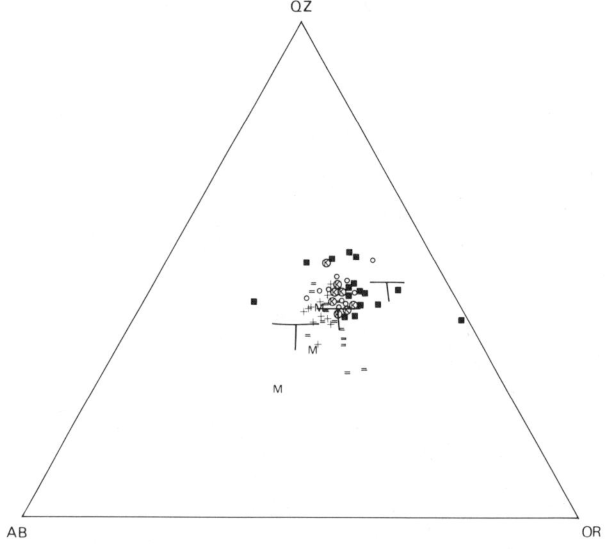

Fig. 5. The QZ-AB-OR ternary minima at 1kb water
pressure (James and Hamilton, 1969). Most granites
plot near one or other cotectic line, indicating
that they were indeed liquids, but deviations
within one particular granitic unit can be important. This situation is attributed to the overwhelming effect of assimilation of quartz and/or
aluminous rocks, the conditions of crystallisation
of the various units being similar as well as their
chemical composition. Ternary minima are indicated
for AN contents of (from left to right) 3, 5 and
7.5 per cent of the sum quartz plus feldspars.
Symbols as in Fig. 3a.

TABLE 1. Rb and Sr Isotopic Data (New Results)

Sample	Rb	Sr	$^{87}Sr/^{86}Sr$		$^{87}Rb/^{86}Sr$
MAKEBUKO GRANITE					
63 866	216	114	0.7941	+ 3	5.53
71 352	222	120	0.7938	+ 2	5.40
LT 4	187	141	0.76968	+ 4	3.86
LT 22	180	131	0.76517	+ 8	4.00
LT 23	177	141	0.76538	+ 4	3.65
LT 26	218	138	0.78276	+ 5	4.60
WC N2	158	158	0.75580	+ 3	2.91
BUKERASAZI GRANITE					
LT 2	183	13.65°	1.40348	+ 16	41.46
LT 7	225	16.34°	1.37049	+ 5	42.45
BURURI MYLONITES					
71 720A	131	50.9	0.87496	+ 5	7.57
71 720B	184	49.9	0.92758	+ 4	10.90
71 720C	162	78.6	0.85149	+ 4	6.05
AKANYARU CHLORITOID SCHISTS					
JPL 190	288	54.0	0.97407	+ 4	15.84
JPL 191	310	57.5	0.98731	+ 5	16.03
JPL 192	290	59.5	0.96363	+ 4	14.46
JPL 193	240	60.2	0.92149	+ 5	11.78
JPL 195	329	54.5	0.99745	+ 6	17.97
JPL 198	183	53.2	0.89988	+ 4	10.14
JPL 200	299	66.3	0.94083	+ 4	13.35

- Rb and Sr concentrations were determined by XRF
  (C. Léger, analyst), except (°) by isotope dilution.
- The errors on the $^{87}Rb/^{86}Sr$ ratios are 2%.
- NBS 987 standard : 0.710235 $\pm$ 0.000026.
- Normalisation for $^{86}Sr/^{88}Sr$ = 0.1194.
- ^{87}Rb = 1.42 $10^{-11}a^{-1}$.
- Isotopic ratios were measured on the VARIAN MAT 260 and
  VARIAN TH5 mass spectrometers of the Centre Belge de
  Géochronologie, Brussels.
- The ages and initial ratios were calculated following
  the method of Williamson (1968).
- All the errors are quoted at the 2σ level.

the age of diagenesis or incipient metamorphism. They are derived from tuffs and tuffites which were composed originally of glass shards which have partly recrystallized during incipient metamorphism.

The Gr 2 granites, which are intruded synkinematically with respect to the D1 deformation, all give similar ages around 1260-1280 Ma (Mutumba, Mugere, Lac : Liégeois et al. 1982, see Fig. 6d; Muramba and Akanyaru : Klerkx et al. 1984, see Fig. 6e and 6f). These ages most likely correspond to the D1 horizontal deformation; the ages represent either the time of synkinematic emplacement of the granites or the age is reset by the deformation climax which occurred shortly after the emplacement of the granitic intrusions.

The Origin of the Gr 1 and Gr 2 Granites

The contamination undergone by the granitic magmas, particularly those at the origin of the Gr 2 granites, restricts a discussion on the nature and the origin of the primary magmas. The strong peraluminous composition which characterizes all these granites has certainly been obtained at least partly when the granitic magma interacted with the Burundian sediments during intrusion (see above and Fig. 3 to 5). An evaluation of the degree of the

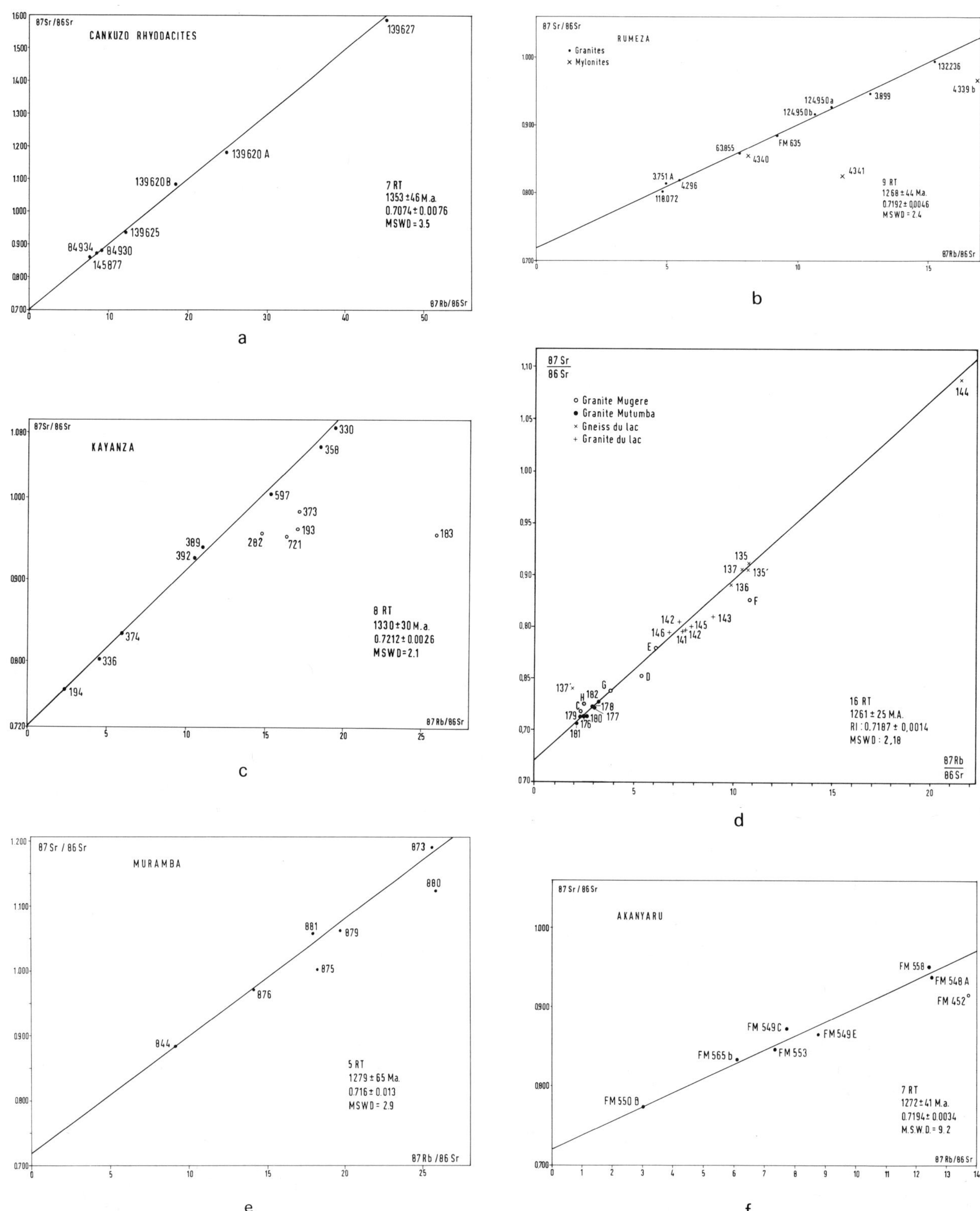

Fig. 6a–6f. Rb-Sr isochrons of early Kibaran magmatic rocks. a. Cankuzo metavolcanics; b. Rumeza Gr 1 granite; c. Kayanza Gr 1 granite; d. Mutumba-Mugere-Lac Gr 2 granite; e. Muramba Gr 2 granite; f. Akanyaru Gr 2 granite. Localities in Figure 2.

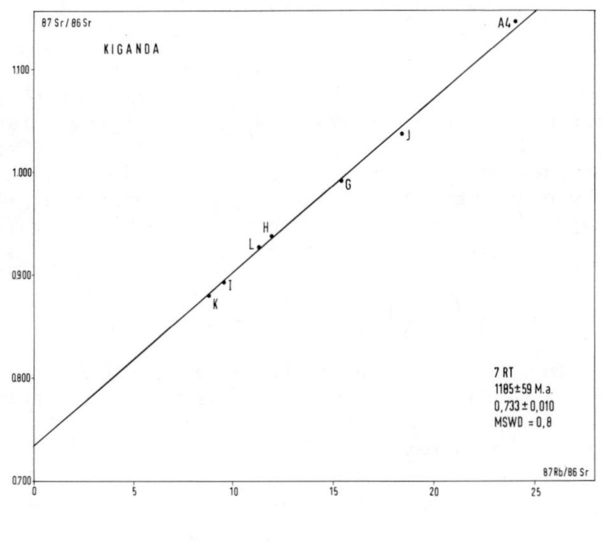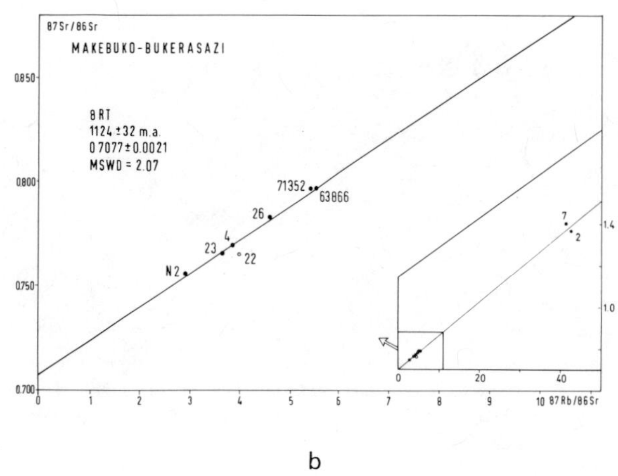

Fig. 7a-7b. Rb-Sr isochrons of late Kibaran rocks and events. a. syn-D2 Kiganda granite; b. Makebuko-Bukerasazi alkaline intrusions (provisional isochron). Localities in Figure 2.

contamination and of its effects on the magma composition has not yet been made. It is therefore not possible to evaluate exactly the original magma composition.

Nevertheless, it appears that the initial $^{87}Sr/^{86}Sr$ ratios of the different granites are relatively low, particularly in view of the highly peraluminous composition which resulted, in part at least, from contamination by crustal material. The Rumeza Gr 1 granite has an initial ratio of 0.707-0.710 (recalculated at 1325 Ma) which corresponds to the ratio of the early Cankuzo volcanics (0.707). A lower crustal origin for these magmas has to be taken into account (Klerkx et al.,1984). The Gr 1 Kayanza granite and the analyzed Gr 2 granites have initial Sr isotopic ratios between 0.716 and 0.721. These ratios are also relatively low considering that these magmas have been strongly contaminated by crustal material. It has been suggested (Klerkx et al.,1984) that the primary magma from which these granites were originated was derived from a low radiogenic source, probably in the lower continental crust, and that they have been progressively contaminated by crustal material, but essentially during their emplacement.

D1 Deformation: Extensional Tectonics or Thrust-Related Deformation?

Characteristics of the D1 deformation. The D1 deformation is most strongly expressed in the more metamorphic western part of the investigated region where Gr 1 and Gr 2 intrusions are abundant. In the less metamorphic eastern part a schistosity parallel to bedding is restricted to the contact zones between the Archaean basement and the Burundian sedimentary cover : a narrow zone of mylonites is developed in the basement adjacent to the cover and in the basal quartzite of the sedimentary cover local cataclastic bands lie parallel with bedding. This deformation is interpreted as a décollement between cover and basement.

Also, in the E, an S1 schistosity is found in phyllitic sediments around the isolated granitic cores and is concordant with the granite contact. This foliation also affects the outer part of granitic intrusions.

The S1 schistosity is ubiquitous in Burundi where it is clearly associated with the granitic intrusions : the schistosity and the metamorphic facies of the metasediments both increase towards the granites; moreover, bedding and schistosity in the metasediments are concordant to the contact and the foliation of the granites.

The main characteristic of this subhorizontal S1 foliation is the scarcity of associated folds. Local isoclinal folds are restricted to metaquartzite xenoliths in the granitic intrusions; in places, where the density of granite bodies is important, structures associated with thin-skinned thrusting are present in the vicinity of the granitic intrusions (Fig. 8) Important thrusting is absent in the sedimentary cover where asymmetric folds present a vergence opposed to the underlying thrust direction.

The local thrust structures are not sufficiently abundant to infer the direction of tectonic transport. Indeed, in two places the transport direction deduced from the vergence of the folds on the thrust ramp is seen in opposite directions, at one place towards SE, at the other place towards the W.

Interpretation of the D1 deformation. The D1 deformation appears then as a regional horizontal schistosity with only minor folds, those accom-

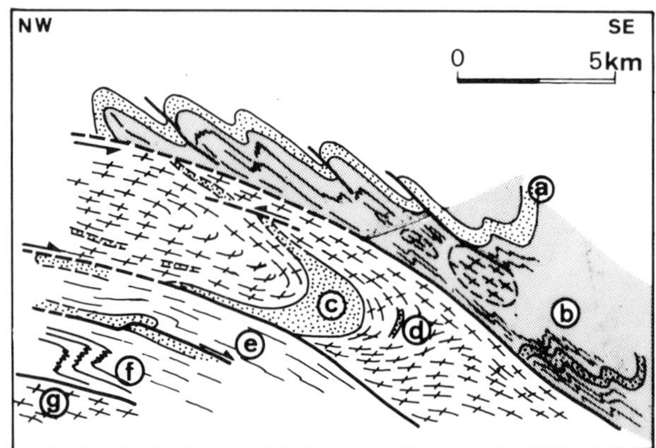

Fig. 8. Thrusting associated with granitoid intrusions. a. Dragfolds in the cover sediments, NW-overturned; b. Isolated granitic intrusions in the cover; c. Remnants of folded metaquartzites in the Gr 2 granites; d. Rests of metasediments in the Gr 2 granites; e. Thin-skinned thrust in a complex of migmatitic gneisses (pre-Kibaran ?) and metaquartzites; f. Recumbent fold; g. Gr 2 granitoids.

panying local thrust structures. Its intensity is spatially linked with the granitic intrusions. In the absence of conclusive information on transport directions, two alternative interpretations are discussed for the origin of the Dl deformation : (1) Dl reflects a regional thrust movement, and the granitic intrusions are emplaced synkinematically with this movement, (2) the Dl deformation is a décollement resulting from early extensional processes; the local thrusting is then connected with the granitic intrusions.

The first explanation implies an entirely compressional type model which considers the cover-basement décollement as a sole thrust. This interpretation also implies that the Dl and D2 deformational phases are contemporaneous : listric thrust faults rising off the décollement surface, steepening, becoming blind and giving way to the upright D2 folds. A major argument against this hypothesis is the age difference between Dl and D2 events obtained from associated granitic intrusions : Dl is dated around 1260-1280 Ma whereas the granites associated with the D2 deformation give ages around 1180 Ma. Both types of granites, either associated with Dl or D2, also are distinctly different in composition.

The second explanation considers that the Dl deformation results form extensional processes. Arguments in support of this hypothesis are the evidence of rifting during Burundian sedimentation and the bimodal nature of extrusive and intrusive magmatic rocks (Klerkx et al. 1984).

In both hypotheses the Gr 2 granites are contemporaneous with the décollement; the increase in the intensity of the foliation and in metamorphic grade near the intrusions is in both cases a result of the synkinematic intrusion of granitic magmas.

The observed isoclinal folds and thin-skinned thrust structures, which are always restricted to the vicinity of granitic intrusions, are not necessarly related to a major thrusting event; they can be explained by the differential movements between parallel intruding granite sheets and intercalated sedimentary sequences. Dragging of the sediments on the granitic sheets or pushing away of the sediments by the intruding granitic bodies may result in local compressive movements which caused local folding or even thrusting.

Evidence for Rifting During Burundian Sedimentation

The junction between the Middle and Upper Burundian sequence corresponds to a transition from mature to immature sediments; the sediments of the Middle Burundian are principally pelites and fine-grained, well sorted quartzites. The Upper Burundian sediments, on the other hand, are arenaceous, coarse-grained, poorly sorted and with angular fragments. The lower part of the Upper Burundian contains conglomerates and subgreywackes rich in fragments derived from the underlying beds, namely fragments of shales. These fragments testify to a phase of erosion prior to, or concomitant with, the deposition of the sediments (Demulder and Theunissen, 1980; Dreesen, 1980). As there is no indication of a phase of folding and erosion before the deposition of the Upper Burundian sediments, it is likely that the erosion of the Middle Burundian shales was due to rifting processes and formation of rift basins in which the Upper Burundian was deposited. These local rift basins possibly correspond to the isolated synclines of Upper Burundian sediments in western Burundi (Fig. 9), whereas a larger basin was formed in the eastern part of the country.

Bimodal Magmatism Related to Extension

Bimodal magmatism is commonly associated with an extensional regime and especially with the development of aulacogens, particularly in the Proterozoic when aulacogens were larger and more frequent (Smith, 1976) : an example is the Athapuscow aulacogen in the Canadian shield (Hoffman, 1973). Bimodal magmatism is also known to be associated with younger extensional regions : the late-Paleozoic plutonic complexes in Morocco are considered to be related to the initial rupture between North America and Africa (Vogel et al. 1976); the association of mafic and acid volcanism during the late Cenozoic of the Western United States is linked to extension (Christiansen and Lipman, 1972); certain parts of modern rift environments are characterized by bimodal magmatism (Barberi et al., 1972; Black et al.,1972; Hart and Walter, 1983).

Igneous activity related to lithospheric extension is considered a result of the intrusion of hot asthenosphere into the continental lithosphere, whatever the process which is ultimately responsible (Sengör and Burke, 1978; Turcotte and Emerman, 1983; Liégeois and Black, 1984). Extrusive magma-

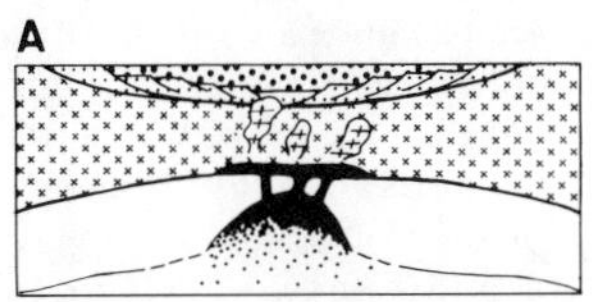

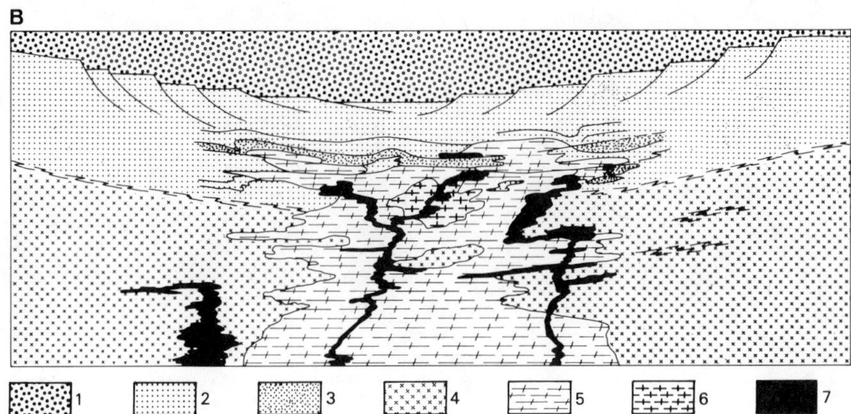

Fig. 9. Section across the Kibaran belt (in Burundi) during the D1 phase (sedimentation, deformation, magmatism): a. At lithospheric scale; b. Across the crustal segment exposed in Burundi. 1. Upper Burundian sediments; 2. Lower and Middle Burundian sediments; 3. Quartzites in 2; 4. Pre-Kibaran basement; 5 and 6. Granitoids intruded during the D1 phase; 7. Mafic magmas intruded during the D1 phase.

tism is not common in the Kibaran belt of Burundi, but there are large amounts of granite and gabbro associated with the D1 phase. Similar bimodal plutonic associations are known in other extensional regions, for example, the late Paleozoic magmatism in Morocco (Vogel et al. 1976). In Morocco, the generation of granitic magmas is considered to have resulted from partial fusion of the lower crust by heat supplied by mafic magmas generated during the extension. Christiansen and McKee (1978) accept a similar process for the origin of the Cenozoic volcanism in the Great Basin of the United States. They propose that the volcanism is connected with thermal effects associated with extension of the continental North American plate. In this region, basaltic magmas generated at different depths within the mantle are intruded in the crust during continuous extension, the heat flow is augmented and rhyolitic magmas are formed in the lower crust by local partial fusion. Hildreth (1981) also considers a primary basaltic source for numerous intermediate and acid magmas. Particularly during a long period of magmatism the intrusion of basaltic magmas is responsible for the partial fusion of crustal rocks so that both acid and mafic magmas will intrude the crust during lithospheric extension.

Although the extrusive igneous activity is limited in the Kibaran belt, it occurs before an intense phase of rifting; this corresponds to an active phase of rifting according to Sengör and Burke (1978), related to thinning of the lithosphere above a mantle plume. This also corresponds to the

views of Condie (1982) who classes the Proterozoic supracrustal assemblages according to their lithologies and considers that assemblages of type II (bimodal volcanites - quartzites - arkoses) correspond to aborted mantle-activated rifts.

In the light of the processes outlined above, we can suggest the following origin for the early Kibaran magmatism. The extension was slow and continuous over a long period of time (at least between 1350 and 1260 Ma ago). This slow extension induced negligible fracturing of the crust, thereby reducing the amount of extrusive activity but facilitating the accumulation of mantle-derived magmas at the base of the crust. The accumulation of heat from these mafic magmas caused partial fusion of crustal rocks, thereby generating magmas of granitic composition. The granite generation could be either continuous or limited to more rapid rates of extension. Indeed, the major episode of granitic intrusion seems to be contemporaneous with the climax of deformation related to extension. Although the granitic magmas intruded during this phase of extension in the upper crust have a strong peraluminous, S-type composition, as discussed above, they probably acquired those peraluminous characteristics by contamination in the upper levels of the crust.

Décollement Tectonics Induced By Crustal Extension

The development of a regional horizontal foliation or ductile shear zones as a consequence of

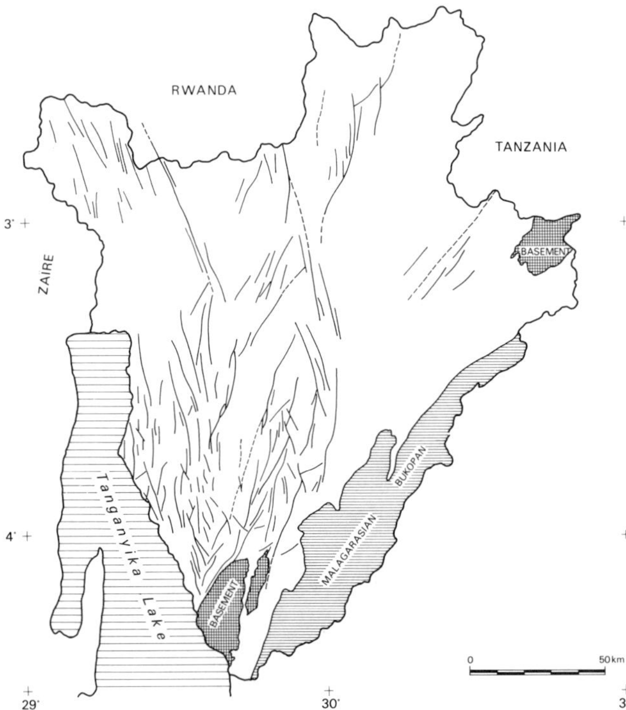

Fig. 10. Distribution of shear zones in Burundi.

crustal extension has been proposed by Oxburgh (1982). As exemplified by the Bay of Biscay (Montadert et al. 1979), the upper part of the crust is thinned by movement along listric faults, whereas the lower part, beneath about 8 km, is deformed by homogeneous and ductile stretching. Considering that the total thickness of the Burundian sediments is between 11 and 14 km, the zone where the granites are intruded and where the D1 structures develop lies in the zone of ductile behaviour. In this zone the extension results in the formation of localized horizontal shear zones, the deep level expression of listric faults; these sites may correspond to localized more intense zones of foliation in the lower Burundian metasediments. Along these horizontal listric faults, the highly fluid granitic magmas could easily intrude the metasediments. The high fluid content, which results in a high fluidity of the magmas, is manifested by the high proportion of hydrated minerals in the granites, their ability to assimilate sedimentary rocks and their high tourmaline content.

The synkinematic emplacement of the granites, intruding the sediments or as elongated bodies, sometimes even as sheets parallel to the décollement zone, was able to create local zones of compression and thrusting. Indeed, the intrusion of highly fluidy magmas can disequilibrate the sedimentary pile and induce differential movements between the granites and the metasediments which are of different competence.

It is possible that the process of extension has been accelerated at the time when the rift basins

of the Upper Burundian were formed; these rifts may be the compensation in the upper part of the crust for the extensional movements which, in a lower part, are compensated by introduction of a greater amount of granitic magmas, under a thickness of 8 to 10 km of sediments which were already deposited at that time.

Compressive Tectonics During a Later Deformation Phase (D2) and D2-Associated Magmatism

The D2 Deformation

The deformation responsible for the morphological signature of the belt produced open upright folds, mostly oriented NE-SW in the investigated area, but which, on a regional scale, swing from a NE-SW direction in the southern part of the Kibaran segment (Shaba, Burundi) towards a NW-SE direction in the northern part (Rwanda, Uganda). This phase of open folding did not result in significant crustal shortening. Although a cleavage is axial planar to the folds, the structures resulting from the D1 deformation and associated magmatism are only partly obliterated by the D2 event.

Granitic intrusions, associated with this phase of deformation, occur in the cores of anticlines and consist of two-mica granites. They are typically intrusive, homogeneous in composition and are not usually foliated; these features distinguish them from the Gr 2 granites. These late intrusions have different compositions (compare Fig. 4 and 6) and also have more typical crustal signature ($^{87}Sr/^{86}Sr$ initial ratio of 0.733). Their Rb-Sr age of 1185 $\pm$ 59 Ma (Klerkx et al., 1984, see Fig. 7a) may be considered to correspond to the age of D2 deformation, as the granites typically occur in the cores of D2 anticlines.

The D2' Shearing Event and Associated Magmatism

The latest Kibaran structural event to affect the region is a shear event, corresponding to the development of shear zones along two conjugated directions, NE and NW (Fig. 9 and 10), which are parallel to both directions of compressive D2 deformation. All the older structures are overprinted by a strongly developed set of NNE and NNW oriented shear zones. They appear as local zones of intense deformation (Fig. 9), characterized by a pervasive cleavage with an increasing intensity towards the centre of the zone. The sedimentary strata generally show moderate to intense folding at the margin of the zones. The centre of the zones consists of cataclastic or even mylonitic rocks. The quartzites in particular show gently plunging or even horizontal stretching and mineral lineations. These shear zones are considered to have behaved as strike-slip zones.

Concerning the sense of displacement along these shear zones, dextral as well as sinistral movements have been observed with a predominance of dextral displacement. Intrusions of alkaline granites

228   KLERKX ET AL.

TABLE 2. Chronological Sequence of Tectonic Events in the
Kibaran Belt and Associated Magmatism

Date	Tectonic Event	Phase of Deformation	
950 ~ 1000 Ma	Post-tectonic granites (Sn-bearing)		COMPRESSION
1100 Ma	Sub Alkaline granites Ultramafic and mafic intrusions	D2′ shear	
1200 Ma	Granitic magmatism	D2 deformation Open, upright, cylindrical folds	
1250 Ma ↑ 1330 Ma	Granitic and associated mafic magmatism	D1 deformation extensional structures (locally thin skinned thrust in granitic environment)	EXTENSION
1350 ~ 1400 Ma	Acid volcanism	Formation of sedimentary basin	

(Tack and De Paepe, 1983; Tack, 1984) are spatially associated with the NE-SW shear direction.

Preliminary age determinations have been performed by the Rb-Sr method on these alkaline intrusions. A preliminary isochron on the Makebuko massif gives an age of 1068 ± 78 Ma (6 whole-rock samples) Ro = 0.7109 ± 0.0044; MSWD = 0.9) (Fig. 7b, Table 1). If two samples of the Bukerasazi pluton, which is a satellite pluton of the Makebuko massif, are added, an age of 1125 ± 25 Ma is obtained (Ro = 0.7077 ± 0.0021, MSWD = 2.0).

The age of the D2' shear is inferred from both the age of the alkaline intrusions which are emplaced around 1100 Ma and which are spacially associated with the shear zones, and from the result, although imprecise, of a provisional Rb-Sr age determination on mylonites from the NE shear zone: 1095 ± 111 Ma (Ro = 0.756 ± 0.022; MSWD = 0.05).

As the alkaline intrusions of Makebuko and Bukerasazi are themselves locally sheared, the maximum age for the shearing event along the NE-SW direction is about 1100 Ma. Moreover, the relatively low $^{87}Sr/^{86}Sr$ initial ratio indicates a deep origin for the alkaline plutons which gives a lithospheric scale to the D2' shear.

More problematic is the significance of the mafic and ultramafic intrusions which, in Burundi, are also aligned NE, parallel both to the major D2 folding direction and to D2' shears. They are part of an important alignment of ultramafic bodies which extends northwards to Lake Victoria in northern Tanzania. These intrusions have also been formerly connected with the shearing event (Klerkx, 1984). However, an age indication (work in progess) suggests that they are older and possibly are contemporaneous with the D2 folding phase. They consist of huge bodies of peridotite with or without associated gabbro, norite and leuconorite. They show clear evidence of intrusions into the Burundian sediments (intrusive contacts, contact metamorphism, deformation of the sediments at the contact). Although these complexes have been only superficially studied, the gabbroic-noritic rocks show evidence of being layered intrusions. The internal structures of the ultramafic rocks, however, e.g. alternating layers of peridotite and

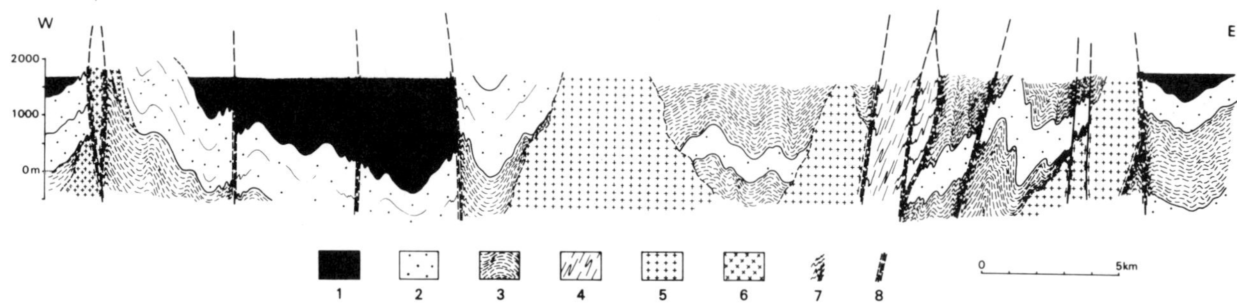

Fig. 11.   Section across sheared zones in central Burundi. 1. Shales; 2. Quartzites; 3. Sandy shales; 4. Sheared rocks; 5. Granitoids; 6. Mafic rocks; 7. Shear folds; 8. Shear faults.

norite, suggest that they have been emplaced as a crystal mush of olivine crystals in a noritic liquid matrix. The mafic intrusions may be considered as derivatives of the ultramafic crystal-liquid mixture.

The processes responsible for the generation of these magmas in the mantle and for their intrusion into the upper crust remain to be studied. Although there is an age difference between the early Kibaran bimodal magmatism that is contemporaneous with the D1 deformation, and the emplacement of the mafic-ultramafic bodies, it may be reasonable to accept that these magmas still belong to the mafic magmatism generated during the early phase of extension.

So called post-tectonic granites in the Kibaran belt have been known for a long time. They are associated with dykes and pegmatites with cassiterite, columbo-tantalite and wolframite mineralisation. They are found mainly in Shaba and in Rwanda and have been dated at about 970-990 Ma (Cahen et al. 1979; Lavreau and Liégeois, 1982). They have typical crustal characteristics, but their genetic significance is not well understood. However, they do not belong to the Kibaran orogeny as these granites cut Pan African Katangan sediments in the Itombwe syncline in Kivu, Zaire (Cahen et al. 1979).

The Northern Kibaran Segment as Part of Kibaran Age Events in Eastern and Southern Africa

Considering the Burundi region to be representative of the northern segment of the Kibaran belt - the Kibaran belt sensu stricto, extending over Shaba, Burundi, NW Tanzania, Rwanda and SW Uganda -, we may infer that the formation of a basin in this area started around 1400 Ma ago and was followed, around 1350 Ma ago by bimodal magmatism, comprising large intrusions of granitoids. This magmatism persisted until 1260 Ma, occurring contemporaneously with a pervasive horizontal deformation, resulting from the décollement of the sedimentary cover over its basement. Décollement and associated magmatism are seen to be related to a process of lithospheric extension rather than to compressive tectonics, and occurred entirely in a intraconti-

nental environment without evidence of crustal rupture. The main structural and magmatic characteristics of the belt were acquired during this phase that ended around 1260 Ma ago.

Subsequent compression, resulting in upright folding (ca. 1180 Ma) and, finally, shearing around 1100 Ma ago have not dramatically modified the earlier features. The significance of D2 folding and particularly of the shearing event will now be discussed in relation to the large scale events which occurred in Kibaran times in the eastern and southern African subcontinent.

The Kibaran belt sensu stricto evolved independently from other subsynchronous, parallel belts in south-eastern Africa. The Kibaran belt s.s. is separated from other Kibaran domains situated more S by the Bangweulu block in Zambia of Lower Proterozoic age, and by the Tanzanian Archaean craton. The two other main belts of Kibaran age are the Irumide belt to the SE and the Malawi-Moçambique belt still farther to the SE. The Irumide belt is mainly composed of supracrustal rocks and presents some analogies with the northern Kibaran segment. It has been quoted previously as an intra-cratonic mobile belt (Shackleton, 1969; Hurley, 1973; Watson, 1976). Recently, however, Daly (1985), with convincing arguments, has interpreted the Irumide belt as the NW-facing foreland fold and thrust belt of the southern Moçambique belt. As the structural evolution of the Irumide belt is much more complex than the northern Kibaran belt, it is not easy to characterize its early development. Nevertheless, the rock sequences and the magmatism are very similar to the northern Kibaran belt. It is thus possible that the Irumide basin evolved by extensional processes in an early stage of its evolution, before involvment in the collision processes which occurred in the Malawi-Moçambique belt. Various authors (Andreoli, 1983; Jourde, 1983; Sacchi, 1983; Daly, 1985) presented arguments for interpreting the orogeny in Moçambique and Malawi in terms of NW-dipping subduction, collision tectonics and accretion.

Ages of about 1100 Ma are reported both for the Irumide belt and for the higher grade Malawi-Moçambique belt. This implies that the younger compressive tectonic events in the northern Kibaran belt

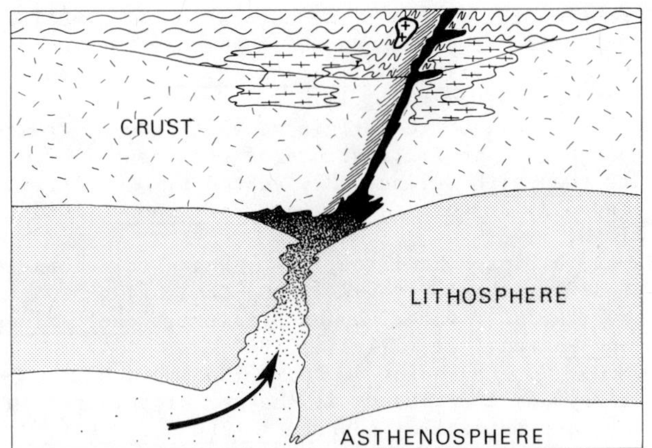

Fig. 12. Strike-slip resulting from delamination of the lithosphere; shear zones develop along the strike-slip zone with associated mafic and ultra-mafic intrusions as well as alkaline intrusions.

are contemporaneous with continental collision and accretion invoked in the southeastern belts.

Daly (1985) has also proposed that the Irumide belt is interrupted at its north eastern margin by a NW-SE directed transform fault to which he assigned a Kibaran age. This transform zone reaffects NW-SE oriented structures of Ubendian age (Lower Proterozoic), the Ubendian belt having been interpreted recently (Daly et al. 1985) as a lateral accretion belt to the Archaean Tanzania craton. The late shearing which occurs in Burundi along a conjugate set of directions, NW-SE and NE-SW, could be considered as a continuation to the north of this Irumide transform zone.

It consequently appears that the compressional tectonics observed in Burundi (from D2 at 1180 Ma to D2' at 1100 Ma) can be attributed to the NW subduction and final collision which may have occurred in the southern Malawi-Moçambique area.

As alkaline intrusions are associated with this shearing, this event can be considered as a major event which has affected the entire lithosphere and may be the result of lithospheric delamination (Kröner, 1983) (Fig. 12). Recent studies favour lithospheric control for alkaline magma genesis (Black et al. 1985). The ultramafic magmas could also be considered the result of the melting of hot asthenosphere injected into the lithosphere.

Conclusions

The northern Kibaran belt, investigated in Burundi, is an example of a linear intracontinental belt which is considered to have originated by a process of crustal extension. Sedimentation began about 1400 Ma ago. A a result of crustal thinning intensive bimodal magmatism - granitic and gabbroic - affected the upper crustal sediments during a period between 1350 and 1280 Ma ago. The extensional tectonics resulted in a décollement of the sedimentary pile over its basement and in widespread horizontal deformation of the lower sedimentary sequence;

this deformation occurred contemporaneously with the intrusion of granitic magmas. The generation of granitic magmas is considered to have result from the fusion of lower crustal rocks by heat transfer from the mafic magmas which intruded the base of the crust during the extensional processes (Fig. 11).

Most of the features which result from the extension have been preserved during later compressive tectonics. The latter culminated around 1180 Ma ago with upright deformation and ended around 1100 Ma ago during a major shearing event. Alkaline granitic intrusions and probably also mafic and ultramafic complexes were intruded along the shear zones which may be related to delamination of the subcrustal mantle lithosphere.

It consequently appears that time relations and structures of the compressional stages in the northern Kibaran belt closely match those in the southern Malawi-Moçambique belt where collision probably occurred. The compressive tectonic phase in the northern segment is considered a product of that collision. The northern Kibaran basin, however, developed initially completely independently of processes in the southern area. It behaved as an aborted rift basin, evolving entirely in intracontinental conditions, whereas continental separation and ocean formation was active in the southern Malawi-Moçambique region.

References

Andreoli, M., Geological evolution of Moçambique Belt granulites in S. Malawi - Uranium mineralisation in the Tete Province of Moçambique. Abstracts Proterozoic '82 Conference, Lusaka, p. 2, 1983.

Barberi, F., Tazieff, H. and Varet, J., Volcanism in the Afar depression : its tectonic and magmatic significance, in East African rift, edited by R.W. Girdler, Tectonophysics, 15, 19-29, 1972.

Black, R., Lameyre, J. and Bonin, B., The structural setting of alkaline complexes, J. Afr. Earth Sci., 3, 5-16, 1985.

Black, R., Morton, V.A. and Varet, J., New data on Afar tectonics (Ethiopia), Nature (Phys. Science), 240, 170-173, 1972.

Cahen, L., Delhal, J. and Deutsch, S., Rubidium-strontium geochronology of some granitic rocks from the Kibaran belt (Central Katanga, Republic of the Congo), Ann. Mus. roy. Afr. centr., Tervuren (Belg.), in-8°, Sci. géol., 59, 65 p.,1967.

Cahen, L., Delhal, J. and Deutsch, S., A comparison of the ages of granites of SW Uganda with those of the Kibaran of Central Shaba (Katanga), Republic Zaire, with some new isotopic and petrogenetical data, Ann. Mus. roy. Afr. centr., Tervuren (Belg.), in-8°, Sci. géol., 73, 45-67, 1972.

Cahen, L. and Ledent, D., Précisions sur l'âge, la pétrogenèse et la position stratigraphique des "granites à étain" de l'est de l'Afrique centrale, Bull. Soc. belg. Géol., 88, 33-49, 1979.

Cahen, L. and Snelling, N.J., The geochronology and

evolution of Africa, <u>Clarendon Press</u>, Oxford, 512 p., 1984.

Cahen, L. and Theunissen, K., The structural evolution of the Kibaran orogeny in Rwanda and Burundi in the light of the presently available radiometric data in the Kibaran belt from Shaba to Uganda, <u>Mus. roy. Afr. centr.</u>, <u>Tervuren</u> (<u>Belg.</u>), <u>Dépt. Géol. Min.</u>, <u>Rapp. ann.</u>1979, 215-217, 1980.

Christiansen, R.L. and Lippman, P.W., Cenozoic volcanism and plate-tectonic evolution of the western United States - Part II, Late Cenozoic, <u>Phil. Trans. R. Soc. London</u>, <u>A 271</u>, 249-284,1972.

Christiansen, R.L. and McKee, E.H., Late Cenozoic volcanic and tectonic evolution of the Great Basin and Columbia Intermontane regions, <u>Geol. Soc. America</u>, <u>Memoir 152</u>, 283-311, 1978.

Condie, K.C., Early and middle Proterozoic supracrustal succession and their tectonic settings, <u>Amer. Journ. Sci.</u>, <u>282</u>, 341-357, 1982.

Daly, M., The Irumide Belt of Zambia and its bearing on collision orogeny during the Proterozoic of Africa, in Collision tectonics, edited by M.P. Coward and A. Ries, <u>Spec. Publ. Geol. Soc. London</u>, 1985 (in press).

Daly, M., Klerkx, J. and Nanyaro, J.T., Early Proterozoic exotic terranes and strike-slip accretion in the Ubendian belt of south west Tanzania, <u>Terra Cognita</u>, 1985 (in press).

de Magnée, I., Coupe géologique des monts Kibara (Katanga), <u>Ann. Soc. géol. Belg.</u>, <u>58</u>, C70-82,1935.

Debon, F. and Le Fort, P., A chemical-mineralogical classification of common plutonic rocks and associations, <u>Trans. R. Soc. Edinburgh</u> : <u>Earth Sciences</u>, <u>73</u>, 135-149, 1983.

Demaiffe, D. and Theunissen, K., Données géochronologiques U-Pb et Rb-Sr relatives au complexe archéen de Kikuka (Burundi), <u>Mus. roy. Afr. centr.</u>, <u>Tervuren</u> (<u>Belg.</u>), <u>Dépt. Géol. Min.</u>,<u>Rapp. ann. 1978</u>, 65-69, 1979.

De Mulder, M. and Theunissen, K., Contribution à l'étude structurale des métasédiments de la Haute-Kitenge (NW Burundi), <u>Mus. roy. Afr. centr.</u>, <u>Tervuren</u> (<u>Belg.</u>), <u>Dépt. Géol. Min.</u>, <u>Rapp. ann. 1979</u>, 185-205, 1980.

Dreesen, R., Shallow-water deposits within the Burundian Proterozoic (Republic of Burundi, East-Africa), <u>Bull. Soc. belg. Géol.</u>, <u>89</u>, 217-238,1980.

Gérards, J. and Ledent, D., Grands traits de la géologie du Rwanda, différents types de roches granitiques et premières données sur les âges de ces roches, <u>Ann. Soc. géol. Belg.</u>, <u>93</u>, 477-489, 1970.

Gérards, J. and Ledent, D., Les réhomogénéisations isotopiques d'âge lufilien dans les granites du Rwanda, <u>Mus. roy. Afr. centr.</u>, <u>Tervuren</u> (<u>Belg.</u>), <u>Dépt. Géol. Min.</u>, <u>Rapp. ann. 1975</u>, 91-103, 1976.

Hart, W.K. and Walter, R.C., Geochemical investigation of volcanism in the west central Afar, Ethiopia, <u>Carnegie Inst. Wash. Year Book</u>, <u>82</u>, 491-497, 1983.

Hildreth, W., Gradients in silicic magma chambers : implications for lithospheric magmatism, <u>Journ. Geoph. Res.</u>, <u>86</u>, 10153-10192, 1981.

Hoffman, P., Evolution of an early Proterozoic continental margin : the Coronation geosyncline and associated aulacogens in the northwestern Canadian Shield, <u>Phil. Trans. R. Soc. London</u>, <u>ser. A 273</u>, 547-581, 1973.

Hurley, P.M., On the origin of 450 $\pm$ 200 my orogenic belts, in Implications of continental of drift <u>to the earth sciences</u>, <u>2</u>, edited by D.H. Tarling and S.K. Runcorn, <u>Academic Press London</u>, 1083-1090, 1973.

James, R.S. and Hamilton, D.L., Phase relations in the System $NaAlSi_3O_8-KAlSi_3O_8-CaAl_2Si_2O_8-SiO_2$ at 1 kilobar water vapour pressure, <u>Contr. Mineral. and Petrol.</u>, <u>21</u>, 111-114, 1969.

Jourde, G., La chaîne du Lurio (Mozambique) : un témoin de l'existence de chaînes kibariennes en Afrique orientale, <u>Abstracts 12th Coll. Afr. Geol.</u>, <u>Brussels</u>, p. 50, 1983.

Klerkx, J., Modèle d'évolution de la chaîne kibarienne, <u>Unesco, Geology for development</u>, <u>Newsletters 3</u>, 43-46, 1984.

Klerkx, J., Lavreau, J., Liégeois, J.-P. and Theunissen, K., Granitoïdes kibariens précoces et tectonique tangentielle au Burundi : magmatisme bimodal lié à une distension crustale, in Géologie africaine-African geology, edited by J. Klerkx and J. Michot, Tervuren, 29-46, 1984.

Kröner, A., Proterozoic mobile belts compatible with the plate tectonic concept, in Proterozoic geology, edited by G.M. Medaris et al., <u>Geol. Soc. Amer. Memoir 161</u>, 59-74, 1983.

Kusznir, N.J. and Park, R.G., A mathematical model of the brittle ductile transition within the lithosphere, <u>J. Geol. Soc. London</u>, <u>140</u>, 1983.

Lachenburch, A.H. and Sass, J.H., Models of an extending lithosphere and heat flow in the Basin and Range province, <u>Geol. Soc. America, Memoir 152</u>, 209-250, 1978.

Lavreau, J. and Liégeois, J.-P., Granites à étain et granito-gneiss burundiens au Rwanda (région de Kibuye) : âge et signification, <u>Ann. Soc. géol. Belg.</u>, <u>105</u>, 289-294, 1982.

Ledent, D., Résultats U/Pb et Rb/Sr obtenus sur des gneiss antérieurs au Burundien au Rwanda et au Burundi, <u>Mus. roy. Afr. centr.</u>, <u>Tervuren</u> (<u>Belg.</u>), <u>Dépt. Géol. Min.</u>, <u>Rapp. ann. 1978</u>, 97-99, 1979.

Le Fort, P., Manaslu leucogranite : a collision signature of the Himalaya. A model for its genesis and emplacement, <u>Journ. Geophys. Res.</u>, <u>16</u>, 10545-10568, 1981.

Liégeois, J.-P. and Black, R., Pétrographie et géochronologie Rb-Sr de la transition calco- alcaline fini pan-africaine dans l'Adrar des Iforas (Mali) : accrétion crustale au Précambrien supérieur, in Géologie africaine-African géology, edited by J. Klerkx and J. Michot, Tervuren, 115-146, 1984.

Liégeois, J.-P., Theunissen, K., Nzojibwami, E. and Klerkx, J., Granitoïdes syncinématiques kibariens au Burundi : étude pétrographique, géochimique et géochronologique préliminaire, <u>Ann. Soc. géol. Belg.</u>, <u>105</u>, 345-356, 1982.

McKenzie, D.P., Some remarks on the development of sedimentary basins, <u>Earth Planet. Sci. Lett.</u>, <u>40</u>, 25-32, 1978.

Montadert, L., Robert, D.G., De Charpal, O. and Guennoc, P., Rifting and subsidence of the Northern Continental Margin of the Bay of Biscay, Init. Rep. Deep Sea Drilling Proj., 48, 1025-1060, 1979.

Ntungicimpaye, A., Le magmatisme basique dans le Burundien de l'ouest du Burundi, Unesco, Geology for development, Newsletters 3, 13-21, 1984a.

Ntungicimpaye, A., Contribution à l'étude du magmatisme basique dans le Kibarien de la partie occidentale du Burundi, Unpublished Ph. D. Thesis, University of Ghent, Belgium, 250 p., 1984b.

Nzojibwami, E., The West Mugere supracrustal complex (Bujumbura) : evidence of an ancient basement complex remobilized during the Kibaran orogeny, Unesco, Geology for Development, Newsletters 3, 5-12, 1984.

Oxburgh, E.R., Heterogeneous lithospheric stretching in early history of orogenic belts, in Mountain building processes, edited by J. Hsü, Academic Press, 85-93, 1982.

Sacchi, R., Late Proterozoic evolution in the southernmost Mozambique belt, Unesco, Geology for development, Newsletters 3, 69-72, 1984.

Shackleton, R.M., Displacement within continents, in Time and space in Orogeny, edited by P.E. Kent G.E. Satternswaite and A.M. Spencer, Spec. Publ. geol. Soc. London, 3, 1-7, 1969.

Sengör, A.M.C. and Burke, K., Relative timing of volcanism on earth and its tectonic implications, Geoph. Res. Letters, 5, 419-421, 1978.

Smith, A.G., Plate tectonics and orogeny : a review, Tectonophysics, 33, 215-285, 1976.

Tack, L., Post-Kibaran intrusions in Burundi, Unesco, Geology for development, Newsletters 3, 45-57, 1984.

Tack, L. and De Paepe, P., Existence de plusieurs massifs granitiques alcalins au Burundi : réflexions préliminaires concernant leur âge et leur signification, Mus. roy. Afr. centr., Tervuren (Belg.), Dépt. Géol. Min., Rapp. ann. 1981, 135-136, 1983.

Tack, L., De Paepe, P., Deutsch, S. and Liégeois, J.-P., The alkaline plutonic complex of the Upper Ruvubu (Burundi) : geology, age, isotopic geochemistry and significance for the regional geology of the western rift, in Géologie africaine-African geology, edited by J. Klerkx and J. Michot, Tervuren, 91-114, 1984.

Theunissen, K., Les principaux traits de la tectonique kibarienne au Burundi, Unesco, Geology for development, Newsletters 3, 25-30, 1984.

Theunissen, K. et Klerkx, J., Considérations préliminaires sur l'évolution tectonique du "Burundien" au Burundi, Mus. roy. Afr. centr., Tervuren (Belg.), Dépt. Géol. Min., Rapp. ann. 1979, 207-214, 1980.

Theunissen, K. and Klerkx, J., Pan-African and late Kibaran tectonics in western Burundi, Abstracts 12th Coll. Afr. geol., Brussels, p. 97, 1983.

Turcotte, D.L. and Emerman, S.T., Mechanisms of active and passive rifting, Tectonophysics, 94, 39-50, 1983.

Vernon-Chamberlain, V.P. and Snelling, N.J., Age and isotope studies on the arena granites of SW Uganda, Ann. Mus. roy. Afr. centr., Tervuren (Belg.), in-8°, Sci. géol., 73, 1-44, 1972.

Vogel, Th.A., Williams, E.R., Preston, J.K. and Walker, B.M., Origin of the late Paleozoic massifs in Morocco, Geol. Soc. America Bull., 87, 1753-1762, 1976.

Waleffe, A., Etude géologique du sud-est du Burundi (Régions du Mosso et du Nkoma), Ann. Mus. roy. Afr. centr., Tervuren (Belg.), in-8°, Sci. géol., 48, 312 p., 1965.

Watson, J.V., Vertical movements in Proterozoic structural provinces, Phil. Trans. R. Soc., A280, 619-640, 1976.

Willems, L., Metamorfe evolutie van het Onder-Burundiaan in Zuid-West Burundi, Unpublished thesis, K.U.Leuven, Belgium, 63 p., 1985.

Williamson, J.H., Least-square fitting of a straight line, Can. J. Phys., 46, 1845-1847, 1968.

PAN-AFRICAN CRUSTAL EVOLUTION IN THE NUBIAN SEGMENT OF NORTHEAST AFRICA

A. Kröner[1], R. Greiling[1], T. Reischmann[1,2], I.M. Hussein[1,3], R.J. Stern[4],
S. Dürr[1], J. Krüger[1] and M. Zimmer[1]

[1]Institut für Geowissenschaften, Universität Mainz, Postfach 3980, 6500 Mainz,
Federal Republic of Germany
[2]Max-Planck-Institut für Chemie, Postfach 3060, 6500 Mainz, F.R.G.
[3]Geological and Mineral Resources Department, P.O. Box 573, Port Sudan, Sudan
[4]Institute of Lithospheric Studies, University of Texas at Dallas,
P.O. Box 688, Richardson, TX 75080, USA

Abstract. Similarities in rock assemblages and broad tectonic features between the late Precambrian to early Paleozoic basement of Arabia, the Eastern Desert of Egypt and the Red Sea Hills of the Sudan have led to evolutionary models that envisage broadly contemporaneous processes of island arc formation and obduction-accretion tectonics for both the Arabian and Nubian segments of this large Pan-African shield. For Arabia such models are based on a wealth of geochemical and isotopic data as well as on systematic regional mapping, while the information on NE Africa is still fragmentary for many regions, and there are much fewer constraints on ages and tectonic settings.

All present models tacitly assume that the juvenile Arabian arc and ophiolite assemblages were accreted onto the African craton between ca. 950 Ma and about 600 Ma ago, but the precise location of the cratonic margin is still unknown. Our work in the southern Eastern Desert of Egypt has revealed the local occurrence of shallow-water, clastic and partly aluminous metasediments that we interpret as passive margin deposits and that were later involved in extensive thrust and nappe tectonics. The recognition of low-angle thrusts that form ramps and duplexes suggest similarities with modern foreland thrust and fold belts. The uppermost nappe unit consists of an ophiolite mélange and well preserved remnants of layered oceanic crust as well as mafic to felsic metavolcanics whose geochemistry suggests formation in marginal and/or intra-arc basins. Voluminous calcalkaline plutonism occurred between ca. 650 Ma and 720 Ma ago, i.e. before and during the above period of thrust tectonics.

We tentatively propose that westward subduction, some 700-750 Ma ago, transformed the attenuated passive margin of the African craton into an active belt whereby marginal basin closure, ophiolite obduction and arc accretion from the east (Arabia) facilitated thrust stacks of juvenile assemblages to override the ancient cratonic edge that may now be buried beneath the SE Desert of Egypt.

In the Red Sea Hills of the Sudan rocks of undisputed continental derivation have only been recognized at one locality, but the extensive carbonate metasediments along and E of the River Nile suggest a stable depositional environment that is not found in the volcano-sedimentary terranes farther E. There, several distinct high-strain belts contain tectonically dismembered ophiolite fragments that range in size from almost complete segments of marginal basin-type oceanic crust to thin lenses of talc-carbonate schist. The ophiolite belts may define sutures between accreted arc terranes, and we have subdivided the Red Sea Hills into five such terranes.

Our age data suggest that arc magmatism occurred from at least 920 Ma ago to about 620 Ma ago. Nd isotopic systematics for several well preserved volcanic suites in the Red Sea Hills demonstrate an intraoceanic environment of formation and derivation of the arc volcanics from a significantly depleted mantle as has previously also been shown for the arc assemblages in Saudi Arabia. Furthermore, our Sr isotopic data confirm the primitive nature of the Red Sea Hills crust.

Although we can delineate several distinct crustal blocks or terranes in the Red Sea Hills that are separated by ophiolite belts or shear-zones, our isotopic results do not support sequential arc accretion models that would imply a general decrease in ages from W to E. We can demonstrate that many of the conventional large-scale lithostratigraphic correlations are erroneous and that there is no general trend from early, primitive arc magmatism to mature, andesite-dominated activity as has been postulated for the Arabian shield. It is therefore difficult at this stage to incorporate the Red Sea Hills into evolutionary models that were proposed for Arabia, although at least 2 of the Nubian ophiolite belts clearly continue into the Arabian shield.

The Pan-African structural domain with low-angle thrusts and ophiolite mélanges extends at least as far W as the River Nile, where the ancient margin of the African craton may be found. The entire domain farther E is characterized by newly accreted magmatic associations of late Precambrian age that may have evolved in settings similar to those presently observed in the Indonesian archipelago.

Introduction

There is now general agreement that the Arabian-Nubian shield represents one of the best documented examples of late Proterozoic to early Paleozoic (Pan-African sensu lato, ∿950-450 Ma, Kröner, 1984) crustal growth through processes of lateral arc and terrane accretion, comparable to the present evolution of the SW Pacific (Roobol et al., 1983; Camp, 1984; Kröner, 1985; Stoeser and Camp, 1985). This model was first developed in Saudi Arabia where a number of volcano-sedimentary/plutonic belts and ophiolite-decorated sutures display internal evolutionary patterns suggesting a development from primitive intraoceanic arcs some 900-950 Ma ago to mature, andesite-dominated arcs some 640-700 Ma ago through processes of oceanic lithosphere subduction, arc collision, ophiolite obduction and magmatic crustal thickening (Brown and Coleman, 1972; Greenwood et al., 1976; Fleck et al., 1980; Gass, 1981; Roobol et al., 1983; Clark, 1985).

Although similar assemblages occur in the Eastern Desert of Egypt, they were first interpreted in terms of a normal stratigraphic succession (Hume, 1934) and following the classical geosynclinal concept (El Ramly, 1972). Garson and Shalaby (1976) were the first to recognize the ophiolitic nature of the widespread mafic-ultramafic complexes in the Eastern Desert, and their identification of porphyry-type and Kuroko-type mineralizations led to a plate tectonic model that proposed arc accretion and marginal basin closure against an older African craton. Subsequent work by Engel et al. (1980), Dixon (1981), Stern (1981) and Stern and Hedge (1985) supported this concept on the basis of geochemical and isotopic data, while Shackleton et al. (1980) and Ries et al. (1983) recognized large nappes of ophiolitic mélange and suggested that the "geosynclinal assemblage" represents a collage of tectonic units whose internal age and stratigraphic relationships are only locally decipherable.

Early work in the Red Sea Hills of the Sudan centered on regional studies N of Port Sudan (Gass, 1955; Ruxton, 1956; Gabert et al., 1960; Kabesh and Lotfi, 1962) and documented volcano-sedimentary and plutonic assemblages. Neary et al. (1976) were the first to recognize the similarity of rock types and their tectonic settings in the NE Red Sea Hills with those in the Eastern Desert of Egypt and in Arabia, and they supported the arc accretion model developed for these regions. Subsequently Vail (1976) attempted to integrate the Red Sea Hills into the geology of the entire Arabian-Nubian shield and suggested that the "greenschist assemblages" were part of the same Pan-African crustal domain as the Eastern Desert of Egypt and the Arabian basement, while he considered the high-grade gneisses at the River Nile to represent the edge of the ancient African craton. Further work led to a first evolutionary model for the Red Sea Hills where 3 separate crustal entities were identified, each reflecting temporally distinct cycles of magmatic activity and separated by ophiolite belts, similar to the situation in Arabia (Vail, 1983, 1985; Embleton et al., 1984).

Detailed mapping, geochemical work, isotopic studies and seismic profiling have enabled workers in Saudi Arabia to modify the original accretion model, particularly after it was recognized that not all crustal segments were of juvenile (i.e. mantle-derived) origin (Stacey and Hedge, 1984; Stoeser and Camp, 1985). In contrast, research in the Sudan and Ethiopian segments of the Nubian shield is still hampered by a lack of detailed, structure-oriented mapping as well as geochemical and geochronologic characterization of most rock units. For this reason, large-scale lithostratigraphic correlations have inhibited the identification of distinct tectonic provinces that can be fitted into a regional tectonic framework and that can be compared with the better known terranes in Egypt and Arabia.

This contribution summarizes the results of field and laboratory studies in selected areas of SE Egypt and the Red Sea Hills of the Sudan that were initiated in 1981 as part of the activities of IGCP Project 164 "Pan-African crustal evolution in the Arabian-Nubian shield". We present and discuss our data in relation to existing models for the evolution of the shield and with particular emphasis on the development of the Nubian segment.

Rock Assemblages in SE Egypt and in the Red Sea Hills

We discuss here the four principal rock associations that make up the bulk of the Pan-African basement along the western flank of the Red Sea between latitudes 18° N and 25° N (Fig. 1). These are: (1) older shelf sequences, (2) arc assemblages, (3) ophiolitic suites, and (4) granitoid intrusives.

The oldest rocks recognized in the Eastern Desert of Egypt are metaquartzites and quartzitic schists that are locally cross-bedded and predominantly feldspathic. These rocks are exposed in the composite dome structure at Hafafit (El Ramly et al., 1984). The aluminous nature of some of these rocks is revealed by the local presence of sillimanite, while primary sedimentary structures were largely destroyed by intense deformation and

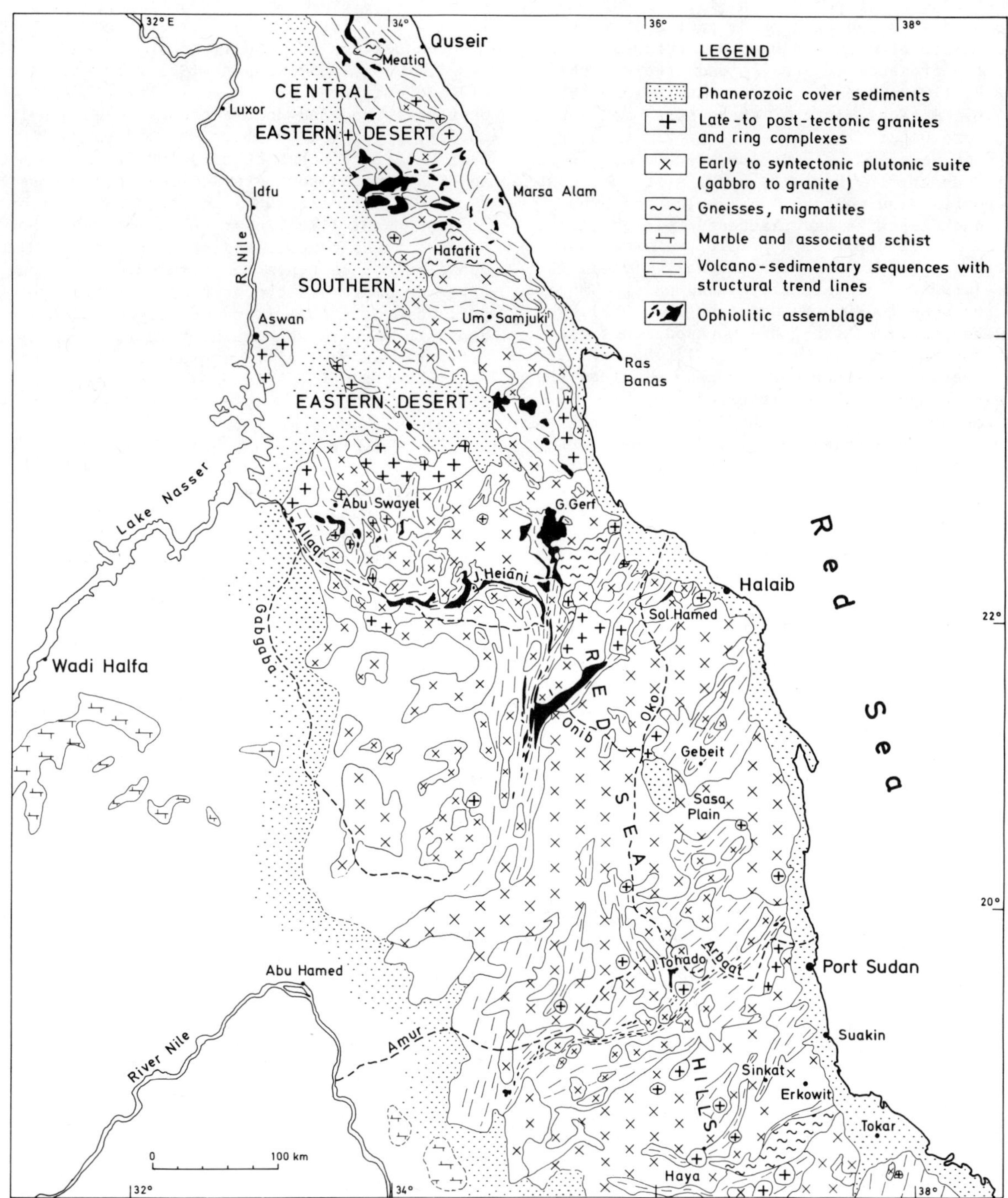

Fig. 1. Simplified geological map of the Eastern Desert of Egypt and the Red Sea Hills of
the Sudan between lat. 18°N and 26°N, showing the principal rock associations and place names
mentioned in the text. Based on Geol. Map of Egypt, 1:2.000.000 (1981), Geol. Map of the Sudan,
1:2.000.000 (1981), Hussein (1986) and Ali (1979), unpublished mapping of the Egyptian Geological Survey and of Geosurvey International Ltd. and on own reconnaissance traverses and Landsat
interpretation.

metamorphism. Similar rocks occur locally around
the Meatiq dome farther N (Fig. 1), although most
of the mylonites and high-grade gneisses of this
structure are of magmatic origin (Sturchio et al.,
1983). The presence of granite and arkose cobbles
with U-Pb zircon ages up to 2060 Ma (Dixon, 1981)
in clastic metasediments of the central and southern Eastern Desert indicates the proximity of ancient continental crust, presumably the margin of
the African craton. However, at Hafafit such clastic deposits also contain local volcanic components, such as tuffs and flattened felsic clasts,
that may be derived from a primitive magma source
(positive $\varepsilon_{Nd}(T)$-value, Harris et al., 1984), perhaps related to initial rifting and passive margin
formation some 800-900 Ma ago (Church, 1982). We
discovered similar high-grade metasediments and
migmatites SE of Jabal Gerf (Fig. 1), but it remains uncertain whether these rocks belong to the
same assemblage as those discussed above.

In the Red Sea Hills of the Sudan, quartzites
with associated marbles were found in small outcrops in the Sasa Plain S of Gebeit and in extensive areas S of Wadi Amur. High-grade, partly aluminous, metasediments also occur near Haya, SSW of
Port Sudan (Fig. 1). In several localities these
rocks appear to occur in the tectonically lowest
positions, and we interpret them to represent continental margin deposits.

Several authors have considered all high-grade
supracrustal rocks of the Red Sea Hills to constitute an older basement included in the "Kashebib
Series" (Gabert et al., 1960; Vail, 1979; Geol.
Map of Sudan, 1981), but at many localities we can
demonstrate from field relationships that the
higher metamorphic grade is due to wide thermal
aureoles around large granitoid batholiths. The
only rocks reflecting high regional metamorphism
are the cordierite-sillimanite-kyanite gneisses of
the "Tolik Series" near Haya (El Samani, 1983).

Tholeiitic as well as calcalkaline volcanic
and volcaniclastic assemblages, known as Shadli
Volcanics or Geosynclinal Metavolcanics in Egypt
(El Ramly, 1972; El Shazly, 1964) and as Nafirdeib
Series in the Sudan (Ruxton, 1956; Ahmed, 1979;
Geol. Map of the Sudan, 1981) are generally less
deformed and metamorphosed than the metasediments
and are therefore considered to be younger. Clear
contacts between these units, however, are rare
and often tectonic, but Shukri and Mansour (1980)
report an unconformity from the southern Eastern
Desert. Geochemically the volcanic rocks have island arc characteristics and are thus comparable
to the assemblages in the central Eastern Desert
and in Saudi Arabia (Engel et al., 1980; Stern,
1981; Hafez and Shalaby, 1983; Vail et al., 1984;
Reischmann and Kröner, 1984), but their genetic,
geographic and chronologic relationships remain
virtually unkown. It is therefore incorrect to regard them as a single, synchronous stratigraphic
unit as will be demonstrated below.

The third major rock assemblage in the Nubian
shield consists of dispersed and tectonically dismembered ophiolite complexes and ophiolitic mé-

langes that occur in large nappes (Shackleton et
al., 1980; Ries et al., 1983; Greiling et al.,
1984) or in distinct shear zones that may mark
sutures between previously separate crustal blocks
(Vail, 1983, 1985; Kröner, 1985). The major ophiolite occurrences are shown in Fig. 1, and the
best preserved and documented examples are in Wadi
Ghadir, Egypt (Elbayoumi, 1980), Jabal Sol Hamed
near Halaib (Fitches et al., 1983), in Wadis Onib
and Sudi in the northern Red Sea Hills (Hussein et
al., 1984; Kröner, 1985) and at Jabal Gerf as well
as along Wadi Allaqi near the Egypt/Sudan border.
Most of these complexes display the full range of
ophiolitic rock types and structures, although
often no longer in the original stratigraphic order. Some of the smaller occurrences lack the diagnostic rock relationships to unequivocally identify them as ophiolitic segments. Furthermore, extensive serpentinization and carbonatization has
profoundly altered many of the ultramafic rocks
to talc-carbonate schist.

Since almost all contacts between the ophiolite
assemblage and other rock units are either tectonic or intrusive, the age relationships are uncertain, and none of the Nubian ophiolites has so far
been dated. Several workers have suggested that
the ophiolite-decorated sutures of the Red Sea
Hills continue into Saudi Arabia (Bakor et al.,
1976; Vail, 1983, 1985; Camp, 1984), but detailed
correlations remain uncertain.

The Shadli-Nafirdeib volcano-sedimentary succession locally contains rare gabbroic and ultramafic fragments in basal conglomerates such as at
Sol Hamed (Fitches et al., 1983) and Jabal Gerf,
and at least in these instances must be younger
than the ophiolites. However, there are probably
also pre-ophiolite arc assemblages in the Red Sea
Hills as shown below.

All the rock types discussed so far were intruded by large masses of gabbroic to granitic
rocks that often constitute composite batholiths.
In Egypt these igneous suites are included in the
"Grey" or "Older" Granitoids (El Ramly and Akaad,
1960) and the Metagabbro-Diorite Complex (Geol.
Map of Egypt, 1981), with ages between 610 Ma and
710 Ma (Dixon, 1981; Stern and Hedge, 1985), while
they are collectively referred to as the "Batholithic Granite" and the gabbro-diorite complex in
the Sudan (Gass, 1955; Ahmed, 1979; Geol. Map of
the Sudan, 1981) with ages between 850 Ma (our
unpubl. data) and 660 Ma (Cavanagh, 1979). Available chemical data suggest these rocks to be generally of I-type and to be related to the evolution of the volcanic arc assemblages as in Saudi
Arabia (Neary et al., 1976; Hussein et al., 1982;
Gillespie and Dixon, 1983; Vail et al., 1984).

A variety of mostly unfoliated, often coarse-
grained and red to pinkish granites post-date all
the above layered and igneous suites and display
A-type characteristics (Hussein et al., 1982).
These rocks are widely distributed in the central
and northern Eastern Desert and in Sinai, where
they are associated with the bimodal, rift-related
Dokhan volcanics (575-595 Ma, Stern and Hedge,

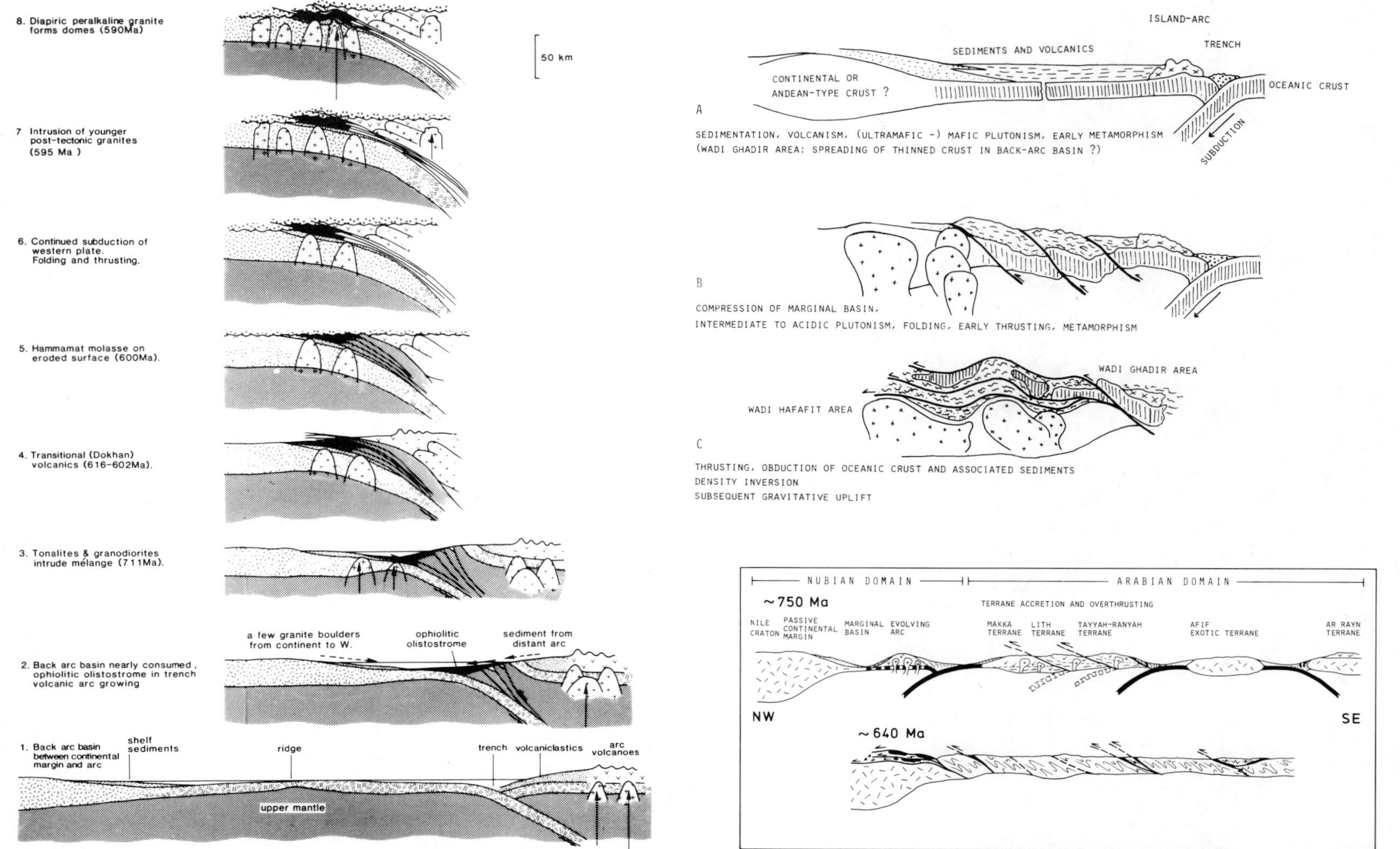

Fig. 2. Contrasting models for the Pan-African tectonic evolution of the Eastern Desert of Egypt (left: Ries et al., 1983; right top: El Ramly et al., 1984) and hypothetical composite NW-SE sections across Nubian and Arabian domains (right bottom) showing suggested evolution between ~750 Ma and ~640 Ma ago (Kröner, 1985).

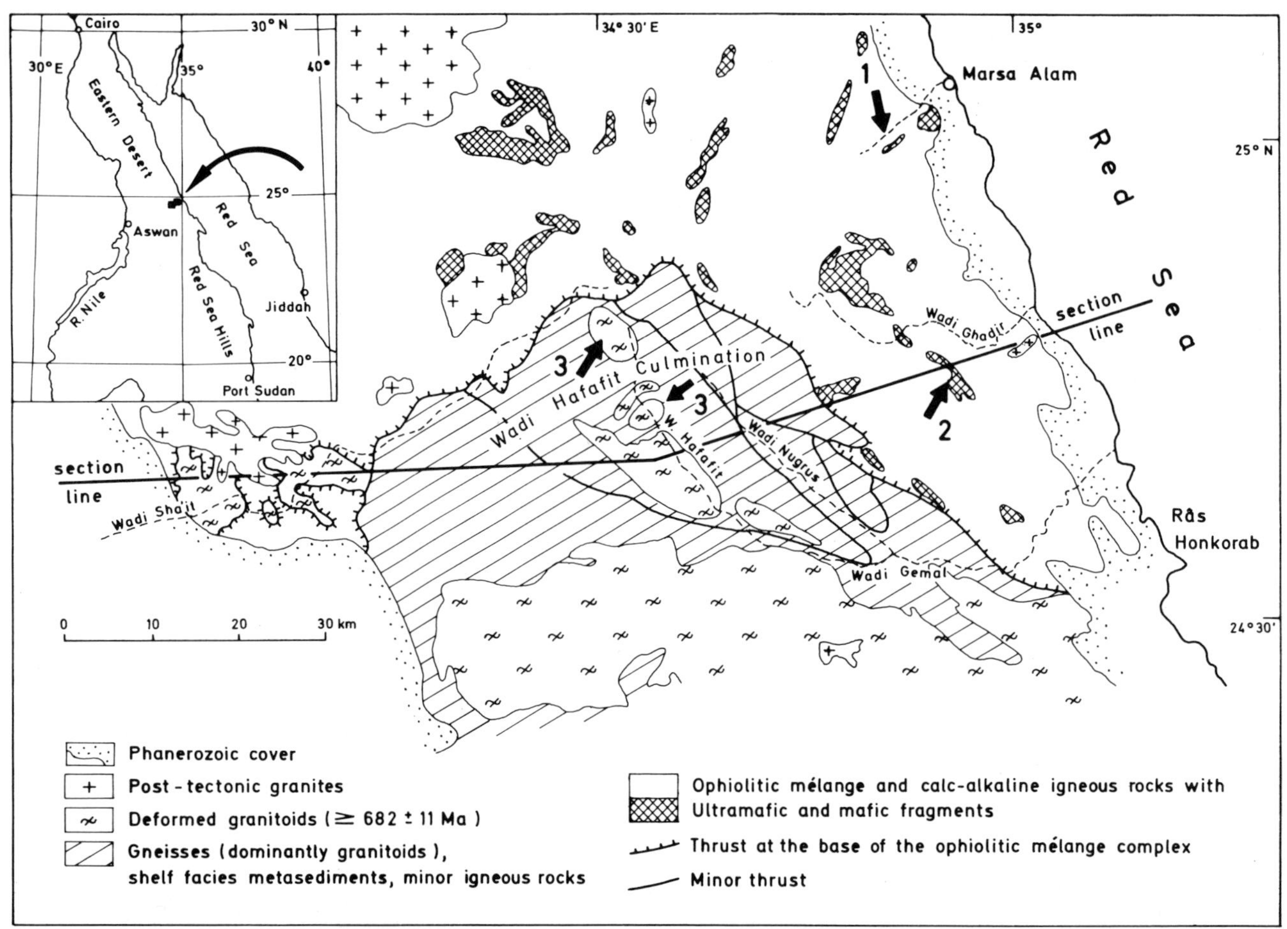

Fig. 3.   Simplified and schematic map of Wadi Hafafit Culmination and environs showing major rock types and tectonic units. Section line refers to Fig. 6. A major (frontal) ramp is assumed below the WHC, parallel to its NE margin. This frontal ramp terminates at its NW end at a corresponding lateral ramp which runs at, and parallel to, the WHC's NW margin. Numbered arrows indicate location of rock suites investigated in detail: 1, Volcanic rocks near Marsa Alam (Zimmer, 1985); 2, Wadi Ghadir ophiolite, sheeted dikes and pillow lavas in Wadi El Beida; 3, Hafafit igneous suite.

1985; Stern, 1985) and are collectively known as "Pink" or "Younger" Granites (El Ramly and Akaad, 1960). Stern et al. (1984) and Stern (1985) relate them to a major episode of NW-SE to N-S directed rifting in northernmost Afro-Arabia. Similar granites occur in the Red Sea Hills where they are volumetrically less important and also indicate significantly older ages around 670 Ma (Cavanagh, 1979; Stern and Kröner, unpubl. data).

Tectonic Evolution of the Southeastern Desert of Egypt

Ries et al. (1983) have attempted to reconstruct the Pan-African evolution of the central Eastern Desert in terms of a two-stage plate tectonic model as shown in Fig. 2a. They suggested that SE-subduction and closure of a back arc basin led to NW-directed ophiolite (marginal basin crust) obduction and nappe formation when an oce-

anic arc collided with the African passive continental margin. El Ramly et al. (1984) and El-bayoumi and Greiling (1984) studied an E-W section in the SE Desert from the Wadi Ghadir ophiolite to the composite Hafafit dome structure (Fig. 3) and, in analogy with the Ries et al. (1983) model, they proposed a passive continental margin development that may have begun between 800 Ma and 1100 Ma ago. In their scenario subduction is to the SW, however, and resulted in the formation of an oceanic arc and marginal basin the rocks of which were subsequently thrust over each other and onto the continent during arc accretion, giving rise to thrust stacks composed of ophiolite and mélange nappes, dismembered arc segments (now found as sheared calcalkaline metavolcanics and volcaniclastic rocks) and continent-derived, predominantly clastic metasediments (now largely psammitic gneisses and migmatites). In this model the intrusion of plutonic rocks began

240   KRONER ET AL.

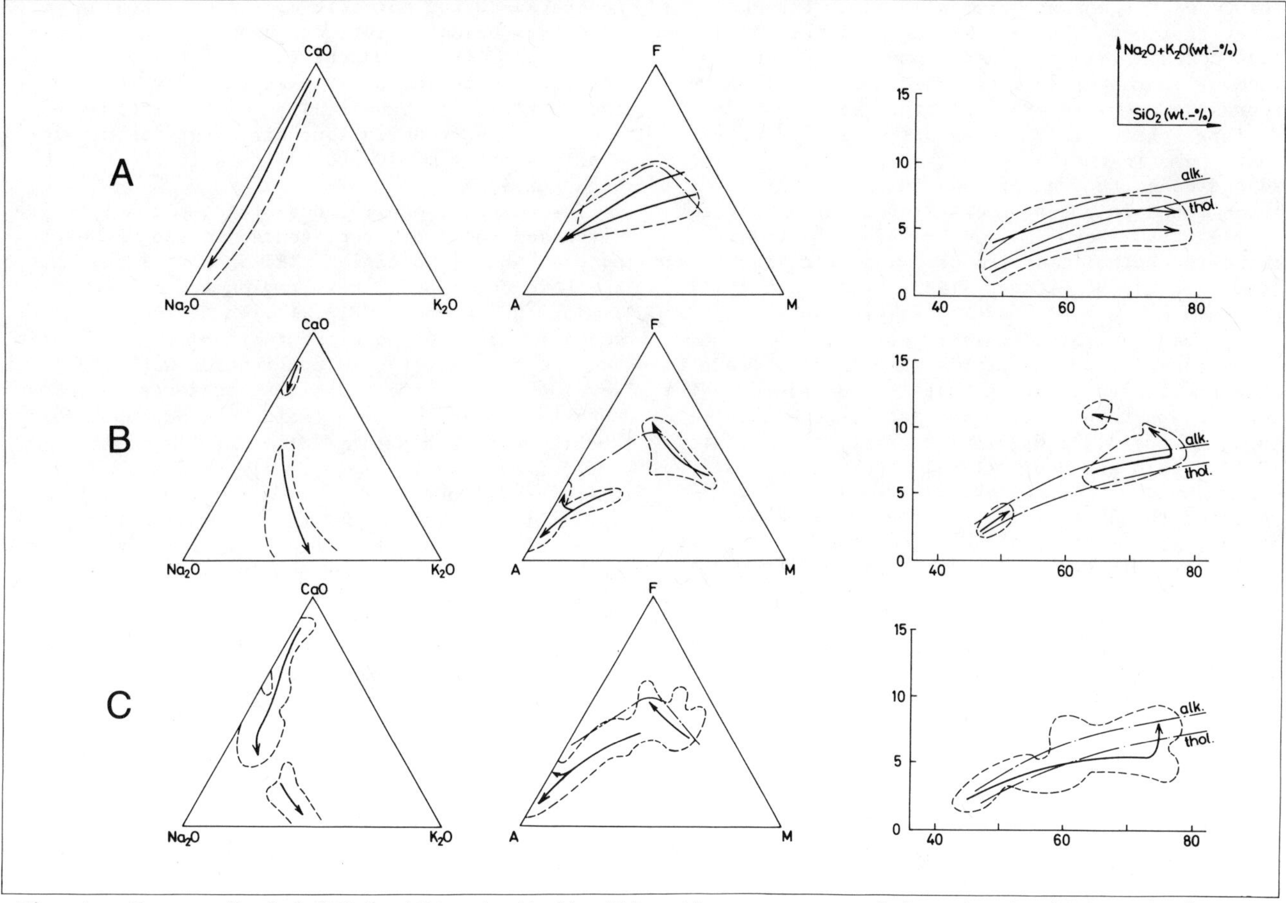

Fig. 4. Ternary Na$_2$O-CaO-K$_2$O, AFM and alkali-silica diagrams summarizing the chemical evolution of magmatic rocks in the Arabian shield (A and B) and the Hafafit area of southeastern Egypt (C).
A: Mantle-derived and subduction-related "primary crust" (andesitic assemblage and dioritic suite, Schmidt and Brown, 1984).
B: Associations produced after collision and "cratonization" (rhyolitic assemblage and granite-gabbro suite, Schmidt and Brown, 1984).
C: Calcalkaline magmatic association of Hafafit, J. Krüger, unpubl. data).

during early deformation of the supracrustal sequences and during the onset of thrust tectonics, i.e. at a time when the previously thinned continental crust was transformed into an active (Andean- or Cordilleran-type) margin (Fig. 2b).

Further support for these models comes from our new geochemical data on a plutonic suite at Hafafit and environs, for metavolcanic rocks near the Red Sea W of Marsa Alam and for the Wadi Ghadir ophiolites, as well as from structural studies around Hafafit and elsewhere in the southeastern Desert.

The granitoid gneisses forming the cores of several conspicuous dome structures in the Wadi Hafafit area (Wadi Hafafit Culmination = WHC) SW of Marsa Alam (Fig. 3) were previously regarded as the oldest rocks in the Eastern Desert and as the basement for the surrounding low-grade and predominantly volcanic rocks (El Ramly and Akaad, 1960). These rocks are intimately associated with

ultramafic, gabbroic, dioritic, tonalitic and trondhjemitic intrusives as well as with a variety of metapelites, now all intensely deformed and partly migmatized (El Ramly et al., 1984). The mafic varieties occur largely as banded amphibolites while the trondhjemitic to granitic members are represented by leucocratic orthogneisses with variable proportions of hornblende, biotite and K-feldspar. Migmatization is most pronounced in the siliceous gneisses. The serpentinized ultramafic bodies are volumetrically insignificant and occur as lens-shaped xenoliths in the younger granitoids.

Intrusive and structural contacts show a temporal relation between these rock types from the oldest ultramafic-gabbroic phases to the youngest granodiorite-granite varieties. The gabbroic to tonalitic members of the suite largely constitute the outer margins of the dome structures at Hafafit while the leucocratic granitoids constitute

the cores but occasionally also occur between the
other gneisses. The central parts of the domes
also contain less strained, competent rocks whose
igneous character is easily recognized and where
intrusive contacts are best preserved.

This systematic field relationship indicates
that these intrusive rocks are genetically related and constitute an igneous suite similar to
those found in the southern Arabian shield (Fleck
et al., 1980; Schmidt and Brown, 1984) and in
composite batholiths of modern magmatic arcs
(Hamilton, 1979; Pitcher, 1983). The field appearance of most of these rocks is also suggestive
of I-type granitoids (Hussein et al., 1982).

We consider the Hafafit granitoids to belong to
the same intrusive suite that is widespread in the
southern Eastern Desert and that is generally
known as the Older Granite (El Ramly and Roufaiel,
1974). Stern and Hedge (1985) report a virtually
concordant U-Pb zircon age of 682±11 Ma for a
foliated tonalite from the northernmost dome
structure at Hafafit that defines the younger age
limit for this suite. Tonalitic plutons from the
region farther SE yielded zircon ages of 710 Ma
(Dixon, 1981; Stern and Hedge, 1985) that may
signify the beginning of active margin plutonism
in this part of the Nubian shield. Greiling et al.
(1984) showed that large-scale recumbent folding,
shearing and nappe formation affected the igneous
suite at Hafafit, and it is speculated below that
the present granitoid-cored large antiformal
structure resulted from westward movement of a
thick pile of nappes over a structural ramp.

The igneous suite at Hafafit was generated
during 4 temporally distinct intrusive episodes
during which increasing differentiation led to
bodies ranging from minor ultramafic dykes and
sills (now serpentinites) to voluminous monzogranites. The AFM-plot (Fig. 4) displays a calcalkaline trend for the whole suite, and the geochemistry with low levels of LIL- and HFS-elements indicates a subduction-related magma generation at depths of less than 50 km and at melting temperatures above 1100°C (Wyllie, 1984). A
direct deriviation of the suite from a mantle reservoir is unlikely in view of the high silica-
content of 68-78% for the trondhjemites and granites, and we suggest that the majority of the
suite was generated from underplated mantle-derived magmas as proposed for Andean batholiths by
Cobbing and Pitcher (1983).

A comparison of the Hafafit igneous suite with
intrusive rocks of the Arabian shield reveals
some important geochemical differences that may
reflect variations in tectonic setting (Fig. 4).
In Arabia a clear separation can be made between
ca. 700-900 Ma old rocks reflecting juvenile
primary crust and younger, 560-660 Ma old rocks
that indicate significant intracrustal melting
during cratonization (Schmidt and Brown, 1984;
Jackson, 1986; Stoeser, 1986; upper 6 diagrams
in Fig. 4). The Hafafit trends (bottom diagrams
in Fig. 4) do not show this distinction and have
similarities with Andean-type suites, i.e. rocks

generated during magmatic crustal thickening
processes below active continental margins (Barker et al., 1981). On the basis of the available
age data it would appear that the time to produce
the variation in genetically related intrusive
rock types as seen in southern Egypt was considerably shorter ($\sim$710-680 Ma) than in the SW Arabian shield.

The volcanic equivalents of the plutonic suite
discussed above are represented in the SE Desert
by the "Shadli Volcanics" (El Shazly, 1964) that
are best developed in the region around the Um
Samjuki Cu-Prospect (Fig. 1). These rocks reflect
several phases of basaltic to rhyolitic volcanism,
and their chemistry, in combination with the poor
development of true andesites, suggests an immature island arc setting (Hafez and Shalaby, 1983).
Rb-Sr whole-rock dating (Stern and Kröner, unpubl.
data) indicates an isochron age of 711±10 Ma for
these rocks, and a low $^{87}Sr/^{86}Sr$ initial ratio
(Sr$_i$) of 0.702 supports derivation from a depleted mantle source. The tholeiitic to calcalkaline
metavolcanics of the Abu Swayel region farther
SW are lithologically similar but significantly
older at 768±31 Ma (Rb-Sr whole rock isochron,
Sr$_i$ = 0.702, Stern and Hedge, 1985).

The metavolcanics near Marsa Alam and N of the
Wadi Ghadir ophiolite nappe (Fig. 3) probably
also belong to the above volcanic association
and are particularly interesting since they do
not display a typical calcalkaline chemistry.
They probably belong to a thrust slice that over-
rode the ophiolite nappes, and the rocks are associated with an olistostrome-turbidite succession (Church, 1982). Pillowed basaltic andesite
with large hornblende pseudomorphs after clinopyroxene phenocrysts are interlayered with two
chemically contrasting basalt types. Of these,
the low-Ti variety has low concentrations of
incompatible elements and is similar to modern
arc tholeiites while the high-Ti type has within-
plate chemical affinities (Table 1). Pearce
(1980) proposed that such rock associations may
be generated above subduction zones in areas
where back-arc rifting has taken place, and the
Marsa Alam volcanics may thus represent an example of intra-arc spreading similar to the situation in the Philippine Sea (Karig, 1971).

When Garson and Shalaby (1976) proposed their
subduction-accretion model for the Eastern Desert
they postulated the existence of marginal basins
solely on the basis of metallogenetic associations, while Stern (1981) recognized the intimate
relationship between pillowed tholeiites and
clastic sediments that indicated nearby continental crust and island arcs during ophiolite formation.

Geochemical data also indicate that the Nubian
ophiolites do not represent normal MOR-type oceanic crust. Trace elements in sheeted dykes and
pillow basalts of the Wadi Ghadir (Egypt) and
Onib (Sudan) complexes reveal a significant enrichment in elements of low ionization potential
relative to modern MORB (Fig. 5), a feature

TABLE 1. Selected Geochemical Features of Representative Volcanic Rocks from the Nubian Shield (all concentrations in ppm)

| | Eastern Desert of Egypt | | Red Sea Hills, Sudan | | | | | | |
| | | | Wadi Abent | | | Gebeit | | | Wadi Arbaat |
Sample	EG 524	EG 523	PS-13	PS-21	PS-28	GE-19	GE-46	GE-72	AR-2
Ti	13608	4256	1858	3657	5096	4137	10731	5875	8993
Zr	189	59.9	16.7	61.9	36.3	22.3	208	79.5	137
Y	39.1	16.3	11.5	18.3	22.9	16.1	47.2	24.9	38.9
Co	31.4	34.0	43.9	20.3	54.2	60	37.8	32.3	44.5
Cr	13.2	52.6	375	53.2	450	491	68.1	150	81.8
V	222	328	290	161	247	321	228	234	238
La	13.7	7.5	1.82	6.75	1.18	5.57	16.5	16.2	17.1
Sm	5.0	5.1	.83	2.63	1.90	2.47	6.83	4.71	6.22
Yb	–	–	1.12	1.91	2.10	1.05	3.70	2.12	3.44
Ti/Zr	72	71	111	59	140	186	51	74	66
Ti/V	61	13	6.4	23	21	13	47	25	38
Zr/Y	4.8	3.7	1.5	3.4	1.6	1.4	4.4	3.2	3.5
Y/Co	1.2	0.5	.26	.90	.42	.27	1.25	.77	.87
$(La/Sm)_n$	1.73	0.93	1.38	1.62	0.39	1.42	1.52	2.17	1.73
$(La/Yb)_n$	–	–	3.53	3.63	0.38	3.58	3.01	5.16	3.36

EG 524: high-Ti basalt (tholeiitic), Marsa Alam
EG 523: low-Ti basalt (tholeiitic), Marsa Alam
PS-13: basaltic andesite (boninitic), Wadi Abent
PS-21: basalt (tholeiitic), Wadi Abent
PS-28: andesite (calcalkaline), Wadi Abent
AR-2: basaltic andesite (calcalkaline), Wadi Abent
GE-19: cpx-phyric basalt (tholeiitic), Gebeit
GE-72: hbl-phyric basalt (calcalkaline), Gebeit
GE-46: pl-phyric andesite (calcalkaline), Gebeit

Data for Eastern Desert from Zimmer (1985) and for Red Sea Hills from Reischmann (1986).

thought to be caused by selective enrichment of the magma source due to aqueous fluids released from the downgoing slab during subduction (Pearce et al., 1981). Although some of the high values in K, Rb and Ba may be due to secondary alteration, immobile elements such as Nb, Ce, P and Zr show a noticeable increase over MORB, similar to the marginal basin basalts in the Scotia Sea and the "Geotimes lavas" of the Oman ophiolite (Pearce et al., 1981; Saunders and Tarney, 1984). The remarkable depletion of Cr in the Wadi Ghadir rocks (Fig. 5) suggests an arc component as was also observed in the Geotimes lavas. Kröner (1985) has provided additional major element distribution patterns for these two Nubian ophiolites that re-

semble those of back-arc basin rocks from post-Mesozoic environments.

Pearce et al. (1984) named rock assemblages that have geochemical characteristics of island arcs but the structure of oceanic crust "supra-subduction zone (SSZ) ophiolites" and relate their origin to seafloor spreading directly above subducted oceanic lithosphere, either in an intra-arc or back-arc basin setting. If this model is applicable to the ophiolites of NE Africa there must have been an important component of westward subduction of oceanic lithosphere below the evolving marginal basin(s) as suggested by El Ramly et al. (1984) and Kröner (1985) while an SSZ-origin is difficult to reconcile with the models of east-

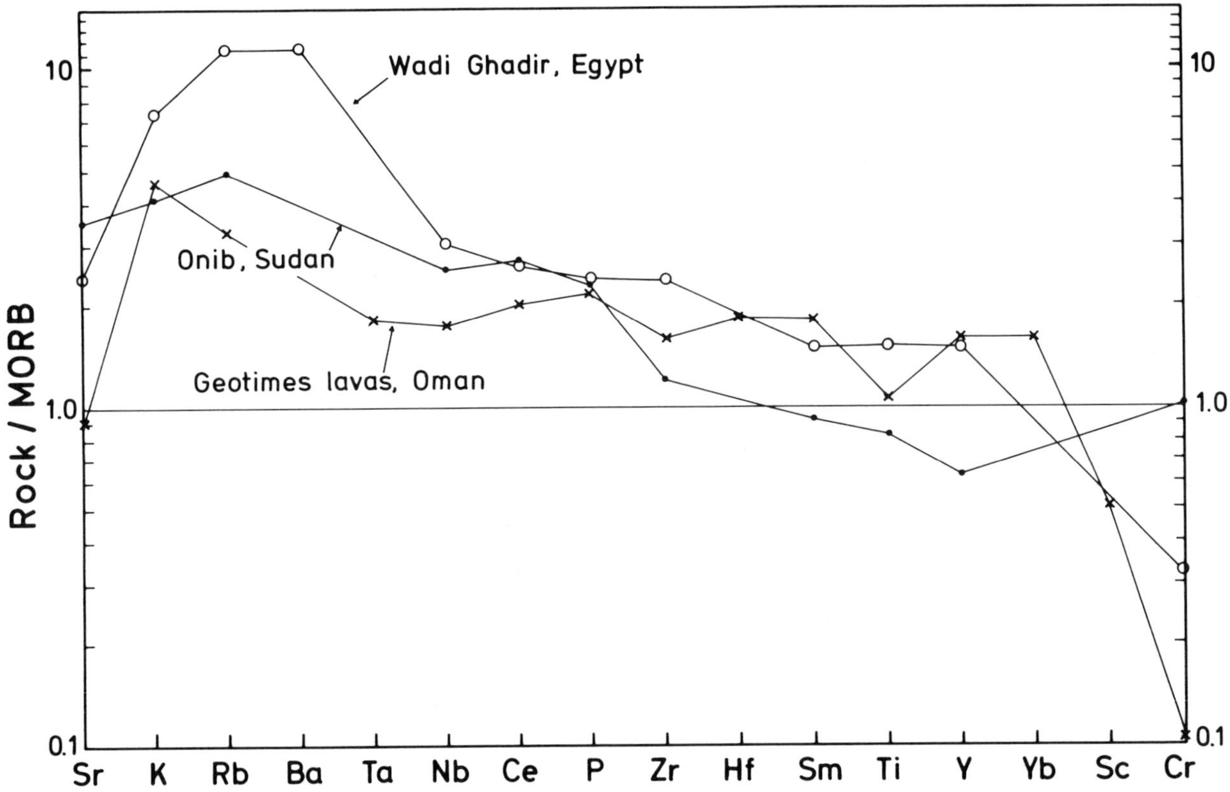

Fig. 5. Geochemical patterns for pillow lavas and sheeted dikes of Wadi Ghadir ophiolite, Egypt (data from Elbayoumi, 1980, and own analyses) and Onib ophiolite, Sudan (data from Hussein, 1986), normalized against modern MORB (normalizing values from Pearce et al., 1981) and compared with pillow basalts of the Oman (Pearce et al., 1981).

ward subduction proposed by Ries et al. (1983) and Camp (1984).

It is equally difficult to generate the Hafafit plutonic suite from mantle reservoirs below a passive continental margin as envisaged by Ries et al. (1983, see also Fig. 2). There are no modern analogues to such granite-forming processes, and we suggest that the Hafafit igneous suite and related plutonic rocks in the SE Desert intruded into an active margin environment close to, or on, the African continent as indicated by the presence of mature arkosic and partly aluminous metasediments. The association of this magmatic arc with a marginal basin setting as inferred above suggests a geographic relationship similar to the Rocas Verdes/Sarmiento Complex in southern Chile (Dalziel et al., 1974) or the Andaman Sea region of the eastern Indian Ocean (Curray et al., 1979).

We propose that westward subduction of oceanic lithosphere under the attenuated African continent first generated a marginal basin behind an evolving oceanic arc. Closure of this basin and subsequent collision of the arc with the continent shifted the arc magmatism westwards. As a result of this accretion process marginal basin crust (Wadi Ghadir ophiolite) was obducted from the E

and thrust over the Wadi Hafafit Culmination (Fig. 2).

Our structural work in a traverse from the Red Sea coast S of Marsa Alam to Hafafit (Elbayoumi and Greiling, 1984) revealed a tectonic pattern that is consistent with the above model. The region can be divided into two distinct structural domains that differ in their tectonic evolution and that are separated by regional low-angle thrusts. They comprise the high-grade Wadi Hafafit Culmination (WHC) in the W and the low-grade ophiolite mélange and associated rocks of the Wadi Ghadir area to the E (Fig. 3). The time of regionally extensive thrusting is bracketed between ca. 680 Ma (sheared tonalite of the WHC, Stern and Hedge, 1985) and the intrusion of post-tectonic granites dated elsewhere in the Eastern Desert at 565-600 Ma (Fullagar, 1980; Stern and Hedge, 1985).

The tectonic evolution of the WHC has been studied by El Ramly et al. (1984) and is summarized in Table 2. Seven local deformation phases (D1-D7) and two metamorphic events occurred prior to the regional thrusting (D8), while only two deformation phases after the mélange-forming event have been recognized in the nappe units to the E

TABLE 2. Sequence of Structural, Metamorphic and Magmatic Events in the Hafafit Area,
Eastern Desert of Egypt (modified after El Ramly et al., 1984)

Magmatic Activity	Metamorphism	Deformation	Minor Features (outcrop and microscopic scale)	Major Structures (map scale)
		D 11	open folds, generally flat-lying axial plane, NNW-SSE to N-S directions locally crenulation cleavage	uplift of granitoid domes
		D 10	open folds, generally steep axial planes, NW-SE to WNW-ESE directions fracture cleavage	
			induced by (D 8-) thrusting above ramps in the footwall (D 10) and in the hangingwall (D 9) of the WHC gneisses	
		D 9	chevron/kink-type folds, NNE-SSW to NE-SW directions	
		D 8	thrusting/faulting, mylonite formation	regional low angle thrusts
		D 7	variable open to tight dragfolds, generally steep axial planes, NW-SE to WNE-ESE directions, crenulation/fracture cleavage, locally intensive, leading over to (D 8) thrusting/faulting	
white pegmatite				
granite intrusion				
		D 6	open folds, generally steep axes and axial planes, NE-SW to E-W directions crenulation cleavage	
red pegmatite				
		D 5	local faults, mylonites	low angle thrusts
		D 4	penetrative foliation, isoclinal folds	open to isoclinal, irregular folds, generally steep axial planes
pegmatitic veins				
granitoid intrusion (Older Granite)	M 2			
		D 3	penetrative foliation, isoclinal folds	
diorite intrusion				
	M 1	D 2	penetrative foliation, metamorphic banding, local migmatization	
gabbro intrusion (locally serpentinite)				
		D 1	planar fabric	
volcanism			sedimentary texture	

The evolution shown here applies only to the central, high-grade domain of the WHC (Elbayoumi and Greiling, 1984) The emplacement of post-tectonic granites (outside the Hafafit domain) occurred during or after D10.

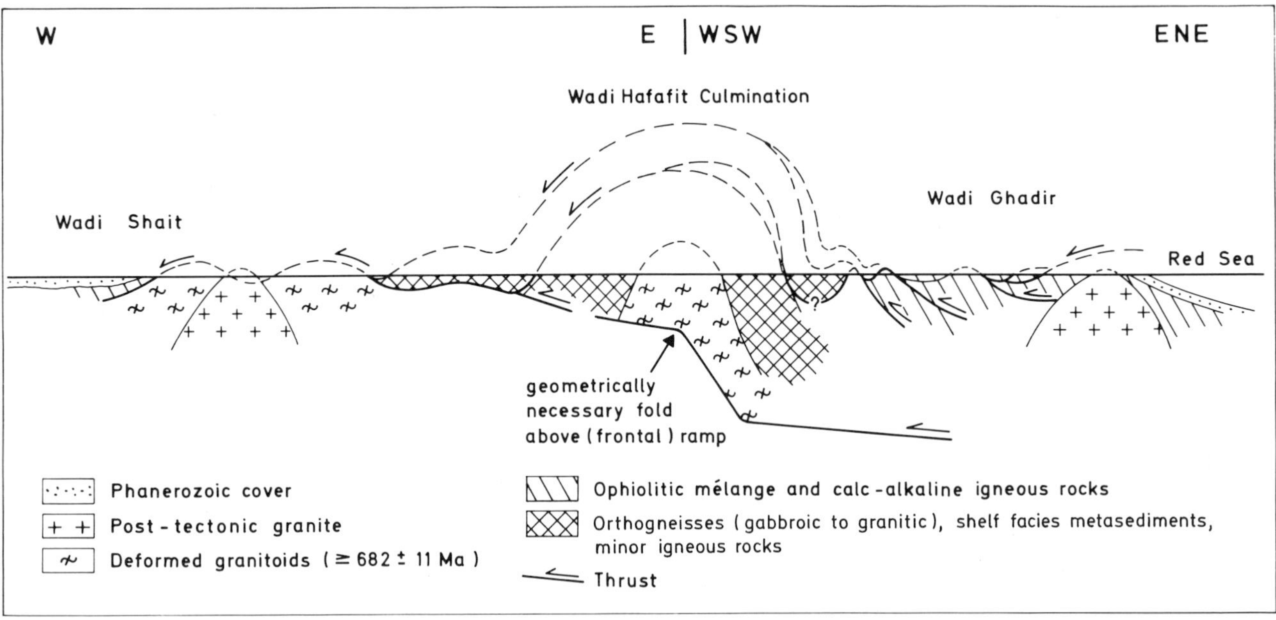

Fig. 6. Simplified cross-section through the Wadi Hafafit Culmination and environs (for location see Fig. 3), showing a thrust-tectonic interpretation of major structures. The NW-SE trending regional folds probably formed above irregularities of a (blind) thrust plane. A major frontal ramp is assumed to underlie the WHC, causing the major antiform. The "blind" thrust is exposed in lower Wadi Shait, where it forms the boundary between the older (?) Wadi Shait granite in the footwall and the Wadi Hafafit gneisses in the hanging wall. The lateral ramp at the NW margin of the WHC is situated north of, and parallel to, the line of section.

of the WHC (Elbayoumi and Greiling, 1984). Here metamorphism is within the greenschist facies and reflects a single prograde event. The major thrust of the region separating the two structural domains dips NE and N below the ophiolitic mélange units of the Wadi Ghadir area (Figs. 3 and 6).

Thrust direction and nappe transport were towards SW as suggested by drag folds, reorientation of older linear elements and branch lines of local thrust units (El Ramly et al., 1984). This is in contrast to the inferred transport direction in the central Eastern Desert where Ries et al. (1983) observed NW-SE stretching lineations. This discrepancy, however, may be resolved if these lineations reflect transport accompanying younger movements on the Nadj fault system as suggested by Stern (1985).

At the exposed low-angle thrusts around the WHC local ramp formation and associated folding have been inferred (Greiling, 1985 and unpubl. data). These thrusts are overprinted by major folds with axes trending NW-SE, i.e. perpendicular to the direction of nappe transport. These folds become progressively tighter from the Wadi Ghadir area towards the northern margin of the WHC (Fig. 6) and, significantly, they are conspicuously absent in the mélange N of the WHC. From this geometrical configuration we infer that these younger folds also originated above a fron-

tal ramp. They terminate at the northern margin of the WHC where this frontal ramp is ending in a lateral ramp (Fig. 6). This interpretation suggests that even the lowermost tectonic unit recognized in the Wadi Hafafit area may be allochthonous and is underlain by a blind thrust. The autochthonous basement may be exposed farther W (Fig. 6).

A minimum amount for the displacement on this floor thrust can be estimated from the distance between the exposed part of the thrust and the position of the assumed frontal ramp, about 38 km. Displacement along the roof thrust of the WHC duplex adds at least 42 km to the total amount of nappe transport (Fig. 6).

The formation of the conspicuous dome structures in the WHC postdated thrusting and folding (El Ramly et al., 1984). Greiling et al. (1984) speculated that low-angle thrusting led to a density inversion that placed the relatively dense nappe units of the eastern domain (mafic metavolcanics, greywacke, ophiolitic mélange) on top of the granitoid-dominated WHC. This inversion also had a thermal blanketing effect and may have given rise to granite diapirism, doming and overprinting of the earlier structures. In spite of this doming after regional thrusting, thrust-related mylonites to the W and N of the WHC were not affected by later metamorphism; we suggest, therefore, that the region between Wadi Ghadir

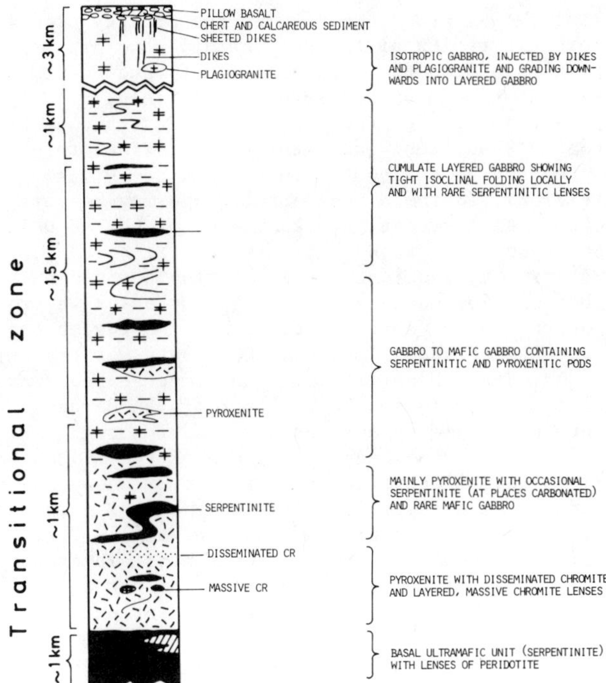

Fig. 7. Schematic section through ophiolite sequence as exposed in Wadis Onib and Sudi, northern Red Sea Hills, Sudan. Note unusually thick transition zone below layered gabbros.

and Wadi Shait (Fig. 6) constitutes part of a foreland thrust and fold belt.

Rock Assemblages and Tectonic Domains in the Red Sea Hills of the Sudan

In contrast to the extensive low-angle thrust regimes in the Eastern Desert of Egypt, major tectonic boundaries in the Red Sea Hills of the Sudan are generally steep and constitute several large-scale shear zones that contain remnants of ophiolites and appear to separate distinct crustal blocks (terranes) whose relationships remain unknown (Vail, 1983, 1985). This configuration is similar to the situation in Saudi Arabia (Kröner, 1985; Stoeser and Camp, 1984), and several of the Sudanese ophiolite-decorated belts appear to continue into the Arabian shield (Vail, 1983, 1985; Camp, 1984), where they have been interpreted as sutures resulting from closure of oceanic domains and collision of island arcs (Bakor et al., 1976; Al-Shanti and Mitchell, 1976). Such an interpretation has also been adopted for the Red Sea Hills (Embleton et al., 1984), and Camp (1984) has proposed an evolutionary model for the central Arabian-Nubian shield that requires the continuation of several distinct tectonic domains and rock assemblages into NE Sudan.

We have undertaken several reconnaissance traverses across the northern Red Sea Hills, and more detailed work concentrated on the geochemistry and geochronology, petrology of selected volcanic sequences as well as the structure and composition of the Onib ophiolite (Fig. 7). The following account is based on these studies, on Landsat and SIR-A radar imagery interpretation and on unpublished work of the Geological and Mineral Resources Dept., Port Sudan. In our analysis we refer to studies in Arabia in order to point out similarities and differences between the Arabian and Nubian segments that must be accommodated in any tectonic model for the Pan-African evolution of the shield.

Ophiolite Belts

Embleton et al. (1984) suggested that two distinct but discontinuous zones of highly sheared rocks with lensoid mafic-ultramafic bodies occur in the Red Sea Hills and mark sutures between accreted arc complexes. The northern and most prominent belt contains the Sol Hamed ophiolite complex near Halaib (Fitches et al., 1983) and its probable southwestern continuation, the Onib ophiolite complex (Hussein et al., 1984; Kröner, 1985). Both are tectonically dismembered but contain all the diagnostic ophiolite components that identify them as obducted remnants of oceanic crust.

The Onib complex contains interlayered gabbros and cumulate ultramafic rocks with podiform chromite lenses that mark an unusually thick (2–3 km) transition from the layered gabbro unit to the basal serpentinites (Fig. 7). Such transition zones and the chromite occurrences seem to be characteristic of ophiolites derived from marginal basins (Hawkins and Evans, 1983) as has now also been documented in the Troodos Complex of Cyprus (Schmincke, pers. comm., 1985). Although high-temperature ductile deformation - probably related to the ocean-floor spreading mechanism - is ubiquitous in these rocks (Kröner, 1985), the primary mineralogy is remarkably well preserved. The cumulate pyroxenites of this transition zone are significantly depleted in TiO_2 relative to $Cr (TiO_2 = 0.01-0.06\%$, Cr = 1500–3000 ppm, I. M. Hussein, unpubl. data) while the sheeted dykes and pillow lavas display a marked enrichment in incompatible elements similar to the rocks in the Wadi Ghadir ophiolite of Egypt (Fig. 5). All these features are considered as indicative of a marginal basin or intra-arc basin setting (Pearce et al., 1981) and typical of so-called supra-subduction zone ophiolites (Pearce et al., 1984).

Dips along the Onib-Sol Hamed belt are steep to vertical, and both complexes face SE with their upper side. We infer from structural data that obduction was from SE to NW and that steepening of the originally low-angle thrusts carrying the ophiolite nappes occurred through backthrusting (as observed at Onib, Hussein et al., 1984) and through large-scale, younger, transcurrent motion along major shear zones as detailed below. The ophiolite belt continues into

the Arabian Yanbu suture (Camp, 1984; see also
Fig. 8) where ophiolites of back-arc origin at
Jabal al Wask and Jabal Ess have yielded Sm-Nd
internal isochron ages of 742±24 Ma and 782±38 Ma
respectively (Claesson et al., 1984). It is thus
likely that the Onib-Sol Hamed marginal basin(s)
also formed about 750 Ma ago.

The terrane N of the above ophiolite belt is
little known, but the information we have from
reconnaissance field traverses indicates that
its rock associations and internal structure are
more reminiscent of the SE Desert of Egypt than
of the Red Sea Hills farther S. At Jabal Gerf we
identified a huge ophiolite nappe complex (Fig. 1)
whose major components consist of sheared ultramafic rocks that were thrust over volcaniclastic
and turbiditic metasediments of probable forearc
provenance as well as the previously mentioned
high-grade metasediments. As in the Hafafit region the sense of thrusting appears to be to the
S or SW. This complex and similar occurrences
that constitute a curvilinear belt SW of Gerf
from Jabal Heiani westwards along Wadi Allaqi
close to Lake Nasser (Fig. 1) probably all belong to one giant ophiolite-mélange nappe and may
mark the southern front of the Egyptian thrust
regime, perhaps even a major suture, against the
NE Red Sea Hills.

The Nakaseib-Amur shear zone in the central
Red Sea Hills (Figs. 1 and 8) may represent the
continuation of the Bir Umq ophiolite belt in
Arabia (Camp, 1984) from which gabbro samples
have yielded zircon ages of 820-870 Ma (Pallister
et al., 1985). It constitutes a well defined
shear belt with discontinuous, serpentinized and
carbonated lenses of ultramafic rocks. The only
preserved, though dismembered, ophiolite occurs
in a nappe complex at Jabal Tohado (Fig. 1) that
was thrust westwards over a turbidite-rich sedimentary sequence containing minor limestone and
dolomite belts.

The Onib-Sol Hamed, the Wadi Allaqi-Jabal
Heiani and the Nakasib-Amur ophiolite belts are
bent and dragged into a broad N-S trending major
transcurrent shear zone in the central part of
the northern Red Sea Hills where large serpentinite lenses "float" within highly deformed and
partly mylonitized schists of mixed volcanic
and volcano-sedimentary derivation (Fig. 8). We
suggest the name "Hamisana shear zone" for this
most prominent linear structure of the Red Sea
Hills that extends from the Wadi Amur region in
the S via Onib into the sand-covered plains of
Wadi Hasium E of Jabal Gerf. Chromiferous ultramafic lenses were found in Wadi Hamisana along
the western part of the shear zone.

The N-S trending Baraka "suture" in the southern Red Sea Hills (Fig. 8) and in NW Eritrea is
little known and was defined by Kazmin (cited in
Mohr, 1979) on the basis of isolated serpentinite
occurrences. It may have its continuation in the
Afaf belt (suture?) of Arabia (Fig. 8), a zone
of strongly deformed rocks that separates the
At Taif-Jiddah from the Asir terrane (Johnson

and Vranas, 1984). In view of the paucity of data
for this region of the Red Sea Hills we are unable to assess the tectonic significance of this
belt that we have not seen in the field.

Accepting the above ophiolite belts and associated shear zones as sutures, 5 distinct terranes can be delineated in the Red Sea Hills
(Fig. 8). In the N the Gerf Terrane may be part
of the same crustal block as most of the SE Desert
of Egypt that extends at least as far as the Wadi
Hafafit Culmination. For the region bounded by
the Onib-Sol Hamed and Nakasib-Amur belts we propose the name Gebeit Terrane, and the region between the Nakasib-Amur and the Baraka sutures may
be named the Haya Terrane. The wedge-shaped area
of the southern Red Sea Hills E of the Baraka
suture is named the Tokar Terrane. The region W
of the Hamisana shear zone and S of the Allaqi-
Jabal Heiani ophiolite belt is poorly known and
largely sand-covered, and we suggest the name
Gabgaba Terrane after the largest wadi in this
region. We shall now discuss the major layered
rock types of these terranes and explore the
question whether they are part of distinct, previously separate crustal segments or whether they
evolved together as part of one large crustal
entity.

Volcano-Plutonic Terranes in the Red Sea Hills

The dominant rock types in all terranes are
the greenschist facies volcano-sedimentary assemblages collectively known in the literature as
Nafirdeib Series and voluminous granitoids of the
Batholithic Granite group (Fig. 8). In addition,
all terranes contain variable proportions of
medium to high-grade quartzofeldspathic, pelitic
and carbonate metasediments, amphibolites, migmatites and gneisses that have traditionally been
considered to constitute the pre-Pan-African basement in the Red Sea Hills and its western margin
on the River Nile (Ruxton, 1956; Gabert et al.,
1960; Vail, 1979). Vail (1983) has recently suggested, however, that at least the metagreywackes
represent an early phase of the volcanic arc
assemblages and constitute a passive margin sedimentary prism laid down on the rifted African
craton. If this assumption is correct, large parts
of the Red Sea Hills should be underlain by older
(i.e. >1000 Ma continental crust).

As stated earlier we interpret most high-grade
layered rocks and migmatites of the Red Sea Hills
as products of Pan-African metamorphism with the
exception, perhaps, of the Tolik Series of the
Haya terrane. Provisional Rb-Sr isotopic results
(Stern and Kröner, unpubl. data) from the Gerf
terrane support this conclusion and also preclude
the possibility that much older continental crust
was involved in the generation of the magmatic
assemblages that we have investigated.

Nevertheless, the volumetrically subordinate
quartzofeldspathic rocks and marbles, particularly those found in the Gabgaba terrane of the NW
Red Sea Hills, may indeed characterize a conti-

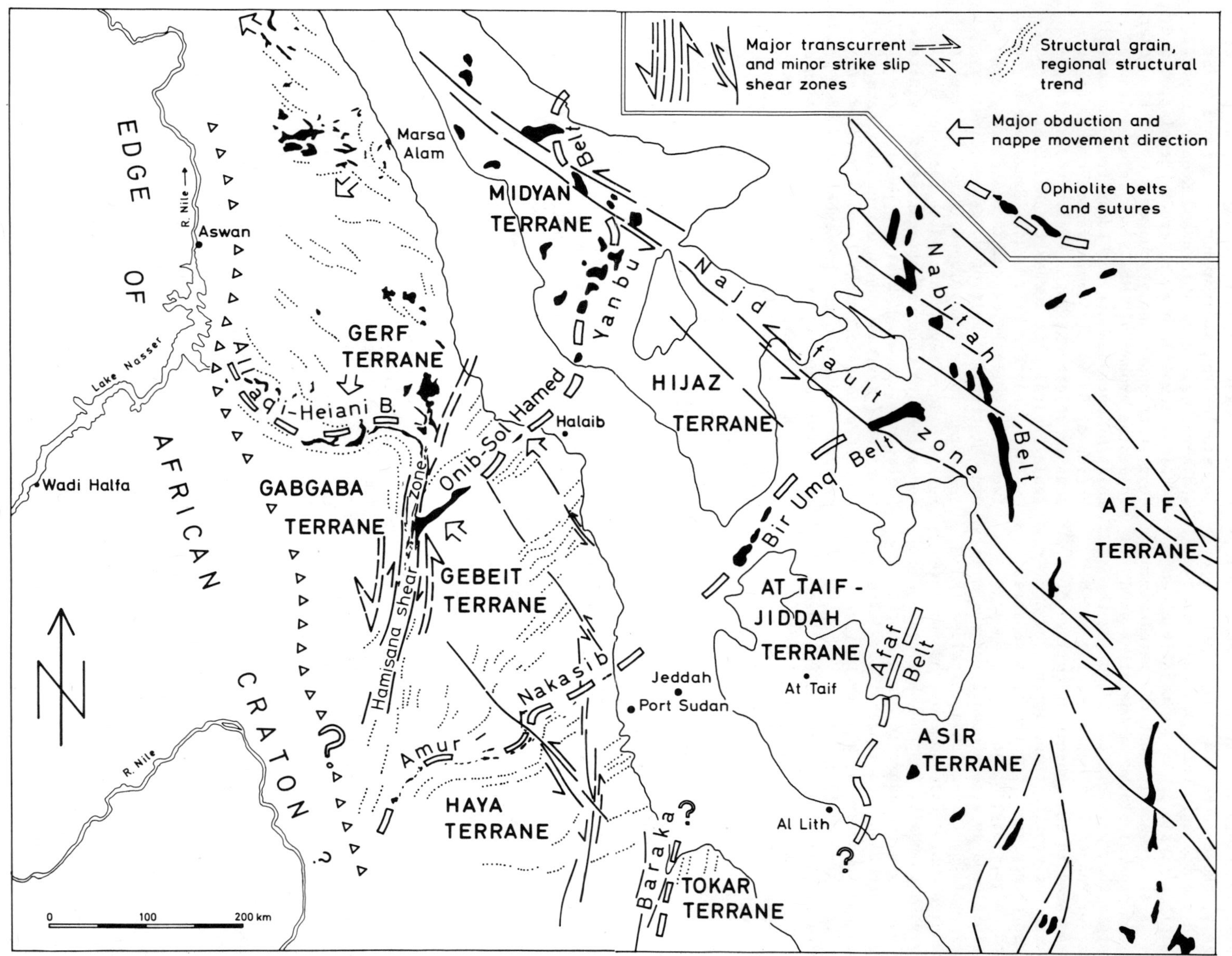

Fig. 8. Simplified and schematic tectonic sketch map of southern Eastern Desert and Red Sea Hills showing ophiolite belts, major structures and lineaments, location of suggested terranes and possible margin of African craton, all juxtaposed against western Arabian shield after closure of Red Sea. Major tectonic units in Arabia as defined by Camp (1984) and Johnson and Vranas (1984).

nental margin environment, the more so since such rock associations have not been identified in the oceanic environment of the Arabian arc complexes. Unfortunately, these rocks are always in tectonic contact with the volcanic units and their original depositional realm cannot be readily ascertained. It is also possible that the metasediments signify small crustal blocks of the attenuated African margin that were rifted off during the early phase of Pan-African ocean opening and became incorporated in the subsequent collage of arc and microcontinent terranes during the accretion process as envisaged for the Arabian shield (Stacey and Hedge, 1984; Kröner, 1985; Stoeser and Camp, 1985) and as also proposed for the modern evolution of the Indonesian archipelago (Pigram and Panggabean, 1984). Some of the Sudanese terranes may be composite in that they contain both passive margin and magmatic arc elements, but we are as yet unable to conclusively demonstrate this and to determine whether they are "exotic", i.e. far-travelled.

Since the geochemistry and age of the low-grade volcano-sedimentary assemblages provide important constraints for crust-formation processes in the region, these are now discussed in the light of recent models for the accretion of the Arabian-Nubian shield.

Roobol et al. (1983) proposed a time-lithostratigraphic model for the Arabian shield that considers immature (i.e. tholeiite-dacite-rhyolite-dominated) island arcs to have formed ∼800 Ma ago while mature (i.e. andesite-dominated) arcs were generated between ∼800 Ma and ∼700 Ma ago. Their concept also indicates a slight younging of the shield from SW to NE. Camp (1984) also included the sedimentary and ophiolitic assemblages in his reconstruction of two late Proterozoic island arcs that trend NE-SW from central Arabia into the Red Sea Hills and that should have collided some 700 Ma ago.

According to these models we should expect ∼700-800 Ma old calcalkaline volcanics in the Gebeit terrane and a ∼700 Ma old arc complex, including a high-grade metamorphic belt (Camp, 1984) in the Haya terrane. The data base currently available makes it possible to identify several distinct volcanic arcs in the Red Sea Hills, and we are able to distinguish different lithological units within the 4 terranes on the basis of our geochemical analyses and provisional geochronology of characteristic rock types.

Starting in the S, the terrane E of the Baraka "suture" is well exposed in the hills S of Tokar (Fig. 1). The volcanic rocks of this region are largely bimodal in composition with amygdaloidal and pillowed basalts and thick units of rhyolite and ignimbrite with associated tuffs forming the dominant lithologies. These rocks and their deformational pattern are similar to the Baish-Baha-Jeddah sequences of the Al-Lith region in Saudi Arabia (Reischmann et al., 1984) that have Rb-Sr whole-rock isochron ages of ∼830 Ma (Kröner et al., 1984). Although a direct correlation is

therefore attractive and has been attempted by Camp (1984) who considered the At-Taif (Saudi Arabia)-Tokar (Sudan) volcanic belt as part of one single frontal arc, our provisional Rb-Sr whole-rock age data (Stern and Kröner, unpubl. results) invalidate such direct correlation and suggest an appreciably younger age for the Tokar volcanics around 670 Ma. Since major volcanic units of this age are not known from the coastal region of SW Saudi Arabia we are unable at present to relate the Tokar region to any large-scale crustal unit across the Red Sea.

In the eastern part of the Haya terrane, between the villages of Suakin and Sinkat, a 40 km wide NE-trending belt of volcano-sedimentary sequences is well developed (Fig. 1). Close to Suakin a low-Ti volcanic suite includes basaltic andesite with chemical affinities to boninites (Reischmann et al., 1986; see Table 1) as well as andesites, dacites and rhyolites. The basaltic andesites contain amphibole pseudomorphs after pyroxene while the felsic members are fine-grained with rare feldspar phenocrysts. An emplacement age of ∼920 Ma is indicated for the felsic members of this suite (Kröner, Compston and Reischmann, unpubl. data). Farther W tholeiitic porphyritic basalts have depleted LREE-patterns and low contents of LIL- and HFS-elements (Reischmann, 1986). Still farther W these rocks grade into an andesite-dominated assemblage with nearly aphyric andesites that are transitional in composition between tholeiitic and calcalkaline rocks and show slight enrichment in LREE. The sediments intercalated with these volcanics consist of intermediate to acidic pyroclastics and volcaniclastics as well as minor, discontinuous beds of dolomite and limestone. Clastic rocks such as quartzite or greywacke were not found in this area.

Plagioclase-phyric andesites, commonly interlayered with basalts and rhyolites, occur in a belt crossed by Wadi Arbaat, some 60 km W of Port Sudan (Fig. 1) and have high concentrations of LIL- and HFS-elements as well as LREE-enriched patterns (Table 1, see also Reischmann, 1986). Clastic sediments are frequently intercalated with these volcanics and contain greywackes and coarse conglomerates with clasts of the associated volcanic rocks and the Batholithic Granite. Limestones and pyroclastic sediments are also common. Although the contrasting assemblages described from these two regions are not in contact with each other, they were intruded by the same large, composite Batholithic Granite (Fig. 1), and we therefore see no reason to consider them as originally separate units.

All these volcanics of the Haya terrane have characteristics of modern island arc settings and may therefore constitute a NE-trending arc complex whose early low-Ti (boninitic) series may suggest a forearc environment in the SE, followed by a tholeiitic sequence to the NW, whereas the mature (i.e. andesite-dominated) suites are exposed still farther NW. This zonal distribution

may be suggestive of a subduction zone dipping
NW. It is possible that this arc represents the
continuation of the Asir arc system in Saudi
Arabia (Camp, 1984), and our limited isotopic data
are consistent with arc formation at least 900 Ma
ago and derivation of the volcanic rocks from a
mantle source.

The large granitoid-dominated area SE of the
above volcanic arc assemblage includes rafts and
xenoliths of amphibolite that are probably con-
tact-metamorphosed remnants of the arc tholeiites.
A well foliated granodiorite also occurs in this
granitoid domain at the Resort of Erkowit, and
U-Pb zircon dating yielded an age of ∿850 Ma
while the Rb-Sr systematics indicate a primitive
source region (Stern, Manton and Kröner, unpubl.
data). This result is compatible with the age of
the large Bidah Pluton in the SW Arabian shield
(Marzouki et al., 1982), and the remarkable similarity in age as well as the almost identical
suite of plutonic rocks in the Erkowit region of
the Sudan and the At Taif region of Saudi Arabia,
ranging from gabbro through diorite, trondhjemite
and granodiorite (Fleck et al., 1980) points to
a genetic and spatial relationship. We therefore
agree with Camp (1984) that at least part of the
central and southern Haya terrane may constitute
the extension of the Arabian Taif frontal arc.

The most extensive exposure of the low-grade
volcano-sedimentary sequences of the Red Sea
Hills occurs in the Gebeit terrane (Fig. 1).
Fitches et al.(1983) reported a Rb-Sr whole-rock
isochron age of 712±58 Ma (Sr$_i$=0.702) for basalts
and andesites S of the Sol Hamed ophiolite complex near the northern margin of the terrane, and
clastic sediments interlayered with these volcanics at Sol Hamed contain rare serpentinite pebbles, implying that the ophiolite complex must be
older than these rocks. Vail et al. (1984) investigated calcalkaline metavolcanics and intrusives near the southern margin of the Gebeit
terrane in the Wadi Oko area for which Klemenic
(1985) reported Rb-Sr isochron ages of 723±6 Ma
and 719±6 Ma respectively with low Sr$_i$ of 0.703.
Since both the metavolcanics at Sol Hamed and of
the Wadi Oko area indicate virtually identical
ages and rock types it is suggestive that the
entire Gebeit terrane constitutes part of a broad
NE-trending arc complex that should represent the
continuation of the ∿700-850 Ma old Hijaz arc of
western Arabia as suggested by Camp (1984).

The geochemistry and isotopic systematics of
metavolcanic rocks from the central part of this
terrane around Gebeit Mine (Fig. 1) were investigated in detail by Reischmann et al. (1985) and
Reischmann (1986). These rocks are separated from
the occurrences farther N by the Batholithic
Granite and from those studied by Vail et al.
(1984) in the S by the higher grade assemblage of
the Sasa Plain (Fig. 1). There are 3 main rock
types at Gebeit: cpx-phyric basalt and basaltic
andesite, hbl-phyric basalt and basaltic andesite
and fsp-phyric andesite. All of these are calcalkaline and have LREE-enriched patterns. The

fsp-porphyries show distinctly higher abundances
of HFS-elements than the cpx- and hbl-phyric
types. Voluminous pyroclastic rocks as well as
greywacke and minor limestone are associated
with these lava flows. These immature sediments
underline the arc character of the entire suite.
In general, the Gebeit Mine volcanics are lithologically and geochemically similar to those
described by Fitches et al. (1983) and Vail et
al. (1984). However, Sm-Nd analyses on mineral
separates and whole-rock samples yield an isochron age of 832±26 Ma ($\epsilon_{Nd}(T)$= +6.8) that we
interpret as the age of eruption of the lavas
(Reischmann et al., 1985). The Rb-Sr isotopic
data obtained for the same samples clearly signify severe disturbance of this system and partial
metamorphic resetting at about 715 Ma ago. This
new result casts doubt on the interpretation of
the ∿720 Ma Rb-Sr ages reported as primary for
other regions of the Red Sea Hills, the more so
since Camp (1984) postulated a major collision
event at about this time in western Arabia and
in NE Sudan that may be reflected by these dates.

Detrital zircons from a metaquartzite of the
Sasa Plain in the central part of the Gebeit
terrane (Fig. 1) yield Pan-African as well as
rare Archean ages (Kröner and Compston, unpubl.
data), and the whole-rock Nd systematics of this
rock are also compatible with derivation from a
mixture of juvenile and ancient crust (Reischmann, unpubl. data). This is our first evidence
for the presence of old continental crust in the
Red Sea Hills, and the Sasa Plain region may thus
constitute a largely concealed microcontinental
fragment in the collage of arcs now surrounding
it, underlining the composite nature of the Gebeit terrane.

The Gabgaba terrane of the westernmost Red Sea
Hills is little known, and no detailed rock descriptions and geochemical data are available.
Our reconnaissance traverses along Wadi Amur have
revealed a N-S sheared belt of mafic metavolcanic rocks with interspersed serpentinite lenses,
and we presume that this constitutes the southward continuation of the Hamisana shear zone although no direct connection has yet been established. Rare serpentinite pebbles in clastic
metasediments immediately W of the Onib Ophiolite
indicate at least some of these rocks to be younger than the mafic-ultramafic complex, a situation similar to that at Sol Hamed. Basaltic and
andesitic metalavas collected just W of the shear
zone at Wadi Amur indicate a provisional age of
∿620 Ma that is significantly younger than those
for the metavolcanics farther E; if this is a
primary emplacement age, the Gabgaba terrane is
one of the youngest provinces in the Red Sea
Hills. Farther W, towards the River Nile, we discovered an extensive sequence of marble with local
intercalations of metaquartzite (Fig. 1), indicating a stable depositional environment not seen in
the terrane farther E. We therefore do not exclude
the possibility that the Gabgaba terrane is part
of the African cratonic margin (Fig. 8).

In summary, field evidence and geochemical data
for the metavolcanic assemblages are broadly in
agreement with the geodynamic concepts proposed
for the Red Sea Hills by Embleton et al. (1984),
Vail (1983, 1985) and Vail et al. (1984) and for
the western Arabian shield by Camp (1984) and
Stoeser and Camp (1985). However, we question
whether all metavolcanic successions in the Red
Sea Hills are of similar age and can be assigned
to only one stratigraphic unit, namely the Nafirdeib Series. In accordance with Embleton et al.
(1984) we suggest to restrict currently used names
to sequences in their respective type areas. The
Sm-Nd age for the Gebeit andesites indicates these
mature arc volcanics to correspond temporally to
the Birak Group bimodal suite of the Hijaz terrane
in Saudi Arabia (Stoeser and Camp, 1985). This
suggests that arc formation was already advanced
in the central Gebeit terrane while the more primitive Birak arc evolved farther E. The ophiolites
of the Onib-Sol Hamed Yanbu belt (Fig. 8) may
represent obducted remnants of intra-arc crust
that formed during a phase of inter-arc spreading
∿740-780 Ma ago.

The isotopic data for the Red Sea Hills confirm
the pattern already well established for the Arabian shield, namely that all dated rock units display uniformly low Sr initial ratios of 0.702-
0.703 that indicate a primitive (mantle) magma
source with no involvement of significantly older
continental crust, a feature now also supported by
the Sm-Nd data (Reischmann et al., 1985). We conclude from these data and from the chemical and
lithological properties of the volcano-sedimentary assemblages of the Red Sea Hills that most of
these rocks were generated in intraoceanic arc
environments and that the early Pan-African continental margin of the Nubian shield must be sought
farther west. This conclusion also forces us to
assume that the high-grade metasedimentary rocks
of the Haya terrane discussed earlier are foreign
to such an oceanic environment, and that their
juxtaposition with the arc complexes is due to
tectonic processes. Whether this justifies the
postulation of exotic terranes in the Red Sea
Hills, however, remains to be established.

Towards a Model for the Pan-African Evolution
of the Nubian Shield

Early models for the development of the Arabian
shield postulated an entirely intraoceanic arc
formation and accretion history between ∿1200 Ma
and ∿700 Ma ago (Greenwood et al., 1976; Gass,
1981), and segments of older continental crust at
the SE edge of the shield were seen as part of an
"Iranian craton" that collided with the Arabian
arc complex some 650-700 Ma ago (Greenwood et al.,
1982). These models tacitly assumed that NE Africa
constituted a passive continental margin early in
the Pan-African history that was eventually converted into an active Andean-type terrane and that
the consolidated Arabian "neocraton" finally col-

lided with the active African margin to form the
Arabian-Nubian shield (see Table 1 in Camp, 1984).

The recognition of older crustal remnants in
the eastern Arabian shield (Afif terrane) and the
implausibly high juvenile crust-generation rates
implied by these models (Reymer and Schubert,
1984) led to some important revision of the simple
accretion scenario as outlined by Stacey and
Stoeser (1983), Camp (1984) and Kröner (1985). In
a departure from previous hypotheses, Stoeser and
Camp (1985) consider the Arabian Afif terrane as
a detached fragment of the African continent and
thus recognize a period of rifting, ∿1200-950 Ma
ago, in NE Africa with crustal attenuation and
subsequent formation of an oceanic domain to the
E. Ensimatic island arc development in a SW Pacif-
ic-type setting between the Afif microplate and
the African margin is postulated for the period
∿950 Ma to ∿715 Ma ago while the main arc-micro-
plate accretion and collision events took place
between ∿715 Ma and about 640 Ma ago.

Our suggested evolution for the SE Desert of
Egypt compares well with the above model although
we have not recognized evidence for extensive
rifting and passive margin formation as early as
1200 Ma ago. However, Cahen et al. (1984) cite
unpublished Rb-Sr data from Egyptian sources that
indicate ages of uncertain significance up to
1200 Ma for clastic metasediments along the western margin of the SE Desert. Whenever the rifting
and crustal attenuation process started, a passive
margin environment seemed to prevail until at
least ∿800 Ma ago, and first arc volcanics and
granitoids signifying conversion into an active
and compressional tectonic regime appear at ∿770-
710 Ma ago. By that time the rifted margin had
evolved into a microplate-arc-back arc basin setting as presently seen in Indonesia (Hamilton,
1979; Pigram and Panggabean, 1984). This is supported by the fact that the Hafafit passive margin
domain occurs to the east of arc-ophiolite assemblages such as found at Abu Swayel and may therefore constitute a rifted-off microplate of continental crust behind which a marginal basin/arc
setting formed.

Similar scenarios can be inferred for the small
areas of the Red Sea Hills with continental-type
metasediments such as the Sasa Plain in the Gebeit
terrane and the Tolik Series of the Haya terrane.
Since undoubted juvenile arc volcanics and ophiolites occur to the W and NW of these suites they
are unlikely to reflect the position of the former
African margin but may constitute rifted fragments
thereof, subsequently incorporated in the arc
accretion development. The only autochthonous continental margin deposits may be those of the western Gabgaba terrane and around Wadi Halfa on the
River Nile (Figs. 1 and 8).

In the central Eastern Desert of Egypt arc
magmatism began considerably later as noted by
Stern and Hedge (1985) and in analogy with the
Midyan terrane of the NW Arabian shield (Clark,
1985; Stoeser and Camp, 1985), and we postulate

from this that the broad age zonation for juvenile
rocks with younging from SE to NW as observed in
the Arabian shield reflects a diachronous estab-
lishment of the subduction-accretion regime in the
Arabian-Nubian shield between ∿900 Ma and ∿600 Ma
ago.

The diachrony is also evident in the terranes
of the Red Sea Hills. A record of the earliest arc
magmatism around 850-920 Ma ago is preserved in
the volcanic and granitoid associations of the
northwestern Haya terrane and correlates well with
the oldest arc complexes in the At Taif-Jiddah
terrane of SW Arabia (Fig. 8). In the Gebeit
terrane, discrete periods of juvenile arc volcanism are indicated by ages of ∿830 Ma and ∿720 Ma
respectively and provide a record of multiple arc
generation. Such episodicity is also evident in
the formerly adjacent Hijaz terrane of Saudi Arabia (Stoeser and Camp, 1985), and there remains
little doubt that the Gebeit-Hijaz block constituted one single crustal domain by about 720 Ma
ago. New age data on the Bir Umq ophiolite in Saudi Arabia (Pallister et al., 1985) imply that the
Gebeit-Hijaz terrane was separated from the Haya-
Asir terrane farther SE by a marginal sea at about
820-870 Ma ago that, on closure, produced the Bir-
Umq-Nakasib-Amur suture some 700-680 Ma ago (Stoeser and Camp, 1985).

This collision event created the dominantly SW-
NE trending structural grain of the northwestern
Red Sea Hills with mega- to meso-scale, often open
and upright, but also tighter, NW- as well as SE-
verging folds and minor thrusts. At about the same
time closure of the marginal seas in SE Egypt initiated the low-angle thrust regime and the foreland thrust and fold belt in the Hafafit region.

The 670 Ma old metavolcanics in the Tokar terrane, the temporally equivalent post-arc rhyolite
suites of the Gebeit and Haya terranes and the
∿620 Ma old basalt-andesite suite at Wadi Amur in
the Gabgaba terrane do not show the SW-NE structures and were presumably generated after the major collision events in the Nubian shield. They
may reflect an extensional tectonic regime that
followed after arc accretion, similar to the situation in the NE Desert of Egypt (Stern et al.,
1984) and to a period of widespread post-colli-
sional intracratonic magmatism 620-680 Ma ago in
the western Arabian shield (Stoeser and Camp,
1985). The prominent Hamisana and other N-S shear
zones affect these rocks and also deform the earlier SW-NE structures in the older terranes. The
dominant movement along these zones has been
sinistral, only locally dextral (Fig. 8). The high
degree of fragmentation of the ophiolite and volcaniclastic assemblages in the Hamisana shear zone
indicates that movement here involved displacements of at least several tens of kilometers.

The cause for these transcurrent movements
could have been the final oblique collision of the
accreting arc terranes to the E (e.g. Ar Rayn
microcontinent of the eastern Arabian shield,
Stoeser et al., 1984; see also Fig. 2) with the

rigid African craton and may have been broadly
synchronous with formation of the N-S trending
Nabitah and Al Amar sutures in the Arabian shield
(Stoeser and Camp, 1985). The conspicuous NW-SE
strike-slip faults in the Red Sea Hills with
sinistral displacements up to some tens of kilometers (Fig. 8) can be mapped on satellite images
(Schönfeld, 1977; Ahmed, 1982) and are probably
related to the left-lateral Najd wrench fault
system of Saudi Arabia that formed 550-630 Ma ago
(Moore, 1979; Stoeser and Camp, 1985; Stern,
1985).

Major remaining uncertainties in all models
concern the precise timing of magmatic and tectono-metamorphic processes in the Arabian-Nubian
shield since the majority of age data, particularly those in the Nubian segment, are based on
Rb-Sr geochronology that does not permit an unequivocal interpretation of primary or reset ages
in rocks with low Rb/Sr ratios. In addition,
seismic reflection profiling would help to place
further constraints on tectonic modelling.

Acknowledgments. This study was initiated by
IGCP Project No. 164 (Pan-African crustal evolution in the Arabian-Nubian shield) and is part
of the Mainz Consortium "Accretion and Differentiation of the Planet Earth". We appreciate
funding by the German Research Council (DFG Kr
590/11), the German Ministries of Science and
Technology (BMFT) and Economic Cooperation (BMZ),
the Volkswagen Foundation, the Max-Planck-Insti-
tut für Chemie and through JPL Grant 956821 to
R. J. Stern. In addition, we received logistic
support during fieldwork from the Geological and
Mineral Resources Department (GMRD), Red Sea
Hills Office, Port Sudan, the BRGM/GMRD Mission
in the Red Sea Hills, the Egyptian Geological Survey and Mining Authority (EGSMA), Cairo, and the
Geological and Mining Centre, Area of Integration,
Asswan. We are particularly indebted to Messrs. S.
Bakheit, M.I. Shaddad and M.N. Bushara, Port
Sudan, A.S. Dawoud, Khartoum, M.F. El Ramly and M.
Helmy, Cairo as well as A.A. Rashwan and M.A. Kar-
fis, Asswan, for assistance in the field and for
discussions on the regional geology. The GMRD and
EGSMA also made available unpublished maps for
which we are grateful.

References

Ahmed, A.A.M., General outline of the geology and
mineral occurrences of the Red Sea Hills, Geol.
Min. Res. Dept., Bull., 30, 63 pp., Khartoum,
1980.

Ali, S.E.M., The geology of the Homogar Group vol-
canic rocks, southern Red Sea Hills, Sudan,
M.Phil. thesis, 161 pp., Portsmouth Polytechnic,
England, 1979.

Al-Shanti, A.M.S. and Mitchell, A.H.G., Late Pre-
cambrian subduction and collision in the Al
Amar-Idsas region, Arabian shield, Kingdom of
Saudi Arabia, Tectonophysics, 30, 41-47, 1976.

Bakor, A.R., I.G. Gass and C.R. Neary, Jabel al
Wask, northwest Saudi Arabia, an Eocambrian
back-arc ophiolite, Earth Planet. Sci. Lett.,
30, 1-9, 1976.

Barker, F., J.G. Arth and T. Hudson, Tonalites in
crustal evolution, Phil. Trans. R. Soc. Lond.,
A301, 293-303, 1981.

Brown, G.F. and R.G. Coleman, The tectonic framework of the Arabian Peninsula, Proc. 24th Int.
Geol. Congr., Montréal, 3, 300-305, 1972.

Cahen, L., N.J. Snelling, J. Delhal and J.R. Vail,
The geochronology and evolution of Africa,
512 pp., Clarendon Press, Oxford, 1984.

Camp, V.E., Island arcs and their role in the evolution of the western Arabian shield, Geol. Soc.
Am. Bull., 95, 913-921, 1984.

Cavanagh, B.J., Rb-Sr geochronology of some pre-
Nubian igneous complexes of central and north-
eastern Sudan, Ph.D. thesis, 239 pp., Univ.
Leeds, England, 1979.

Church, W.R., The northern Appalachians and the
Eastern Desert of Egypt (abstract), Precambrian
Res., 16, A13, 1982.

Claesson, S., J.S. Pallister and M. Tatsumoto,
Samarium-neodymium data on two late Proterozoic
ophiolites of Saudi Arabia and implications for
crustal and mantle evolution, Contrib. Mineral.
Petrol., 85, 244-252, 1984.

Clark, M.D., Late Proterozoic crustal evolution
of the Midyan region, northwestern Saudi Arabia,
Geology, 13, 611-615, 1985.

Cobbing, E.J. and W.S. Pitcher, Andean plutonism
in Peru and its relationship to volcanism and
metallogenesis at a segmented plate edge, Geol.
Soc. Am. Mem., 159, 277-291, 1983.

Curray, J.R., D.G. Moore, L.A. Lawwer, F.J. Emmel,
R.W. Raitt, M. Henry and R. Kiekhefer, Tectonics
of the Andaman Sea and Burma, Am. Assoc. Petr.
Geol. Mem., 29, 189-198, 1979.

Dalziel, I.W.D., M.J. de Wit and K.F. Palm-
er, Fossil marginal basin in the southern
Andes, Nature, 250, 291-294, 1974.

Dixon, T.H., Age and chemical characteristics of
some pre-Pan-African rocks in the Egyptian
shield, Precambrian Res., 14, 119-133, 1981.

Elbayoumi, R.M.A., Ophiolites and associated
rocks of Wadi Ghadir, east of Gebel Zabara,
Eastern Desert, Egypt, Ph.D. thesis, 227 pp.,
Cairo Univ., Egypt, 1980.

Elbayoumi, R.M.A. and R. Greiling, Tectonic evolution of a Pan-African plate margin in southeastern Egypt - a suture zone overprinted by
low angle thrusting, in Géologie africaine -
African geology, edited by J. Klerkx and J.
Michot, pp. 47-56, Musée royal de l'Afrique
centrale, Tervuren, 1984.

El Ramly, M.F., A new geological map for the
basement rocks in the Eastern and South-Western
Deserts of Egypt, scale 1:1.000.000, Ann. geol.
Surv. Egypt, 2, 2-18, 1972.

El Ramly, M.F. and M. K. Akaad, The basement complex in the central Eastern Desert of Egypt between lat. 24°30' and 25°40'N, Geol. Surv.
Egypt, paper no. 8, 35 pp., 1960.

El Ramly, M.F. and G.S.S. Roufaiel, Sodic-silica
metasomatism and introduced zircon in the
Migif-Hafafit gneisses, Eastern Desert, Egypt.
J. Geol., 18, 119-126, 1974.

El Ramly, M.F., R. Greiling, A. Kröner and A.A.
Rashwan, On the tectonic evolution of the Wadi
Hafafit area and environs, Eastern Desert of
Egypt, Fac. Earth Sci., Univ. Jeddah, Bull., 6,
113-126, 1984.

El Samani, Y., Contribution à l'étude géologique
et minéralogique des minéralisations Cu, Pb,
Zn, Ba, Mn de la plaine d'Alikaleib (Soudan),
Thesis 3ème cycle, 239 pp., Univ. Orléans,
France, December 1983.

El Shazly, E.M., On the classification of the
Precambrian and other rocks of magmatic affili-
ation in Egypt, U.A.R., Proc. 22nd Int. Geol.
Congr., New Delhi, India, Part 10, 88-101, 1964.

Embleton, J.C.B., D.J. Hughes, P.M. Klemenic, S.
Poole and J.R. Vail, A new approach to the
stratigraphy and tectonic evolution of the Red
Sea Hills, Sudan, Fac. Earth Sci., Univ. Jeddah,
Bull., 6, 101-112, 1984.

Engel, A.E.J., T.H. Dixon and R.J. Stern, Late
Precambrian evolution of Afro-Arabian crust from
ocean arc to craton, Geol. Soc. Am. Bull., 91,
699-706, 1980.

Fitches, W.R., R.H. Graham, I.M. Hussein, A.C.
Ries, R.M. Shackleton and R.C. Price, The late
Proterozoic ophiolite of Sol Hamed, NE Sudan,
Precambrian Res., 19, 385-411, 1983.

Fleck, R.J., W.R. Greenwood, D.G. Hadley, R.E.
Anderson and D.L. Schmidt, Rubidium-strontium
geochronology and plate tectonic evolution of
the southern part of the Arabian shield, U.S.
Geol. Surv., Prof. Paper 1131, 38 pp., 1980.

Fullagar, P.D., Pan-African age granites of
northeastern Africa: new or reworked sialic
material?, in Geology of Libya, edited by M.J.
Salem and M.T. Bucrevich, v.3, pp. 1051-1058,
Academic Press, London, 1980.

Gabert, G., B.P. Ruxton and H. Venzlaff, Über
Untersuchungen im Kristallin der nördlichen
Red Sea Hills im Sudan, Geol. Jahrb., 77, 241-
270, 1960.

Garson, M.S. and I.M. Shalaby, Precambrian-Lower
Paleozoic plate tectonics and metallogenesis in
the Red Sea region, Geol. Ass. Canada, Spec.
Paper, 14, 573-596, 1976.

Gass, I.G., The geology of the Dunganab area,
Anglo-Egyptian Sudan, M.Sc. thesis, 112 pp,
Univ. Leeds, England, 1955.

Gass, I.G., Pan-African (Upper Proterozoic) plate
tectonics of the Arabian-Nubian shield, in
Precambrian plate tectonics, edited by A. Kröner, pp. 387-405, Elsevier, Amsterdam, 1981.

Geological Map of Egypt, 1: 2 000 000, Egyptian
Geol. Survey and Mining Authority, Cairo, 1981.

Geological Map of the Sudan, 1: 2 000 000, Geological and Mineral Resources Dept., Khartoum,
1981.

Gillespie, J.G. and T.H. Dixon, Pb isotope systematics of some granitic rocks from the Egyptian shield, Precambrian Res., 20, 63-77, 1983.

Greenwood, W.R., D.G. Hadley, R.E. Anderson, R. J. Fleck and D.L. Schmidt, Late Proterozoic cratonization in southwestern Saudi Arabia, Phil. Trans. R. Soc. Lond., A280, 517-527, 1976.

Greenwood, W.R., D.B. Stoeser, R.J. Fleck and J.S. Stacey, Late Proterozoic island-arc complexes and tectonic belts in the southern part of the Arabian shield, Kingdom of Saudi Arabia, Ministry of Petr. Min. Res., Jiddah, Open-file report USGS-OF-02-8, 46 pp., 1982.

Greiling, R., Thrust tectonics in Pan-African rocks of SE Egypt (abstract), Terra Cognita, 5, 257, 1985.

Greiling, R., A. Kröner and M.F. El Ramly, Structural interference patterns and their origin in the Pan-African basement of the southeastern Desert of Egypt, in Precambrian tectonics illustrated, edited by A. Kröner and R. Greiling, pp. 401-412, E. Schweizerbart'sche Verlagsbuchhandlung, Stuttgart, 1984.

Hafez, A. and I.M. Shalaby, On the geochemical characteristics of the volcanic rocks at Umm Samiuki, Eastern Desert, Egypt, Egypt. J. Geol., 27, 73-92, 1983.

Hamilton, W., Tectonics of the Indonesian region, U.S. Geol. Surv., Prof. Paper 1078, 345 pp., 1979.

Harris, N.B.W., C.J. Hawkesworth and A.C. Ries, Crustal evolution in northeast and east Africa from model Nd ages, Nature, 309, 773-776, 1984.

Hawkins, J.W. and C.A. Evans, Geology of the Zambales Range, Luzon, Philippine islands: ophiolite derived from an island arc-back arc basin pair, in The tectonic and geologic evolution of southeast Asian seas and islands, part 2, edited by D.E. Hayes, pp. 95-123, AGU Geophys. Monogr. 27, 1983.

Hume, W.F., Geology of Egypt, 2, pt. 1: The metamorphic rocks, Govt. Press, Cairo, 293 pp., 1934.

Hussein, A.A.A., M.A Ali and M.F. El Ramly, A. proposed new classification of the granites of Egypt, J. Volcanol. Geotherm. Res., 14, 187-198, 1982.

Hussein, I.M., Tectonic evolution of the northern Red Sea Hills, Sudan, with particular reference to the geology west of Port Sudan and the Wadi Onib ophiolite complex, Ph.D. thesis, Univ. Mainz, West Germany, 1986.

Hussein, I.M., A. Kröner and St. Dürr, The Wadi Onib mafic-ultramafic complex of the northern Red Sea Hills, Sudan: a dismembered Pan-African ophiolite (abstract), 27th Int. Geol. Congr., Moscow, 4, 326, 1984.

Hussein, I.M., A. Kroner and St. Durr, Wadi Onib - a dismembered Pan-African ophiolite in the Red Sea Hills of Sudan, Fac. Earth Sci., Univ. Jeddah, Bull., 6, 319-328, 1984.

Jackson, N.J., Petrogenesis and evolution of Arabian felsic plutonic rocks, J. Afr. Earth Sci., 4, 47-59, 1986.

Johnson, P.R. and G.J. Vranas, The origin and development of late Proterozoic rocks of the Arabian shield: an analysis of terranes and mineral environments, Kingdom of Saudi Arabia, Deputy Ministry for Min. Res., Jiddah, Open-file report RF-OF-04-32, 96 pp., 1984.

Kabesh, M.L. and M. Lotfi, On the basement complex of the Red Sea Hills, Sudan, Bull. El Mataria, Le Caire, 12, 1-19, 1962.

Karig, D.E., Structural history of the Mariana island arc system, Geol. Soc. Am. Bull., 82, 323-344, 1971.

Klemenic, P.M., New geochronological data on volcanic rocks from northeast Sudan and their implication for crustal evolution, Precambrian Res., 30, 263-276, 1985.

Kröner, A., Late Precambrian plate tectonics and orogeny: a need to redefine the term Pan-African, in Géologie africaine - African geology, edited by J. Klerkx and J. Michot, pp. 23-28, Musée royal de l'Afrique centrale, Tervuren, 1984.

Kröner, A., Ophiolites and the evolution of tectonic boundaries in the late Proterozoic Arabian-Nubian shield of northeast Africa and Arabia, Precambrian Res., 27, 277-300, 1985.

Kröner, A., M. Halpern and A. Basahel, Age and significance of metavolcanic sequences and granitoid gneisses from the Al-Lith area, southwestern Arabian shield, Fac. Earth Sci., Univ. Jeddah, Bull., 6, 379-388, 1984.

Marzouki, F.M.H., N.J. Jackson and C.R. Ramsay, Composition, age and origin of two Proterozoic diorite-tonalite complexes in the Arabian shield, Precambrian Res., 19, 31-50, 1982.

Mohr, P., Lithology and structure of the Precambrian rocks of Eritrea, in Evolution and mineralization of the Arabian-Nubian shield, vol.1, edited by S.A. Tahoun, pp. 7-16, Pergamon Press, Oxford, 1979.

Moore, J.M., Tectonics of the Najd transcurrent fault system, Saudi Arabia, J. Geol. Soc. Lond., 136, 441.454, 1979.

Neary, C.R., I.G. Gass and B.J. Cavanagh, Granitic association of northeastern Sudan, Geol. Soc. Am. Bull., 87, 1501-1512, 1976.

Pallister, J.S., J.S. Stacey and L.B. Fischer, Precambrian ophiolites of Arabia: geochronology and implications for continental accretion (abstract), Eos, AGU, 66, 420, 1985.

Pearce, J.A., Geochemical evidence for the genesis and eruptive setting of lavas from Tethyan ophiolites, in Ophiolites, edited by A. Panayiotou, pp. 261-272, Geol. Survey Dept., Cyprus, 1980.

Pearce, J.A., T. Alabaster, A.W. Shelton and M.P. Searle, the Oman ophiolite as a Cretaceous arc-basin complex: evidence and implication, Phil. Trans. R. Soc. Lond., A300, 299-317, 1981.

Pearce, J.A., S.J. Lippard and S. Roberts, Characteristics and tectonic significance of supra-subduction zone ophiolites, in Marginal basin geology, edited by B.P. Kokelaar and M.F. Howells, pp. 77-94, Geol. Soc. London, Spec. Publ. 16, 1984.

Pigram, C.J. and H. Panggabean, Rifting of the

northern margin of the Australian continent and
the origin of some microcontinents in eastern
Indonesia, Tectonophysics, 107, 331-353, 1984.

Pitcher, W.S., Granite type and tectonic environment, in Mountain building processes, edited by
K. Hsü, pp. 19-40, Academic Press, London, 1983.

Reischmann, T., Geologie and Genese spätprotero-
zoischer Vulkanite der Red Sea Hills, Sudan,
Ph.D. thesis, 202 pp., Univ. Mainz, Germany,
April 1986.

Reischmann, T. and A. Kröner, Geochemistry of
late Proterozoic metavolcanic rocks in the Red
Sea Hills, NE Sudan (abstract), Terra Cognita,
4, 93, 1984.

Reischmann, Th., A. Kröner and A. Basahel, Petrography, geochemistry and tectonic setting of
metavolcanic sequences from the Al Lith area,
southwestern Arabian shield, Fac. Earth Sci.,
Univ. Jeddah, Bull., 6, 365-378, 1984.

Reischmann, T., A. Kröner and A.W. Hofmann, Isotope geochemistry of Pan-African volcanic rocks
from the Red Sea Hills, Sudan (abstract), Terra
Cognita, 5, 288, 1985.

Reischmann, T., A.W. Hofmann and A. Kröner, Precambrian boninites from the Red Sea Hills, NE
Sudan (abstract), Eos, 68, in press, 1986.

Reymer, A. and G. Schubert, Phanerozoic addition
rates to the continental crust and crustal
growth, Tectonics, 3, 63-78, 1984.

Ries, A.C., R.M. Shackleton, R.H. Graham and W.R.
Fitches, Pan-African structures, ophiolites and
mélange in the Eastern Desert of Egypt: a traverse at 26°N, J. geol. Soc. London, 140, 75-95,
1983.

Robol, M.J., C.R. Ramsay, N.J. Jackson and D.P.F.
Darbyshire, Late Proterozoic lavas of the central Arabian shield - evolution of an ancient
volcanic arc system, J. geol. Soc. London, 140,
185-202, 1983.

Ruxton, B.P., Major rock groups of the northern
Red Sea Hills, Sudan, Geol. Mag., 93, 314-310,
1956.

Saunders, A.D. and J. Tarney, Geochemical characteristics of basaltic volcanism within back-arc
basins in Marginal basin geology, edited by B.P.
Kokelaar and M.F. Howells, pp. 59-76, Geol.
Soc. London, Spec. Publ. 16, 1984.

Schmidt, D.L. and G.F. Brown, Major-element
chemical evolution of the late Proterozoic
shield of Saudi Arabia, Fac. Earth Sci., Univ.
Jeddah, 6, 1-21, 1984.

Schönfeld, M., Scherzonen im Kristallin der Red
Sea Hills im nordöstlichen Sudan und ihre Be-
deutung für die Entwicklung des Roten Meeres -
erste Ergebnisse einer Auswertung von LANDSAT-
Aufnahmen, Geotekt. Forsch., 53, 107-121, 1977.

Shackleton, R.M., A.C. Ries, R.H. Graham und W.R.
Fitches, Late Precambrian ophiolitic mélange
in the eastern desert of Egypt, Nature, 285,
472-474, 1980.

Shukuri, N.M. and M.S. Mansour, Lithostratigraphy
of Um Samiuki District, Eastern Desert of Egypt,

in Evolution and mineralization of the Arabian-
Nubian shield, v. 4, edited by P.G. Cooray and
S.A. Tahoun, pp. 83-93, Pergamon Press, Oxford,
1980.

Stacey, J.S. and D.B. Stoeser, Distribution of
oceanic and continental leads in the Arabian-
Nubian shield, Contrib. Mineral. Petrol., 84,
91-105, 1983.

Stacey, J.S. and C.E. Hedge, Geochronologic and
isotopic evidence for early Proterozoic crust
in the eastern Arabian shield, Geology, 12,
310-313, 1984.

Stern, R.J., Petrogenesis and tectonic setting
of late Precambrian ensimatic volcanic rocks,
central Eastern Desert of Egypt, Precambrian
Res., 16, 195-230, 1981.

Stern, R.J., The Najd fault system, Saudi Arabia
and Egypt: a late Precambrian rift-related
transform system?, Tectonics, 4, 497-511, 1985.

Stern, R.J., D. Gottfried and C.E. Hedge, Late
Precambrian rifting and crustal evolution in
the Northeastern Desert of Egypt, Geology, 12,
168-172, 1984.

Stern, R.J. and C.E. Hedge, Geochronologic and
isotopic constraints on late Precambrian crustal evolution in the Eastern Desert of Egypt,
Am. J. Sci., 285, 97-127, 1985.

Stoeser, D.B., Distribution and tectonic setting
of plutonic rocks of the Arabian shield, J.
Afr. Earth Sci., 4, 21-46, 1986.

Stoeser, D.B. and V.E. Camp, Pan-African microplate accretion of the Arabian shield, Geol.
Soc. America Bull., 96, 817-826, 1985.

Stoeser, D.B., J.S. Stacey, W.R. Greenwood and
L.B. Fisher, U/Pb zircon geochronology of the
southern portion of the Nabitah mobile belt and
Pan-African continental collision in the Saudi
Arabian shield, Ministry of Petr. Min. Res.,
Jiddah, Technical Record USGS-Tr-04-5, 88 pp.,
1984.

Sturchio, N.C., M. Sultan and R. Batiza, Geology
and origin of Meatiq dome, Egypt: a Precambrian
metamorphic core complex?, Geology, 11, 72-76,
1983.

Vail, J.R., Outline of the geochronology and tectonic units of the basement complex of northeast Africa, R. Soc. Lond., Proc., A350, 127-
141, 1976.

Vail, J.R., Outline of geology and mineralization
of the Nubian shield east of the Nile valley,
Sudan, in Evolution and mineralization of the
Arabian-Nubian shield, vol. 1, edited by S.A.
Tahoun, pp. 97-107, Pergamon Press, Oxford,
1979.

Vail, J.R., Pan-African crustal accretion in
north-east Africa, J. afr. Earth Sci., 1, 285-
294, 1983.

Vail, J.R., Pan-African (late Precambrian) tectonic terrains and reconstruction of the
Arabian-Nubian shield, Geology, 13, 839-842,
1985.

Vail, J.R., D.C. Almond, D.J. Hughes, P.M.

Klemenic, S. Poole, S.E.M. Nour and J.C.B.
Embleton, Geology of the Wadi Oko-Khor Hayet
area, Red Sea Hills, Sudan, Geol. and Mineral
Resources Dept., Khartoum Bull. 34, 20 pp.,
1984.
Wyllie, P.J., Constraints imposed by experimental
petrology on possible and impossible magma
sources and products, Phil. Trans. R. Soc.
Lond., A310, 439-456, 1984.
Zimmer, M., Petrographie und Geochemie spätpro-
terozoischer Metavulkanite und Ophiolith-Ge-
steine der südöstlichen Eastern Desert, Ägyp-
ten, Diploma thesis, 120 pp., University of
Mainz, July 1985.

PROTEROZOIC CRUSTAL DEVELOPMENT IN THE PAN-AFRICAN REGIME OF NIGERIA

A. C. Ajibade[1] and M. Woakes

Department of Geology, Ahmadu Bello University, Zaria, Nigeria

M. A. Rahaman

Department of Geology, University of Ife, Ife, Nigeria

Abstract. The evolution of the Nigerian basement, which lies within the Pan-African mobile
belt to the east of the West African Craton, is
discussed in the light of the new data from some
critical areas of the basement. The Nigerian
basement can be divided into two provinces: (1)
the Western Province, approximately west of latitude $8^{\circ}E$, is characterised by narrow, sediment-
dominated, N-S trending, low-grade schist belts in
a predominantly migmatite-gneiss 'older' basement,
and the whole is intruded by Pan-African granitic
plutons; (2) the Eastern Province comprises mainly
a migmatite-gneiss complex intruded by larger volumes of Pan-African granites and the Mesozoic ring
complexes of Central Nigeria. Geochronological
work has demonstrated that the Nigerian basement
includes rocks of Liberian (Archean), Eburnean
(early Proterozoic), middle Proterozoic and Pan-
African (late Proterozoic) ages. The Pan-African
belt east of the West African Craton is believed
to have evolved by plate tectonic processes. The
evolution of the Nigerian basement during the late
Proterozoic is considered to be related to the
activities taking place at the plate boundary.
Initial crustal extension and continental rifting
at the West African craton margin, about 1000 Ma
ago, led to the formation of graben-like structures in Western Nigeria and the subsequent deposition of the rocks of the schist belts. Closure of
the ocean at the cratonic margin about 600 Ma ago
and crustal thickening in the Dahomeyan led to the
deformation of the sediments, the reactivation of
pre-existing rocks and the emplacement of the Pan-
African granites. The Kibaran ($\sim$ 1100 Ma) event
is not considered to have been a major tectonic
event in Nigeria.

Introduction

Nigeria lies in the extensive region east of
the West African craton and northwest of the Congo
craton which was affected by the Pan-African orogeny about 600 Ma ago. The Nigerian Basement lies
to the south of the Tuareg shield (Figure 1). Evidence from the eastern and northern margins of the
West African craton indicates that the Pan-African
belt evolved by plate tectonic processes which involved the collision between the passive continental margin of the West African craton and the active continental margin (Pharusian belt) of the
Tuareg shield about 600 Ma ago (Burke and Dewey,
1972; Leblanc, 1976; 1981; Black et al., 1979;
Caby et al., 1981). The collision at the plate
margin is believed to have led to the reactivation
of the internal region of the belt. The Nigerian
basement complex lies in the reactivated part of
the belt.

Radiometric ages indicate that the Nigerian
basement is polycyclic and includes rocks of
Liberian (2700 $\pm$ 200 Ma), Eburnean (200 $\pm$ 200 Ma),
Pan-African (600 $\pm$ 150 Ma) and, questionably,
Kibaran (1100 $\pm$ 200 Ma) ages. The most obvious
effects of the Pan-African Orogeny in Nigeria is
the emplacement of large volumes of granitoids and
the resetting of mineral ages in virtually all
rock types in the basement. However, little is
known about the nature of the Pan-African event in
Nigeria, or indeed, about any of the earlier
events.

A 400 km wide zone of low-grade schist belts
occurs in the western part of Nigeria. These
schist belts have been variously considered to be
Archean in age (Russ, 1957), middle Proterozoic
(Oyawoye, 1964, 1972) and Pan-African (McCurry,
1973). More recently, Grant (1978) and Turner
(1983) have suggested that there are two generations of schist belts, one Kibaran, the other Pan-
African. The schist belts are considered critical
to the understanding of the evolution of the
Nigerian basement.

This paper discusses the geology of the Nigerian basement in the light of our recent work in
some critical areas and then discusses the evolution of the basement during the late Proterozoic.

---

[1] Now at Federal University of Technology,
Minna, Nigeria

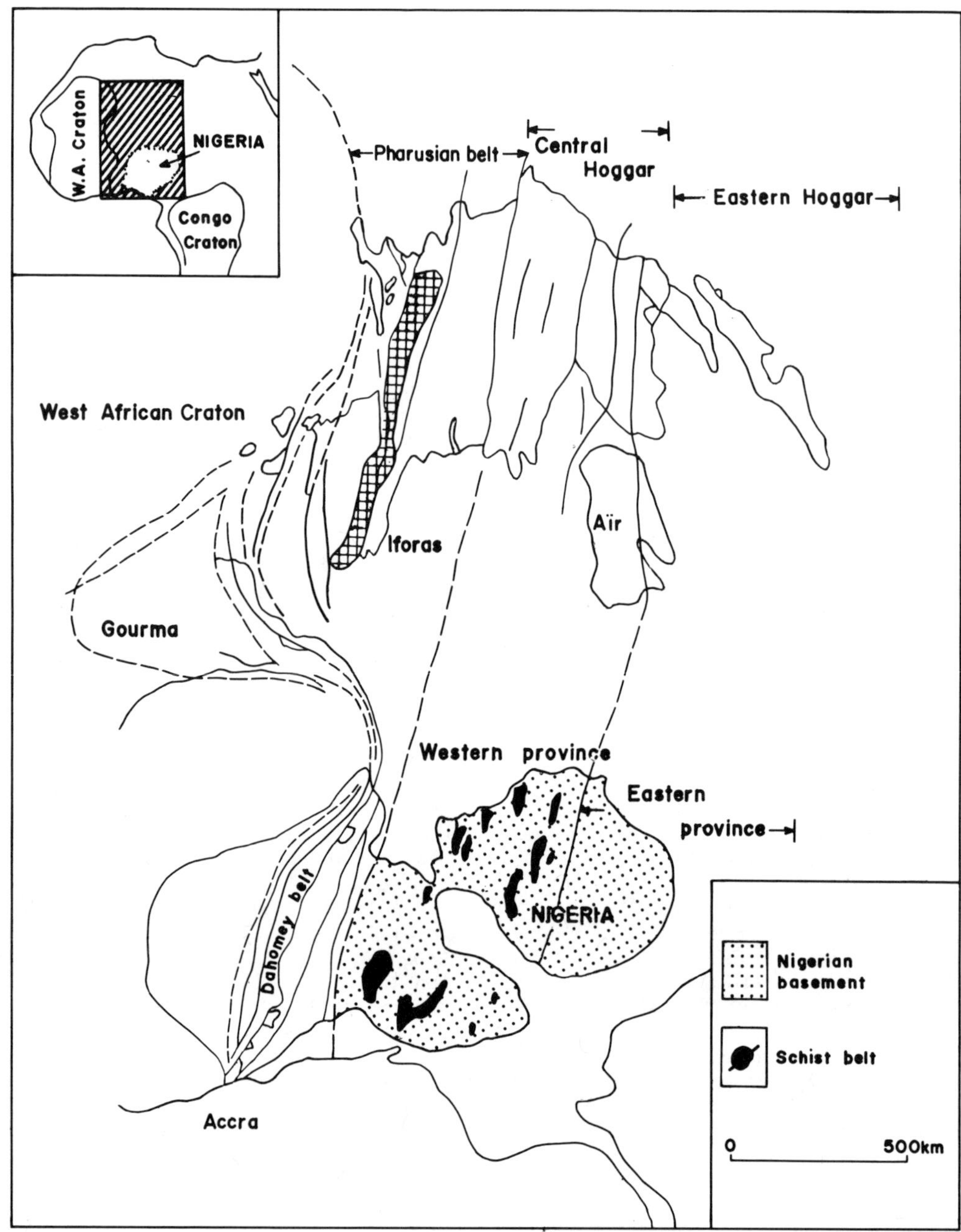

Fig. 1. Generalised geological map of the Pan-African belt east of the West African Craton showing the relative positions of the Nigerian basement and the Tuareg shield. Broken vertical lines show possible connection between the two regions. (Modified from Caby et al., 1981).

It also aims at highlighting some of the problems of the Nigerian basement complex.

General Geology

Three broad lithological groups are usually distinguished within the Nigerian basement complex:

1) A polymetamorphic migmatite-gneiss complex which is composed largely of migmatites and gneisses of various compositions and amphibolites. Relict metasedimentary rocks represented by medium-to-high grade calcareous, pelitic and quartzitic rocks occur within the migmatites and gneisses, and they have been described as "Ancient Metasediments" (Oyawoye, 1972). The migmatite-gneiss

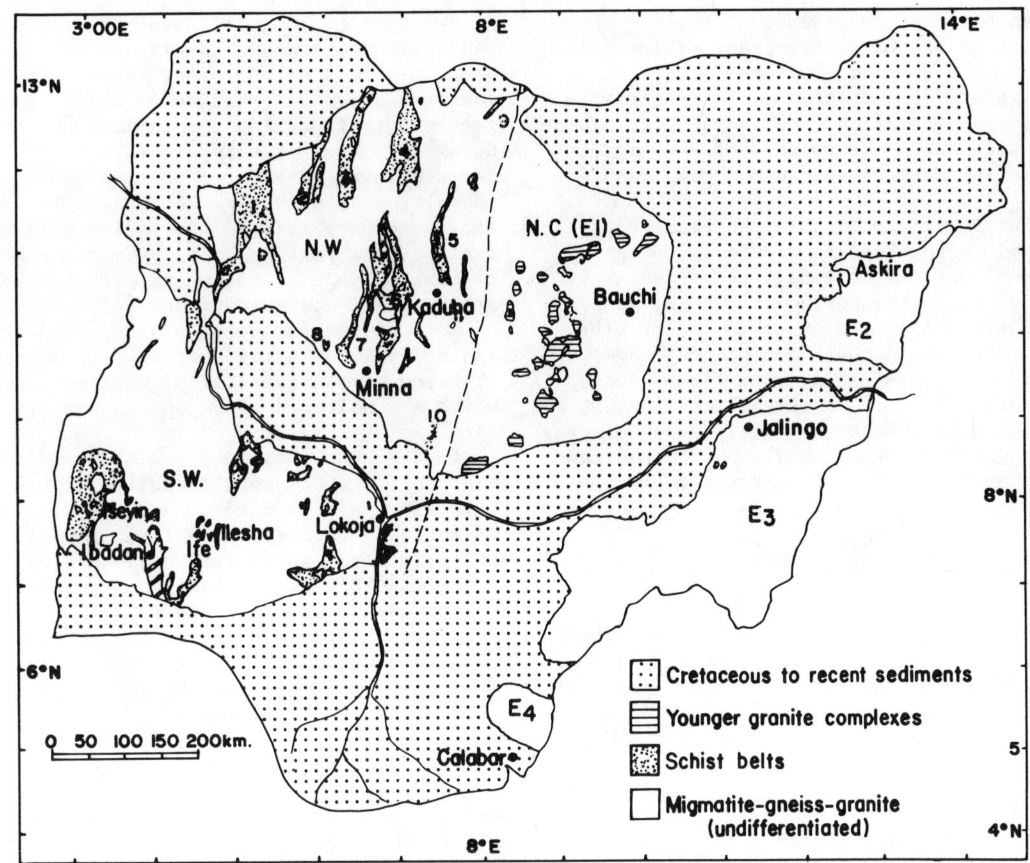

Fig. 2. Simplified geological map of Nigeria. The schist belts of N W Nigeria are numbered 1-10. 1 Zuru, 2 Anka, 3 Maru, 4 Wonaka, 5 Malumfashi, 6 Kushaka, 7 Birnin Gwari, 8 Ushama, 9 Kazaure, 10 Toto belts. Vertical line represents the limit of the schist belts. E_1 to E_4 are areas where basement rocks are exposed in the Eastern Province.

complex is considered to be the basement sensu stricto, and isotopic ages varying from Liberian to Pan-African have been obtained from the rocks. The Pan-African ages have been interpreted as due to isotopic rehomogenization in preexisting rocks during the Pan-African orogeny.

2) Low-grade sediment dominated schists which form narrow belts in the eastern half of the country (Figure 2) have been described as "Newer Metasediments" (Oyawoye, 1972), "Younger Meta-sediments" (McCurry, 1976) and "Unmigmatised to Slightly Migmatised Schists" (Rahaman, 1976). The schist belts are believed to be relicts of a supracrustal cover which was infolded into the migmatite gneiss complex (Russ, 1957; McCurry, 1973). The schist belts are intruded by Pan-African granitoids. The age of the schist belts is discussed in later sections.

3) Syntectonic to late tectonic granitic rocks which cut both the migmatite-gneiss complex and the schist belts. The granitoids include rocks varying in composition from granite to tonalite and charnockite with smaller bodies of syenite and gabbro. The granitoids have yielded radiometric ages in the range of 750-500 Ma which lie within the Pan-African age spectrum. These Pan-African granitoids are called the Older Granites in Nigeria to distinguish them from the Mesozoic, tin-bearing granite complexes of Central Nigeria which are referred to as the Younger Granites.

Our work in different parts of the basement has indicated that the three-fold grouping of the basement rocks based on lithology and inferred age is too simple. Both the migmatite-gneiss complex and the rocks of the schist belts have parallel N-S structures and have conformable field relations. Many of the larger Pan-African granites are foliated and are elongated parallel to the country rocks. The parallel and conformable relations within the basement rocks has made it difficult to separate rocks that were formed during the Pan-African event from those formed during the earlier events in many parts of Nigeria. It is hazardous, therefore, to make general statements on any of the rock groups or to correlate a sequence worked out in one area with that of another without adequate data.

Two distinct provinces can be distinguished in

the Nigerian basement: the Western Province which
is characterised by narrow low-grade schist belts
and in which each belt is separated from the
others by migmatites and gneisses or granites; and
the Eastern Province which comprises mostly migmatites, gneisses and large volumes of Pan-African
granites and is intruded by the Mesozoic Younger
Granites.

The evolution of the Nigerian basement can best
be understood in the Western Province. Detailed
work has been done in two widely separated areas
of this province in the northwest and the southwest, and correlation from one area to the other
is often uncertain. Much of our present knowledge
of the sequence of pre-Pan-African events comes
from the southwest while the field relationship of
the low-grade schist belts are best studied in the
northwest. Only when the two are considered together can the evolution of the province be considered.

Northwestern Nigeria

The first detailed work in northwestern Nigeria
was carried out by Truswell and Cope (1963) who
mapped the now classical Kusheriki area. They
were so impressed by the parallel structures and
transitional contacts between the different metamorphic units in the area that they concluded that
the gneissic complex, the schist formations and
the Older Granites in the area were formed during
a single orogenic event. On the basis of the then
available isotopic data, mainly K-Ar and Rb-Sr
mineral and whole-rock dates which lay between 480
Ma and 540 Ma, Truswell and Cope (1963) concluded
that the basement rocks in the area could not be
older than Cambrian. The interpretation of the
geology of the other areas was based on this conclusion in the subsequent publications of the
Geological Survey of Nigeria (Carter et al., 1963;
Jones and Hockey, 1964; MacLeod et al., 1971).

Subsequent geochronological work on the gneissic complex, particularly by Grant (1970) and
Grant et al., (1972) has shown that the Nigerian
basement is polycyclic. However, no field evidence
of a structural break between the gneissic complex
and the schist belts was provided.

One of us (A.C.A.) has extended the mapping of
the Kusheriki area into the adjacent Zungeru/Minna
area (Figure 3) and re-examined some of the type
localities described by Truswell and Cope (1963).

Field and Age Relations of the Gneissic Complex

Truswell and Cope (1963) divided the metamorphic rocks of the Kusheriki area (Figure 3) into
three formations: the Kusheriki, the Kushaka
Schist and the Birnin Gwari Schist Formations. The
Kusheriki Formation includes migmatites and
gneisses, a quartzo-feldspathic rock which was
named the Zungeru Granulite Member, and small
bodies of kyanite-and sillimanite bearing metasedimentary rocks. The Birnin Gwari and Kushaka
Schist Formations occur in low-grade schist belts.

Two types of migmatites and gneisses were mapped by Truswell and Cope (1963) in the area, and
both were considered to have been formed by metasomatism and migmatisation of adjacent metasedimentary rocks during the emplacement of the Older
Granites. The first type occurs in the western
parts and was considered to have been derived from
psammitic hosts, while the second type occurs in
the eastern parts, and was believed to have been
derived from pelitic hosts. Contacts between all
the metamorphic rocks, where seen in the field,
are parallel and gradational.

Work in the adjacent Zungeru/Minna region
(Ajibade, 1980; Ajibade et al., 1979) confirms the
presence of two types of migmatites and gneisses
which are believed to belong to two different
ages.

In the west (Figure 3), a supposed metasedimentary unit, the Zungeru Granulite Member, lies between the gneissic complex and the Birnin Gwari
Formation. This unit was later identified as cataclastic rocks which were derived from the adjacent migmatites and gneisses and was renamed the
Zungeru Mylonites (Ajibade et al., 1979). The cataclastic deformation of the migmatite-gneiss complex was considered to have taken place during the
initial folding (D_1) of the Birnin Gwari Schist
Formation. Both the Zungeru Mylonites and the
Birnin Gwari Schist Formation were deformed together during subsequent movements (D_2-D_3). The
identification of the Zungeru Mylonites explains
the gradational contacts between the mylonites and
the gneissic complex and the parallel structures
between the mylonites and the Birnin Gwari Schist
Formation. The recognition of the Zungeru Mylonites also provided the first evidence of a structural break between the gneissic complex and the
schist belts.

In the eastern part, the Kushaka Schist Formation was also seen to lie parallel to gneisses
and migmatitic gneisses. Recent detailed field
work at contacts between the Kushaka Schist Formation and the gneisses (Ajibade, unpublished 1985)
reveals that the "gneisses" are foliated tonalites
and granodiorites which are intrusive into the
Kushaka Schist Formation. Field relations show
that the foliated granitic rocks are part of the
Older (Pan-African) Granites. Age determination
on these rocks is currently in progress.

Although there are no reliable radiometric data
on the migmatites and gneisses, our work confirms
the presence of at least two generations of gneisses in the Kusheriki Zungeru-Minna areas. The
first, occurs in the western parts and was probably formed during an earlier episode of high-
grade metamorphism and migmatisation prior to the
deposition of the sediments and volcanic rocks of
the schist belts. The early high-grade metamorphism led to the formation of kyanite-and sillimanite-bearing quartzites and the migmatites.
The second generation of migmatites and gneisses
occurs in the eastern part and was formed as a
result of extensive migmatisation of the Kushaka
Schist Formation by an early phase of the Pan-

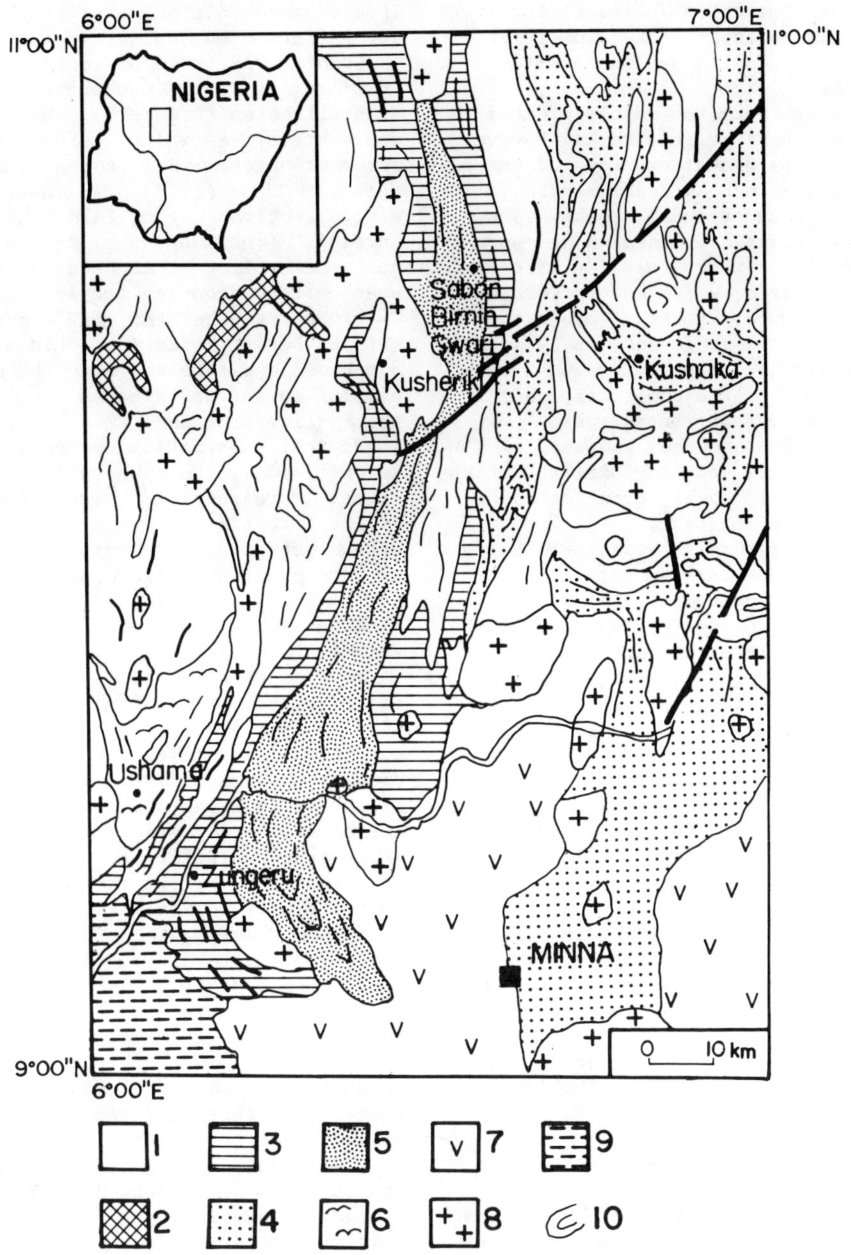

Fig. 3. Geological map of the Kusheriki/Minna–Zungeru areas. 1 Migmatites and gneisses (pre–Pan-African), 2 Kyanite and sillimanite bearing quartzites, 3 Zungeru Mylonites, 4 Kushaka Formation, 5 Birnin Gwari Formation, 6 Ushama Schist Formation, 7 Foliated tonalites and granodiorites (early Pan-African), 8 Late Pan-African granites, 9 Cretaceous to Recent Sediments, 10 Photogeological trends.

African granites which were also deformed during the Pan-African event.

The Schist Belts

The schist belts of northwestern Nigeria occupy narrow N–S trending synformal troughs and form strike ridges which break the monotonous landscape of the Northern Plains. Ten main schist belts have so far been described in this region (Figure 2). Each belt is separated from the adjacent belts by migmatites, gneisses or granites. The outcrop pattern has been compared to that of Archean greenstone belts (Wright and McCurry,

1970; Turner, 1983). The schist belts differ from
greenstone belts only in their bulk composition
which is predominantly clastic sediments with
minor volcanic rocks.

The schist belts are composed mainly of metamorphosed pelitic and semi-pelitic rocks. However,
each belt differs in the amount and type of associated minor lithologies such as conglomerates,
greywackes, quartzites and volcanic rocks. The
Anka belt is unique because it contains an appre-
ciable amount of volcanic rocks, varying in composition from acid to basic, which are interlayered
with the clastic sedimentary rocks. In addition
there are late tectonic volcanic and hypabyssal
rocks, mainly rhyolites and dacites, as well as
small serpentinite-amphibolite complexes, which
overlie or intrude the metamorphosed rocks. A
molasse-type sedimentary sequence, composed of
unmetamorphosed greywackes and conglomerates which
include clasts of volcanic rocks, granites and
phyllites also occurs in the belt.

The lithological differences between the belts
have been used as evidence for suggesting that the
schist belts were not deposited in the same basin
and for suggesting the presence of two generations
of schist belts (Grant, 1978; Turner, 1983;
Fitches et al., 1985).

Evidence from granitic terrains separating two
of the belts, the Kushaka and the Birnin Gwari
Schist Formation in the Kusheriki/Zungeru – Minna
area, shows that relicts of schist are widespread
in deformed granites (Ajibade, unpublished 1985).
This suggests that each belt was not confined to
its present map limits. The same is believed to
be true of the other belts. However, there is at
present no evidence to decide whether the lithological differences reflect deposition in different sedimentary basins or, as suggested by
Fitches et al., (1985) whether the differences
reflect lateral volcanic and facies changes within
a complex volcano-sedimentary basin.

All the schist belts have been metamorphosed in
the greenschist facies, but lower amphibolite facies grade is locally attained.

<u>Structural and Age Relations of the Schist Belts</u>

The age of the schist belts is critical to any
discussion of the evolution of the Nigerian basement during the Proterozoic. Very few detailed
structural studies have been carried out on the
schist belts, and there are fewer reliable geochronological data. The few data available have
been interpreted in different ways, thus leading
to considerable controversy on the age(s) of the
schist belts.

McCurry (1973; 1976) identified two phases of
folding in each of four of the schist belts and
concluded that they were relicts of a once continuous Katangan (900-1000 Ma) sedimentary cover
which became downfolded into the gneissic complex
during the Pan-African event. This conclusion was
based on K-Ar mica and whole-rock data on phyll-

ites from the different belts which varied from
600 Ma to 450 Ma (Harper et al., 1973).

Grant (1978) recognised two contrasting structural styles in the two schist belts of the
Kusheriki area (Figure 3). He described small
scale structures which indicate four phases of
deformation (F_1 – F_4) in the Kushaka Schist Formation). Grant (1978) also identified only a
single gently plunging fold in the "Gneissic
Envelope" (the Zungeru Mylonites) and suggested
that the earlier structures in the gneisses have
been destroyed during the infolding of the Birnin
Gwari Schist Formation. He argued that the
Kushaka Schist Formation could have been metamorphosed and infolded into the migmatite-gneiss
complex during the Kibaran ($\sim$ 1100 Ma) or even in
the earlier Eburnean (2200 Ma) envent. The Birnin
Gwari Schist Formation was considered to have been
deformed and metamorphosed during the Pan-African
event. More recently Turner (1983), using Grant's
(1978) criteria and the assumption that some belts
(the older group) are extensively invaded by Pan-
African granites while others (the younger) are
cut by Pan-African granites only near their margins, assigned each belt throughout Nigeria to
either the Kibaran or Pan-African episodes/events.

Ajibade et al., (1979) have demonstrated in the
better exposed Zungeru/Minna area that the Birnin
Gwari Formation and the Zungeru Mylonites are as
complexly deformed as the Kushaka Formation. A
recent traverse across most of the schist belts of
northwestern Nigeria by us revealed that each belt
has several generations of small-scale structures
which appear similar to those of the other belts.
We also noted that in many areas of poor exposure
only a single foliation is seen in each belt. It
is our conclusion, therefore, that the schist
belts can not be distinguished on the basis of
minor structures. It is also our view that the
amount of granitic intrusion can not be used to
separate rocks which can be shown to predate the
granitic rocks into two age groups. The differences in the amount of granitic intrusion in each
belt may reflect differences of crustal levels
only.

The first reliable radiometric data for the
schist belts was reported by Ogezi (1977) who obtained a Rb-Sr isochron age of 1064 $\pm$ 64 Ma from
phyllites from the Maru belt. He interpreted this
age as dating the time of metamorphism of the belt
and concluded that there was a major Kibaran tectonic event in northwestern Nigeria. This critical age was confirmed by re-analysis of the same
samples by Ajibade (1980). Ajibade (1980) also
obtained a Rb-Sr isochron age of 566 $\pm$ 17 Ma on
phyllites from the Birnin Gwari Formation. He
argued that the schist belts are all of the same
age and that if the 1100 Ma age obtained from the
Maru belt is correct, then the 566 Ma age obtained
from the Birnin Gwari Formation must represent a
later date recording the time of Sr isotope rehom-
ogenisation resulting from mild reheating during
the emplacement of the Pan-African granites.

TABLE 1. Suggested Sequence of Events in the Ibadan Area (Modified from Grant, 1970 and Burke et al., 1976).

Event	Age
6. Emplacement of dolerite dykes.	480-500 Ma Late Pan-African.
5. Emplacement of granite veins and pegmatites.	500-600 Ma Pan-African thermo-tectonic event.
4. F_2 folding and emplacement of Ibadan granite gneiss.	2200 Ma Eburnean
3. Emplacement of Semi-concordant aplite sheets in the banded gneisses.	2750 Ma
2. F_1 folding, high amphibolite facies metamorphism resulting in the formation of banded gneiss, quartzite and amphibolite.	?
1. Deposition of shale, greywacke, sandstone and basalt.	?

Preliminary Rb-Sr whole-rock data from the Zungeru Mylonites (Grant and Ajibade, unpublished 1985) have yielded ages varying from 600 Ma – 700 Ma which suggests a major Pan-African deformation of the region.

Holt (1982) also obtained a Rb-Sr age of 762 Ma on tonalites which are intercalated with phyllites and volcanoclastics in the Anka belt. This age is interpreted as the age of the emplacement of the tonalites and places a minimum age on the deposition of the sediments in the belt.

On the basis of the available evidence, including the new data from the Zungeru Mylonites, we are of the opinion that the schist belts of northwestern Nigeria contain late proterozoic sediments and volcanics which were deformed and metamorphosed during the Pan-African event about 600 Ma ago. The Kibaran 'age' is discussed in the concluding sections.

Southwestern Nigeria

The southwestern segment differs from the northwest in that the schists in the southwest do not form well-defined belts like those in the northwest and are poorly exposed. In addition to the low-grade schists, a series of ancient metasedimentary and metavolcanic rocks belonging to the migmatite-gneiss complex is rather well developed in the southwest whereas these rocks occur only as small relicts in the northwest. In the absence of detailed field and geochronological data it is often impossible to differentiate the older schist from the younger in many parts of the southwest. Detailed field and geochronological data from the Ibadan area (Grant, 1970; Burke et al., 1976) provided the most complete picture of the pre-Pan-African history of the Nigerian basement so far.

Pre-Pan-African History

The Ibadan area is underlain by a migmatite-gneiss complex which consists of banded gneiss with associated quartzite, quartz schist, mica schist and amphibolite layers. Small bodies of granite gneiss are emplaced in the sequence. The quartzites form prominent topographic features around Ibadan, and their outcrop pattern indicates a refolded fold. The association of the banded gneiss with quartzite and mica schist, which are believed to have been derived from sandstones and shales, led Jones and Hockey (1964) to suggest that the banded gneiss were derived primarily from sedimentary rocks, probably shales and greywackes. Some of the amphibolite bands were found to be tholeiitic in composition, and Freeth (1971) inferred that the Ibadan sequence included basalts.

The first reliable radiometric evidence for a pre-Pan-African event in Nigeria was provided by Grant (1970) who obtained a Rb-Sr isochron age of 2200 $\pm$ 70 Ma on the Ibadan Granite Gneiss (Burke et al., 1976) which was previously mapped as part of the Older Granites. The sequence of events established in the Ibadan area is summarized in Table 1. Similar sequences, calibrated by geochronology, have not been established in any other part of the southwest.

Grant et al., (1972) obtained a Rb-Sr isochron age of 1120 $\pm$ 124 Ma on the Ile-Ife granite gneiss. This age was interpreted as a metamorphic age and has been widely (but mistakenly) quoted by some workers as evidence for Kibaran igneous activity in Nigeria (Hubbard, 1975; Ogezi, 1977; Ajibade, 1980; Holt, 1982; Turner, 1983). The granite gneiss has recently yielded an upper concordia intercept age of 1825 $\pm$ 27 Ma and a lower intercept age of 567 $\pm$ 26 Ma, based on a U-Pb zircon dating, (Lancelot and Rahaman, in press) The 1825 Ma upper intercept age has been interpreted as dating the actual time of emplacement of the granite gneiss while the lower intercept age was interpreted as a minimum age for the main Pan-African event which caused radiogenic lead loss.

Rahaman et al., (1983) also obtained a Rb-Sr whole rock isochron age of 1904 + 18 Ma on the Igbetti granite gneiss body north of Iseyin, (Figure 2).

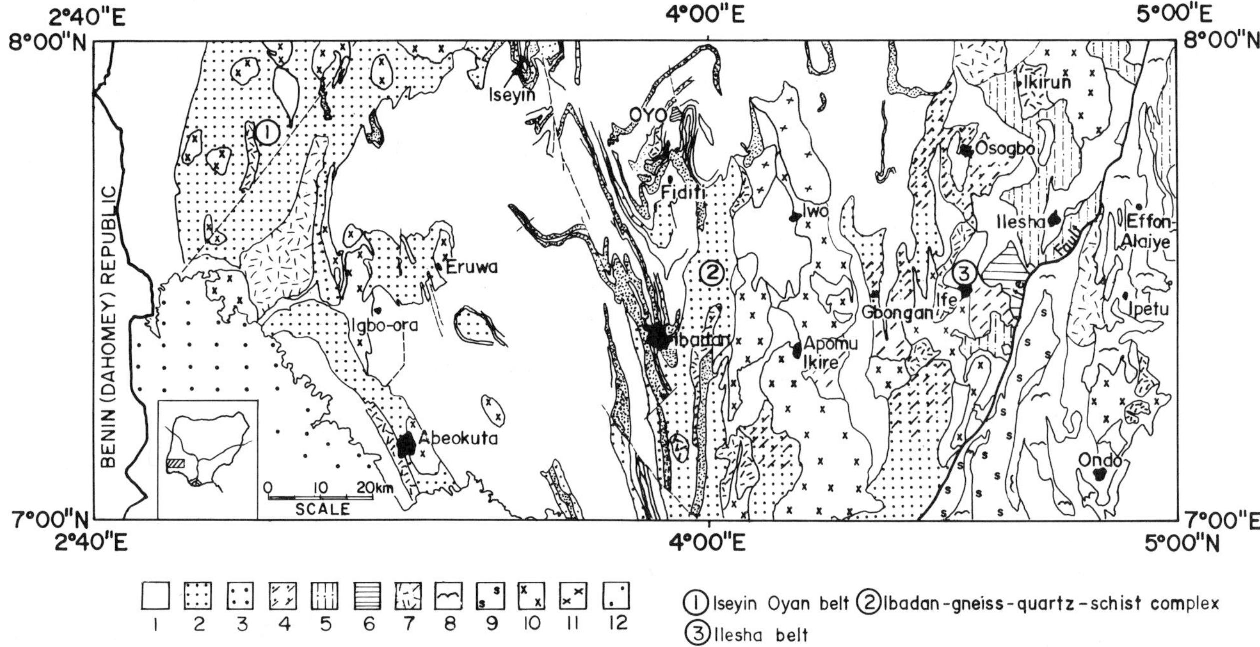

Fig. 4. Geological map of parts of southwestern Nigeria. 1 Migmatites and gneisses with intercalated amphibolites, 2 Gneiss and schist complex, 3 Quartzite and schist, 4 Pegmatised schist, 5 Schist and epidiorite complex, 6 Epidiorite (Archean and Eburnean), 7 Granite gneiss (Late Lower Proterozoic), 8 Quartzite, 9 Feldspathic quartzite, 10 Granites, 11 Quartz syenite and charnockites (Late Proterozoic), 12 Cretaceous to Recent sedimentary rocks.

The Schist Belts

The schist belts of southwestern Nigeria are extremely poorly exposed because of tropical climatic conditions and rain forest vegetation in that region. Three schist belts were distinguished in the region by Turner (1983, Figure 4). These are the Iseyin-Oyan River, the Ilesha and Igarra-Kabba-Lokoja belts. Turner (1983) correlated the schist belts with those of the northwest and assigned each of the belt to either the Kibaran or the Pan-African episodes/events.

The Iseyin-Oyan River belt is continuous into the Ibadan area and appears to form part of the Archean sequence of banded gneiss-quartzite-schist of the area and was so mapped by Jones and Hockey (1964) and Burke et al., (1976). Metamorphism is also higher in this belt than in the belts of the northwest. Rahaman (1976) described pelitic rocks containing garnet, staurolite and, locally, sillimanite.

The Igarra-Kabba-Lokoja belt on the other hand is composed of upper greenschist facies metapelites with interlayered quartzites, marble, and metaconglomerates. The metaconglomerates include clasts that are derived from the adjacent migmatite-gneiss complex (Odeyemi, 1976).

The Ilesha belt is of particular interest because it contains unusual suites of rocks and there are some geochronological constraints on its age.

Two structural units with contrasting lithologies were distinguished in this belt which lies in the Iwo sheet (Hubbard, 1975). The two units are separated by the NNE-trending Ife Fault Zone (Figure 4). The western unit, which contains gold mineralization, consists of massive amphibolites ("epidorites"), amphibole schist, talc chlorite schist, talc-tremolite rock and pelitic schists with associated granite, gneiss and pegmatite. They have been metamorphosed in the lower amphibolite facies (Hubbard, 1975; Olade and Elueze, 1978). Hubbard (1968) initially suggested that the belt might represent a reworked Archaean greenstone belt, but later (Hubbard, 1975) reinterpreted the belt as supracrustal rocks which were deformed during the Kibaran episode/event on the basis of the 1120 Ma age obtained by Grant et al., (1972) on the Ile-Ife granite gneiss. Hubbard (1975), however, noted that the Ibadan sequence continues into the Iwo area and observed "Before Grant and others' (1972) determination of a 1150 Ma Kibaran age for the Ife granite gneiss, it perhaps seemed most reasonable to relate the main schist belts of the Iwo sheet to the Ibadan schist sequence ... Grant's age does not of course necessarily preclude the correlation with the pre- or early Eburnean schist of the Ibadan area". Klemm

266   AJIBADE ET AL.

et al., (1984) presented some geochemical data on the basic and ultrabasic rocks from the belt and concluded that the rocks might represent an Archean greenstone belt. Olade and Elueze (1978) also presented detailed petrochemical data on the amphibolites and associated talc-tremolite schists to suggest that they are products of metamorphosed basalts and peridotites. The rocks have low K_2O, Rb, Rb/Sr, CaO/Al_2O_3 and high K/Rb, Na_2O/K_2O suggesting a rather primitive nature and probable dervation from a subcrustal source region. However, Olade and Elueze (1978) consider the rocks to be part of the late Proterozoic sequence that was deformed and metamorphosed during the Pan-African.

The new U-Pb zircon age of 1825 Ma for the Ile-Ife granite gneiss (Lancelot and Rahaman, in press) which is emplaced into the belt seems to confirm Hubbard's interpretation of the schists as probably representing a reworked greenstone belt of pre-Eburnean or early Proterozoic age. If this is correct, it may not be possible to correlate the western unit of the Ilesha belt with the schist belts in northwestern Nigeria which we think are late Proterozoic in age. The eastern unit, east of the Ife fault system (Figure 4) is composed of quartzite, quartz schist, quartzo-feldspathic gneiss, minor iron-rich schist and quartzites, and amphibolites. The quartzite overlies amphibole schist, as is the case to the west (Hubbard, 1975), and indicates that the contact may be stratigraphic superposition or an overthrust relationship. This unit is considered younger than the western unit and probably represents a late Proterozoic shelf or deltaic deposit (Turner, 1983).

In summary, southwestern Nigeria is underlain by two generations of schist belts: one which belongs to the migmatite-gneiss complex sequence, and of probable Archean to early Proterozoic age, the other belongs to the late Proterozoic sequence. Except in the eastern belts, it is difficult to identify one generation from the other. The Iseyin-Oyan River and the western half of the Ilesha belt are believed to belong to the older sequence but may also contain elements of the younger sequence.

The Eastern Province

Basement complex rocks are exposed in four areas of this province: a large area of north-central Nigeria which continues into the north-west and three smaller areas in the eastern parts of the country (Figure 2). The different areas are separated by Cretaceous to Recent sediments of the Chad Basin and the Benue Valley. The north-central block is intruded by the tin-bearing Younger Granite complexes. The basement in this province is the least studied in the Nigerian basement as attention in the province has been directed to the Younger Granites and the sedimentary formations, both of which are mineralized. From the available information on the province, the basement complex consists of a migmatite-gneiss complex which has been extensively intruded by granitic and charnockitic rocks of Pan-African age. The migmatite-gneiss complex is composed predominantly of migmatites, banded gneisses and granite gneisses. Relicts of metasedimentary and metavolcanic rocks are widely distributed in the migmatite-gneiss complex, and they form part of the succession.

MacLeod et al., (1971) recognized a field succession of a pre-migmatitic granulitic gneiss, migmatite, granite gneiss and granite but considered them to form a single petrogenetic unit on the basis of a K-Ar biotite age of 510 ± 20 Ma obtained from a biotite granulite. All the radiometric data, which include whole rock Rb-Sr isochron ages so far have been obtained mainly on foliated granitic and charnockitic rocks, and have given Pan-African ages (van Breeman et al., 1977), but no final conclusion can be reached until more field and geochronological work has been done in the province. The migmatite gneiss complex is considered here to be similar to that of other parts of the basement complex.

The Older (Pan-African) Granites

The "Older Granites" of Nigeria include a wide spectrum of rocks varying in composition from tonalite through granodiorite to granite, syenite and charnockitic rocks. The granitoids have been emplaced into both the migmatite-gneiss complex and the schist belts, and they occur in all parts of the Nigerian basement. They give K-Ar and Rb-Sr ages in the range 750 Ma to 450 Ma, and they are the only undisputed products of the Pan-African orogeny in Nigeria. Recent observations in the migmatite gneiss complex, described earlier, indicate that the extent of Pan-African plutonism is still largely unknown. An important question in the study of the Older Granites and, by implication, the nature of the Pan-African orogeny in Nigeria, is to what extent the granitoids were derived from melting of a pre-existing crust or represent new additions to the crust from the mantle?

Recent work in different parts of the Nigerian basement has shown that the Older Granites are high-level intrusions emplaced by stoping and diapiric processes (Fitches et al., 1985). The most gneissic of the rocks contain angular xenoliths of schists from the adjacent belts, while narrow contact aureoles have been mapped in the country rocks (Russ, 1957; Egbuniwe, 1982; Ajibade, 1982). Evidence from cross-cutting pegmatites and microgranites shows that the foliation in the granites is a result of post-emplacement deformation, because early pegmatites and microgranites have parallel foliations to those of the granites while later one cut across the foliation.

The charnockitic member of the Older Granite suite deserves special mention as its occurrence in parts of the country has been used as evidence for Archean granulite facies metamorphism in the Nigerian basement (Hubbard, 1975; Dempster, unpublished MS 1965). The term "charnockitic"

was suggested by Cooray (1977) to describe the
dark rocks which contain the common charnockitic
minerals, quartz + plagioclase + orthopyroxene $\pm$
alkali feldspar $\pm$ clinopyroxene $\pm$ biotite $\pm$ hornblende $\pm$ fayalite. The charnockitic rocks vary in
composition from norite, diorite to enderbite and
true charnockite. They occur either associated
with the other members of the Older Granite suite
or as discrete bodies within the migmatite-gneiss
complex. Complex contact relations between the
charnockitic rocks and the surrounding rocks have
led to different interpretations of the age and
origin of the charnockitic rocks. On the basis of
field relations Hubbard (1968; 1975) identified
two generations of charnockites in the Iwo area.
The earlier one occurs within the migmatite-gneiss
complex and is considered to have crystallized in
a granulite facies environment. The other is
believed to have formed during the Pan-African
plutonism, either by crystallization from magma or
by metasomatism of pre-existing rocks (Hubbard,
1968).

All the radiometric data, including whole rock
Rb-Sr and U-Pb zircon ages so far obtained on the
charnockitic rocks give Pan-African ages (van
Breeman et al., 1977; Tunbosun, 1983; Tunbosun et
al., 1984; Rahaman, unpublished 1984). Field relations show that the charnockitic rocks are intrusive rocks which have been emplaced into a predominantly upper greenschist to amphibolite facies
environment (Cooray, 1972; 1977; Eborall, 1976;
van Breeman, 1977).

Origin and Tectonic Setting of the Granitic Rocks

A number of geochemical and isotopic data have
recently become available on the granitoids and
are being used to answer the question whether the
Older Granites are juvenile additions to the crust
or represent remobilised crustal rocks (van
Breeman et al., 1977; Holt, 1982; Fitches et al.,
1985; Rahaman et al., 1983).

The major and trace element geochemistry of
most of the Older Granites investigated show calcalkaline to alkaline on an alkalinity index diagram (Fitches et al., 1985). Some of the granitoids show an iron-enriched fractionation pattern
which is closer to the trend shown by tholeiitic
rocks. The granitoids belong to the I-type of
White and Chappell (1974). The trace elements
plot in the arc and/or collisional fields on several discrimination diagrams except for a late,
mildly peralkaline quartz syenite in the northwest
which plots in a field that overlaps the collisional and mid-plate fields (Fitches et al., 1985).

Rb-Sr isotopic measurements on the Older Granite from different parts give initial $^{87}Sr/^{86}Sr$
ratios which range from 0.7065 to 0.7125. These
initial ratios are significantly above the
"expected" upper mantle values 600 Ma ago.

The granitoids of the Adrar des Iforas (Mali),
close to the Pan-African suture, by contrast all
have low initial $^{87}Sr/^{86}Sr$ ratios of 0.7035 to
0.705 which have been interpreted as indicating a

mantle origin with only minor crustal contamination (Leigeois and Black, 1983). The geological
setting, the expanded composition (Pitcher, 1979)
and the trace element geochemistry of the Nigerian
Pan-African granites indicate that the granitic
magmas were mainly derived from mantle sources but
were heavily contaminated by crustal materials,
and also by partial melting of the lower crust.

Summary and Conclusions

Pre-Pan-African Events

Field and geochronological evidence has confirmed the polycyclic nature of the Nigerian basement. Poor exposure and the overprinting effect
of the last tectonic event to affect the basement,
the Pan-African, have made it extremely difficult
to differentiate the pre-Pan-African rocks in most
places. This is more so because the Pan-African
metamorphism is in the upper greenschist and
lower amphibolite facies, close to the grade attained by the pre-existing rocks which is predominantly in the middle to upper amphibolite
facies. In spite of these limitations, it is now
possible to make general statements about the pre-
Pan-African history of the Nigerian basement. The
evolution of the basement during the Pan-African
is then discussed in the regional context of the
Pan-African belt of West Africa.

The migmatite-gneiss complex is considered to
contain rocks of Archean age which have been deformed and modified several times prior to the
Pan-African orogeny. Evidence from the Ibadan
area indicates that the Archean rocks included
metasedimentary and metavolcanic rocks which were
deformed before the emplacement of the Eburnean
granite gneiss. The granite gneiss is also considered to have been derived from pre-existing
rocks by partial fusion (Burke et al., 1976). If
the interpretation of Hubbard (1968; 1975) of the
Ilesha schist belt as a reworked Archean greenstone belt is correct, then the Nigerian gneissic
complex contains components of granite-greenstone
belts which now occur only as relicts in most
parts.

The early Proterozoic Eburnean event was probably accompanied by sedimentation, deformation
metamorphism and syntectonic igneous activity.
However, the only reliable evidence for this event
comes from the Ibadan granite gneiss which has
yielded an age of 2200 Ma. The nature of the
Eburnean event and the full extent of Eburnean
plutonism will not be known until more detailed
field and geochronological data become available
from other parts of the basement.

The Ile-Ife granite gneiss and the Igbetti
gneiss have yielded a U-Pb zircon age of 1825 $\pm$ 27
Ma (Lancelot and Rahaman, in press) and a Rb-Sr
isochron age of 1904 $\pm$ 18 Ma. (Rahaman et al.,
1983) respectively. These ages are now considered
to be too young for the Eburnean event. They correspond to similar ages that have been reported
from orthogneisses in the Iforas region of Mali by

Caby and Andrepoulus-Renauld (1983). The orthogneisses are considered to belong to a shield-
scale anorogenic magmatism. The 1800-1900 Ma
granite gneisses of Nigeria may also belong to
this series and were deformed during the Pan-
African event. This may also explain the difficulty in differentiating the early Proterozoic
gneiss from the foliated Pan-African granites in
the field in Nigeria.

Geochronological evidence for a Kibaran event
in Nigeria came from a Rb-Sr isochron age of 1120
± 140 Ma with initial $^{87}Sr/^{86}Sr$ ratio of 0.739
0.014 from the Ile-Ife granite gneiss (Grant et
al., 1972) and the phyllites of the Maru belt
(Ogezi, 1977). Because of the elevated initial
ratio, Grant et al., (1972) concluded that the
Ile-Ife gneiss was much older, a conclusion now
confirmed by the upper intercept zircon age of
1825 ± 27 Ma (Lancelot and Rahaman, in press).
The only remaining evidence for a Kibaran event
is from the Maru phyllites. As in the case of
the Tuareg shield, the Kibaran event is not well
documented. Similar Kibaran ages have been obtained only from metasedimentary rocks but not on
intrusive granites (Black, 1980; 1984).

It has been argued in the earlier sections that
the schist belts of northwestern Nigeria including
the Maru belt, are all of the same age and that
they were deformed and metamorphosed for the first
time during the Pan-African event. It is our conclusion that there was no major Kibaran tectonic
event in Nigeria.

The Pan-African Orogeny

The evolution of the Nigerian basement during
the Pan-African can best be discussed in the regional context of the Pan-African belt of West
Africa (Figure 1). Geological and geophysical
evidence from the western province of the belt and
the cratonic margin has been used to erect geodynamic models for the evolution of the belt, (Grant,
1969; Burke and Dewey, 1972, Bertrand and Caby,
1978; Caby et al.,1981; Black et al., 1979).
Essentially, the evolution of the belt is seen as
a collision-type orogeny with an eastward-dipping
subduction zone.

According to Burke and Dewey (1972) and
Trompette (1979), the subduction zone in the
southern sector of the belt (Ghana-Togo-Benin)
may have been shallow-dipping, extending a long
way under the Dahomeyan shield and therefore resulting in steep thermal gradients as far east as
Nigeria. The collision at the cratonic margin
led to the reactivation of the Dahomeyan basement
including Nigeria. The Dahomeyan was thickened to
form a Tibetan-type crust. Partial melting in the
lower parts of the thickened crust led to the formation of granitic magmas. While this model may
explain the emplacement of some of the Older
Granites, it does not explain the evolution of the
schist belts.

Since the recognition of the suture along the
eastern margin of the West African craton and the

nature of the orogeny, attempts have been made to
relate the schist belts to the subduction processes in the cratonic margin. Vaniman (1976), Holt
(1982) and Turner (1983) consider that the schist
belts may have been deposited in a back-arc basin
developed after the onset of subduction at the
cratonic margin. However, the distance of the
nearest of the Nigerian schist belts from the site
of subduction is at least 200-250 km. This is in
excess of 100-150 km from arc to back-arc basins
in present day arc systems (Gass, 1981). In addition, only the Anka belt has characteristics that
can be compared to those of an island arc.

The possibility that the schist belts may represent additional microcontinents separating pre-
existing microcontinents have also been suggested
(McCurry and Wright, 1977; Turner, 1983; Rahaman,
unpublished). While the volcanic rocks in some of
the belts have chemical characteristics that indicate possible ocean floor derivation, none of
the belts has the typical Wilson Cycle signature
(Kroner, 1983) which includes sutures, ophiolites,
calcalkaline volcanism and blue schist assemblages.

In our view, crustal thinning which was important in the Western Nigerian province is probably
related to initial crustal extension and continental rifting at the craton margin about 1000Ma ago.
This led to the formation of graben-like structures in a broad zone in Nigeria and the deposition
of predominantly clastic sediments with minor volcanic rocks. We agree with Turner (1983) that the
present synformal belts are not equivalent to the
original basins of deposition as they show no
zonation of facies parallel to their margins.

Closure of the ocean at the craton margin,
about 600 Ma ago, and crustal thickening to the
east led to the deformation and metamorphism of
the sediments, partial melting of the upper mantle
and lower crust and the emplacement of the Older
Granites in Nigeria. Uplift of the region around
Anka led to the post-tectonic deposition of the
coarse clastic sediments of the region with the
associated rhyolitic and dacitic volcanic rocks
which have been dated at 500 Ma.

The work of French geologists across the Pan-
African belt in the much better exposed Tuareg
shield provides a basis for comparing the geology
of that shield to that of the Togo-Benin-Nigeria
sector (Figure 1). Nigeria lies to the south of
the central and eastern Hoggar-Air domains. Both
regions are polycyclic and are of similar distances from the craton margin. They are characterised by large volumes of Pan-African granitoids. The central Hoggar and the Western Province
of the Nigerian basement both contain N-S schist
belts that are parallel to the craton margin but
those of the central Hoggar are of higher grade.
A major difference between the Nigerian basement
and indeed the whole of the southern sector of the
belt and the Tuareg shield is the occurrence of
extensive shear and mylonite zones in the Tuareg
shield which divide the whole shield into structural provinces. Although it is improbable that

such structures occur in the Nigerian basement,
some N-S mylonite and fault zones have been recognised in the Nigerian basement (Ajibade et al.,
1979; Hubbard, 1975). The low-grade schist belts
of the central Hoggar which are assumed to be of
late Proterozoic age are interpreted as intracratonic troughs bounded and controlled by the shear
zones (Black, 1984). In Nigeria there is no direct evidence for fault bounded troughs to the
schist belts but the Ile-Ife Fault Zone appears to
have controlled the deposition of the eastern unit
of the Ilesha schist belt. The Pan-African granites in both regions may ultimately have been
derived from the same source, mainly the upper
mantle (Black, 1984; Liegeois and Black, 1983;
Fitches et al., 1985), but fusion related to
crustal thickening may have played an important
role in the Nigerian basement.

Acknowledgements. The authors would like to
thank A.Kroner, N.K.Grant and other referees for
critically reading the manuscript. Field work was
supported by Research Grant from Ahmadu Bello
University (A.C.A. and M.W.) and Ife University
(M.A.R.) We are grateful to Pascal Affaton for
stimulating discussion in the field during a
traverse across the schist belts of northwestern
Nigeria and the showing two of us (A.C.A. and
M.A.R.) the geology across the orogenic front in
Togo.

References

Ajibade, A. C., Geotectonic evolution of the
    Zungeru Region, Nigeria. Unpubl. Ph.D. thesis,
    Univ. of Wales (Aberystwyth), 1980.
Ajibade, A. C., The Origin of the Older Granites
    of Nigeria: some evidence from the Zungeru
    region. Nigerian Jour. Min. and Geol., 19, 223-
    230, 1982.
Ajibade, A. C., W. R. Fitches, and J. B. Wright,
    The Zungeru Mylonites, Nigeria: recognition of a
    major tectonic unit. Rev. de Geol. Phys., 21(5),
    359-363, 1979.
Bertrand, J. M. L. and R. Caby, Geodynamic evolution
    of the Pan-African orogenic belt. A new inter-
    pretation of the Hogger Shield (Algerian
    Sahara). Geol. Rundsch., 67, 357-388, 1978.
Black, R., Precambrian of West Africa. Episodes, 4,
    3-8, 1980.
Black, R, H. Ba, E. Ball, J. M. L. Bertrand, A. M.
    Boullier, R. Caby, I. Davison, J. Fabre, M.
    Leblanc, and L. I. Wright, Outline of the Pan-
    African geology of Adrar des Iforas (rep. of
    Mali). Geol. Rundsch. 68(2) 543-564, 1979.
Burke, K. C. and J. F. Dewey, Orogeny in Africa,
    in African Geology, edited by T.F.J.Dessauvagie
    and A.J.Whiteman, Univ. Ibadan, pp. 583-608,
    1972.
Burke, K., S. J. Freeth and N. K. Grant, The
    structure and sequence of geological events in
    the basement complex of the Ibadan area,
    Western Nigeria. Precambrian Res., 3, 537-545,
    1976.

Caby, R., J. M. L. Bertrand and R. Black, Pan-
    African Ocean Closure and Continental Collision
    in the Hoggar-Iforas segment. Central Sahara, in
    Precambrian Plate Tectonics, edited by A.Kroner,
    Elsevier Amsterdam, 407-434, 1981.
Caby, R., and U. Andreopoulous-Renauld, Age a 1800
    Ma du Magmatisme sub-alcalin associe aux meta-
    sediments monocycliques dans la Chaine Pan-
    Africane du Sahara Central. Jl. Afri. Earth
    Sci., 2 193-197, 1984.
Carter, J. D., W. Barber, and E. A. Tait, The
    Geology of parts of Adamawa, Bauchi and Bornu
    Provinces in North-Eastern Nigeria. Geol.
    Surv. Nigeria Bull., 30, 1963.
Chappell, B. W. and A. J. R. White, Two cont-
    rasting granite types, Pacific Geol., 8, 173-
    174, 1974.
Cooray, P. G. A note on the charnockites of the
    Ado-Ekiti area, Western State, Nigeria, in
    Geology of Africa, edited by T.F.J. Dessauvagie
    and A.J.Whiteman, pp 45-54, Univ. Ibadan, 1972.
Cooray, P. G., Classification of the charnockitic
    rocks of Nigeria, Nigeria Jour. Min. Geol. 14,
    1-6, 1977.
Eborall, M. I., Intermediate Rocks from Older
    Granite Complexes of the Bauchi area, Northern
    Nigeria. in Geology of Nigeria, edited by C.A.
    Kogbe, pp. 65-74, Elizabethan Publ. Co., Lagos,
    1976.
Egbuniwe, I. G., Geotectonic evolution of the Maru
    Belt, N.W.Nigeria, Unpubl. Ph.D. thesis, Univ.
    of Wales, Aberystwyth, 1982.
Elueze, A. A., Dynamic metamorphism and oxidation
    of amphibolites of Tegina area, northwest
    Nigeria, Precambrian Res., 14, 379-388, 1981.
Fitches, W. R., A. C. Ajibade, I. G. Egbuniwe, R.
    W. Holt and J. B. Wright., Late Proterozoic
    Schist belts and Plutonism in N.W.Nigeria.
    J. Geol. Soc. London, 142, 319-337, 1985.
Freeth, S. J., Geochemical and related studies of
    West African igneous and metamorphic rocks.
    Ph.D. thesis, Univ. Ibadan, 1971.
Gass, I. G., Pan-African (Upper Proterozoic) plate
    tectonics of the Arabian-Nubian shield, in
    Precambrian Plate Tectonics, edited by A. Kroner,
    pp 387-402, Elsevier, Amsterdam, 1981.
Grant, N. K., Geochemistry of Precambrian basement
    rocks from Ibadan, Southwestern Nigeria, Earth
    Planet. Sci. Lett., 10, 29-38, 1970.
Grant, N. K., Structural distinction between a
    meta-sedimentary cover and an underlying base-
    ment in the 600 Ma old Pan-African domain of
    North-western Nigeria, Bull. Geol. Soc. Amer.,
    89, 50-58, 1978.
Grant, N. K., M. Hickman, F. R. Burkholder and
    J. L. Powell, Kibaran metamorphic belt in the
    Pan-African domain of W.Africa, Nature Phys.
    Sci. 238, 90-91, 1972.
Harper, C. T., G. Sherrer, P. McCurry and J. B.
    Wright, K-Ar retention ages from the Pan-
    African of Northern Nigeria, Bull. Geol.
    Soc. Amer., 84, 919-926, 1973.
Holt, R., Geotectonic evolution of the Anka belt
    in the Precambrian basement complex of N.W.

Nigeria, Ph.D. Thesis, Open University, England, 1982.

Hubbard, F. H., The association Charnockite-Older Granite in Southwestern Nigeria, Nigerian J. Min. Geol. 3, 25-32, 1968.

Hubbard, F. H., Precambrian crustal development in Western Nigeria, indications from the Iwo Region, Bull. Geol. Soc. Amer., 86, 548-554, 1975.

Jones, H. A. and R. D. Hockey, The geology of part of Southwestern Nigeria, Geol. Surv. Nigeria Bull., 30 1964.

Klemm, D., W. Schneider and B. Wagner, The Precambrian metavolcanic-sedimentary sequence east of Ife and Ilesha, S.W.Nigeria, A Nigerian 'greenstone belt'? Jl. Afr. Earth Sci., 2:2, 161-176, 1984.

Kroner, A., Proterozoic mobile belts compatible with the plate-tectonics concept, in Proterozoic Geology, edited by G.M.Medaris Jr., Geol. Soc. Amer. Memoir 161, 59-74, 1983.

Lancelot, J. R., and M. A. Rahaman, A Propos de l'evenement a 1000 Ma en Afrique ocidentale: 2°/L'orthogneiss d'Ile-ife (Nigeria), C. R. Somm. Soc. Geol. Fr., in press.

Leblanc, M., The Late Proterozoic ophiolites of Bou Azzer (Morocco) evidence for Pan-African plate tectonics, in Precambrian Plate Tectonics, edited by A.Kroner, Elsevier, Amsterdam: pp 435-451, 1981.

Leigeois, J. P. and R.Black, Preliminary results on the geology and geochemistry of the late Pan-African composite batholith of western Iforas (Mali), Abstracts, 12th Coll. Afr. Geol. Ternuren, p 62, 1983.

MacLeod, W. N., D. C. Turner and E. P. Wright, The Geology of the Jos Plateau - General Geology, Geol. Surv. Nigeria Bull., 32(1), 1971.

McCurry, P., Geology of degree sheet 21 (Zaria), Overseas Geol. Min. Res., 45, H.M.S.O., London, 1973.

McCurry, P., The Geology of the Precambrian to Lower Palaeozoic rocks of Northern Nigeria, a review, in Geology of Nigeria, edited by C.A. Kogbe, pp. 15-39, Elizabethan Pub. Co. Lagos, 1976.

McCurry, P., Geology of degree sheets 10 (Zuru), 20 (Chafe) and part of 9 (Katsina) Nigeria, Overseas Geol. Min. Res., 53, H.M.S.O. London, 1978.

McCurry, P. and J. B. Wright. Geochemistry of calc-alkaline volcanics in Northwestern Nigeria, and a possible Pan-African suture zone, Earth Planet. Sci. Lett., 37, 90-96, 1977.

Odeyemi, I. B., Preliminary report on the field relationships of the basement complex rocks around Igarra, Midwest Nigeria, in Geology of Nigeria, edited by C.A.Kogbe, pp. 59-63, Elizabethan Publ. Co. Lagos, 1976.

Ogezi, A. E. O., Geochemistry and Geochronology of basement rocks from Northwestern Nigeria, Ph.D. thesis, Leeds Univ., 1977.

Olade, M. A. and A. A. Elueze, Petrochemistry of the Ilesha amphibolites and Precambrian crustal evolutions in the Pan-African domain of SW Nigeria, Precambrian Res., 8, 303-318, 1978.

Oversby, V. M., Lead isotopic study of aplites from the Precambrian basement rocks near Ibadan, Southwestern Nigeria, Earth Planet. Sci. Lett., 27, 177-180, 1975.

Oyawoye, M. O., The contact relationship of the Charnockite and biotite gneiss at Bauchi, Charnockite and biotite gneiss at Bauchi, Northern Nigeria, Geol. Mag., 101(2), 138-144, 1964.

Oyawoye, M. O., The basement complex of Nigeria, in African Geology, edited by T.F.J.Dessauvagie and A.J.Whiteman, pp. 66-102, Univ. Ibadan, 1972.

Pitcher, W. S., The nature ascent and emplacement of granitic magmas, Jl. Geol. Soc. Lond., 136, 627-662, 1979.

Rahaman, M. A., Review of the basement geology of Southwestern Nigeria, in Geology of Nigeria, edited by C.A.Kogbe, pp. 41-58, Elizabethan Publ. Co., Lagos, 1976.

Rahaman, M. A., W. O. Emofureta and M. Caen Vachette, The Potassic-granites of the Igbetti area: further evidence of the Polycyclic evolution of the Pan-African belt in Southwestern Nigeria, Precambrian Res. 22, 75-92, 1983.

Trompette, R., The Pan-African dahomeyide fold belt : a collision orogeny? (abstract), 10th Colloque de Geologie Africaine, Montpellier, 72-73, 1979.

Truswell, J. F. and R. N. Cope, The geology of parts of Niger and Zaria provinces, Northern Nigeria, Bull. Geol. Surv. Nigeria, 29, 1963.

Tunbosun, I. A., Geochronologie U/Pb de socle precambrian de Sud Quest Nigeria, Thesis Doct Zeme cycle, Univ. Montpellier, 1983.

Tunbosun, I. A, J. R. Lancelot, M. A. Rahaman and O. Ocan., U/Pb Pan-African ages of two charnockite-granite associations from Southwestern Nigeria, Contr. Min. Petrol.,88, 188-195, 1984.

Turner, D. C., Upper Proterozoic Schist Belts in the Nigerian Sector of the Pan-African province of West Africa, Precambrian Res., 21, 55-79, 1983.

van Breeman, O, R. T. Pidgeon and P. Bowden, Age and isotopic studies of some Pan-African granites from North central Nigeria, Precambrian Res., 4, 307-319, 1977.

Vaniman, D. T., The Godani granodiorite plutons. Nigeria, Petrology and regional setting, Ph.D. Thesis, Univ. California, Santa Cruz, 1976.

Wright, J. B, and P. McCurry, A reappraisal of some aspects of Precambrian shield geology: Discussion, Bull. Geol. Soc. Amer.,81, 3491-3492, 1970.